ANALYTICAL MODELS OF THERMAL STRESSES IN COMPOSITE MATERIALS I

Analytical Models of Thermal Stresses in Composite Materials I

Ladislav Ceniga

Nova Science Publishers, Inc.
New York

For permission to use material from this book please contact us:
Telephone 631-231-7269; Fax 631-231-8175
Web Site: http://www.novapublishers.com

Library of Congress Cataloging-in-Publication Data
Available upon request.

ISBN 978-1-60456-085-5

Published by Nova Science Publishers, Inc. ✦ New York

Contents

Preface

This book is dedicated with love to my dearest parents.

This book is the first volume of the trilogy *Analytical models of thermal stresses in composite materials I, II, III*, presenting, in each of the volumes, original results only, created by the author. The fact that the author proceeds from fundamental equations of Mechanics of Solid Continuum presented in Sections 3.1.3–3.1.6, 3.1.8 confirms originality of the results and accordingly establishment of new scientific school with an interdisciplinary character belonging to the scientific branch Applied Mechanics. As an imagination considered for the analytical models, an **elastic** solid continuum is represented by a multi-particle-(envelope)-matrix system consisting of components represented by **spherical** particles periodically distributed in an **infinite** matrix, without or with a **spherical** envelope on the surface of each of the spherical particles. The multi-particle-(envelope)-matrix system with different distribution of the spherical particles is considered as a model system for the determination of the thermal stresses in real composite materials with finite dimensions included in the categories as presented in Section 2.1 (see Items 1–4). Dependent on the particle and envelope radii, R_1 and $R_1 < R_2$, respectively, resulting in the envelope thickness $t = R_2 - R_1$, on the inter-particle distance d and on the particle volume fraction v, representing parameters of a real composite material, the thermal stresses originating during a cooling process are a consequence of the difference in thermal expansion coefficient of the components, as well as a consequence of the difference in dimensions of isotropic cubic crystalline lattices resulting from a phase transformation originating at least in one of the components.

As usual in Mechanics of Solid Continuum representing a fundamental basis of Applied Mechanics, the state of stress of a system can be determined using different mathematical techniques resulting in different differential equations for an investigated quantity, e.g. radial displacement, radial stress. Consequently, integration constants included in solutions of each of the different differential equations are derived regarding defined boundary conditions. With regard to the Castigliano's theorem (see Section 3.1.9) and to the different solutions describing the state of stress of the system, such solution is considered to exhibit minimal energy of the system.

With regard to Volume I, the thermal stresses acting in the isotropic multi-particle-(envelope)-matrix system, represented by the isotropic components, are determined using eight different mathematical techniques (see Sections 5.1–5.3, 6.1, 7.1, 7.2, 8.1) applied to the transformed equilibrium equations (3.35), (3.36) and consequently resulting in eight different linear differential equations with **non-zero right sides** related to the radial displacement of an arbitrary point in the isotropic elastic solid continuum, considering boundary conditions defined in Section 4.1.

With regard to specific conditions for the isotropic multi-particle-envelope-matrix system defined in Section 9.1, Equation (3.35) is not considered, and using ten different mathematical techniques presented in Section 9.2, Equation (3.36) is transformed to ten different linear differential equations with **zero right sides**, exhibiting solutions suitable for the determination of the thermal stresses in the spherical particle and envelope. Additionally, with regard to completeness of the presented topic, as presented in Sections 5.7, 6.5 and 9.5, the solutions to result from the eight and ten different linear differential equations with and without right sides, respectively, are used for the determination of the thermal stresses in an isotropic one-particle-(envelope)-matrix system containing, in contrast to the isotropic multi-particle-(envelope)-matrix, one spherical particle only, considering boundary conditions defined in Section 4.2. Finally, Section 9.6.1 presents the mathematical techniques resulting in fifteen different linear differential equations with **zero right sides** related to the radial stress acting in an arbitrary point in the isotropic elastic solid continuum, in contrast to Sections 5.1–5.3, 6.1, 7.1, 7.2, 8.1, 9.2 related to the radial displacement.

Along with two different mathematical techniques, applied to **both** Equations (3.35), (3.36) and resulting in two different differential equations with **zero right sides** presented in Volume II, the analytical models of the thermal stresses acting in the **isotropic** components of the isotropic multi-particle-(envelope)-matrix system, determined in Volumes I, II as extensive as possible regarding a physical point of view, exhibit accordingly permanent validity without a need of the updating in future.

Devoted to an anisotropic elastic solid continuum, Volume III presents analytical models of the thermal stresses acting in the one-axial and triaxial anisotropic components of the anisotropic multi-particle-(envelope)-matrix system. Accordingly, combinations of the analytical models related to the isotropic, one-axial and triaxial anisotropic components presented in the three volumes result in analytical models related to the isotropic-one-axial-anisotropic, isotropic-triaxial-anisotropic and one-triaxial anisotropic multi-particle-(envelope)-matrix systems.

In addition to the thermal stresses acting in each of the components, all volumes of the trilogy present formulae for **corresponding quantities** represented by the elastic energy density w, the elastic energy W and the thermal stresses σ_1, σ_2, σ_3 acting in three mutually perpendicular directions and inducing the elastic energy density w_1, w_2, w_3, respectively, used for the determination of thermal-stress induced phenomena. Presented in Volume III in forms of general formulae to include the parameters w, σ_i, w_i $(i=1,2,3)$ related to each of the isotropic, one-axial and triaxial anisotropic components, surface and curve integrals of w, σ_i, w_i over a corresponding surface and along a corresponding curve in the isotropic, one-axial and triaxial anisotropic elastic solid continuum, respectively, result in the thermal-stress induced phenomena including

1. elastic energy fluctuations to represent energy barriers,
2. thermal-stress induced strengthening or weakening,
3. crack formation conditions along with critical values of the parameters R_1, R_2, v of composite materials,
4. analytical models of thermal-stress induced cracks to form in brittle composite materials.

Along with a transformation of the thermal stresses to stresses originating in a crystalline lattice as a consequence of the presence of a central substitutive atom presented in Chapter 11 of Volume I for cubic crystalline lattices and in Volume III, the trilogy *Analytical models of thermal stresses in composite materials I, II, III* represents **an integrated scientific work** with regard to **elasticity** of the multi-particle-(envelope)-matrix system, to the **spherical** particle and the **spherical** envelope, and to the **infinite** matrix.

With regard to the selected topics of Applied Mathematics in Sections 12.4.1–12.4.5 presenting the mathematical knowledge to be required for the study of this book, Volume I as well as the trilogy, exhibiting an interdisciplinary character with respect to Items 1–4, serve several categories of readers, including senior undergraduates and PhD. students in Mechanical Engineering, Applied Mechanics, Applied Physics, Materials Engineering, along with researchers and practicing engineers working at universities, scientific institutes and in industry. Readers of the categories are recommended to study Section *1.8.2 Categories of readers* in detail[1].

[1]This work was supported by the Slovak Research and Development Agency under the contract No. COST-0022-06.

About the author

Born in 1965 in Košice, Slovak Republic, graduated from the Mechanical Engineering Faculty of the Technical University in Košice (1988) (Department of Mechanical Engineering Technology), and from the Faculty of Sciences of the P. J. Šafárik University in Košice (1993) (Department of Physics of Solids), both with distinction, awarded the prize of the Chancellor of the Technical University in Košice for excellent study results, aimed at heat and chemical treatment and magnetic properties of amorphous alloys (1994–2000), defending a PhD. thesis in Physics of Condensed Matters and Acoustics at the Institute of Experimental Physics of the Slovak Academy of Sciences in Košice (1999), since 2000 employed at the Institute of Materials Research of the Slovak Academy of Sciences in Košice, Dr. Ladislav Ceniga currently works on analytical models of thermal stresses and related thermal-stress induced phenomena in composite materials[2].

Dr. Ladislav Ceniga

Košice, Slovak Republic
March 2006

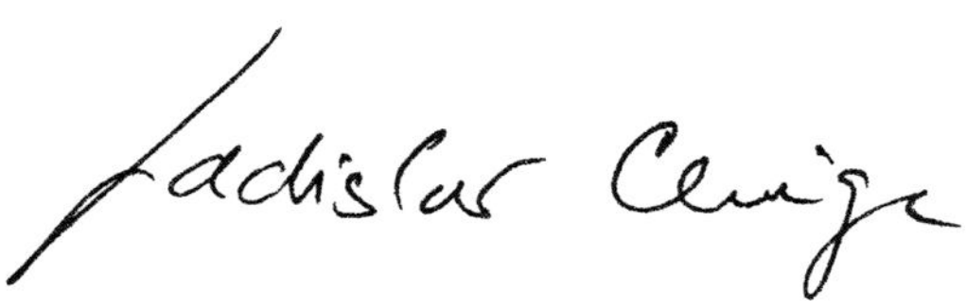

[2]E-mail addresses: lceniga@yahoo.com, ladislav_ceniga@yahoo.com, lceniga@hotmail.com, ladislav_ceniga@hotmail.com, ceniga@post.sk, lceniga@imr.saske.sk

Chapter 1

Outline of principles

1.1 Cell model

To derive thermal stresses in a multi-particle-(envelope)-matrix system with infinite dimensions replacing the types of real composite materials with finite dimensions as presented in Section 2.1 (see Items 1-4), an infinite matrix is **imaginarily** divided into identical cells, and a shape of the cell containing a central spherical particle with the particle centre O corresponds to particles distribution, and without or with a spherical envelope on the surface of each of the spherical particles (see Fig. 2.1a,b), where the spherical particles, the spherical envelopes and the infinite matrix represent components of the multi-particle-envelope-matrix system. The thermal stresses are consequently investigated within the cell, and additionally, such imaginary dividing of the infinite matrix is required regarding the particles distribution that the cells can fulfil the infinite matrix perfectly, and accordingly rectangular-based and hexagonal-based prismatic cells are considered in this book. Corresponding mathematical techniques similar to those presented in Sections 2.2, 2.3 can be applied to such cells to fulfil the infinite matrix perfectly, exhibiting a different shape than the rectangular-based and hexagonal-based prismatic cells.

The cell dimension along the axis x_i of the Cartesian system $(Ox_1x_2x_3)$ is a function of the inter-particle distance d_i $(i = 1,2,3)$ of the particle volume fraction $v \in (0, v_{max})$ and the particle radius R_1, representing, along with radii of the spherical envelope $R_1 < R_2$, material parameters of a composite system, where v_{max} depends on the particles distribution (see Section 2.2). Resulting from the matrix infinity, analytical models of the thermal stresses in a certain cell are identical with those in any cell with the same shape.

1.2 Mathematical techniques

With regard to an arbitrary point with a position determined by a system of suitable coordinates, an infinitesimal spherical cap in the point P with a position determined by the spatial polar coordinates $[r, \varphi, \nu]$ (see Fig. 3.1) represents an infinitesimal part of a solid continuum within which the state of deformation and stress is investigated, as usual in Mechanics of Solid Continuum, where $r = |OP|$. Considering elasticity and isotropy of the solid continuum, the determination of the state of deformation and stress results from the Cauchy's, Saint-Venant's and equilibrium equations, and from the Hooke's laws, representing fundamental equations of Mechanics of Solid Continuum as a basis of Applied Mechanics presented in Section 3.1.

With regard to an experience of the author, the thermal stresses in the **isotropic** multi-particle-(envelope)-matrix system consisted of **isotropic components** can be derived using such mathematical techniques to lead to differential equations for **the radial displacement** u_{1q} of the infinitesimal spherical cap in each of the isotropic components [1]. Accordingly, the Cauchy's equations (see Eqs. (3.1)–(3.5)), representing relationships between the strain ε_{ijq} $(i,j=1,2,3)$ and the radial displacement u_{1q} along with the derivations $\partial u_{1q}/\partial\varphi$, $\partial u_{1q}/\partial\nu$, are substituted to the Hooke's laws (see Eqs. (3.26)–(3.30)), and the stress σ_{ijq} is consequently derived as a function of u_{1q}, $\partial u_{1q}/\partial\varphi$, $\partial u_{1q}/\partial\nu$ (see Eqs. (3.31)–(3.34)). The dependence $\sigma_{ijq} = \sigma_{ijq}\left(u_{1q}, \partial u_{1q}/\partial\varphi, \partial u_{1q}/\partial\nu\right)$ is substituted to the equilibrium equations (3.15)–(3.17), representing relationships between the radial and tangential and shear stresses, σ_{11q} and σ_{22q}, σ_{33q} and σ_{12q}, σ_{13q}, respectively. Finally, the equilibrium equations are accordingly transformed to a system of differential equations, representing a relationship between the derivations $\partial^3 u_{1q}/\partial r\partial\varphi^2$, $\partial^3 u_{1q}/\partial r\partial\nu^2$, $\partial^2 u_{1q}/\partial\varphi^2$, $\partial^2 u_{1q}/\partial\nu^2$ (see Eq. (3.35)), and a relationship between u_{1q}, $\partial u_{1q}/\partial r$, $\partial^2 u_{1q}/\partial r^2$, $\partial^2 u_{1q}/\partial\varphi^2$, $\partial^2 u_{1q}/\partial\nu^2$ (see Eq. (3.36)).

As presented in Sections *Mathematical techniques* (see Sections 5.1–5.3, 6.1, 7.1, 7.2, 8.1), different mathematical techniques applied to the differential equations (3.35), (3.36) lead to different linear differential equations with non-zero right sides (see Eqs. (5.5), (5.13), (5.18), (6.3), (7.2), (7.11), (8.2), (8.3)) exhibiting solutions suitable for the determination of the thermal stresses in the spherical particle (see Sections 5.1–5.3), the spherical envelope (see Sections 5.1–5.3, 6.1, 7.1, 7.2, 8.1) and the cell matrix (see Sections 5.1–5.3, 6.1, 7.1, 7.2), considering the boundary conditions defined in Section 4.1.

With regard to the multi-particle-envelope-matrix system provided that $\beta_p \neq \beta_e = \beta_m$ and $\varepsilon_{11te} - \varepsilon_{11tp} \neq f(\varphi,\nu)$ (Eqs. (3.48)–(3.50), (12.23), (12.24)), Equation (3.35) is not considered, and Equation (3.36) is transformed to the linear differential equations (9.2), (9.6), (9.10), (9.13), (9.17), (9.21), (9.24), (9.28), (9.32), (9.36) with zero right side exhibiting solutions suitable for the determination of the thermal stresses in the spherical particle and envelope (see Section 9.2), where the coefficient β_q includes the thermal expansion coefficient α_q and the phase-transformation induced radial strain ε_{11tq}, the latter dependent or independent on the angles φ, ν of a system of the spatial polar coordinates $[r,\varphi,\nu]$. Additionally, the thermal stresses can be determined from the linear differential equations (9.126), (9.130), (9.134), (9.137), (9.141), (9.144), (9.147), (9.151), (9.155), (9.159), (9.162), (9.166), (9.174), (9.179) for the radial stress σ_{11q} with zero right sides, where

[1]With regard to a component consisted of anisotropic crystalline lattices with mutually different orientation of the crystalline lattice axes, the poly-crystalline component is considered to be isotropic [1, p 5–15].

the initial linear differential equation (9.126) is derived by the substitution of the Saint-Venant's equation (3.6) and the Hooke's laws (see Eqs. (3.18)–(3.20)) to the equilibrium equation (3.15) (see Section 9.6.1). Finally, to determine the thermal stresses in the multi-particle-envelope-matrix system provided that $\beta_p \neq \beta_e = \beta_m$ for $\varepsilon_{11te} - \varepsilon_{11tp} \neq f(\varphi, \nu)$, solutions for the spherical particle, for the spherical envelope and for the cell matrix are related to one of the linear differential equations with zero right side (see Eqs. (9.2), (9.6), (9.10), (9.13), (9.17), (9.21), (9.24), (9.28), (9.32), (9.36)) and to one of the different linear differential equations with non-zero right sides (see Eqs. (5.5), (5.13), (5.18), (6.3), (7.2), (7.11)), respectively.

The state of deformation and stress in one of the components is determined from a solution related to one of the different linear differential equations. Accordingly, solutions related to one or two, and to one, two or three different linear differential equations are considered for the multi-particle-matrix and multi-particle-envelope-matrix systems, respectively. With regard to the Castigliano's theorem (see Section 3.1.9), such combination of solutions resulting from the different linear differential equations is considered to result in minimal thermal-stress induced elastic energy of the cell, $W_c = W_p + W_m$ and $W_c = W_p + W_e + W_m$ (see Eqs. (3.44), (3.45)), related to the multi-particle-matrix and multi-particle-envelope-matrix systems, respectively, where W_p, W_e and W_m, as corresponding quantities, represent thermal-stress induced elastic energy of the spherical particle, the spherical envelope and the cell matrix (see Eqs.(3.42)–(3.45)), respectively.

Additionally, with regard to completeness of the presented topic and to the analysis in Section 2.2.2, provided that $\varepsilon_{11tm} - \varepsilon_{11tp} = f(\varphi, \nu)$, $\varepsilon_{11te} - \varepsilon_{11tp} = f(\varphi, \nu)$, $\varepsilon_{11tm} - \varepsilon_{11te} = f(\varphi, \nu)$ and $\varepsilon_{11tm} - \varepsilon_{11tp} \neq f(\varphi, \nu)$, $\varepsilon_{11te} - \varepsilon_{11tp} \neq f(\varphi, \nu)$, $\varepsilon_{11tm} - \varepsilon_{11te} \neq f(\varphi, \nu)$, Sections 5.7, 6.5 and 9.5 present analytical models of the thermal stresses in an **isotropic** one-particle-(envelope)-matrix system containing, in contrast to the multi-particle-(envelope)-matrix, one spherical particle only, respectively, considering the boundary conditions defined in Section 4.2.

Representing corresponding quantities, the thermal stress σ_{iq} along the axis x_i ($i = 1,2,3$; $q = p,e,m$) inducing the elastic energy density w_{iq}, and the thermal-stress induced elastic energy density w_q for the multi- and one-particle-(envelope)-matrix systems (see Section 3.1.8) are used for the determination of related phenomena (see Items $1, \ldots, 4$ in *Preface*), presented along with the thermal stresses in anisotropic multi-particle-(envelope)-matrix systems in Volume III.

Finally, with regard to the transformations presented in Chapter 11, formulae for the thermal stresses in the multi-particle-matrix system for the particles distribution with the inter-particle distance $d_i = d$ ($i = 1,2,3$) (see Fig. 2.1a) are transformable to stresses originating in a crystalline lattice as a consequence of presence of a central substitutive atom.

1.3 Reason of thermal stresses

As presented in Section 3.2.1, with regard to the multi-particle-matrix and multi-particle-envelope-matrix systems, thermal stresses originate as a consequence of the differences $\alpha_m - \alpha_p \neq 0$ and $\alpha_e - \alpha_p \neq 0$, $\alpha_m - \alpha_e \neq 0$, as well as $\varepsilon_{11tm} - \varepsilon_{11tp} \neq 0$ and $\varepsilon_{11te} - \varepsilon_{11tp} \neq 0$, $\varepsilon_{11tm} - \varepsilon_{11te} \neq 0$ (see Eqs. (12.23), (12.24)), respectively, in the temperature-dependent or temperature-independent thermal expansion coefficients α_p, α_e, α_m, as well as in the radial strains ε_{11tp}, ε_{11te}, ε_{11tm} induced by a phase transformation (see Section 12.1) originating at least in one of components of the multi-particle-(envelope)-matrix system, where α_q and ε_{11tq} are included in the coefficient β_q $(q = p,e,m)$ (see Eqs. (3.48)–(3.50)). The phase-transformation induced radial strain ε_{11tq} results from the difference in dimensions of mutually transforming cubic crystalline lattices of the spherical particle, the spherical envelope, the cell matrix (see Section 12.1), thus for $q = p,e,m$, respectively.

In case of the non-zero difference in the thermal expansion coefficients, the thermal stresses originate in the temperature interval $\langle T_f, T_r \rangle$, where the temperature T_r of the thermal stress generation is minimal temperature from the sets $\{T_{rp}, T_{rm}\}$ and $\{T_{rp}, T_{re}, T_{rm}\}$ related to the multi-particle-matrix and multi-particle-envelope-matrix systems, respectively, T_f is final temperature of a cooling process. and T_{rq} is the recovery temperature below that the stress relaxation as a consequence of thermal-activated processes does not occur in the spherical particle, the spherical envelope, the cell matrix, thus for $q = p,e,m$, respectively. In case of the zero difference in the thermal expansion coefficients and provided that $T_{tq} \in \langle T_f, T_r \rangle$, the thermal stresses originate at the temperature T_t representing maximal temperature from the sets $\{T_{tp}, T_{tm}\}$ and $\{T_{tp}, T_{te}, T_{tm}\}$ related to the multi-particle-matrix and multi-particle-envelope-matrix systems, respectively, where $\langle T_f, T_t \rangle \subset \langle T_f, T_r \rangle$.

The same analysis as presented in Section 1.3 is also valid for the one-particle-(envelope)-matrix system.

1.4 Radial stresses p_1, p_2

The thermal stresses and the corresponding quantities are functions of the radial stresses p_1 and p_1, p_2 (see Section 10.1), acting on the particle-matrix and particle-envelope, matrix-envelope boundaries, representing common parameters included in all formulae related to the multi-particle-matrix and multi-particle-envelope-matrix systems, respectively, where $p_{1+i} > 0$ and $p_{1+i} < 0$ $(i = 0,1)$ are compressive and tensile, respectively.

Considering the multi-particle-matrix system, the radial stress p_1, as a consequence of the condition $\beta_m - \beta_p \neq 0$, is a function of the coefficients ϱ_{1p} and ϱ_{1m} derived by the tangential strains in the spherical particle and in the cell matrix, $(\varepsilon_{22p})_{r=R_1}$ and $(\varepsilon_{22m})_{r=R_1}$, respectively, on the particle-matrix boundary, thus for $r = R_1$ (see Eqs. (3.56), (3.58)). As presented in Section 1.2, the tangential strains

$(\varepsilon_{22p})_{r=R_1}$ and $(\varepsilon_{22m})_{r=R_1}$ are derived both by one solution or by two different solutions, related to one or two of the different linear differential equations, respectively. With regard to the determination of the radial stress p_1, and to the Castigliano's theorem (see Section 3.1.9), such solution or such combination of solutions, resulting in such coefficients ϱ_{1p} and ϱ_{1m}, are considered to result in minimal thermal-stress induced elastic energy of the cell, $W_c = W_p + W_m$ (see Eq. (3.44)). Consequently, considering one or two different linear differential equations, the tangential strains $(\varepsilon_{22p})_{r=R_1}$ and $(\varepsilon_{22m})_{r=R_1}$ are derived, using uniform notations.

Similarly, considering the multi-particle-envelope-matrix system, the radial stresses p_1 and p_2 as consequences of the conditions $\beta_e - \beta_p \neq 0$ and $\beta_m - \beta_e \neq 0$ related to the particle-envelope and matrix-envelope boundaries (see Eqs. (3.59), (3.60)), respectively, are functions of the coefficients

1. ϱ_{1p}, $\varrho_{1+i1+je}$ $(i,j=0,1)$, ϱ_{2m} related to the radial stresses p_1, p_2 on the condition $\beta_p \neq \beta_e \neq \beta_m$, transformed to the condition $\alpha_p \neq \alpha_e \neq \alpha_m$ provided that $T > T_{tq}$ or $T_{tq} \notin \langle T_f, T_r \rangle$ $(q=p,e,m)$ (see Eqs. (3.49), (3.50)),

2. ϱ_{1p}, ϱ_{1e}^{me} related to the radial stress p_1 on the condition $\beta_p \neq \beta_e = \beta_m$, transformed to the condition $\alpha_p \neq \alpha_e = \alpha_m$ provided that $T > T_{tq}$ or $T_{tq} \notin \langle T_f, T_r \rangle$ $(q=p,e,m)$,

3. ϱ_{2m}, ϱ_{2e}^{pe} related to the radial stress p_2 on the condition $\beta_p = \beta_e \neq \beta_m$, transformed to the condition $\alpha_p = \alpha_e \neq \alpha_m$ provided that $T > T_{tq}$ or $T_{tq} \notin \langle T_f, T_r \rangle$ $(q=p,e,m)$.

The coefficients ϱ_{1p} and $\varrho_{1+i1+je}$ $(i,j=0,1)$, ϱ_{1+ie}^{me}, ϱ_{1+ie}^{pe} and ϱ_{2m} are derived from the tangential strains in the spherical particle and in the spherical envelope and in the cell matrix, $(\varepsilon_{22p})_{r=R_1}$ and $(\varepsilon_{22e})_{r=R_{1+i}}$ and $(\varepsilon_{22m})_{r=R_2}$, respectively, where ϱ_{1e}^{pe}, ϱ_{2e}^{me} are included in formulae for the thermal stresses and for the corresponding quantities related to the spherical particle, the cell matrix provided that $\beta_p = \beta_q \neq \beta_m$, $\beta_p \neq \beta_q = \beta_m$, respectively. As presented in Section 1.2, the tangential strains $(\varepsilon_{22p})_{r=R_1}$, $(\varepsilon_{22e})_{r=R_{1+i}}$ $(i=0,1)$ and $(\varepsilon_{22m})_{r=R_2}$ are derived by solutions related to one, two or three of the different linear differential equations. With regard to the determination of the radial stresses p_1, p_2, and to the Castigliano's theorem (see Section 3.1.9), such solution or such combination of solutions, resulting in such coefficients ϱ_{1p}, $\varrho_{1+i1+je}$ $(i,j=0,1)$, ϱ_{1+ie}^{me}, ϱ_{1+ie}^{pe}, ϱ_{2m}, are considered to result in minimal thermal-stress induced elastic energy of the cell, $W_c = W_p + W_e + W_m$ (see Eq. (3.45)). Consequently, considering one, two or three different linear differential equations, the tangential strains $(\varepsilon_{22p})_{r=R_1}$, $(\varepsilon_{22e})_{r=R_{1+i}}$ $(i=0,1)$ and $(\varepsilon_{22m})_{r=R_2}$ are derived, using uniform notations.

With regard to the conditions $\beta_p = \beta_m$ and $\beta_p = \beta_e = \beta_m$ (see Eqs. (3.58), (3.63), (3.64)), the thermal stresses in the multi-particle-matrix and multi-particle-envelope-matrix systems do not occur, respectively.

Finally, the same analysis as presented in Section 1.4 is also valid for the one-particle-(envelope)-matrix system with the thermal-stress induced elastic energy $W^{(1)} = W_p^{(1)} + \left(W_e^{(1)}\right) + W_m^{(1)}$ [2] (see Section 2.2.2.1).

1.5 Temperature range

As presented in Sections 1.3, 3.2.1, induced by a phase transformation only, originating at least in one of components of the multi-particle-(envelope)-matrix system, the thermal stresses originate at the temperature T_t of the phase transformation. Induced by the differences in the thermal expansion coefficients α_p, α_e, α_m, the thermal stresses originate during a cooling process between the recovery temperature T_r, below that the stress relaxation as a consequence of thermal-activated processes does not occur in a composite material [1, p 38–40], and the final temperature T_f.

With regard to a temperature dependence of the elastic thermal-stress loading, an analysis concerning the critical final temperature of a cooling process, T_{ys} and T_{as}, related to the yield stress in compression or tension of the spherical particle, the spherical envelope, the cell matrix, σ_{yscq} or σ_{ystq}, provided that $p_{1+i} > 0$ or $p_{1+i} < 0$ $(i = 0{,}1)$, for $q = p{,}e{,}m$, and to the adhesion strength σ_{as}^{q} of the particle-matrix, particle-envelope, matrix-envelope boundaries, provided that $p_{1+i} < 0$, for $q = pm{,}pe{,}me$, respectively, is presented in Section 10.3.

1.6 Finite matrix

With regard to a finite matrix, the analytical model is required to consider a shape and dimensions of the finite matrix, and a position of each cell in the finite matrix. Additionally, boundary conditions, related to the radial displacement u_{1m} or to the radial stress σ_{11m} both on the surfaces of each cell, are required to be determined for each cell separately.

Accordingly, even though the boundary conditions are defined, an application of the analytical model to a real composite material, resulting in numerical dependencies of the thermal stresses and the corresponding quantities on the parameters v, R_1, R_2 of the real composite material, would be probably time-consuming.

1.7 Subscripts and notation

As an example for the understanding of a calculation of subscripts presented in this book, the subscript $1+i \;\; 1+j \;\; 1+k\varphi m$, related to the angle φ and to material parameters of the cell matrix (m), is transformed to the subscripts $112\varphi m$ and $212\varphi m$ for $i, j = 0$, $k = 1$ and $i, k = 1$, $j = 0$, respectively. Analogically, the subscript

[2]The superscript (1) is related to the *one*-particle-(envelope)-matrix system.

$1 + i \quad 1 + jm$ is transformed to the subscripts $11m$ and $21m$ for $i, j = 0$ and $i = 1$, $j = 0$, respectively.

Similarly, the subscript $i1 + j \quad 1 + ke$ related to material parameters of the spherical envelope (e) is transformed to the subscripts $312e$ and $121e$ for $i = 3$, $j = 0$, $k = 1$ and $i, j = 1$, $k = 0$, respectively. Analogically, the subscript $1 + i + 2je$ is transformed to the subscripts $1e$ and $3e$ and $4e$ for $i, j = 0$ and $i = 0$, $j = 1$ and $i, j = 1$, respectively.

Finally, the subscript $1 + ijq$ related to the material parameters of the spherical particle $(q = p)$, the spherical envelope $(q = e)$ and the cell matrix $(q = m)$ is transformed to the subscript $12q$ and $23q$ for $i = 0$, $j = 2$, and $i = 1$, $j = 3$, respectively. Additionally, the subscript $11 + iq$ is transformed to the subscripts $11q$ and $12q$ for $i = 0$ and $i = 1$, respectively.

With regard to the identical right sides of the conditions $\varepsilon_{11tm} - \varepsilon_{11tp} = f(\varphi, \nu)$, $\varepsilon_{11te} - \varepsilon_{11tp} = f(\varphi, \nu)$, $\varepsilon_{11tm} - \varepsilon_{11te} = f(\varphi, \nu)$, the differences on the left sides are defined only as functions of the variables φ, ν. Finally, the differences $\varepsilon_{11tm} - \varepsilon_{11tp}$, $\varepsilon_{11te} - \varepsilon_{11tp}$, $\varepsilon_{11tm} - \varepsilon_{11te}$ are not required to be identical mutually.

1.8 Recommendations of author

1.8.1 Combinations of solutions

The thermal stresses and the corresponding quantities in the isotropic multi-particle-matrix system are represented by s_p and s_m sets of formulae for the spherical particle and for the cell matrix, respectively, where $s_p = 1$, $s_m = 12$, and accordingly, the state of stress and deformation is defined by $s_p \times s_m = 12$ combinations. As presented in Section 1.4, the radial stress p_1, acting on the particle-matrix boundary and representing a common parameter in formulae of the $s_p \times s_m$ combinations, is a function of the coefficients ϱ_{1p} and ϱ_{1m} (see Section 10.1), considering 1 set of formulae for the spherical particle and one of 12 sets of formulae for the cell matrix, respectively.

Regarding or disregarding the phase-transformation induced radial strain ε_{11tq} given by Eqs. (12.23), (12.24), the thermal stresses and the corresponding quantities in the isotropic multi-particle-envelope-matrix system are represented by

1. s_p and s_p^{pe} sets of formulae for the spherical particle provided that $\beta_p \neq \beta_e \neq \beta_m$ or $\beta_p \neq \beta_e = \beta_m$, and $\beta_p = \beta_e \neq \beta_m$ (see Eqs. (3.48)–(3.50)), respectively, where $s_p = s_p^{pe} = 1$,

2. s_e and s_e^{me} and s_e^{pe} sets of formulae for the spherical envelope provided that $\beta_p \neq \beta_e \neq \beta_m$ and $\beta_p \neq \beta_e = \beta_m$ and $\beta_p = \beta_e \neq \beta_m$, respectively, where $s_e = 10$, $s_e^{me} = 5$ for $\varepsilon_{11te} - \varepsilon_{11tp} = f(\varphi, \nu)$, $s_e^{me} = 1$ for $\varepsilon_{11te} - \varepsilon_{11tp} \neq f(\varphi, \nu)$, $s_e^{pe} = 5$,

3. s_m and s_m^{me} sets of formulae for the cell matrix provided that $\beta_p \neq \beta_e \neq \beta_m$ or $\beta_p = \beta_e \neq \beta_m$, and $\beta_p \neq \beta_e = \beta_m$, respectively, where $s_m = s_m^{me} = 12$.

Similarly, the state of stress and deformation in the isotropic multi-particle-envelope-matrix system is defined by $s_p \times s_e \times s_m = 120$, and $s_p \times s_e^{me} \times s_m^{me} = 60$ or $s_p \times s_e^{me} \times s_m^{me} = 12$ for $\varepsilon_{11te} - \varepsilon_{11tp} = f(\varphi, \nu)$ or $\varepsilon_{11te} - \varepsilon_{11tp} \neq f(\varphi, \nu)$, and $s_p^{pe} \times s_e^{pe} \times s_m = 60$ combinations of the sets of formulae for the spherical particle, for the spherical envelope and for the cell matrix, provided that $\beta_p \neq \beta_e \neq \beta_m$ and $\beta_p \neq \beta_e = \beta_m$ and $\beta_p = \beta_e \neq \beta_m$, respectively.

As presented in Section 1.4, the radial stresses p_1 and p_2, acting on the particle-envelope and matrix-envelope boundaries and representing common parameters in formulae of the $s_p \times s_e \times s_m$ and $s_p \times s_e^{me} \times s_m^{me}$ and $s_p^{pe} \times s_e^{pe} \times s_m$ sets, are functions of the coefficients

4. ϱ_{1p}, $\varrho_{1+i1+je}$ $(i, j = 0,1)$, ϱ_{2m}, related to the radial stresses p_1, p_2 provided that $\beta_p \neq \beta_e \neq \beta_m$, considering 1 set of formulae for the spherical particle, one of 10 sets of formulae for the spherical envelope, and one of 12 sets of formulae for the cell matrix,

5. ϱ_{1p}, ϱ_{1e}^{me}, related to the radial stress p_1 provided that $\beta_p \neq \beta_e = \beta_m$, considering one set of formulae for the spherical particle, and one of 5 sets or 1 set of formulae for the spherical envelope provided that $\varepsilon_{11te} - \varepsilon_{11tp} = f(\varphi, \nu)$ or $\varepsilon_{11te} - \varepsilon_{11tp} \neq f(\varphi, \nu)$, respectively,

6. ϱ_{2m}, ϱ_{2e}^{pe}, related to the radial stress p_2 provided that $\beta_p = \beta_e \neq \beta_m$, considering one of 12 sets of formulae for the cell matrix, and one of 5 sets of formulae for the spherical envelope.

As presented in Section 1.4, the radial stresses p_1, p_2 represent common parameters included in all formulae for the thermal stresses and for the corresponding quantities derived by the different mathematical techniques presented in Sections 5.1–5.3, 6.1, 7.1, 7.2, 8.1, 9.2). With regard to the Castigliano's theorem (see Section 3.1.9), such combination of the sets of formulae is considered to result in minimal thermal-stress induced elastic energy of the cell, $W_c = W_p + W_m$ and $W_c = W_p + W_e + W_m$ (see Eqs. (3.44), (3.45)), related to the multi-particle-matrix and multi-particle-envelope-matrix systems, respectively. When $\{u_{1p}\}$, $\left\{u_{1m}^{(1)}, \ldots, u_{1m}^{(n)}\right\}$, for $n = s_m$, are sets of the radial displacements of the infinitesimal spherical cap in the spherical particle and in the cell matrix derived by the different mathematical techniques, and u_{1p}, $u_{1m}^{(i)} \subset \left\{u_{1m}^{(1)}, \ldots, u_{1m}^{(n)}\right\}$ represent such combination to result in minimal elastic energy of the cell, $W_c = W_p + W_m$, the radial stress p_1 acting on the particle-matrix boundary is derived, regarding Eq. (3.58), from the condition

$$\left[\varepsilon_{22m}^{(i)}\right]_{r=R_1} - (\varepsilon_{22p})_{r=R_1} = \beta_m - \beta_p\,, \tag{1.1}$$

resulting in a dependence of p_1 (see Eq. (10.1)) on the coefficients ϱ_{1p} and $\varrho_{1m}^{(i)}$ included in $(\varepsilon_{22p})_{r=R_1}$ and $\left[\varepsilon_{22m}^{(i)}\right]_{r=R_1}$ related to the solutions u_{1p} and $u_{1m}^{(i)}$, respectively.

Similarly, the radial stresses p_1 and p_2 are derived, regarding Eqs. (3.63), (3.64), from the conditions

$$\left[\varepsilon_{22e}^{(i)}\right]_{r=R_1} - (\varepsilon_{22p})_{r=R_1} = \beta_e - \beta_p \,, \tag{1.2}$$

$$\left[\varepsilon_{22m}^{(j)}\right]_{r=R_2} - \left[\varepsilon_{22e}^{(i)}\right]_{r=R_2} = \beta_m - \beta_e \,, \tag{1.3}$$

resulting in dependencies of p_1 and p_2 (see Eqs. (10.2), (10.3)) on the coefficients ϱ_{1p}, ϱ_{1+ke} $(k=0,1)$, ϱ_{2m} included in $(\varepsilon_{22p})_{r=R_1}$, $\left[\varepsilon_{22e}^{(i)}\right]_{r=R_{1+k}}$, $\left[\varepsilon_{22m}^{(j)}\right]_{r=R_2}$ related to the solutions u_{1p}, $u_{1e}^{(i)} \subset \left\{u_{1e}^{(1)}, \ldots, u_{1e}^{(n)}\right\}$, $u_{1m}^{(j)} \subset \left\{u_{1m}^{(1)}, \ldots, u_{1m}^{(l)}\right\}$ for $n = s_e$, $l = s_m$, respectively, representing such combination resulting in minimal elastic energy of the cell, $W_c = W_p + W_e + W_m$, for the multi-particle-envelope-matrix system provided that $\beta_p \neq \beta_e \neq \beta_m$.

Finally, the conditions (1.2) and (1.3) are considered for the determination of the radial stresses p_1 and p_2 related to the multi-particle-envelope-matrix system provided that $\beta_p \neq \beta_e = \beta_m$ and $\beta_p = \beta_e \neq \beta_m$ for $n = s_e^{me}$ and $l = s_m^{pe}$, respectively.

With regard to a physical point of view, the searching of such combination to result in minimal thermal-stress induced elastic energy of the cell is physically correct but time-consuming. Considering practicability, with regard to the quantitative determination of the elastic thermal-stress loading in a real composite material, the recommendations of the author for readers of Categories 1b, 1c, 2 presented in Section 1.8.2 are required to be study in detail.

The same analysis as presented for the multi-particle-(envelope)-matrix system is also valid for the one-particle-(envelope)-matrix system provided that $\varepsilon_{11tm} - \varepsilon_{11tp} = f(\varphi, \nu)$, $\varepsilon_{11te} - \varepsilon_{11tp} = f(\varphi, \nu)$, $\varepsilon_{11tm} - \varepsilon_{11te} = f(\varphi, \nu)$ and $\varepsilon_{11tm} - \varepsilon_{11tp} \neq f(\varphi, \nu)$, $\varepsilon_{11te} - \varepsilon_{11tp} \neq f(\varphi, \nu)$, $\varepsilon_{11tm} - \varepsilon_{11te} \neq f(\varphi, \nu)$, considering $s_p = s_p^{pe} = 1$, $s_e = 10$, $s_e^{me} = s_e^{pe} = 5$, $s_m = s_m^{me} = 12$ and $s_p = s_p^{pe} = 7$, $s_e = 19$, $s_e^{me} = 2$, $s_e^{pe} = 8$, $s_m = s_m^{me} = 1$ (see Sections 9.3.1–9.3.3), respectively.

1.8.2 Categories of readers

With regard to practicability, this book as well as the trilogy is devoted to the following categories of readers:

1. senior undergraduates, PhD students and senior researchers in
   (a) Mechanical Engineering and Applied Mechanics,
   (b) Applied Physics,

   (c) Materials Engineering,

2. practicing engineers in Materials Engineering, working at universities, scientific institutes and in industry.

Exhibiting an interdisciplinary character resulting from an application of principles of Mechanics of Solid Continuum to isotropic composite materials, this book is intended to be used as a textbook on Applied Mechanics regarding readers of **Category 1a**. The readers are consequently able to derive analytical models of thermal stresses in the isotropic multi-particle-**multi-envelope**-matrix system, considering the conditions $\beta_p \neq \beta_e \neq \beta_m$, $\beta_p \neq \beta_e = \beta_m$, $\beta_p = \beta_e \neq \beta_m$ (see Eqs. (3.48)–(3.50)), and consequently to define corresponding boundary conditions to determine integration constants included in solutions of linear differential equations with non-zero right sides and related to the spherical particle, to the spherical envelope and to the cell matrix (see Section 4.1). The same is also valid for the isotropic one-particle-**multi-envelope**-matrix system with the boundary conditions defined in Section 4.2. Additionally, the presented mathematical techniques can be used for the determination of the state of stress in other systems representing an isotropic elastic solid continuum. Accordingly, Chapters 2, 3, 4, 10 and Sections 1.2, 1.4, 1.6, 1.7, 1.8.1, 5.1–5.4, 6.1, 6.2, 7.1–7.3, 8.1, 9.1–9.3, 9.6, 12.1 are recommended to be studied in detail.

The differential equations (5.5), (5.13), (5.18), (6.3), (7.2), (7.11), (8.2), (8.3) are solved using the Wronskian's method resulting in the solutions given by Eqs. (5.7), (5.15), (5.20), (6.4), (7.3), (7.12), (8.4), (8.5), respectively. Accordingly, the selected topics of Applied Mathematics presented in Section 12.4 represents the mathematical knowledge to be required for the study of the book regarding readers of **Category 1a**, including Section 12.4.5 to explain the Wronskian's method illustratively applied to the differential equation (5.5).

In addition to **Category 1a**, formulae presented in Sections 5.5, 5.6 and 5.7; 6.3, 6.4 and 6.5; 7.4, 7.5; devoted to the multi-particle-matrix, multi-particle-envelope-matrix and one-particle-(envelope)-matrix systems, result from Equations (5.7), (5.15), (5.20); (6.4); (7.3), (7.12); respectively, including integration constants determined by the boundary conditions presented in Sections 4.1.1, 4.1.2, 4.1.3 related to the spherical particle ($q = p$), spherical envelope ($q = e$), the cell matrix ($q = m$), respectively.

With regard to **Category 1b**, this book is intended to be used as well as a handbook of formulae by which the elastic energy W induced by stresses originating as a consequence of the presence of a central substitutive atom in cubic crystalline lattices can be determined numerically, representing a considerable quantity in the scientific branch Physics of Transport Phenomena (see Chapter 11). Additionally, formulae for the thermal-stress induced elastic energy density w in components of the isotropic multi-particle-(envelope)-matrix system derived in Volume I can be substituted to general formulae for the elastic energy fluctuation representing an energy barrier derived in Volume III, and consequently the general formulae can be incorporated into analytical models created by physicists and describing an interaction of energy barriers with dislocations or magnetic domain walls.

Similarly, with regard to **Categories 1c, 2**, this book is additionally intended to be used as well as a handbook of formulae by which the elastic thermal-stress loading in the real composite materials included in Items 1–4 presented in Section 2.1 can be quantitatively determined. Additionally, formulae for the parameters w, σ_i, w_i ($i=1,2,3$) related to components of the isotropic multi-particle-(envelope)-matrix system derived in Volume I, substituted to general formulae describing thermal-stress induced phenomena resulting from surface and curve integrals of w, σ_i, w_i derived in Volume III, enable to determine thermal-stress induced strengthening or weakening as well as critical values of the particle radius R_1, the envelope radius R_2, as functions of the particle volume fraction v, resulting in the critical state of a real composite material regarding a crack formation.

Considering the elastic thermal-stress loading, readers of **Categories 1c, 2** usually require to determine average values $\overline{p_1}$ and $\overline{p_2}$ of the radial stresses p_1 and p_2 (see Eqs. (10.1)–(10.6)) dependent on v, R_1, R_2, acting on the particle-envelope and matrix-envelope boundaries, respectively. Consequently, with regard to the multi-particle-envelope-matrix system, one of the average values $\overline{p_1}$, $\overline{p_2}$ or both $\overline{p_1}$, $\overline{p_2}$ can exhibit zero values at the parameters v_c, R_{1c}, R_{2c} at which one of the radial stresses $\overline{p_1}$, $\overline{p_2}$ or both $\overline{p_1}$, $\overline{p_2}$ are transformed from compressive to tensile or vice versa. The compressive or tensile radial stresses $\overline{p_1} > 0$ or $\overline{p_1} < 0$ and $\overline{p_2} > 0$ or $\overline{p_2} < 0$ result in compressive or tensile thermal stresses in the spherical particle and the cell matrix, accordingly representing resistance against compressive or tensile mechanical loading.

Consequently, the parameters v_c, R_{1c}, R_{2c} represent critical values at which the thermal stresses change from positive to negative or vice versa, and a resistive effect against mechanical loading is transformed to a contributory effect or vice versa, respectively. The resistive and contributory effects result in thermal-stress induced strengthening and weakening, respectively, derived by mathematical techniques presented in Volume III. With regard to the critical parameters v_c, R_{1c}, R_{2c}, determined from numerically generated dependencies of the average values $\overline{p_1}$, $\overline{p_2}$ on the variables v, R_1, R_2 and related to a real composite material, readers of **Categories 1c, 2** are able to design and develop microstructure of a real composite material with such material parameters v, R_1, R_2 resulting in the thermal-stress induced strengthening against required mechanical loading. Additionally, provided that the time-dependent development of the microstructure to be tailored regarding the required mechanical loading is defined analytically or experimentally, readers of **Categories 1c, 2** are able to determine the time τ_c when the real composite material exhibits the critical parameters v_c, R_{1c}, R_{2c}.

As an example, microstructure to exhibit usually the unstable equilibrium regarding the enthalpy of a real composite material is significantly transformed at the temperature T higher than the room temperature, 20°C [1, p 38–40]. On the condition of mechanical loading at $T \gg 20°C$, the time τ_c is required to be determined, accordingly representing the lifetime of the real composite material with regard to a transformation of the resistive effect to the contributory effect resulting from a transformation of the microstructure with the critical parameters v_c, R_{1c}, R_{2c}.

Finally, the similar analysis concerning the multi-particle-envelope-matrix system presented in Section 10.2 is considered for the multi-particle-matrix system regarding the average value $\overline{p_1}$ of the radial stress p_1 acting on the particle-matrix boundary and dependent on the material parameters v, R_1, and regarding the critical parameters v_c, R_{1c} at which $\overline{p_1}$ is extremal. Additionally, the average value $\overline{p_1}$ is not assumed to be transformed from positive to negative or vice versa.

As presented in Section 1.8.1, the searching of such combination of solutions to result in minimal thermal-stress induced elastic energy of the cell is time-consuming due to large number of the combinations (see Section 1.8.1). Additionally, with regard to the function $W_c = W_c(v, R_1, R_2)$, different numerical values of the parameters v, R_1, R_2 of a real composite material can result in different combinations of solutions to result in minimal elastic energy of the cell. Representing a hypothetic assumption with regard to the quantitative determination of the elastic thermal-stress loading, the combinations of solutions are assumed to exhibit mutually negligible the differences in the elastic energy of the cell.

Along with the parameters w, σ_i, w_i $(i=1,2,3)$, formulae for the thermal stresses in the spherical particle are presented in Section 5.5.1 related to the multi-particle-matrix and multi-particle-envelope-matrix systems, the latter provided that $\beta_p \neq \beta_e \neq \beta_m$, as well as in Sections 5.6.1.2 and 9.4.1 related to the multi-particle-envelope-matrix system provided that $\beta_p = \beta_e \neq \beta_m$ and $\beta_p \neq \beta_e = \beta_m$, respectively.

Formulae for the thermal stresses in the spherical envelope are presented in Sections 5.6.2.1, 6.4.1.1, 7.5.1.1; 5.6.2.2, 6.4.1.2, 7.5.1.2, 9.4.2; 5.6.2.3, 6.4.1.3, 7.5.1.3, provided that $\beta_p \neq \beta_e \neq \beta_m$; $\beta_p \neq \beta_e = \beta_m$; $\beta_p = \beta_e \neq \beta_m$, respectively. Consequently, with regard to readers of **Categories 1b, 1c, 2**, the formulae from one of Sections 5.6.2.1, 6.4.1.1, 7.5.1.1; from one of Sections 5.6.2.2, 6.4.1.2, 7.5.1.2, 9.4.2; from one of Sections 5.6.2.3, 6.4.1.3, 7.5.1.3, related to the conditions $\beta_p \neq \beta_e \neq \beta_m$; $\beta_p \neq \beta_e = \beta_m$; $\beta_p = \beta_e \neq \beta_m$ are considered to be applied to a real isotropic composite material, respectively.

Formulae for the thermal stresses in the cell matrix are presented in Sections 5.5.2, 6.3.1, 7.4.1 and 5.6.3, 6.4.2, 7.5.2, related to the multi-particle-matrix and multi-particle-envelope-matrix systems, respectively, the latter provided that $\beta_p \neq \beta_e \neq \beta_m$, $\beta_p \neq \beta_e = \beta_m$ (see Eqs. (3.48)–(3.50)). Similarly, readers of **Categories 1b, 1c, 2** are recommended to apply the formulae from one of Sections 5.5.2, 6.3.1, 7.4.1 and from one of Sections 5.6.3, 6.4.2, 7.5.2 to a real isotropic composite material for the quantitative determination of the elastic thermal-stress loading.

Additionally, with regard to completeness of the presented topic and to the analysis in Section 2.2.2, analytical models of the thermal stresses acting in the isotropic one-particle-(envelope)-matrix system are presented in Sections 5.7, 6.5, 9.5.

Finally, along with the sections to include formulae related to components of the isotropic multi-particle-(envelope)-matrix system, readers of **Categories 1b, 1c, 2** are recommended to study Chapters 2, 10, 11 and Sections 1.2–1.4, 1.7, 1.8.3, 3.2.1, 3.3, 12.1.2, 12.2, 12.3 in detail. Considering readers of **Categories 1, 2**, regarding an application of the analytical models of the thermal stresses to a real isotropic composite material, Section 2.4 is recommended to be studied in detail.

Considering the programming techniques presented in Section 1.8.3, using a suitable programming language (Fortran, Pascal) for the creation of a software containing formulae for the thermal stresses and for the corresponding quantities, the latter represented by w, W, σ_i, w_i $(i=1,2,3)$, numerical dependencies generated for a real isotropic composite material by the software can be consequently compared with experimental results. To facilitate in the software creation, the sections presenting formulae for the thermal stresses and for w, σ_i, w_i related to components of the isotropic multi-particle-(envelope)-matrix system are written in an uniform style. Considering surface and curve integrals presented in Volume III in a form of general formulae to result in the determination of the physical quantities defined in Items 1–4 in Chapter *Preface*, the parameters w, σ_i, w_i are derived in the simplest forms regarding terms with variable r included in w, σ_i, w_i. Additionally, formulae of indefinite integrals for the r-dependent terms are presented in Section 12.4.4.

Finally, the monographs, presented in Section *Bibliography* and written in Slovak and Czech, represent common textbooks of the scientific branches Material Science [1], Mechanics of Solid Continuum [2]–[4] and Applied Mathematics [5]–[7], and accordingly equivalent monographs are offered by any public library as well as libraries at universities and scientific institutes.

1.8.3 Programming techniques

With regard to the software creation, a derivation of the parameter Q regarding the variable η can be calculated numerically by the formula

$$\frac{\partial Q}{\partial \eta} \approx \frac{Q(\eta) - Q(\eta + \Delta\eta)}{\Delta\eta}, \tag{1.4}$$

where $Q = \kappa, \psi$, and $\eta = \varphi, \nu$. The angle steps $\Delta\varphi$, $\Delta\nu$ can be taken as small as possible, e.g. $\Delta\varphi = \Delta\nu = 10^{-6}$ [deg].

Similarly, with regard to the elastic energy W representing a definite integral of the coefficient Ω as a function of the angles $\varphi, \nu \in \langle 0, \pi/2 \rangle$, the definite integral can be calculated numerically by the formula

$$W = \int_0^{\pi/2}\int_0^{\pi/2} \Omega \, d\varphi \, d\nu \approx \sum_{j=0}^{m} \left(\sum_{i=0}^{n} W\left(i \times \Delta\varphi; j \times \Delta\nu\right) \Delta\varphi \right) \Delta\nu, \tag{1.5}$$

where n, m are integral parts of the real numbers $\pi/(2\,\Delta\varphi)$, $\pi/(2\,\Delta\nu)$, respectively. With regard to an experience of the author, the steps $\Delta\varphi = \Delta\nu = 0.01 - 0.1$ [deg], dependent on a computer hardware, is sufficient.

With regard to the determination of the particle radius R_1, in case of the R_1-independent and φ, ν-dependent function $(\varepsilon_{22q})_{r=R_1}$ $(q=p,m)$, the definite integral in Eq. (3.65) can be calculated numerically by the formula

$$\int_0^{\pi/2}\int_0^{\pi/2} (\varepsilon_{22q})_{r=R_1}\, d\varphi\, d\nu \approx \sum_{j=0}^{m}\left(\sum_{i=0}^{n} \left[\varepsilon_{22q}\left(i \times \Delta\varphi; j \times \Delta\nu\right)\right]_{r=R_1} \Delta\varphi\right)\Delta\nu, \quad (1.6)$$

considering the same conditions concerning the steps $\Delta\varphi$, $\Delta\nu$ as related to Eq. (1.5).

With regard to the quantitative determination of the elastic thermal-stress loading and the critical parameters v_c, R_{1c}, R_{2c} (see Section 10.2), the average values $\overline{p_1}$ and $\overline{p_2}$ of the radial stresses p_1, p_2 given by Eqs. (10.1)–(10.3), (10.5) are recommended to be derived. Considering the same conditions concerning the steps $\Delta\varphi$, $\Delta\nu$ as related to Eq. (1.5), the definite integral given by Eq. (10.6) can be calculated numerically by the formula

$$\overline{p_{1+i}} = \left(\frac{2}{\pi}\right)^2 \int_0^{\pi/2}\int_0^{\pi/2} p_{1+i}\, d\varphi\, d\nu \approx \left(\frac{2}{\pi}\right)^2 \sum_{k=0}^{m}\left(\sum_{j=0}^{n} p_{1+i}\left(j \times \Delta\varphi; k \times \Delta\nu\right)\Delta\varphi\right)\Delta\nu,$$
$$i = 0, 1. \quad (1.7)$$

Chapter 2

Cell model

2.1 Geometric boundary condition for cell matrix

The analytical models of thermal stresses and the corresponding quantities presented in this book are applicable to real composite materials represented by

1. grains and precipitates, the latter aperiodically distributed in the grains or on the grain boundaries without or with a continuous phase on the surface of each of the precipitates, where the grains and the precipitates along with the continuous phase exhibit mutually different properties,

2. two types of aperiodically distributed grains with different properties and without continuous phases on the surface of the grains of both types,

3. two types of aperiodically distributed grains with identical properties and one type of the grains only is covered by a continuous phase with different properties than those of the grains of both types,

4. one type of grains covered by a continuous phase with different properties than those of the grains.

With regard to the analytical models of the thermal stresses, the real composite materials with finite dimensions are replaced by a multi-particle-(envelope)-matrix system with infinite dimensions, consisted of components represented by periodically distributed spherical particles without or with a spherical envelope on the particle surface, respectively, embedded in an infinite matrix. To derive the thermal stresses acting in the multi-particle-(envelope)-matrix system, the infinite matrix is **imaginarily** divided into identical cells with the shape to correspond to particles distribution, and consequently the thermal stresses are investigated within the cell representing a part of the solid continuum related to one spherical particle. Considering the matrix infinity, formulae for thermal stresses related to a certain cell are consequently valid for any cell in the infinite matrix (see Fig. 2.1a,b).

As presented in Items 1, 2 and 3, the precipitates, one of the two types of the grains, and the grains covered by a continuous phase are considered to represent the periodically distributed spherical particles. Consequently, with regard to Items 1, 2 and 3, the grains embedding the precipitates, the other type of the grains, and the grains without a continuous phase are considered to represent the infinite matrix resulting in the cell matrix. With regard to Item 4, the grain covered by the continuous phase represents the spherical particle covered by the spherical envelope with such envelope radius that the spherical envelope fills the cell maximally, and the rest of the cell, i.e. the cell matrix, is represented by the grains, neglecting the presence of the continuous phase on their surfaces (see Section 2.4). Finally, regarding Items 1–4, boundaries of the grains representing the infinite matrix, and consequently the cell matrix, are considered to represent a coherent solid continuum, and the continuous phase is considered to represent the spherical envelope.

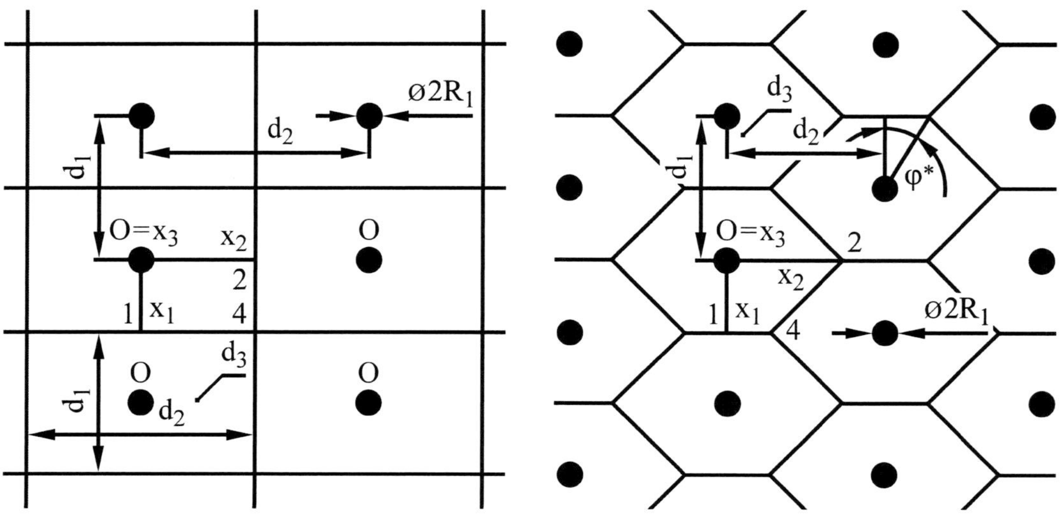

a **b**

Figure 2.1: The infinite matrix divided into the imaginary rectangular-based (a) and hexagonal-based prismatic (b) cells with the central spherical particles of the radius R_1 in the point O of the Cartesian system $(Ox_1x_2x_3)$, and the inter-particle distance d_i along the axis x_i $(i = 1,2,3)$. Along with the inter-particle distance d_i, the hexagonal-based prismatic cell is additionally determined by the angle $\varphi^* \in (0, \pi/2)$. Considering the multi-particle-envelope-matrix system, the spherical envelope on the particle surface with the radii $R_1 < R_2$ is not presented regarding simplification of the figure.

With regard to the periodical distribution of the spherical particles with the radius R_1 shown in Fig. 2.1a,b, the infinite matrix is **imaginarily** divided into rectangular-based (a) and hexagonal-based (b) prismatic cells, both defined by the inter-particle distance d_i along the axis x_i $(i = 1,2,3)$ and the latter additionally defined by the angle $\varphi^* \in (0, \pi/2)$. Moreover, due to symmetry of the multi-particle-(envelope)-matrix system resulting from the matrix infinity and from the periodical distribution, the thermal stresses are sufficient to be investigated in the interval

$\varphi, \nu \in \langle 0, \pi/2 \rangle$ regarding the spatial polar coordinates $[r, \varphi, \nu]$ (see Figs. 2.1a,b, 3.1). The dimensions R_1, R_2, d_i $(i=1,2,3)$ in Fig. 2.1a,b are related to the temperature T_r or T_t, the latter provided that the thermal stresses are induced by a phase transformation only (see Sections 3.3, 10.3).

The spherical particles in the individual planes $x_1'x_2'$, mutually parallel with the inter-plane distance d_3 along the axis x_3', are mutually located along the axis x_3'.

Considering zero displacement of the points O, the centres *1*, *2*, *4* of the abscissae *O1O*, *O2O*, *O4O* in the rectangular-based prismatic cell, respectively, exhibit zero displacement along these abscissae, as also valid for the arbitrary point C on the cell surface along the axis x_1' representing a radial direction (see Figs. 2.2, 3.1). Assuming non-zero radial displacement on the surface of a certain cell, the same condition is also valid on the surface of any cell, resulting from the matrix infinity, as well as on the surface of neighbouring cells. Accordingly, the space between neighbouring cells is replaced by the vacuum what is physically unacceptable. The same conditions are also valid for the points *1*, *2*, *4* of the abscissae *O1O*, *O22O*, *O4*, respectively, and for the arbitrary point C on the surface of the hexagonal-based prismatic cell (see Figs. 2.1b, 2.3). The same physically unacceptable situation arises provided that cubic and regular-hexagonal-based prismatic cells, transformed from the rectangular-based and hexagonal-based prismatic cells on the conditions $d_i = d$ $(i=1,2,3)$ and $d_1 = d_3 = d$, $d_2 = d\sqrt{3}/2$, $\varphi^* = \pi/6$, respectively, are replaced by spherical cells connected in one point related to a pair of the spherical cells, and the matrix between three spherical cells is replaced by the vacuum.

The analytical model of the thermal stresses acting in the multi-particle-(envelope)-matrix system with a finite matrix is required to consider a position of each cell in the finite matrix as well as a shape and dimensions of each cell. Additionally, boundary conditions, related to the radial displacement u_{1m} or to the radial stress σ_{11m} both on the surfaces of each cell, are required to be defined for each cell separately. Accordingly, even though the boundary conditions are defined, an application of the analytical model to a real composite material, resulting in numerical dependencies of the thermal stresses and the corresponding quantities on the particle volume fraction v and on the particle and envelope radii, R_1 and $R_1 < R_2$, respectively, would be probably time-consuming.

On the one hand, as a hypothetic conclusion resulting from the tendency of a system to exhibit the state corresponding to minimal energy [4, p 219], provided that the multi-particle-(envelope)-matrix system is acted by the thermal stresses only, the system represented by the real composite materials included in Items 1–4 might be assumed to exhibit such particles distribution of those shown in Fig. 2.1a,b, which results in lower thermal-stress induced elastic energy of the cell, W_c (see Eqs. (3.44), (3.45)). On the other hand, with regard to the quantitative determination of the elastic thermal-stress loading in the real composite materials, such particles distribution of those shown in Fig. 2.1a,b is considered within the presented analytical models, which results in lower thermal-stress induced elastic energy of the cell.

2.1.1 Summary

Accordingly, the following conclusions, including a geometric boundary condition for the cell matrix defined in Item 1, are considered:

1. zero radial displacement of an arbitrary point in the cell matrix on the cell surface, $(u_{1m})_{r=r_c} = 0$, accordingly represents a geometric boundary condition for the cell matrix, simultaneously valid for a cell of such shape to fill the infinite matrix perfectly, where the φ, ν-dependent distance r_c as a function of the experimentally determined parameters v, R_{1ex}, d_{1ex}, d_{2ex}, d_{3ex} (see Section 2.2.1) is derived in Section 2.3,

2. conversely, such imaginary dividing of the infinite matrix is required regarding particles distribution that cells can fulfil the infinite matrix perfectly,

3. corresponding mathematical techniques similar to those presented in Sections 2.3.1, 2.3.2 can be applied to such cells to fulfil the infinite matrix perfectly, exhibiting a different shape than the rectangular-based and hexagonal-based prismatic cells.

2.2 Particle volume fraction

2.2.1 General formulae

With regard to a real composite material, provided that the particles and the envelopes, exhibiting general shapes and aperiodic distribution, are both considered to exhibit the spherical shape and periodic distribution, the multi-particle-(envelope)-matrix system is characterized at the final temperature T_f of a cooling process by the average particle and envelope radii, R_{1ex} and $R_{1ex} < R_{2ex}$, respectively, by the average inter-particle distance d_{iex} ($i = 1,2,3$) along the axis x_i of the Cartesian system $(Ox_1x_2x_3)$ (see Fig. 2.1a,b), and by the particle volume fraction v, where at least four of the parameters R_{1ex}, d_{1ex}, d_{2ex}, d_{3ex}, v are determined by experimental methods, e.g. using microscopy[1]. Provided that the inter-particle distance d_{iexT} ($i = 1,2,3$) is determined at the temperature $T \in \langle T_f, T_r \rangle$, the inter-particle distance d_{iex} at the temperature T_f has the form

$$d_{iex} = d_{iexT} \left(1 + \beta_{imT}\right), \quad i = 1, 2, 3, \tag{2.1}$$

where the coefficients β_{1mT}, β_{2mT} and β_{3mT} are equal to β_m (see Eqs. (3.48)–(3.50)) for $\varphi = 0$, $\nu = \pi/2$, for $\varphi = \nu = \pi/2$ and for $\varphi = \nu = 0$, respectively, replacing the temperature T_r in the integrals of Eqs. (3.48)–(3.50) by the temperature T. The angles φ, ν are related to the φ, ν-dependent radial strain ε_{11tm} (see Eq. (12.23)),

[1]With regard to the microscopy, the envelope thickness t is usually determined, and thus the envelope radius has the form $R_{2ex} = R_{1ex} + t$ (see Eq. (2.5)).

included in the coefficient β_m and induced by a phase transformation originating in the cell matrix at the temperature $T_{tm} \in \langle T_f, T_r \rangle$, where the relaxation temperature T_r is minimal temperature of the sets $\{T_{rp}, T_{rm}\}$ and $\{T_{rp}, T_{re}, T_{rm}\}$ related to the multi-particle-matrix and multi-particle-envelope-matrix systems (see Section 10.3), respectively.

With regard to four parameters of R_{1ex}, d_{1ex}, d_{2ex}, d_{3ex}, v determined by experimental methods, the fifth parameter can be derived from the formula

$$v = \frac{V_p}{V_c} = \frac{4\pi R_{1ex}^3}{3V_c} \in (0, v_{max}), \tag{2.2}$$

where $V_p = 4\pi R_{1ex}^3/3$ is volume of the spherical particle, and V_c is volume of the rectangular-based and hexagonal-based prismatic cells resulting from the particles distribution in Fig. 2.1a,b. The hexagonal-based prismatic cell is additionally defined by the angle φ^*.

Consequently, the parameters R_{1ex}, R_{2ex}, d_{1ex}, d_{2ex}, d_{3ex}, v can exhibit different numerical values related to cells in different areas of the multi-particle-(envelope)-matrix system, provided that the cells fill the infinite matrix perfectly, as expressed in Item 2 in Section 2.1.1.

With regard to a physical point of view, the multi-particle-(envelope)-matrix system 'is transformed' on the condition $v \longrightarrow 0$ to the one-particle-(envelope)-matrix system containing one spherical particle only to be embedded in the infinite matrix.

2.2.2 Analysis of interval boundaries $v \longrightarrow 0$, $v \longrightarrow v_{max}$

As presented in Section 3.1.8, elastic energy of the infinite multi-particle-(envelope)-matrix system is represented by the thermal-stress induced elastic energy of the cell, W_c (see Eqs. (3.44), (3.45)).

The dependence $W_c - v$, resulting from dependencies of the thermal stresses on v, is required to exhibit limits $W_{c0} = \lim_{v \longrightarrow 0} W_c$, $W_{cmax} = \lim_{v \longrightarrow v_{max}} W_c$, where v_{max} is given by Eqs. (2.4), (2.7), (2.10) considering $t = 0$ for the multi-particle-matrix system, and W_c is the thermal-stress induced elastic energy of the cell (see Eqs. (3.44), (3.45)). Accordingly, the analytical model of the thermal stresses inducing the elastic energy $W_{c0} = \lim_{v \longrightarrow 0} W_c$ related to the multi-particle-(envelope)-matrix system defines, regarding $v \longrightarrow 0$, the state of deformation and stress in the one-particle-(envelope)-matrix system.

The limits $W_{c0} = \lim_{v \longrightarrow 0} W_c$, $W_{cmax} = \lim_{v \longrightarrow v_{max}} W_c$ are meant to be determined numerically, considering material parameters of a real composite material, and consequently the former is compared with a numerical value of $W^{(1)}$ related to the same real composite material. Additionally, the limit $W_{c0} = \lim_{v \longrightarrow 0} W_c$ exists provided that the positive dependence $W_c - r_c$ is a decreasing function of the v-dependent term $r_c = r_c(v)$, or the term $r_c \longrightarrow \infty$ for $v \longrightarrow 0$ (see Section 2.3) is reduced in fractions included in a formula for W_c, otherwise $\lim\limits_{v \to 0} W_c \to \infty$. With regard to a real composite material with $v \nrightarrow 0$, an investigation concerning the condition $\lim\limits_{v \to 0} W_c \to \infty$ is irrelevant.

With regard to the Castigliano's theorem (see Section 3.1.9), such solution or such combination of solutions for components of the multi-particle-(envelope)-matrix system are considered to result in minimal energy W_c. When $W_{c0}^{(i)}$ $(i = 1, \ldots, \mathrm{n})$ is minimal elastic energy of the set $\left\{W_{c0}^{(1)}, \ldots, W_{c0}^{(n)}\right\}$ related to n combinations of solutions, and provided that $W_{c0}^{(i)} \leq W^{(1)}$, consequently the combination of solutions to result in the elastic energy $W_{c0}^{(i)} = \lim\limits_{v \to 0} W_c^{(i)}$ is simultaneously considered for the determination of the elastic thermal-stress loading in the one-particle-(envelope)-matrix system, regarding $v \to 0$. Accordingly, the analytical model for the one-particle-(envelope)-matrix system arises from that for the multi-particle-(envelope)-matrix system.

On the one hand, the condition $W_{c0}^{(i)} > W^{(1)}$ results in a discontinuity of solutions related to the multi-particle-(envelope)-matrix system for $v \to 0$ and to the one-particle-(envelope)-matrix system. On the other hand, with regard to a real composite material with $v \nrightarrow 0$, an investigation of the discontinuity is irrelevant, such as in the case $\lim\limits_{v \to 0} W_c \to \infty$.

2.2.2.1 Elastic energy $W^{(1)}$

With regard to the elastic energy $W^{(1)} = W_p^{(1)} + \left(W_e^{(1)}\right) + W_m^{(1)}$[2] accumulated in the one-particle-(envelope)-matrix system, considering the φ, ν-dependent differences, $\varepsilon_{11tm} - \varepsilon_{11tp} = f(\varphi, \nu)$, $\varepsilon_{11te} - \varepsilon_{11tp} = f(\varphi, \nu)$ and $\varepsilon_{11tm} - \varepsilon_{11te} = f(\varphi, \nu)$ (see Eq. (12.23)), resulting in the φ, ν-dependent differences $\beta_m - \beta_p$, $\beta_e - \beta_p$ and $\beta_m - \beta_e$ (see Eqs. (3.48)–(3.50)), respectively, and consequently in the φ, ν-dependent radial stresses p_1 and p_2 acting on the particle-matrix, particle-envelope and matrix-envelope boundaries, respectively,

1. formulae for the elastic energy $W_m^{(1)}$ in the infinite matrix are presented in Sections 5.7.2, 6.5.2,

2. formulae for the elastic energy $W_p^{(1)}$ and $W_e^{(1)}$ in the spherical particle and in the spherical envelope are identical with those for W_p and W_e presented in Sections 5.5.1, 5.6.1.2 and 5.6.2, 6.4.1, 7.5.1, respectively. Section 5.7.1 is related to the one-particle-matrix and one-particle-envelope-matrix systems regarding the conditions $\beta_p \neq \beta_m$ and $\beta_p \neq \beta_e \neq \beta_m$, $\beta_p \neq \beta_e = \beta_m$, respectively.

Considering the φ, ν-independent differences, $\varepsilon_{11tm} - \varepsilon_{11tp} \neq f(\varphi, \nu)$, $\varepsilon_{11te} - \varepsilon_{11tp} \neq f(\varphi, \nu)$ and $\varepsilon_{11tm} - \varepsilon_{11te} \neq f(\varphi, \nu)$ (see Eq. (12.24)), resulting in the

[2]The superscript (1) is related to the *one*-particle-(envelope)-matrix system.

φ, ν-independent differences $\beta_m - \beta_p$, $\beta_e - \beta_p$ and $\beta_m - \beta_e$ (see Eqs. (3.48)–(3.50)), respectively, and consequently in the φ, ν-independent radial stresses p_1, p_2,

3. formulae for $W_m^{(1)}$ are presented in Section 9.5.3,

4. formulae for $W_p^{(1)}$ and $W_e^{(1)}$ are presented and identical with those for W_p and W_e in Sections 9.4.1, 9.5.1 and 9.4.2, respectively.

With regard to Items 1–4, the analysis concerning the coefficients ϱ_{1p}, ϱ_{1+im}, $\varrho_{1+i1+je}$ $(i, j=0,1)$, ϱ_{1e}^{me}, ϱ_{2e}^{pe} included in formulae for the radial stresses p_1, p_2 (see Eqs. (10.1)–(10.5)) related to the one-particle-(envelope)-matrix system is presented in Section 10.1.2.

2.2.3 Rectangular-based prismatic and cubic cells

Considering the volume $V_c = d_{1ex}d_{2ex}d_{3ex}$ of the rectangular-based prismatic cell, the particle volume fraction v has the form

$$v = \frac{4\pi R_{1ex}^3}{3d_{1ex}d_{2ex}d_{3ex}} \in (0, v_{max}), \tag{2.3}$$

where the parameter v_{max}, determined from the condition $R_{1ex} = (d_{ex}/2) - t$, is derived as

$$v_{max} = \frac{4\pi}{3d_{1ex}d_{2ex}d_{3ex}} \left(\frac{d_{ex}}{2} - t \right)^3, \tag{2.4}$$

where d_{ex} is the minimum of the set $\{d_{1ex}, d_{2ex}, d_{3ex}\}$, considering the thickness t of the spherical envelope in the form

$$t = R_{2ex} - R_{1ex}, \tag{2.5}$$

and accordingly $t = 0$ for the multi-particle-matrix system. Similarly, on the condition $d_{iex} = d_{ex}$ $(i=1,2,3)$ corresponding to a cubic cell, we get

$$v = \frac{4\pi}{3} \left(\frac{R_{1ex}}{d_{ex}} \right)^3 \in (0, v_{max}), \tag{2.6}$$

$$v_{max} = \frac{\pi}{6} \left(1 - \frac{2t}{d_{ex}} \right)^3 = \frac{\pi}{6} \left(1 - \frac{t}{R_{1ex} + t} \right)^3, \tag{2.7}$$

transforming to $v_{max} = \pi/6$ for the multi-particle-matrix system.

2.2.4 Hexagonal-based prismatic cell

Considering the volume $V_c = d_{1ex}d_{2ex}d_{3ex}$ of the hexagonal-based prismatic cell[3], the particle volume fraction v and the parameter v_{max} are given by Eqs. (2.3) and (2.4), respectively, where d_{ex} is the minimum of the set $\{d_{1ex}, d_{24}, d_{3ex}\}$. The parameter d_{24} represents altitude of the triangle *O24* related to the basis *24* derived in the form[4]

$$d_{24} = \frac{d_{1ex}\left(2d_{2ex} - d_{1ex}\tan\varphi^*\right)}{2\left[d_{1ex}^2 + 4\left(d_{2ex} - d_{1ex}\tan\varphi^*\right)^2\right]^{1/2}}. \tag{2.8}$$

On the conditions $d_{1ex} = d_{3ex} = d_{ex}$, $d_{2ex} = d_{ex}\sqrt{3}/2$ corresponding to a regular hexagon, thus for $\varphi^* = \pi/6$, we get

$$v = \frac{4\pi\sqrt{3}}{9}\left(\frac{R_{1ex}}{d_{ex}}\right)^3 \in (0, v_{max}), \tag{2.9}$$

where the parameter v_{max}, determined from the condition $R_{1ex} = (d_{ex}/2) - t$, is derived as

$$v_{max} = \frac{\pi\sqrt{3}}{18}\left(1 - \frac{2t}{d_{ex}}\right)^3 = \frac{\pi\sqrt{3}}{18}\left(1 - \frac{t}{R_{1ex} + t}\right)^3, \tag{2.10}$$

transforming to $v_{max} = \pi\sqrt{3}/18$ for the multi-particle-matrix system, thus for $t = 0$, where the envelope thickness t is given by Eq. (2.5).

2.3 Determination of distance $r_c = r_c(\varphi, \nu)$

With regard to the geometric boundary condition $(u_{1m})_{r=r_c} = 0$ defined in Item 1 in Section 2.1.1, the parameter r_c is a φ, ν-dependent distance from the particle centre O to the point C on the cell surface (see Figs. 2.2, 2.3) is required to be determined, where C is a point of intersection of the axis x_1', representing a radial direction, with the cell surface. Finally, the φ, ν-dependent distance $r_c = |OC|$, representing length of the abscissa OC in a radial direction, is derived as a function of the experimentally determined parameters v, R_{1ex}, d_{1ex}, d_{2ex}, d_{3ex} (see Section 2.2.1) in Sections 2.3.1 and 2.3.2, concerning the rectangular-based and hexagonal-based prismatic cells, respectively, and resulting from the formula for the particle volume fraction given by Eq. (2.3).

[3]Replacing d_i by d_{iex} $(i = 1,2,3)$ (see Fig. 2.1b), the volume V_c of the hexagonal-based prismatic cell is given by $V_c = 4d_{3ex}\left(S_{O14} + S_{O24}\right)$, where $S_{O14} = |O1| \times |14|/2$ and $S_{O24} = |O1| \times |O2|/2$ [7, p 117] represent surface area of the triangles *O14* and *O24*, respectively, considering the lengths $|O1| = d_{1ex}/2$, $|O2| = d_{2ex} - |14|$, $|14| = \left(d_{1ex}\tan\varphi^*\right)/2$ of the abscissae *O1*, *O2*, *14*, respectively. To avoid misunderstanding with regard to Fig. 2.1b, the abscissa *O4* is not drawn.

[4]Considering the surface area $S_{O24} = d_{24} \times |24|/2 = |O1| \times |O2|/2$ of the triangle *O24* with the basis $|24|$ and the altitude d_{24}, we get $d_{24} = |O1| \times |O2| / |24|$, where $|24|^2 = |O1|^2 + \left(d_{2ex} - 2|14|\right)^2$.

2.3.1 Rectangular-based prismatic and cubic cells

The φ, ν-dependent distance r_c between the particle centre O and the points C on the surfaces *3657* and *1456*, *2457* of the rectangular-based prismatic cell (see Fig. 2.2) is derived as

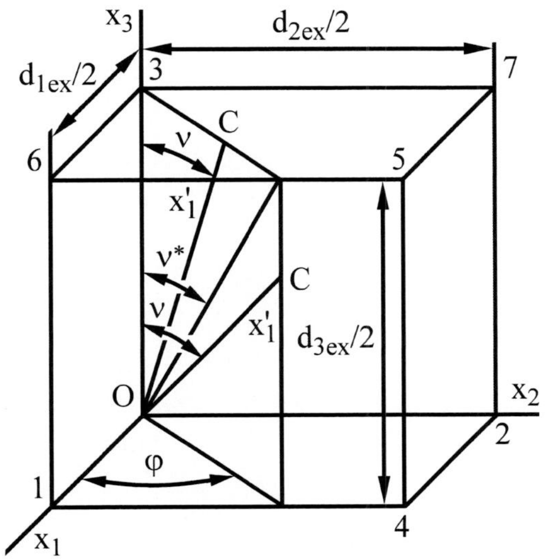

Figure 2.2: The points C as intersections of the axis x_1', representing a radial direction, with the surfaces *3657* and *1456* of one eighth of the rectangular-based prismatic cell (see Fig. 2.1a) for $\nu \in \langle 0, \nu^* \rangle$ and $\nu \in \langle \nu^*, \pi/2 \rangle$, respectively.

$$r_c = \frac{d_{3ex}}{2\cos\nu}, \qquad \nu \in \langle 0, \nu^* \rangle\,, \tag{2.11}$$

$$r_c = \frac{1}{2c_\varphi \sin\nu}, \qquad \nu \in \left\langle \nu^*, \frac{\pi}{2} \right\rangle, \tag{2.12}$$

where the angle ν^* and the coefficient c_φ have the forms

$$\nu^* = \arctan\left(\frac{1}{c_\varphi d_{3ex}}\right), \tag{2.13}$$

$$c_\varphi = \frac{\cos\varphi}{d_{1ex}}, \qquad \varphi \in \langle 0, \varphi^* \rangle\,, \tag{2.14}$$

$$c_\varphi = \frac{\sin\varphi}{d_{2ex}}, \qquad \varphi \in \left\langle \varphi^*, \frac{\pi}{2} \right\rangle, \tag{2.15}$$

$$\varphi^* = \arctan\left(\frac{d_{2ex}}{d_{1ex}}\right). \tag{2.16}$$

On the condition $d_{iex} = d_{ex}$ ($i = 1,2,3$) corresponding to a cubic cell, regarding Eq. (2.6), we get

$$r_c = \frac{d_{ex}}{2\cos\nu} = \frac{R_{1ex}}{2\cos\nu}\left(\frac{4\pi}{3v}\right)^{1/3}, \qquad \nu \in \langle 0, \nu^* \rangle\,, \tag{2.17}$$

$$r_c = \frac{d_{ex}}{2c_\varphi \sin\nu} = \frac{R_{1ex}}{2c_\varphi \sin\nu}\left(\frac{4\pi}{3v}\right)^{1/3}, \quad \nu \in \left(\nu^*, \frac{\pi}{2}\right\rangle, \tag{2.18}$$

where the angle ν^* and the coefficient c_φ for the cubic cell are derived as

$$\nu^* = \arctan\left(\frac{1}{c_\varphi}\right), \tag{2.19}$$

$$c_\varphi = \cos\varphi, \quad \varphi \in \left\langle 0, \frac{\pi}{4}\right\rangle, \tag{2.20}$$

$$c_\varphi = \sin\varphi, \quad \varphi \in \left(\frac{\pi}{4}, \frac{\pi}{2}\right\rangle. \tag{2.21}$$

2.3.2 Hexagonal-based prismatic cell

The φ, ν-dependent distance r_c between the particle centre O and the points C on the surfaces *3657* and *1456* of the hexagonal-based prismatic cell with the parameters d_{1ex}, d_{2ex}, d_{3ex}, φ^* (see Fig. 2.3) is derived by Eqs. (2.11)–(2.14), considering the coefficient c_φ for $\varphi \in (\varphi^*, \pi/2\rangle$ in the form

$$c_\varphi = \frac{d_{1ex}\sin\varphi + 2\left(d_{2ex} - d_{1ex}\tan\varphi^*\right)\cos\varphi}{d_{1ex}\left(2d_{2ex} - d_{1ex}\tan\varphi^*\right)}, \quad \varphi \in \left(\varphi^*, \frac{\pi}{2}\right\rangle. \tag{2.22}$$

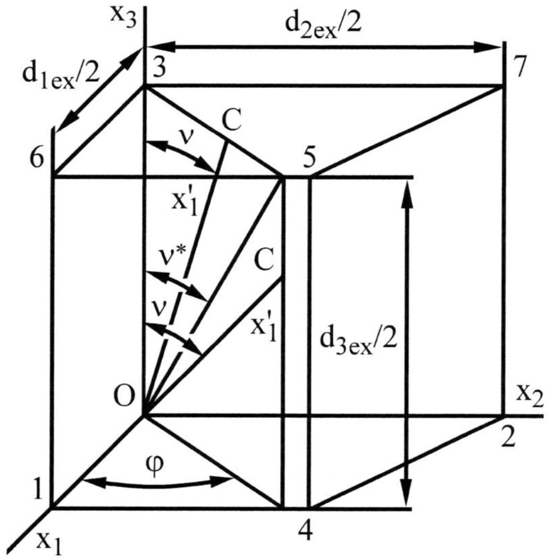

Figure 2.3: The points C as intersections of the axis x_1', representing a radial direction, with the surfaces *3657* and *1456* of one eighth of the hexagonal-based prismatic cell (see Fig. 2.1b) for $\nu \in \langle 0, \nu^*\rangle$ and $\nu \in \langle \nu^*, \pi/2\rangle$, respectively.

On the conditions $d_{1ex} = d_{3ex} = d_{ex}$, $d_{2ex} = 2d_{ex}\sqrt{3}/3$ corresponding to a regular hexagon, thus for $\varphi^* = \pi/6$, regarding Eq. (2.9), we get

$$r_c = \frac{d_{ex}}{2\cos\nu} = \frac{R_{1ex}}{2\cos\nu}\left(\frac{4\pi\sqrt{3}}{9v}\right)^{1/3}, \quad \nu \in \langle 0, \nu^*\rangle, \tag{2.23}$$

$$r_c = \frac{d_{ex}}{2c_\varphi \sin\nu} = \frac{R_{1ex}}{2c_\varphi \sin\nu}\left(\frac{4\pi\sqrt{3}}{9v}\right)^{1/3}, \quad \nu \in \left(\nu^*, \frac{\pi}{2}\right), \tag{2.24}$$

where the angle ν^* is given by Eq. (2.19), considering the coefficient c_φ derived as

$$c_\varphi = \cos\left(\varphi - \frac{z\pi}{3}\right), \quad \varphi \in \left\langle \frac{\pi(2z-1)}{6}, \frac{\pi(2z+1)}{6} \right\rangle, \quad z = 0, 1. \tag{2.25}$$

2.4 Real composite material

Provided that particles in a real composite material are represented by precipitates without or with a continuous phase on the surface of each of the precipitates as defined in Item 1 in Section 2.1, both considered to be spherical with the radii R_{1ex} and $R_{1ex} < R_{2ex}$, respectively, in contrast to the particles distribution characterized by d_{1ex}, d_{2ex}, d_{3ex} (see Fig. 2.1a,b), the real composite material is characterized by one inter-particle distance $d_{ex} = d_{1ex} = d_{2ex} = d_{3ex}$, and accordingly the cubic cell and the φ, ν-dependent distance r_c for $v \in (0, v_{max})$ given by Eqs. (2.7), (2.17)–(2.21) are recommended to be considered in formulae for the thermal stresses and the corresponding quantities, where R_{1ex}, R_{2ex}, d_{ex} are determined by experimental methods. With regard to Items 2, 3 in Section 2.1, the same conditions are also valid provided that a real material is represented

1. by two types of grains, A and B, with the volume fractions v_A and $v_B = 1 - v_A$, respectively, characterized by different properties without continuous phases on the surface of the grains of both types,

2. by two types of grains with identical properties and one type of the grains is covered by a continuous phase with different properties than those of the grains of both types.

With regard to Item 1 and to the analytical models of the thermal stresses, provided that $v_A < v_{max}$ and simultaneously $v_B < v_{max}$, the grains A, B can represent the spherical particle, the infinite matrix, respectively, or vice versa, resulting in the elastic energy of the cell, $W_{cAB} = W_{pA} + W_{mB}$ or $W_{cBA} = W_{pB} + W_{mA}$ (see Eq. (3.44)), substituting $v = v_A$ or $v = v_B$ in Eqs. (2.17), (2.18), respectively. Considering the Castigliano's theorem (see Section 3.1.9) [4, p 219], and provided that $W_{cAB} < W_{cBA}$ or $W_{cAB} > W_{cBA}$, the $A - B$ or $B - A$ multi-particle-matrix system is considered, respectively. Finally, provided that $v_Q > v_{max}$ $(Q = A, B)$, the grains Q are considered to represent the infinite matrix.

Provided that a real composite material is represented by one type of grains covered by a continuous phase with different properties than those of the grains as defined in Item 4 in Section 2.1, both considered to be spherical with the radii R_{1ex}

and $R_{2ex} = R_{1ex}+t$, respectively, the prismatic cell with the regular-hexagonal basis and the φ, ν-dependent distance r_c for $v \to v_{max}$ given by Eqs. (2.10), (2.23)–(2.25) are recommended to be considered for the analytical models of the thermal stresses.

Chapter 3

Thermal stresses in elastic solid continuum

3.1 Selected topics of Mechanics of Solid Continuum

Along with the analysis concerning the radial displacement of an infinitesimal spherical cap and the elastic energy (see Sections 3.1.2, 3.1.8), Section 3.1, presenting fundamental equations of a solid continuum including the Castigliano's theorem (see Sections 3.1.3–3.1.5, 3.1.9)), represents a short introduction to the scientific branch Mechanics of Solid Continuum, assuming elasticity of the solid continuum resulting from the Hooke's laws to consider a linear strain-stress dependence (see Section 3.1.6). Finally, the differential equations (3.35), (3.36), resulting from the Cauchy's and equilibrium equations, and the Hooke's laws, represent initial equations to which the mathematical techniques presenting in Sections 5.1–5.3, 6.1, 7.1, 7.2, 8.1, 9.2 are applied.

3.1.1 Coordinate system

As usual in Mechanics of Solid Continuum representing a fundamental basis of Applied Mechanics, the state of deformation and stress of a solid continuum is investigated within an infinitesimal part of the solid continuum in an arbitrary point with a position determined by a suitable system of coordinates. Additionally, the infinitesimal part has a suitable shape regarding geometry of the solid continuum. An influence of the solid continuum on the infinitesimal part is represented by displacements of the infinitesimal part along individual directions, and by forces acting along individual directions on surfaces of the infinitesimal part [2, p 12–14].

Thermal stresses are investigated in the arbitrary point P of the solid continuum along the axes x_1', x_2', x_3' of the Cartesian system $(Px_1'x_2'x_3')$ (see Fig. 3.1). With regard to the spatial polar coordinates $[r, \varphi, \nu]$, the axes $x_1' = x_r$ and $x_2' = x_\varphi \parallel x_1x_2$, $x_3' = x_\nu$ represent radial and tangential directions related to the spherical surface in the point P, respectively, where $r = |OP|$ is radius of the spherical surface, and $P = [r, \varphi, \nu]$ for $r \in \langle 0, r_c \rangle$, $\varphi, \nu \in \langle 0, \pi/2 \rangle$ (see Section 2.1). Finally, the infinitesimal part of the solid continuum in the point P is represented by an infinitesimal spherical cap described by the parameters dr, $d\varphi$, $d\nu$ (see Figs. 3.2, 3.3).

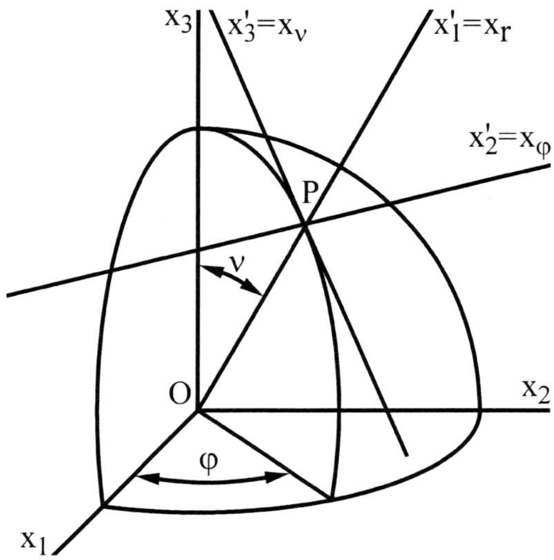

Figure 3.1: The radial and tangential axes $x_1' = x_r$ and $x_2' = x_\varphi \parallel x_1x_2$, $x_3' = x_\nu$, respectively, and the point P with a position determined by the spatial polar coordinates $[r, \varphi, \nu]$ regarding the Cartesian system $(Ox_1x_2x_3)$, where O is the centre of the spherical particle, and x_2', x_3' are tangents to the spherical surface with the radius $r = |OP|$.

3.1.2 Radial displacement

3.1.2.1 Multi-particle-matrix system

With regard to the multi-particle-matrix system, the thermal stresses originate as a consequence of the condition $\beta_m - \beta_p \neq 0$ (see Eq. (3.56)), and as presented in Section 3.2.1, the coefficient β_q $(q = p, e, m)$ is a function of the thermal expansion coefficient α_q and the radial strain ε_{11tq} (see Eqs. (3.48)–(3.50), (12.23), (12.24)), the latter as a function of the difference in dimensions of cubic crystalline lattices considered in a radial direction, transforming mutually at the temperature $T_{tq} \in \langle T_f, T_r \rangle$ of a phase transformation originating in the spherical particle $(q = p)$ and/or the cell matrix $(q = m)$, where the temperature T_r is the recovery temperature below that the stress relaxation as a consequence of thermal-activated processes does not occur in a composite material [1, p 38–40], representing minimal temperature from the set $\{T_{rp}, T_{rm}\}$, and T_f is final temperature of a cooling process (see Sections 10.3, 12.1).

As presented in Section 3.3, on the condition $\alpha_p \neq \alpha_m$ for $T \in \langle T_f, T_r \rangle$, the parameter R_1 represents a particle radius at the temperature T_r. Provided that the phase transformation inducing the radial strain ε_{11tq} (see Eqs. (12.23), (12.24)) originates at the temperature $T_{tq} \in \langle T_f, T_r \rangle$ at least in one of components of the multi-particle-matrix system, and on the conditions $\alpha_p = \alpha_m$ valid at least in the temperature interval $\langle T_{tq}, T_r \rangle$ and simultaneously for $\varepsilon_{11tm} - \varepsilon_{11tp} \neq 0$, the parameter R_1 represents particle radius at the temperature T_t as maximal temperature from the set $\{T_{tp}, T_{tm}\}$, where $\langle T_f, T_t \rangle \subset \langle T_f, T_r \rangle$ (see Section 10.3).

As presented in Section 2.1, and resulting from the matrix infinity, an investigation of the thermal stresses related to a certain cell is also valid for any cell in the infinite matrix. Considering $\beta_m - \beta_p > 0$ or $\beta_m - \beta_p < 0$, the cell matrix on the particle-matrix boundary is pushed or pulled by the spherical particle in a radial direction, respectively, and vice versa. The pulling or the pushing in an arbitrary point on the particle-matrix boundary, thus for $r = R_1$, is realized in a direction of the normal to the infinitesimal surface with the dimensions $R_1\,d\varphi \times R_1\,d\nu$ in the arbitrary point.

Considering the arbitrary point P to be a point on the particle-matrix boundary (see Fig. 3.1), and since the axis x_1' of the Cartesian system $(Px_1'x_2'x_3')$ in the arbitrary point P represents a radial direction regarding the Cartesian system $(Px_1'x_2'x_3')$ and simultaneously the normal to the particle-matrix boundary in the point P, the displacement of the infinitesimal spherical cap with the dimensions $dr \times r\,d\varphi \times r\,d\nu$ (see Fig. 3.2) in the cell matrix and in the spherical particle on the particle-matrix boundary, thus for $r = R_1$, is realized along the axis x_1', i.e. in a radial direction only.

Considering an infinitesimal part of a solid continuum represented by fibres with the lengths R_1 and $r_c - R_1$, and with the dimensions $r\,d\varphi \times r\,d\nu$ of the surface perpendicular to the normal at the radii $r \in \langle 0, R_1 \rangle$ and $r \in \langle R_1, r_c \rangle$ in the spherical particle and in the cell matrix in a direction of the axis x_1', respectively, any infinitesimal part of the fibres at the radius r and with the dimensions $dr \times r\,d\varphi \times r\,d\nu$ exhibits the radial displacement u_{1q} only. The φ, ν-dependent distance r_c as a function of the experimentally determined parameters v, R_{1ex}, d_{1ex}, d_{2ex}, d_{3ex} (see Section 2.2.1) is derived in Section 2.3.

Finally, with regard to the Cartesian system $(Px_1'x_2'x_3')$, the infinitesimal spherical cap exhibits the radial displacement u_{1q} only, along the axis x_1' representing the normal to the particle-matrix boundary.

Resulting from zero radial displacements in the particle centre O and on the cell surface, regarding the geometric boundary conditions (4.3) and (4.11) for the spherical particle and for the cell matrix (see Section 2.1), and from the maximal values $\left|(u_{1p})_{r=R_1}\right|$ and $\left|(u_{1m})_{r=R_1}\right|$ on the particle-matrix boundary, thus for $r = R_1$, the radial displacements u_{1p} and u_{1m} in the spherical particle and in the cell matrix are functions of the variable $r \in \langle 0, R_1 \rangle$ and $r \in \langle R_1, r_c \rangle$, resulting in the differential changes $(\partial u_{1p}/\partial r)\,dr$ and $(\partial u_{1m}/\partial r)\,dr$ corresponding to the change dr of the variable r, and exhibiting zero values in the particle centre O and on the cell surface (see Eqs. (4.3), (4.11)), thus for $r = 0$ and $r = r_c$, respectively. Accordingly, $|u_{1p}|$ and $|u_{1m}|$ represent increasing and decreasing functions of the variable $r \in \langle 0, R_1 \rangle$ and $r \in \langle R_1, r_c \rangle$, respectively.

With regard to the φ, ν-dependence of the radial displacement u_{1m} in the cell matrix resulting from the φ, ν-dependence of the parameter r_c (see Section 2.3), the φ, ν-dependent compressive or tensile radial stress, $p_1 > 0$ or $p_1 < 0$, is accordingly induced on the particle-matrix boundary, consequently resulting in the φ, ν-dependence of the radial displacement u_{1p} in the spherical particle. Consequently, resulting from the $u_{1q} - \varphi$ and $u_{1q} - \nu$ dependencies $(q = p, m)$, the shear strains, stresses, ε_{12q}, σ_{12q} and ε_{13q}, σ_{13q}, are induced, respectively. Additionally, the difference $(\Delta u_{1mp})_{r=R_1} = R_1\,(\beta_m - \beta_p) \neq 0$ (see Eq. (3.56)) is φ, ν-independent or φ, ν-dependent as a consequence of the φ, ν-independent or φ, ν-dependent difference, $\varepsilon_{11tm} - \varepsilon_{11tp} \neq f(\varphi, \nu)$ or $\varepsilon_{11tm} - \varepsilon_{11tp} = f(\varphi, \nu)$ (see Eqs. (12.23), (12.24)), respectively, the former including $\varepsilon_{11tm} - \varepsilon_{11tp} = 0$, respectively.

Additionally, considering infinitesimal fibres in the radial direction in the spherical particle and in the cell matrix with the dimensions $R_1 \times R_1\, d\varphi \times R_1\, d\nu$ and $(r_c - R_1) \times R_1\, d\varphi \times R_1\, d\nu$, respectively, any infinitesimal fibre part with dimensions $dr \times R_1\, d\varphi \times R_1\, d\nu$ in the spherical particle and in the cell matrix is shifted in the radial direction as the normal to the particle-matrix boundary in the point P and exhibits zero radial shift in the particle centre O and on the cell surface (see Figs. 2.1, 2.2), respectively, regarding the geometric boundary condition for the cell matrix (see Section 2.1).

Similarly, any infinitesimal spherical cap with the dimensions $dr \times r\, d\varphi \times r\, d\nu$ for $r \in \langle 0, R_1 \rangle$ and $r \in \langle R_1, r_c \rangle$ (see Figs. (3.2)–(3.5)) is shifted in a radial direction and exhibits zero radial shift in the particle centre O and on the cell surface, respectively.

The same analysis is also valid for the one-particle-matrix system containing one spherical particle, and in contrast to the multi-particle-matrix system, the φ, ν-dependencies of the radial displacements u_{1p}, u_{1m} result from the φ, ν-dependent difference $\varepsilon_{11tm} - \varepsilon_{11tp} = f(\varphi, \nu)$ only (see Eqs. (12.23), (12.24)).

3.1.2.2 Multi-particle-envelope-matrix system

The same analysis as presented in Section 3.1.2.1 and concerning the radial displacement u_{1q} along with the dependencies $u_{1q} - \varphi$, $u_{1q} - \nu$ in the spherical particle and in the cell matrix as components of the multi-particle-matrix system, i.e. $q = p, m$, is considered for the multi-particle-envelope-matrix system, i.e. $q = p, e, m$. Additionally, the radial displacement u_{1e} in the spherical envelope is a function of the variable $r \in \langle R_1, R_2 \rangle$, and at the very most, a constant function regarding the variable r.

As presented in Section 3.3, on the conditions $\alpha_p \neq \alpha_e \neq \alpha_m$, $\alpha_p \neq \alpha_e = \alpha_m$, $\alpha_p = \alpha_e \neq \alpha_m$ for $T \in \langle T_f, T_r \rangle$, the parameters R_1, R_2 represent radii at the temperature T_r. Provided that a phase transformation inducing the radial strain ε_{11tq} (see Eqs. (12.23), (12.24)) originates at the temperature $T_{tq} \in \langle T_f, T_r \rangle$ at least in one of components of the multi-particle-envelope-matrix system represented by the spherical particle $(q = p)$, the spherical envelope $(q = e)$ and the cell matrix $(q = m)$, and on the condition $\alpha_p = \alpha_e = \alpha_m$ valid at least in the temperature interval $\langle T_{tq}, T_r \rangle$ and simultaneously for $\varepsilon_{11te} - \varepsilon_{11tp} \neq 0$, $\varepsilon_{11tm} - \varepsilon_{11te} \neq 0$, the parameters R_1, R_2 represent radii at the temperature T_t representing maximal temperature from the set $\{T_{tp}, T_{te}, T_{tm}\}$, and T_r is minimal temperature from the set $\{T_{rp}, T_{re}, T_{rm}\}$, where $\langle T_f, T_t \rangle \subset \langle T_f, T_r \rangle$ (see Section 10.3).

Accordingly, the compressive or tensile radial stresses $p_1 > 0$ and $p_2 > 0$ or $p_1 < 0$ and $p_2 < 0$, acting on the particle-envelope and matrix-envelope boundaries, are induced as consequences of the conditions $\beta_e - \beta_p \neq 0$ and $\beta_m - \beta_e \neq 0$ (see Eqs. (3.59), (3.60), (3.63), (3.64)), respectively.

As only exception resulting from the analysis presented in Section 9.1 and concerning the conditions $\beta_p \neq \beta_e = \beta_m$ and $\varepsilon_{11te} - \varepsilon_{11tp} \neq f(\varphi, \nu)$, the radial displacements u_{1p}, u_{1e} and consequently the radial stress p_1 are not functions of the variables φ, ν, and accordingly the shear strain and stress, $\varepsilon_{12+iq} = 0$ and $\sigma_{12+iq} = 0$ $(i=0,1;\ q=p,e)$ (see Eqs. (3.4), (3.5), (3.33), (3.34); see Sections 9.4.1, 9.4.2), respectively, as well as $(u_{1m})_{r=R_2} \neq f(\varphi, \nu)$ considering the boundary condition (4.12).

The analysis presented in Section 3.1.2.1 is also valid for the one-particle-matrix system, and in contrast to the multi-particle-matrix system, the φ, ν-dependencies of the radial displacements u_{1p}, u_{1e}, u_{1m} result from the φ, ν-dependent difference in the phase-transformation induced radial strains ε_{11tp}, ε_{11te}, ε_{11tm} only.

3.1.2.3 Summary

The main results of the analyses presented in Sections 3.1.2.1, 3.1.2.2 are as follows:

1. With regard to the elastic thermal-stress loading to generate on the particle-matrix, particle-envelope and matrix-envelope boundaries, the infinitesimal spherical cap as an infinitesimal part of a solid continuum represented by the multi-particle-(envelope)-matrix system, is shifted along the normal to the boundaries only.

2. The normal is identical with the axis x_1' representing a radial direction determined by the angles φ, ν and perpendicular to tangential directions represented by the axes x_2', x_3' of the Cartesian system $(Px_1'x_2'x_3')$ with a position given by the spatial polar coordinates $[r, \varphi, \nu]$ regarding the Cartesian system $(Ox_1x_2x_3)$, where $r = |OP|$ and O is the particle centre (see Fig. 3.1).

3. Accordingly, with regard to the Cartesian system $(Px_1'x_2'x_3')$, the infinitesimal spherical cap exhibits the radial displacement u_{1q} only $(q=p,e,m)$.

4. The absolute values $|u_{1p}|$ and $|u_{1m}|$ related to the spherical particle and the cell matrix represent increasing and decreasing functions of the variable $r \in \langle 0, R_1 \rangle$ and $r \in \langle R_1, r_c \rangle$, $r \in \langle R_2, r_c \rangle$, exhibiting zero values in the particle centre O and on the cell surface (see Eqs. (4.3), (4.11)), thus for $r = 0$ and $r = r_c$, respectively, where the intervals $\langle R_1, r_c \rangle$ and $\langle R_2, r_c \rangle$ are related to the multi-particle-matrix and multi-particle-(envelope)-matrix systems, respectively.

5. Additionally, the radial displacement u_{1e} in the spherical envelope is a function of the variable $r \in \langle R_1, R_2 \rangle$, and at the very most, a constant function regarding the variable r. Although the assumption for the radial displacement u_{1e} as a constant function regarding the variable r may be improbable, nevertheless it is considered, with regard to completeness of the presented topic, in Sections 6.4.1.2.C, 6.4.1.3.B and 9.4.2, 9.5.2.3, the last two for $\lambda_e = 0$ (see Section 9.3.2).

6. With regard to the multi-particle-matrix and multi-particle-envelope-matrix systems, the φ, ν-dependencies of the radial displacements u_{1p}, u_{1e}, u_{1m} result from the φ, ν-dependent differences in the phase-transformation induced radial strains, $\varepsilon_{11tm} - \varepsilon_{11tp} = f(\varphi, \nu)$, and $\varepsilon_{11te} - \varepsilon_{11tp} = f(\varphi, \nu)$ or $\varepsilon_{11tm} - \varepsilon_{11te} = f(\varphi, \nu)$ (see Eq. (12.23)), respectively, as well as from the φ, ν-dependent distance r_c as a function of the experimentally determined parameters v, R_{1ex}, d_{1ex}, d_{2ex}, d_{3ex} (see Sections 2.2.1, Eq. (2.3)). With regard to the one-particle-(envelope)-matrix system, the φ, ν-dependencies of the radial displacements u_{1p}, u_{1e}, u_{1m} result from the differences in ε_{11tp}, ε_{11te}, ε_{11tm} only. As only exception resulting from the analysis presented in Section 9.1 and concerning the conditions $\beta_p \neq \beta_e = \beta_m$ and $\varepsilon_{11te} - \varepsilon_{11tp} \neq f(\varphi, \nu)$ (see Eqs. (3.48)–(3.50)), the radial displacements u_{1p}, u_{1e}, $(u_{1m})_{r=R_2}$ are not functions of the variables φ, ν (see Section 9.1).

3.1.3 Cauchy's equations

The Cauchy's equations, representing geometric equations, define a relationship between displacements and strains of an infinitesimal part of a solid continuum. With regard to the infinitesimal spherical cap (see Figs. 3.2, 3.3), the radial strain $\varepsilon_{11} = \varepsilon_r$, the tangential strains $\varepsilon_{22} = \varepsilon_\varphi$, $\varepsilon_{33} = \varepsilon_\nu$, and the shear strains $\varepsilon_{12} = \varepsilon_{r\varphi}$, $\varepsilon_{13} = \varepsilon_{r\nu}$ are derived as[1]

$$\varepsilon_{11} = \varepsilon_r = \frac{|1'''1'| - |1''1|}{|1''1|} = \frac{1}{dr}\left[\left(dr + \frac{\partial u_1}{\partial r} dr\right) - dr\right] = \frac{\partial u_1}{\partial r}, \tag{3.1}$$

$$\varepsilon_{22} = \varepsilon_\varphi = \frac{|1'2'| - |12|}{|12|} = \frac{(u_1 + r)\, d\varphi - r\, d\varphi}{r\, d\varphi} = \frac{u_1}{r}, \tag{3.2}$$

$$\varepsilon_{33} = \varepsilon_\nu = \frac{|1'3'| - |13|}{|13|} = \frac{(u_1 + r)\, d\nu - r d\nu}{r d\nu} = \frac{u_1}{r}, \tag{3.3}$$

$$\varepsilon_{12} = \varepsilon_{21} = \varepsilon_{r\varphi} = \varepsilon_{\varphi r} = \tan\left[\angle\left(|12|, |12^*|\right)\right] = \frac{1}{r\, d\varphi}\left(\frac{\partial u_1}{\partial \varphi} d\varphi\right) = \frac{1}{r}\frac{\partial u_1}{\partial \varphi}, \tag{3.4}$$

$$\varepsilon_{13} = \varepsilon_{31} = \varepsilon_{r\nu} = \varepsilon_{\nu r} = \tan\left[\angle\left(|13|, |13^*|\right)\right] = \frac{1}{r d\nu}\left(\frac{\partial u_1}{\partial \nu} d\nu\right) = \frac{1}{r}\frac{\partial u_1}{\partial \nu}, \tag{3.5}$$

where ε_{ij} $(i, j = 1,2,3)$ is strain of the surface of the infinitesimal spherical cap with the normal x'_j realized along the axis x'_i, and $\varepsilon_{ij} = \varepsilon_{ji}$ [3, p 14][2].

With regard to the analysis presented in Sections 3.1.2.1, 3.1.2.2, and accordingly as a consequence of radial loading along the normal to the particle-matrix and particle-envelope, matrix-envelope boundaries, the strain ε_{23} is equal to zero.

[1]The subscripts p, e and m related to the spherical particle, the spherical envelope and the cell matrix, respectively, are omitted in Sections 3.1.3–3.1.7.

[2]The points 1, 1′, 1″, 1‴ in Figs. 3.2, 3.3 are in the plane $x'_1x'_2$ as well as in the plane $x'_1x'_3$,

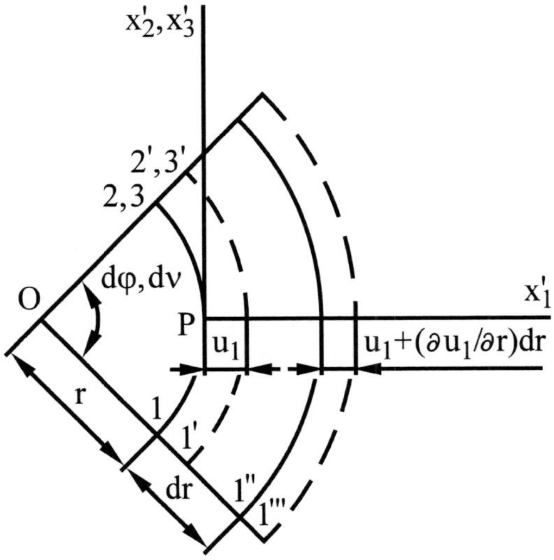

Figure 3.2: The radial displacements u_1 and $u_1 + (\partial u_1/\partial r)\,dr$ of the infinitesimal spherical cap in the point P at the radii r and $r + dr$, respectively, in the planes $x'_1x'_2$ or $x'_1x'_3$ (see Fig. 3.1), effected by the radial and tangential stresses, $\sigma_{11} = \sigma_r$ and $\sigma_{22} = \sigma_\varphi$ or $\sigma_{33} = \sigma_\nu$ (see Figs. 3.4, 3.5), respectively.

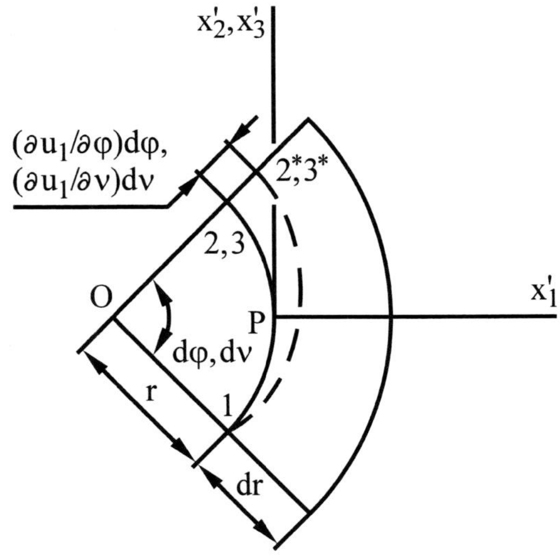

Figure 3.3: The φ, ν-dependent radial displacement u_1 of the infinitesimal spherical cap in the point P in the planes $x'_1x'_2$ and $x'_1x'_3$ (see Fig. 3.1), effected by the shear stresses $\sigma_{12} = \sigma_{21} = \sigma_{r\varphi} = \sigma_{\varphi r}$ and $\sigma_{13} = \sigma_{31} = \sigma_{r\nu} = \sigma_{\nu r}$ (see Figs. 3.4, 3.5), respectively, where the distances $|22^*| = (\partial u_1/\partial \varphi)\,d\varphi$ and $|33^*| = (\partial u_1/\partial \nu)\,d\nu$.

3.1.4 Saint-Venant's equations

As a principle valid in Mechanics of Solid Continuum, a system being continuous before deformation is required to exhibit the continuity after deformation as well, and accordingly individual strains are required to be mutually compatible [4, p 162]. The Saint-Venant's equations, representing equations of compatibility between individual strains, are derived in the following way. Performing $\partial \mathrm{Eq.}\,(3.2)\,/\partial r$ and considering Eq. (3.1), we get

$$\varepsilon_{11} - \varepsilon_{22} - r\,\frac{\partial \varepsilon_{22}}{\partial r} = 0. \tag{3.6}$$

Performing $\partial \mathrm{Eq.}\,(3.1)\,/\partial \varphi$, $\partial \mathrm{Eq.}\,(3.4)\,/\partial r$, and $\partial \mathrm{Eq.}\,(3.1)\,/\partial \nu$, $\partial \mathrm{Eq.}\,(3.5)\,/\partial r$, we get

where $x'_1 = x_r$, $x'_2 = x_\varphi$, $x'_3 = x_\nu$ (see Fig. 3.1). The points 2, $2'$, 2^* and 3, $3'$, 3^* in Figs. 3.2, 3.3 along with the infinitesimal angles $d\varphi$ and $d\nu$ are in the planes $x'_1x'_2$ and $x'_1x'_3$, respectively.

As presented in this book, the conventional subscripts r, φ, ν, $r\varphi$, $r\nu$, $\varphi\nu$ in connection with the spatial polar coordinates $[r, \varphi, \nu]$ (see Fig. 3.1) are replaced by the subscripts 11, 22, 33, 12, 13, 23, respectively.

$$\frac{\partial \varepsilon_{11}}{\partial \varphi} - \varepsilon_{12} - r\,\frac{\partial \varepsilon_{12}}{\partial r} = 0, \tag{3.7}$$

$$\frac{\partial \varepsilon_{11}}{\partial \nu} - \varepsilon_{13} - r\,\frac{\partial \varepsilon_{13}}{\partial r} = 0. \tag{3.8}$$

Performing $\partial \mathrm{Eq.}\,(3.2)\,/\partial \varphi$ and considering Eq. (3.4), and performing $\partial \mathrm{Eq.}\,(3.2)\,/\partial \nu$ and considering Eq. (3.5), we get

$$\frac{\partial \varepsilon_{22}}{\partial \varphi} - \varepsilon_{12} = 0, \tag{3.9}$$

$$\frac{\partial \varepsilon_{22}}{\partial \nu} - \varepsilon_{13} = 0. \tag{3.10}$$

Performing $\partial \mathrm{Eq.}\,(3.4)\,/\partial \nu$, $\partial \mathrm{Eq.}\,(3.5)\,/\partial \varphi$, we get

$$\frac{\partial \varepsilon_{12}}{\partial \nu} = \frac{\partial \varepsilon_{13}}{\partial \varphi}. \tag{3.11}$$

3.1.5 Equilibrium equations

As usual in Mechanics of Solid Continuum, stresses acting on an infinitesimal part with a suitable shape in an arbitrary point of a solid continuum are required to be derived. Equations of the equilibrium of individual forces acting on the infinitesimal spherical cap along the axes x_1', x_2', x_3' (see Figs. 3.4, 3.5) are derived as

$$\begin{aligned}
&\left(\sigma_{11} + \frac{\partial \sigma_{11}}{\partial r}\,dr\right)(r+dr)\,d\varphi\,(r+dr)\,d\nu + \left(\sigma_{12} + \frac{\partial \sigma_{12}}{\partial \varphi}\,d\varphi\right)\cos\frac{d\varphi}{2}\,r\,d\nu\,dr \\
&+ \left(\sigma_{13} + \frac{\partial \sigma_{13}}{\partial \nu}\,d\nu\right)\cos\frac{d\nu}{2}\,r\,d\varphi\,dr - \left[\sigma_{11} r\,d\varphi\,r\,d\nu + \left(\sigma_{22} + \frac{\partial \sigma_{22}}{\partial \varphi}\,d\varphi\right)\sin\frac{d\varphi}{2}\,r\,d\nu\,dr\right. \\
&+ \left(\sigma_{33} + \frac{\partial \sigma_{33}}{\partial \nu}\,d\nu\right)\sin\frac{d\nu}{2}\,r\,d\varphi\,dr + \sigma_{22}\sin\frac{d\varphi}{2}\,r\,d\nu\,dr + \sigma_{33}\sin\frac{d\nu}{2}\,r\,d\varphi\,dr \\
&\left. + \,\sigma_{12}\cos\frac{d\varphi}{2}\,r\,d\nu\,dr + \sigma_{13}\cos\frac{d\nu}{2}\,r\,d\varphi\,dr\right] = 0,
\end{aligned} \tag{3.12}$$

$$\begin{aligned}
&\left(\sigma_{22} + \frac{\partial \sigma_{22}}{\partial \varphi}d\varphi\right)\cos\frac{d\varphi}{2}\,r\,d\nu\,dr + \left(\sigma_{21} + \frac{\partial \sigma_{21}}{\partial r}dr\right)(r+dr)\,d\varphi\,(r+dr)\,d\nu \\
&+ \left(\sigma_{12} + \frac{\partial \sigma_{12}}{\partial \varphi}d\varphi\right)\sin\frac{d\varphi}{2}\,r\,d\nu\,dr + \sigma_{12}\sin\frac{d\varphi}{2}\,r\,d\nu\,dr \\
&- \left(\sigma_{22}\cos\frac{d\varphi}{2}\,r\,d\nu\,dr + \sigma_{21}\,r\,d\varphi\,r\,d\nu\right) = 0,
\end{aligned} \tag{3.13}$$

$$\left(\sigma_{33} + \frac{\partial\sigma_{33}}{\partial\nu}\,d\varphi\right)\cos\frac{d\nu}{2}\,r\,d\varphi\,dr + \left(\sigma_{31} + \frac{\partial\sigma_{31}}{\partial r}\,dr\right)(r+dr)\,d\varphi\,(r+dr)\,d\nu$$
$$+\left(\sigma_{13} + \frac{\partial\sigma_{13}}{\partial\nu}\,d\nu\right)\sin\frac{d\nu}{2}\,r\,d\varphi\,dr + \sigma_{13}\sin\frac{d\nu}{2}\,r\,d\varphi\,dr$$
$$-\left(\sigma_{33}\cos\frac{d\nu}{2}\,r\,d\varphi\,dr + \sigma_{31}\,r\,d\varphi\,r\,d\nu\right) = 0, \tag{3.14}$$

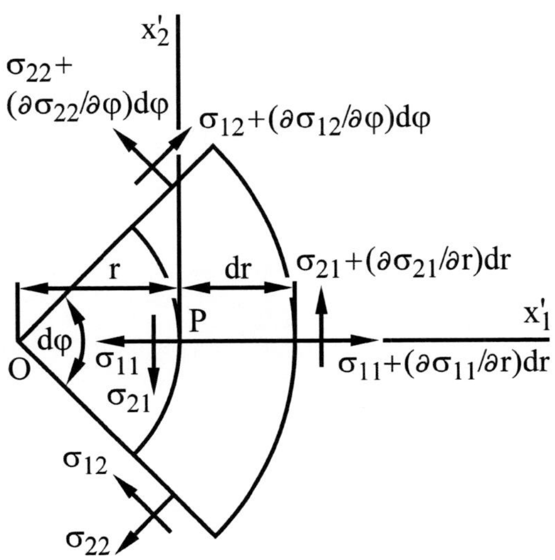

Figure 3.4: The radial, tangential, shear stresses in the plane $x_1'x_2'$, $\sigma_{11} = \sigma_r$, $\sigma_{22} = \sigma_\varphi$, $\sigma_{12} = \sigma_{21} = \sigma_{r\varphi} = \sigma_{\varphi r}$, respectively, acting on surfaces of an infinitesimal spherical cap in the point P (see Fig. 3.1), where $x_1' = x_r$, $x_2' = x_\varphi$.

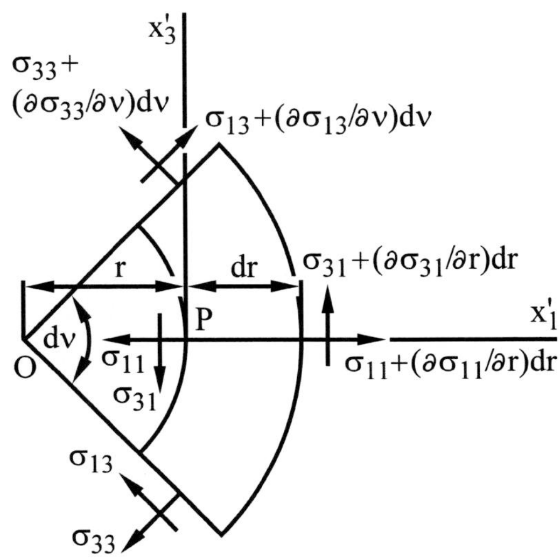

Figure 3.5: The radial, tangential, shear stresses in the plane $x_1'x_3'$, $\sigma_{11} = \sigma_r$, $\sigma_{33} = \sigma_\nu$, $\sigma_{13} = \sigma_{31} = \sigma_{r\nu} = \sigma_{\nu r}$, respectively, acting on surfaces of an infinitesimal spherical cap in the point P (see Fig. 3.1), where $x_1' = x_r$, $x_3' = x_\nu$.

where the stress σ_{ij} $(i,j = 1,2,3)$ acts along the axis x_i' on the surface of the infinitesimal spherical cap with the normal x_j', and $\sigma_{ij} = \sigma_{ji}$ [3, p 13]. Accordingly, $\sigma_{11} = \sigma_r$ and $\sigma_{22} = \sigma_\varphi$, $\sigma_{33} = \sigma_\nu$ and $\sigma_{12} = \sigma_{21} = \sigma_{r\varphi} = \sigma_{\varphi r}$, $\sigma_{13} = \sigma_{31} = \sigma_{r\nu} = \sigma_{\nu r}$ represent radial and tangential and shear stresses, respectively.

With regard to the analysis presented in Sections 3.1.2.1, 3.1.2.2, and accordingly as a consequence of radial loading along the normal to the particle-matrix and particle-envelope, matrix-envelope boundaries, the stress σ_{23} is equal to zero.

Substituting $\sin d\varphi/2 \approx d\varphi/2$, $\sin d\nu/2 \approx d\nu/2$, $\cos d\varphi/2 = \cos d\nu/2 \approx 1$, $(dr)^2 = (d\varphi)^2 = (d\nu)^2 = 0$, for $d\varphi \approx 0$, $d\nu \approx 0$, $dr \approx 0$ [2, p 125–126], we get

$$2\sigma_{11} - \sigma_{22} - \sigma_{33} + r\,\frac{\partial \sigma_{11}}{\partial r} + \frac{\partial \sigma_{12}}{\partial \varphi} + \frac{\partial \sigma_{13}}{\partial \nu} = 0, \tag{3.15}$$

$$\frac{\partial \sigma_{22}}{\partial \varphi} + 3\sigma_{12} + r\,\frac{\partial \sigma_{12}}{\partial r} = 0, \tag{3.16}$$

$$\frac{\partial \sigma_{33}}{\partial \nu} + 3\sigma_{13} + r\,\frac{\partial \sigma_{13}}{\partial r} = 0. \tag{3.17}$$

3.1.6 Hooke's laws

The Hooke's laws, representing a relationship between stresses and strains of an infinitesimal part of an elastic solid continuum, are derived as [3, p 30]

$$\varepsilon_{11} = s_{11}\sigma_{11} + s_{12}\left(\sigma_{22} + \sigma_{33}\right), \tag{3.18}$$

$$\varepsilon_{22} = s_{12}\left(\sigma_{11} + \sigma_{33}\right) + s_{11}\sigma_{22}, \tag{3.19}$$

$$\varepsilon_{33} = s_{12}\left(\sigma_{11} + \sigma_{22}\right) + s_{11}\sigma_{33}, \tag{3.20}$$

$$\varepsilon_{12} = s_{44}\sigma_{12}, \tag{3.21}$$

$$\varepsilon_{13} = s_{44}\sigma_{13}, \tag{3.22}$$

and the elastic moduli s_{11}, s_{12}, s_{44} for an isotropic elastic solid continuum have the forms

$$s_{11} = \frac{1}{E}, \tag{3.23}$$

$$s_{12} = -\,\frac{\mu}{E}, \tag{3.24}$$

$$s_{44} = \frac{2\left(1 + \mu\right)}{E}, \tag{3.25}$$

where E and μ are the Young's modulus of elasticity and the Poisson's number, respectively.

3.1.7 Differential equations of radial displacement

Considering Eqs. (3.1)–(3.5), the Hooke's laws for an isotropic elastic solid continuum are transformed to the forms

$$s_{11}\sigma_{11} + s_{12}(\sigma_{22} + \sigma_{33}) = \frac{\partial u_1}{\partial r}, \tag{3.26}$$

$$s_{11}\sigma_{22} + s_{12}(\sigma_{11} + \sigma_{33}) = \frac{u_1}{r}, \tag{3.27}$$

$$s_{11}\sigma_{33} + s_{12}(\sigma_{11} + \sigma_{22}) = \frac{u_1}{r}, \tag{3.28}$$

$$s_{44}\sigma_{12} = \frac{1}{r}\frac{\partial u_1}{\partial \varphi}, \tag{3.29}$$

$$s_{44}\sigma_{13} = \frac{1}{r}\frac{\partial u_1}{\partial \nu}, \tag{3.30}$$

and consequently, we get

$$\sigma_{11} = (c_1 + c_2)\frac{\partial u_1}{\partial r} - 2c_2\frac{u_1}{r}, \tag{3.31}$$

$$\sigma_{22} = \sigma_{33} = -c_2\frac{\partial u_1}{\partial r} + c_1\frac{u_1}{r}, \tag{3.32}$$

$$\sigma_{12} = \frac{1}{s_{44}r}\frac{\partial u_1}{\partial \varphi}, \tag{3.33}$$

$$\sigma_{13} = \frac{1}{s_{44}r}\frac{\partial u_1}{\partial \nu}, \tag{3.34}$$

where the condition $\sigma_{22} = \sigma_{33}$ results from $\varepsilon_{22} = \varepsilon_{33}$ (see Eqs. (3.2), (3.3)), and the coefficients c_1, c_2, c_3 are given by Eqs. (3.38)–(3.40).

Substituting Eqs (3.31)–(3.34) to $[\partial\text{Eq.}(3.16)/\partial\varphi] + [\partial\text{Eq.}(3.17)/\partial\nu]$ and Eq. (3.15), the equilibrium equations (see Eqs. (3.15)–(3.17)) are transformed to the forms

$$r\frac{\partial U_1}{\partial r} = c_3 U_1, \tag{3.35}$$

$$r^2\frac{\partial^2 u_1}{\partial r^2} + 2r\frac{\partial u_1}{\partial r} - 2u_1 + \frac{U_1}{s_{44}(c_1 + c_2)} = 0, \tag{3.36}$$

where the function U_1 is derived as

$$U_1 = \frac{\partial^2 u_1}{\partial \varphi^2} + \frac{\partial^2 u_1}{\partial \nu^2}. \tag{3.37}$$

The substitution of Eqs. (3.26)–(3.30) to the Saint-Venant's equations (3.6)–(3.11) represents tautology not leading to suitable differential equations.

As presented in Sections 5.1–5.3, 6.1, 7.1, 7.2, 8.1, the different mathematical techniques applied to a system of differential equations (3.35), (3.36) lead to the different linear differential equations (5.5), (5.13), (5.18), (6.3), (7.2), (7.11), (8.2), (8.3) with non-zero right sides.

As only exception resulting from the analysis presented in Section 9.1 and concerning the conditions $\beta_p \neq \beta_e = \beta_m$ and $\varepsilon_{11te} - \varepsilon_{11tp} \neq f(\varphi, \nu)$ to result in $U_1 = 0$ (see Eq. (3.37)), Equation (3.36) is transformed to the different linear differential equations (9.2), (9.6), (9.10), (9.13), (9.17), (9.21), (9.24), (9.28), (9.32), (9.36) with zero right sides.

Appendix. Considering the elastic moduli s_{11q}, s_{12q}, s_{44q} for an isotropic particle $(q=p)$, for an isotropic envelope $(q=e)$ and for an isotropic matrix $(q=m)$ (see Eqs. (3.23)–(3.25)), regarding the Young's modulus of elasticity E_q and the Poisson's number μ_q, the coefficients c_{1q}, c_{2q}, c_{3q} have the forms

$$c_{1q} = \frac{s_{11q}}{s_{11q}\left(s_{11q} + s_{12q}\right) - 2s_{12q}^2} = \frac{E_q}{\left(1 + \mu_q\right)\left(1 - 2\mu_q\right)}, \tag{3.38}$$

$$c_{2q} = \frac{s_{12q}}{s_{11q}\left(s_{11q} + s_{12q}\right) - 2s_{12q}^2} = -\frac{\mu_q E_q}{\left(1 + \mu_q\right)\left(1 - 2\mu_q\right)}, \tag{3.39}$$

$$c_{3q} = \frac{s_{44q}c_{1q} + 2}{s_{44q}c_{2q} - 1} = -4\left(1 - \mu_q\right). \tag{3.40}$$

With regard to the Poisson's number of an elastic solid continuum and real materials, $\mu_q = 0.25$ [3, p 17] and $\mu_q < 0.5$ [1, p 59], respectively, the coefficient $c_{3q} < 0$, and additionally, an investigation concerning the condition $\mu_q = 0.5$ to result in $c_{1q} \to \infty$, $c_{2q} \to -\infty$ and consequently in possible discontinuities is irrelevant.

3.1.8 Elastic energy

Induced by elastic stresses acting in an arbitrary point of an elastic solid continuum, the elastic energy density w_q accumulated in the arbitrary point has the form [3, p 24]

$$w_q = \frac{1}{2}\left(\sum_{i=1}^{3} \sigma_{iiq}\varepsilon_{iiq} + \sum_{i,j=1;\ i\neq j}^{3} \sigma_{ijq}\varepsilon_{ijq}\right), \tag{3.41}$$

and consequently the elastic energy W_q, accumulated in the volume V_q of the elastic solid continuum is derived, using the spatial polar coordinates $[r, \varphi, \nu]$ (see Fig. 3.1) and considering the φ, ν-dependent elastic energy density, $w_q = w_q(\varphi, \nu)$, in the form

$$W_q = \int_{V_q} w_q \, dV_q = 8 \int_0^{\pi/2} \int_0^{\pi/2} \int_{r_1}^{r_2} w_q \, r^2 \, dr \, d\varphi \, d\nu, \tag{3.42}$$

where $dV_q = r^2 \, dr \, d\varphi \, d\nu$ is the volume of the infinitesimal spherical cap for the angles $\varphi, \nu \in \langle 0, \pi/2 \rangle$. With regard to the multi-particle-matrix and multi-particle-envelope-matrix systems, the variable r is from the intervals $\langle 0, R_1 \rangle$, $\langle R_1, r_c \rangle$ and $\langle 0, R_1 \rangle$, $\langle R_1, R_2 \rangle$, $\langle R_2, r_c \rangle$, related to the spherical particle $(q = p)$, the cell matrix $(q = m)$ and to the spherical particle, the spherical envelope $(q = e)$, the cell matrix, respectively. With regard to the one-particle-matrix and one-particle-envelope-matrix systems, the variable r related to the infinite matrix is from the intervals $\langle R_1, \infty)$ and $\langle R_2, \infty)$, respectively. Provided that $w_q \neq w_q(\varphi, \nu)$, the elastic energy W_q is derived as

$$W_q = \int_{V_q} w_q \, dV_q = \int_0^{4\pi} \int_{r_1}^{r_2} w_q \, r^2 \, dr \, d\Phi = 4\pi \int_{r_1}^{r_2} w_q \, r^2 \, dr, \tag{3.43}$$

where $dV_q = r^2 \, dr \, d\Phi$ for $\Phi \in \langle 0, 4\pi \rangle$ (see Eqs. (9.47), (9.60), (9.72), (9.82), (9.96), (9.107), (9.117)).

With regard to the elastic energy of the spherical particle and the cell matrix, W_p and W_m, respectively, the elastic energy of the infinite multi-particle-matrix system is represented by the elastic energy of the cell in the form

$$W_c = W_p + W_m. \tag{3.44}$$

Similarly, the elastic energy of the cell of the infinite multi-particle-envelope-matrix system is derived as

$$W_c = W_p + W_e + W_m, \tag{3.45}$$

where W_e is the elastic energy of the spherical envelope.

Considering the determination of the thermal-stress induced strengthening presented in Volume III, the stress σ_{iq} resulting in the strain $\varepsilon_{iq} = s_{11q}\sigma_{iq}$, both in a direction of the axis x_i $(i = 1,2,3)$ (see Fig. 2.1a,b), and the elastic energy density w_{iq} accumulated in an arbitrary point of an elastic solid continuum and induced by the stress σ_{iq}, are required to be derived as

$$\sigma_{iq} = \sum_{j=1}^{3} a'_{ji}\sigma_{jjq} + \sum_{j,k=1;\, j \neq k}^{3} a'_{ji}\sigma_{jkq}, \quad i = 1, 2, 3, \tag{3.46}$$

$$w_{iq} = \frac{\varepsilon_i \sigma_{iq}}{2} = \frac{s_{11q}\sigma_{iq}^2}{2}, \quad i = 1, 2, 3, \tag{3.47}$$

where $a'_{ji}\sigma_{jkq}$ represents a coordinate of the stress σ_{jkq} in a direction represented by the axis x_i, and the coefficient $a'_{ji} = \cos\left[\angle\left(x'_j, x_i\right)\right]$ given by Eqs. (12.26)–(12.34) is a direction cosine of the angle of the axes x_i, x'_j $(i, j, k = 1,2,3)$.

3.1.9 Castigliano's theorem

As presented in Sections 5.1–5.3, 6.1, 7.1, 7.2, 8.1, 9.2, the thermal stresses in the isotropic multi-particle-(envelope)-matrix system are derived by different mathematical techniques applied to the differential equations (3.35), (3.36) and consequently leading to different differential equations resulting in different solutions. Resulting from the tendency of a system to exhibit minimal energy, the Castigliano's theorem is formulated as follows [4, p 219]:

> Assuming different states of stress to fulfil equilibrium and boundary conditions, the state of stress to exhibit minimal energy of a system is realized.

Accordingly, such solution or such combination of solutions for components of the isotropic multi-particle-(envelope)-matrix system is considered to result in minimal thermal-stress induced elastic energy of the cell, W_c (see Eqs. (3.44), (3.45)). The same is also valid for the isotropic one-particle-(envelope)-matrix system, considering minimal thermal-stress induced elastic energy of the system, $W^{(1)} = W_p^{(1)} + \left(W_e^{(1)}\right) + W_m^{(1)}$[3] (see Items 1-4 in Section 2.2.2.1).

3.2 Reason of thermal stresses

3.2.1 Coefficients β_p, β_e, β_m

With regard to the multi-particle-matrix and multi-particle-envelope-matrix system, the thermal stresses originate

1. as a consequence of the differences $\alpha_m - \alpha_p \neq 0$ and $\alpha_e - \alpha_p \neq 0$, $\alpha_m - \alpha_e \neq 0$ in the temperature-dependent or temperature-independent thermal expansion coefficients α_p, α_e, α_m,

2. as a consequence of the differences $\varepsilon_{11tm} - \varepsilon_{11tp} \neq 0$ and $\varepsilon_{11te} - \varepsilon_{11tp} \neq 0$, $\varepsilon_{11tm} - \varepsilon_{11te} \neq 0$ in the radial strains ε_{11tp}, ε_{11te}, ε_{11tm} (see Eqs. (12.23), (12.24)) induced by a phase transformation in the spherical particle, the spherical envelope, the cell matrix at the temperature T_{tp}, T_{te}, T_{tm}, respectively,

where α_q and ε_{11tq} are included in the coefficient β_q $(q = p, e, m)$ (see Eqs. (3.48)–(3.50)).

[3]The superscript (1) is related to the *one*-particle-(envelope)-matrix system.

With regard to Item 2, the thermal stresses originate as a consequence of the radial strain ε_{11tq} on the condition of a phase transformation originating at least in one of components of the multi-particle-(envelope)-matrix system. The phase-transformation induced radial strain ε_{11tq} results from the difference in dimensions of mutually transforming cubic crystalline lattices of the spherical particle, the spherical envelope, the cell matrix (see Section 12.1), thus for $q = p, e, m$, respectively.

In case of the non-zero difference in the thermal expansion coefficients, the thermal stresses originate in the temperature interval $\langle T_f, T_r\rangle$, where the temperature T_r of the thermal stress generation is minimal temperature from the sets $\{T_{rp}, T_{rm}\}$ and $\{T_{rp}, T_{re}, T_{rm}\}$ related to the multi-particle-matrix and multi-particle-envelope-matrix systems, respectively, and T_f is final temperature of a cooling process. The recovery temperature T_{rq}, below that the stress relaxation as a consequence of thermal-activated processes does not occur in the components, is defined approximately by the relationship $T_{rq} = (0.35 - 0.4) \times T_{mq}$ [1, p 39] and exactly by an experiment, where T_{mq} is the melting temperature. In case of the zero difference in the thermal expansion coefficients and provided that $T_{tq} \in \langle T_f, T_r\rangle$, the thermal stresses originate at the temperature T_t representing maximal temperature from the sets $\{T_{tp}, T_{tm}\}$ and $\{T_{tp}, T_{te}, T_{tm}\}$ related to the multi-particle-matrix and multi-particle-envelope-matrix systems, respectively, where $\langle T_f, T_t\rangle \subset \langle T_f, T_r\rangle$.

Accordingly, the coefficient β_q has the forms

$$\beta_q = \varepsilon_{11tq} + \int_{T_f}^{T_{tq}} \alpha_{Iq}\, dT + \int_{T_{tq}}^{T_r} \alpha_{IIq}\, dT, \quad T_{tq} \in \langle T_f, T_r\rangle, \quad T \le T_{tq}, \tag{3.48}$$

$$\beta_q = \int_{T}^{T_r} \alpha_{IIq}\, dT, \quad T_{tq} \in \langle T_f, T_r\rangle, \quad T > T_{tq}, \tag{3.49}$$

$$\beta_q = \int_{T_f}^{T_r} \alpha_q\, dT, \quad T_{tq} \notin \langle T_f, T_r\rangle, \tag{3.50}$$

where α_{Iq}, α_{IIq} are temperature-dependent or temperature-independent thermal expansion coefficients at the temperature $T \le T_{tq}$, $T \ge T_{tq}$ provided that $T_{tq} \in \langle T_f, T_r\rangle$, respectively, and α_q is a temperature-dependent or temperature-independent thermal expansion coefficient on the condition $T_{tq} \notin \langle T_f, T_r\rangle$.

Provided that the thermal stresses originate as a consequence of the differences in the thermal expansion coefficients α_p, α_e, α_m and as well as in the phase-transformation induced radial strains ε_{11tp}, ε_{11te}, ε_{11tm} (see Items 1, 2), contributions of the individual differences to the thermal stresses can be determined by the substitution either $\varepsilon_{11tq} = 0$ or $\alpha_q = 0$ ($q = p, e, m$), respectively.

Considering $T = T(\tau)$ to represent a time-dependent function, the total differential dT has the form [4]

$$dT = \frac{\partial T}{\partial \tau} d\tau, \tag{3.51}$$

and the temperature intervals $\langle T_f, T_{tq} \rangle$, $\langle T_{tq}, T_r \rangle$, $\langle T, T_r \rangle$, $\langle T_f, T_r \rangle$ in integrals of Eqs. (3.48)–(3.50) are replaced by the time intervals $\langle \tau_f, \tau_{tq} \rangle$, $\langle \tau_{tq}, \tau_r \rangle$, $\langle \tau, \tau_r \rangle$, $\langle \tau_f, \tau_r \rangle$, where τ_f, τ_{tq}, τ_r, τ are determined numerically from the conditions $T_f = T(\tau_f)$, $T_{tq} = T(\tau_{tq})$, $T_r = T(\tau_r)$, $T = T(\tau)$, respectively.

3.2.2 Dimension changes of components

Cooling down the multi-particle-matrix system from the relaxation temperature T_r to the temperature $T \in \langle T_f, T_r \rangle$, the spherical particle and the spherical hole in the cell matrix, both with the radius R_1 at the temperature T_r, tend to exhibit the radii R_{1p} and R_{1m}, respectively, derived as

$$R_{1p} = R_1 (1 - \beta_p), \tag{3.52}$$

$$R_{1m} = R_1 (1 - \beta_m). \tag{3.53}$$

Similarly, with regard to the multi-particle-envelope-matrix system, the spherical envelope and the spherical hole in the cell with the radii $R_1 < R_2$ and R_2 at the temperature T_r, tend to exhibit the radii $R_{1e} < R_{2e}$ and R_{2m}, respectively, derived as

$$R_{1+ie} = R_{1+i} (1 - \beta_e), \quad i = 0, 1, \tag{3.54}$$

$$R_{2m} = R_2 (1 - \beta_m), \tag{3.55}$$

and the radius R_{1p} of the spherical particle in the multi-particle-envelope-matrix system is given by Eq. (3.52).

An influence of a phase transformation is represented by the φ, ν-dependent or, regarding simplification, φ, ν-independent radial strain ε_{11tq} given by Eqs. (12.23) or (12.24), respectively. Considering the φ, ν-dependent radial strain ε_{11tq} $(q = p, e, m)$ (see Eq. (12.23)), the φ, ν-dependent parameters R_{1p}, R_{1m}, R_{1e} and R_{2e}, R_{2m} given by Eqs. (3.52)–(3.55) represents distances in a radial direction, represented by the axis x_1' and determined by the angles φ, ν, from the particle centre O to a point on the particle-matrix, particle-envelope and matrix-envelope boundaries (see Fig. 3.1), respectively.

The thermal stresses in the multi-particle-matrix system originate as a consequence of the condition $(\Delta u_{1mp})_{r=R_1} \neq 0$, and accordingly the radial stress p_1 is

[4] The parameter $v_T = \partial T / \partial \tau$ is cooling rate.

induced on the particle-matrix boundary, where the parameter $(\Delta u_{1mp})_{r=R_1}$, regarding Eqs. (3.52), (3.53), has the form

$$(\Delta u_{1mp})_{r=R_1} = R_{1p} - R_{1m} = R_1 (\beta_m - \beta_p), \tag{3.56}$$

and consequently the condition $(\Delta u_{1mp})_{r=R_1} \neq 0$ results from $\beta_p \neq \beta_m$. Consequently, the distance R_{1T} in a radial direction at the temperature $T \in \langle T_f, T_r \rangle$ from the particle centre O to a point on the particle-matrix boundary is derived as

$$R_{1T} = R_{1p} + (u_{1p})_{r=R_1} = R_{1m} + (u_{1m})_{r=R_1}, \tag{3.57}$$

where $(u_{1p})_{r=R_1}$ and $(u_{1m})_{r=R_1}$ are radial displacements in the spherical particle and in the cell matrix on the particle-matrix boundary, thus for $r = R_1$, respectively, induced by the radial stress p_1. Considering Eqs. (3.52), (3.53), (3.57) and the Cauchy's equation $u_1 = r\,\varepsilon_{22}$ (see Eq. (3.2)), the compressive or tensile radial stress $p_1 > 0$ or $p_1 < 0$, respectively, acting on the particle-matrix boundary, can be determined from the condition

$$(\varepsilon_{22m})_{r=R_1} - (\varepsilon_{22p})_{r=R_1} = \beta_m - \beta_p, \tag{3.58}$$

where the tangential strains ε_{22p}, ε_{22m} are related to one or two different solutions related to one or two of the different linear differential equations. With regard to a physical point of view, such solution or such combination of two different solutions is considered to result in minimal thermal-stress induced elastic energy of the cell (see Eqs. (3.44), (3.45)).

The radial stresses p_1 and p_2 are induced on the particle-envelope and matrix-envelope boundaries as consequences of the conditions $(\Delta u_{1ep})_{r=R_1} \neq 0$ and $(\Delta u_{1me})_{r=R_2} \neq 0$, respectively, where the parameters $(\Delta u_{1ep})_{r=R_1}$ and $(\Delta u_{1me})_{r=R_2}$, regarding Eqs. (3.52), (3.54), (3.55), are derived as

$$(\Delta u_{1ep})_{r=R_1} = R_{1p} - R_{1e} = R_1 (\beta_e - \beta_p), \tag{3.59}$$

$$(\Delta u_{1me})_{r=R_2} = R_{2e} - R_{2m} = R_2 (\beta_m - \beta_e), \tag{3.60}$$

and consequently the conditions $(\Delta u_{1ep})_{r=R_1} \neq 0$ and $(\Delta u_{1me})_{r=R_2} \neq 0$ result in $\beta_p \neq \beta_e$ and $\beta_m \neq \beta_e$, respectively. Consequently, the distances R_{1T} and R_{2T} in a radial direction and at the temperature $T \in \langle T_f, T_r \rangle$ from the particle centre O to a point on the particle-envelope and matrix-envelope boundaries, respectively, have the forms

$$R_{1T} = R_{1p} + (u_{1p})_{r=R_1} = R_{1e} + (u_{1e})_{r=R_1}, \tag{3.61}$$

$$R_{2T} = R_{2e} + (u_{1e})_{r=R_2} = R_{2m} + (u_{2m})_{r=R_2}, \tag{3.62}$$

where $(u_{1p})_{r=R_1}$, $(u_{1e})_{r=R_1}$ are radial displacements in the spherical particle and in the spherical envelope on the particle-envelope boundary, thus for $r = R_1$, respectively, induced by the radial stress p_1, and $(u_{1e})_{r=R_2}$ and $(u_{1m})_{r=R_2}$ are radial displacements in the spherical envelope and in the cell matrix on the matrix-envelope boundary, thus for $r = R_2$, respectively, induced by the radial stress p_2. Considering Eqs. (3.52), (3.54), (3.55), (3.61), (3.62) and the Cauchy's equation $u_1 = r\,\varepsilon_{22}$ (see Eq. (3.2)), the compressive or tensile radial stresses $p_1 > 0$ or $p_1 < 0$ and $p_2 > 0$ or $p_2 < 0$, respectively, are derived from the conditions

$$(\varepsilon_{22e})_{r=R_1} - (\varepsilon_{22p})_{r=R_1} = \beta_e - \beta_p, \tag{3.63}$$

$$(\varepsilon_{22m})_{r=R_2} - (\varepsilon_{22e})_{r=R_2} = \beta_m - \beta_e, \tag{3.64}$$

where the tangential strains ε_{22p}, ε_{22e}, ε_{22m} are related to one, two or three different solutions related to one, two or three of the different linear differential equations. Similarly, regarding a physical point of view, such solution or such combination of two or three different solutions is considered to result in minimal thermal-stress induced elastic energy of the cell (see Eqs. (3.44), (3.45)).

Resulting from φ, ν-dependencies of the radial displacements $(u_{1p})_{r=R_1}$, $(u_{1m})_{r=R_{1+i}}$, $(u_{1e})_{r=R_{1+i}}$ $(i=0,1)$ as a consequence of the φ, ν-dependent radial stresses p_1, p_2, the distances R_{1T}, R_{2T} given by Eqs. (3.57), (3.59), (3.60) are φ, ν-dependent. As only exception resulting from the analysis presented in Section 9.1 and concerning the conditions $\beta_p \neq \beta_e = \beta_m$ and $\varepsilon_{11te} - \varepsilon_{11tp} \neq f(\varphi, \nu)$, the radial displacements u_{1p}, u_{1e} and consequently the radial stress p_1 are φ, ν-independent, and consequently the distance R_{1T} given by Eq. (3.61) is not functions of the variables φ, ν.

Finally, with regard to the conditions $\beta_p = \beta_m$ and $\beta_p = \beta_e = \beta_m$, considering Eqs. (3.58), (3.63), (3.64), the thermal stresses in the multi-particle-matrix and multi-particle-envelope-matrix systems do not occur, respectively.

3.2.3 Temperature dependence of thermal stresses

Temperature dependencies of the thermal stresses is represented by temperature dependencies of the radial stresses $p_1 = f(T)$, $p_2 = f(T)$ to be functions of the coefficient β_q $(q=p,e,m)$ (see Eqs. (3.48)–(3.50)) as well as the elastic moduli s_{11q}, s_{12q}, s_{44q} included in the tangential strain ε_{22q}, where β_q and ε_{22q} are included in the conditions (3.58), (3.63), (3.64) considered for the determination of p_1, p_2 (see Eqs. (3.48)–(3.50)). On the conditions $\alpha_p \neq \alpha_m$, and $\alpha_p \neq \alpha_e \neq \alpha_m$ or $\alpha_p \neq \alpha_e = \alpha_m$ or $\alpha_p = \alpha_e \neq \alpha_m$ related to the multi-particle-matrix and multi-particle-envelope-matrix systems, respectively, temperature dependencies of the thermal stresses results from temperature intervals in the integrals related included in the coefficient β_q, as well as from temperature dependencies of the elastic moduli s_{11q}, s_{12q}, s_{44q} in the temperature intervals. Provided that the thermal expansion coefficients α_{Iq}, α_{IIq}, α_q included in β_q are not functions of the temperature T, the integrals in the coefficient β_q are transformed to the forms $\alpha_{Iq}(T_{tq} - T_f)$, $\alpha_{IIq}(T_r - T_{tq})$, $\alpha_{IIq}(T_r - T)$, $\alpha_q(T_r - T_f)$ [6, p 471].

On the conditions $\alpha_p = \alpha_m$ and $\alpha_p = \alpha_e = \alpha_m$, provided that $\varepsilon_{11tm} - \varepsilon_{11tp} \neq 0$, and $\varepsilon_{11te} - \varepsilon_{11tp} \neq 0$ or $\varepsilon_{11tm} - \varepsilon_{11te} \neq 0$ (see Eqs. (12.23), (12.24)), respectively, a phase transformation originating at least in one of components of the multi-particle-(envelope)-matrix system is a reason of the thermal stresses originating at the temperature $T_{tq} \in \langle T_f, T_r \rangle$. Accordingly, temperature dependencies of the thermal stresses results only from temperature dependencies of the elastic moduli s_{11q}, s_{12q}, s_{44q} in the temperature interval $\langle T_f, T_t \rangle$, and provided that $s_{11q} \neq f(T)$, $s_{12q} \neq f(T)$, $s_{44q} \neq f(T)$, the thermal stresses are temperature-independent, where T_t represents maximal temperature from the sets $\{T_{tp}, T_{tm}\}$ and $\{T_{tp}, T_{te}, T_{tm}\}$ related to the multi-particle-matrix and multi-particle-envelope-matrix systems, respectively, where $\langle T_f, T_t \rangle \subset \langle T_f, T_r \rangle$, and consequently the radii R_1, R_2 are related to the temperature T_t (see Section 3.3).

3.3 Determination of radii R_1, R_2

Applying the analytical models of the thermal stresses and the corresponding quantities presented in Chapters 5–9 to a real composite material, numerical values of the particle and envelope radii, R_1 and $R_1 < R_2$, respectively, are required to be determined.

3.3.1 Multi-particle-matrix system

With regard to the multi-particle-matrix system provided that $\alpha_p \neq \alpha_m$ for $T \in \langle T_f, T_r \rangle$, the parameter R_1 represents particle radius at the temperature T_r representing minimal temperature from the set $\{T_{rp}, T_{rm}\}$. Provided that a phase transformation originates at the temperature $T_{tq} \in \langle T_f, T_r \rangle$ at least in one of components of the multi-particle-matrix system, thus for $q = p$ or $q = m$, and on the condition $\alpha_p = \alpha_m$ valid at least in the temperature interval $\langle T_{tq}, T_r \rangle$, the parameter R_1 represents particle radius at the temperature T_t representing maximal temperature from the set $\{T_{tp}, T_{tm}\}$, where $\langle T_f, T_t \rangle \subset \langle T_f, T_r \rangle$. Considering this analysis concerning the temperature T_r or T_t, with regard to Eq. (3.57) and the Cauchy's equation $u_1 = r\,\varepsilon_{22}$ (see Eq. (3.2)), and provided that R_{1ex} represents average particle radius determined for a real composite material by experimental methods at the temperature $T < T_r$ or $T < T_t$, the particle radius R_1 at the temperature T_r or T_t, respectively, is derived by numerical methods from the equation

$$R_{1ex} = \left(\frac{2}{\pi}\right)^2 \int_0^{\pi/2}\int_0^{\pi/2} R_{1T}\, d\varphi\, d\nu = R_1 \left\{ 1 + \left(\frac{2}{\pi}\right)^2 \int_0^{\pi/2}\int_0^{\pi/2} \left[\left(\varepsilon_{22q}\right)_{r=R_1} - \beta_q \right] d\varphi\, d\nu \right\},$$

$$q = p, m, \tag{3.65}$$

where the right side of Eq. (3.65) represents an average value of the φ, ν-dependent distance R_{1T} from the particle centre O to a point on the particle-matrix boundary in a radial direction determined by the angles φ, ν and represented by the axis x_1' (see Fig. 3.1). The tangential strain $(\varepsilon_{22q})_{r=R_1}$ $(q=p,m)$ is or is not a function of R_1, as dependent on the reduction of the term R_1 in fractions included in $(\varepsilon_{22q})_{r=R_1}$. In case of the R_1-dependent function $(\varepsilon_{22q})_{r=R_1}$, the particle radius R_1 representing a variable of Eq. (3.65) is derived by numerical methods for the determination of roots of transcendental equations, e.g. by the Newton's method of tangents, the method Regula falsi, the iteration method [7, p 970–971]. In case of the R_1-independent function $(\varepsilon_{22q})_{r=R_1}$, the integral in Eq. (3.65) is derived for a real composite material regarding Eq. (1.6). Finally, to derive the root R_1 of Eq. (3.65), one of the subscripts p, m in Eq. (3.65) is sufficient to be considered.

On the condition $\varepsilon_{11tq} = 0$ for $T \in \langle T_f, T_r \rangle$ $(q=p,m)$, or provided that the radial strain ε_{11tq} is considered to be φ, ν-independent (see Eq. (12.24)), both resulting in $\beta_q \neq f(\varphi, \nu)$, Equation (3.65) is transformed to the form

$$R_{1ex} = \left(\frac{2}{\pi}\right)^2 \int_0^{\pi/2}\int_0^{\pi/2} R_{1T}\, d\varphi\, d\nu = R_1 \left\{ 1 - \beta_q + \left(\frac{2}{\pi}\right)^2 \int_0^{\pi/2}\int_0^{\pi/2} (\varepsilon_{22q})_{r=R_1}\, d\varphi\, d\nu \right\}, \quad q = p, m. \tag{3.66}$$

Equations (3.65), (3.66) are considered when R_{1ex} is determined by experimental methods related to particles embedded in a matrix without their separation from the matrix, e.g. using microscopy.

With regard to experimental methods using the separation of particles from a matrix of a real composite material, the particle radius R_{1ex} determined by the experimental methods at the temperature $T < T_r$ or $T < T_t$ is equal to an average value of the φ, ν-dependent parameter R_{1p} given by Eq. (3.52) as a function of the variables φ, ν, due to the φ, ν-dependent radial strain $\varepsilon_{11tp} \neq 0$ (see Eq. (12.23)) for $T \in \langle T_f, T_r \rangle$, and consequently the particle radius R_1 at the temperature T_r or T_t, respectively, has the form

$$R_1 = R_{1ex} \left[1 - \left(\frac{2}{\pi}\right)^2 \int_0^{\pi/2}\int_0^{\pi/2} \beta_p\, d\varphi\, d\nu \right]^{-1}, \tag{3.67}$$

and similarly, on the conditions $\varepsilon_{11tp} = 0$ for $T \in \langle T_f, T_r \rangle$ or $\varepsilon_{11tp} \neq f(\varphi, \nu)$ (see Eq. (12.24)), both resulting in $\beta_p \neq f(\varphi, \nu)$, we get

$$R_1 = \frac{R_{1ex}}{1 - \beta_p}. \tag{3.68}$$

3.3.2 Multi-particle-envelope-matrix system

With regard to the multi-particle-envelope-matrix system provided that $\alpha_p \neq \alpha_e \neq \alpha_m$, $\alpha_p \neq \alpha_e = \alpha_m$, $\alpha_p = \alpha_e \neq \alpha_m$ for $T \in \langle T_f, T_r \rangle$, the parameters R_1, R_2 represent radii at the temperature T_r. Provided that a phase transformation originates at the temperature $T_{tq} \in \langle T_f, T_r \rangle$ at least in one of components of the multi-particle-envelope-matrix system, thus for $q = p$ or $q = e$ or $q = m$, and on the condition $\alpha_p = \alpha_e = \alpha_m$ valid at least in the temperature interval $\langle T_{tq}, T_r \rangle$, the parameters R_1, R_2 represent radii at the temperature T_t representing maximal temperature from the set $\{T_{tp}, T_{te}, T_{tm}\}$, where $\langle T_f, T_t \rangle \subset \langle T_f, T_r \rangle$. As in case of the multi-particle-matrix system (see Eqs. (3.65), (3.67)), and with regard to Eqs. (3.61), (3.62), the radii R_1, R_2 are derived on the conditions $\beta_p \neq \beta_e \neq \beta_m$ or $\beta_p = \beta_e \neq \beta_m$ (see Eqs. (3.48)–(3.50)) from the equation

$$R_{1+iex} = \left(\frac{2}{\pi}\right)^2 \int_0^{\pi/2}\int_0^{\pi/2} R_{1+iT}\, d\varphi\, d\nu = R_{1+i}\left\{1 + \left(\frac{2}{\pi}\right)^2 \int_0^{\pi/2}\int_0^{\pi/2} \left[\left(\varepsilon_{22q}\right)_{r=R_{1+i}} - \beta_q\right] d\varphi\, d\nu\right\},$$
$$i = \begin{cases} 0 & \text{for} \quad q = p, e, \\ 1 & \text{for} \quad q = m, e, \end{cases} \tag{3.69}$$

transforming, on the conditions $\varepsilon_{11tq} = 0$ for $T \in \langle T_f, T_r \rangle$ or $\varepsilon_{11tq} \neq f(\varphi, \nu)$ (see Eq. (12.24)), both resulting in $\beta_q \neq f(\varphi, \nu)$, to the form

$$R_{1+iex} = \left(\frac{2}{\pi}\right)^2 \int_0^{\pi/2}\int_0^{\pi/2} R_{1+iT}\, d\varphi\, d\nu = R_{1+i}\left\{1 - \beta_q + \left(\frac{2}{\pi}\right)^2 \int_0^{\pi/2}\int_0^{\pi/2} \left(\varepsilon_{22q}\right)_{r=R_{1+i}} d\varphi\, d\nu\right\},$$
$$i = \begin{cases} 0 & \text{for} \quad q = p, e, \\ 1 & \text{for} \quad q = m, e. \end{cases} \tag{3.70}$$

With regard to the analysis presented at the beginning of Chapter 9 concerning the conditions $\beta_p \neq \beta_e = \beta_m$ and $\varepsilon_{11te} - \varepsilon_{11tp} \neq f(\varphi, \nu)$, the radial displacements u_{1p}, u_{1e}, $\left(u_{1m}\right)_{r=R_2} \neq f(\varphi, \nu)$ and the tangential strains ε_{22p}, ε_{22e}, $\left(\varepsilon_{22m}\right)_{r=R_2}$ are not functions of the variables φ, ν, regarding the Cauchy's equation $u_1 = r\,\varepsilon_{22}$ (see Eq. (3.2)). Accordingly, Equation 3.69 is transformed to the form

$$R_{1+iex} = R_{1+i}\left[1 + (\varepsilon_{22q})_{r=R_{1+i}} - \left(\frac{2}{\pi}\right)^2 \int_0^{\pi/2}\int_0^{\pi/2} \beta_q \, d\varphi \, d\nu\right],$$

$$i = \begin{cases} 0 & \text{for} \quad q = p, e, \\ 1 & \text{for} \quad q = m, e, \end{cases} \tag{3.71}$$

and similarly, on the conditions $\varepsilon_{11tq} = 0$ for $T \in \langle T_f, T_r \rangle$ or $\varepsilon_{11tq} \neq f(\varphi, \nu)$ (see Eq. (12.24)), both resulting in $\beta_q \neq f(\varphi, \nu)$, we get

$$R_{1+iex} = R_{1+i}\left[1 + (\varepsilon_{22q})_{r=R_{1+i}} - \beta_q\right], \quad i = \begin{cases} 0 & \text{for} \quad q = p, e, \\ 1 & \text{for} \quad q = m, e. \end{cases} \tag{3.72}$$

Finally, to derive the roots R_1, R_2 of Eqs. (3.69)–(3.72) by numerical methods for the determination of roots of transcendental equations [7, p 970–971], one of the subscripts p, e and m, e for $i=0$ and $i=1$, respectively, is sufficient to be considered.

With regard to experimental methods using the separation of particles from a matrix of a real composite material, the equation related to the multi-particle-envelope-matrix system and corresponding to Eq. (3.67) is derived as

$$R_{1+i} = R_{1+i\,ex}\left[1 - \left(\frac{2}{\pi}\right)^2 \int_0^{\pi/2}\int_0^{\pi/2} \beta_q \, d\varphi \, d\nu\right]^{-1}, \quad i = \begin{cases} 0 & \text{for} \quad q = p, \\ 1 & \text{for} \quad q = e, \end{cases} \tag{3.73}$$

transforming, on the conditions $\varepsilon_{11tq} = 0$ for $T \in \langle T_f, T_r \rangle$ or $\varepsilon_{11tq} \neq f(\varphi, \nu)$ (see Eq. (12.24)), both resulting in $\beta_q \neq f(\varphi, \nu)$, to the form

$$R_{1+i} = \frac{R_{1+i\,ex}}{1 - \beta_q}, \quad i = \begin{cases} 0 & \text{for} \quad q = p, \\ 1 & \text{for} \quad q = e. \end{cases} \tag{3.74}$$

3.3.3 Recommendations of author

Provided that R_{1ex} and R_{1ex}, R_{2ex}, related to the multi-particle-matrix and multi-particle-envelope-matrix systems, respectively, are determined by experimental methods at the temperature T_r or $T_t \in \langle T_f, T_r \rangle$ without (with) the separation of particles from a matrix of a real composite material, the radii R_1, R_2 are equal to R_{1ex}, R_{2ex}, respectively.

The determination of the radii R_1, R_2, included in Eqs. (3.65)–(3.65), (3.69)–(3.73), using numerical methods to solve roots of the transcendental equations (3.65), (3.66), (3.69)–(3.72) including numerical integration [7, p 970–971], is exact but time-consuming. Additionally, with regard to the multi-particle-envelope-matrix system provided that $\beta_p \neq \beta_e \neq \beta_m$, the roots R_1, R_2 related to Eqs. (3.69), (3.70) are required to be derived simultaneously from two equations considering the subscript i, i.e. from Eq. (3.69) for $i = 0,1$, or from Eq. (3.70) for $i = 0,1$. With regard to practicability, the same conditions, i.e. $R_1 = R_{1ex}$, $R_2 = R_{2ex}$, are considered, neglecting an influence of the terms β_q and ε_{22q} in Eqs. (3.65)–(3.74).

Chapter 4

Boundary conditions

4.1 Multi-particle-(envelope)-matrix system

Considering material parameters of the spherical particle ($q=p$), the spherical envelope ($q=e$) and the cell matrix ($q=m$), the radial displacement u_{1q} and the thermal stress σ_q, the latter representing radial, tangential and shear stresses, are generally derived as

$$u_{1q} = \sum_{i=1}^{n} C_{iq} a_{iq} u_{irq}, \tag{4.1}$$

$$\sigma_q = \sum_{i=1}^{n} C_{iq} b_{iq} \sigma_{irq}, \tag{4.2}$$

where $a_{1q}u_{1rq}, \ldots, a_{nq}u_{nrq}$ represent particular solutions of one of the different linear differential equations for the radial displacement u_{1q} with non-zero or zero right sides derived by one of the presented different mathematical techniques. The integration constant C_{iq} related to the particular solution $a_{iq}u_{irq}$ is determined from boundary conditions, and the term u_{irq} ($i=1,\ldots,$ n) is a function of the variable r or a constant function regarding the variable r. Consequently, the term σ_{irq} as a function of u_{irq} depends on the variable r, and accordingly, $b_{iq}\sigma_{irq}$ is derived by the solution $a_{iq}u_{irq}$ (see Eqs. (3.31)–(3.34)). Finally, the radial displacement u_{1p}, as a total solution of one of the different linear differential equations with non-zero right sides, represents a linear combination of the solutions $a_{1q}u_{1rq}, \ldots, a_{nq}u_{nrq}$.

With regard to the spherical envelope provided that $\beta_p \neq \beta_e \neq \beta_m$, the integration constants C_{1e}, C_{2e} derived by two boundary conditions (see Section 4.1.2) are determined using the Cramer's rule defined in Section 12.4.1 (see Eqs. (12.41), (12.42), (12.44)–(12.46)). Similarly, with regard to the cell matrix, the Cramer's rule is used for the determination of the integration constants C_{1m}, C_{2m} or C_{1m}, C_{2m}, C_m derived by two or three boundary conditions (see Sections 4.1.3.1, 4.1.3.2), respectively.

4.1.1 Spherical particle

With regard to the spherical particle with the radius R_1 and with the centre O in the beginning of the Cartesian system $(Ox_1x_2x_3)$, considering the spatial polar coordinates $[r, \varphi, \nu]$ (see Figs. 2.1, 3.1), the thermal stresses σ_{iip}, σ_{12+jp} $(i = 1,2,3;$ $j = 0,1)$, and the radial displacement u_{1p} are required to fulfil the boundary conditions determined as

$$(u_{1p})_{r=0} = 0, \tag{4.3}$$

$$(\sigma_{iip})_{r\to 0} \not\to \pm\infty, \quad (\sigma_{12+jp})_{r\to 0} \not\to \pm\infty, \quad i = 1,2,3; \quad j = 0,1, \tag{4.4}$$

$$(\sigma_{11p})_{r=R_1} = -p_1, \tag{4.5}$$

where (4.3) and (4.4), (4.5) represent geometric and stress boundary conditions for the spherical particle, respectively. The boundary conditions (4.4)–(4.3) are related to the multi-particle-matrix and multi-particle-envelope-matrix systems, on the conditions $\beta_p \neq \beta_m$, and $\beta_p \neq \beta_e \neq \beta_m$ or $\beta_p \neq \beta_e = \beta_m$ (see Eqs. (3.48)–(3.50)) resulting in the radial stress p_1 acting on the particle-matrix and particle-envelope boundaries (see Section 1.4), respectively.

Accordingly, the corresponding function u_{irp} $(i = 1, \ldots, \mathrm{n})$ of the set $\{u_{1rp}, \ldots, u_{nrp}\}$ is considered, and additionally, the absolute value $|u_{1p}|$ of the radial displacement u_{1p} of the infinitesimal spherical cap in the spherical particle is required to represent an increasing function of the variable $r \in \langle 0, R_1\rangle$, exhibiting a maximal value on the particle-matrix and particle-envelope boundaries, thus for $r = R_1$, related to the multi-particle-matrix and multi-particle-envelope-matrix systems, respectively, and accordingly the dependence $u_{1p} - r$ is required to be monotonous (see Section 12.4.3).

As examples of a suitable choice, the term $u_{1rp} = r^{\lambda_{1p}}$ and consequently the radial displacement $u_{1p} = C_{1p}\, a_{1p}\, r^{\lambda_{1p}}$ for $\lambda_{1p} > 1$ fulfil the boundary conditions for the spherical particle (see Eqs. (4.4)–(4.3)), regarding Eqs. (4.2), (3.31)–(3.34). Conversely, the term $u_{2rp} = r^{\lambda_{2p}}$ for $\lambda_{2p} < 0$ does not fulfil the boundary condition for the spherical particle (see Eq. (4.3)), due to $(u_{2rp})_{r\to 0} = \left(C_{2p}\, a_{2p}\, r^{\lambda_{2p}}\right)_{r\to 0} \to \infty$ (see Section 5.4.1).

With regard to the multi-particle-envelope-matrix system, provided that $\beta_p = \beta_e \neq \beta_m$ resulting in $(u_{1p})_{r=R_1} = (u_{1e})_{r=R_1}$, and considering the Cauchy's equation $u_1 = r\,\varepsilon_{22}$ (see Eq. (3.2)), the boundary condition (4.5) is replaced by the strain boundary condition determined as

$$(\varepsilon_{22p})_{r=R_1} = (\varepsilon_{22e})_{r=R_1} = \frac{(u_{1e})_{r=R_1}}{R_1} = -p_2\, \varrho_{1e}^{pe}, \tag{4.6}$$

where $(\varepsilon_{22q})_{r=R_1}$ is the tangential strain in the spherical particle $(q = p)$ and in the spherical envelope $(q = e)$ on the particle-envelope boundary, thus for $r = R_1$, and the radial stress p_2 acting on the matrix-envelope boundary arises from the condition $\beta_e \neq \beta_m$ (see Section 1.4). Consequently, the thermal stresses and the corresponding quantities for the spherical particle are function of the coefficient ϱ_{1e}^{pe} (see Section 1.4). The absolute value $|u_{1p}|$ is also required to represent an increasing function of the variable $r \in \langle 0, R_1 \rangle$, exhibiting the maximal value $\left|(u_{1p})_{r=R_1}\right| = |p_2\, \varrho_{1e}^{pe} R_1|$ on the particle-envelope boundary, thus for $r = R_1$.

Finally, the boundary conditions (4.3) and (4.4), the latter regarding Eq. (3.31), define a course of the dependence $u_{1p} - r$ to represent a monotonous function of the variable $r \in \langle 0, R_1 \rangle$ (see Section 12.4.3), and consequently Equations (4.5) or (4.6) determine one integration constant only. Except for Eq. (4.5), no additional boundary conditions exist for the stresses σ_{iip}, σ_{12+jp} ($i = 1,2,3$; $j = 0,1$) in the particle centre O (see Figs. 2.1a,b, 3.1), thus for $r = 0$, and consequently, the dependencies $\sigma_{iip} - r$, $\sigma_{12+jp} - r$ result from the dependence $u_{1p} - r$ only. Additionally, the monotonousness of the dependence $u_{1p} - r$ needs not to result in the monotonousness of the dependencies $\sigma_{iip} - r$, $\sigma_{12+jp} - r$.

4.1.2 Spherical envelope

Provided that $\beta_p \neq \beta_e \neq \beta_m$ (see Eqs. (3.48)–(3.50)), the thermal radial stress σ_{11e} in the spherical envelope with the radii $R_1 < R_2$ is required to fulfil the stress boundary conditions determined as

$$(\sigma_{11e})_{r=R_1} = -p_1, \tag{4.7}$$

$$(\sigma_{11e})_{r=R_2} = -p_2. \tag{4.8}$$

Accordingly, one of the terms u_{ire}, u_{jre} ($i \neq j$; $i, j = 1, \ldots, \mathrm{n}$) from the set $\{u_{1re}, \ldots, u_{nre}\}$ is required to represent a function of the variable $r \in \langle R_1, R_2 \rangle$.

Provided that $\beta_p \neq \beta_e = \beta_m$, considering the boundary condition (4.7) for the spherical envelope, the absolute value $|u_{1e}|$ exhibits a maximal value on the particle-envelope boundary, thus for $r = R_1$, and accordingly, $|u_{1e}|$ is required to represent a decreasing function of the variable $r \in \langle R_1, R_2 \rangle$ or, at the very most, a constant function regarding the variable r. Finally, as presented in Section 3.1.2.3 (see Item 5), the assumption $u_{1e} \neq f(r)$ may be improbable, nevertheless it is considered, with regard to completeness of the presented topic, in Sections 6.4.1.2.C, 6.4.1.3.B and 9.4.2, 9.5.2.3, the last two for $\lambda_e = 0$ (see Section 9.3.2). As an example, the terms $u_{2re} = r^{\lambda_{2e}}$ or $u_{3re} = 1$, the former representing a decreasing function of the variable r for $\lambda_{2e} < 0$, fulfil the requirements (see Eqs. (5.4), (5.10), (6.8)).

Similarly, $\beta_p = \beta_e \neq \beta_m$, considering the boundary condition (4.8) for the spherical envelope, the absolute value $|u_{1e}|$ exhibits a maximal value on the matrix-envelope boundary, thus for $r = R_2$, and accordingly, $|u_{1e}|$ is required to represent an increasing function of the variable $r \in \langle R_1, R_2 \rangle$ or, at the very most, a constant function regarding the variable r. As an example, the terms $u_{1re} = r^{\lambda_{1e}}$ or $u_{3re} = 1$, the former representing an increasing function of the variable r for $\lambda_{1e} > 1$, fulfil the requirements (see Eqs. (5.4), (5.9), (6.8)).

Finally, provided that $\beta_p \neq \beta_e = \beta_m$ or $\beta_p = \beta_e \neq \beta_m$, the dependence $u_{1e} - r$ is required to represent a monotonous function of the variable $r \in \langle R_1, R_2 \rangle$ (see Section 12.4.3).

4.1.3 Cell matrix

The thermal stresses and the corresponding quantities in the cell matrix are derived regarding the standard boundary conditions (4.9)–(4.12) to be mandatory, without or with consideration of the additional boundary condition (4.14). As presented in Section 4.1.3.2, the additional boundary condition to assume the minimum or the maximum of the $u_{1m} - r$ dependence on the cell surface, thus for $r = r_c$, regarding $u_{1m} > 0$ or $u_{1m} < 0$, respectively, might be considered as a possible state of deformation of a solid continuum on cell boundaries. Accordingly, the condition (4.14) is applied to Eqs. (6.4), (7.3) to represent solutions for the radial displacement u_{1q} containing three integration constants C_{1q}, C_{2q}, C_{3q} (see Sections 6.3.1.D, 6.4.2.1.D, 6.4.2.2.D, 7.4.1.D, 7.5.2.1.D, 7.5.2.2.D). Finally, the φ, ν-dependent distance r_c as a function of the experimentally determined parameters v, R_{1ex}, d_{1ex}, d_{2ex}, d_{3ex} (see Section 2.2.1) is derived in Section 2.3.

4.1.3.1 Mandatory boundary conditions

With regard to the cell matrix, the thermal radial stress σ_{11m} and the radial displacement u_{1m} are required to fulfil the boundary conditions determined as

$$(\sigma_{11m})_{r=R_1} = -p_1, \tag{4.9}$$

$$(\sigma_{11m})_{r=R_2} = -p_2, \tag{4.10}$$

$$(u_{1m})_{r=r_c} = 0, \tag{4.11}$$

where (4.9), (4.10) and (4.11) represent stress and geometric boundary conditions for the cell matrix, respectively. The pairs (4.9), (4.11) and (4.10), (4.11) of the boundary conditions are related to the multi-particle-matrix and multi-particle-envelope-matrix systems, on the conditions $\beta_p \neq \beta_m$, and $\beta_p \neq \beta_e \neq \beta_m$ or $\beta_p \neq \beta_e = \beta_m$ (see Eqs. (3.48)–(3.50)) resulting in the radial stresses p_1 and p_1, p_2 acting on the particle-matrix and particle-envelope, matrix-envelope boundaries (see Section 1.4), respectively.

With regard to the multi-particle-envelope-matrix system, provided that $\beta_p \neq \beta_e = \beta_m$ resulting in $(u_{1m})_{r=R_2} = (u_{1e})_{r=R_2}$, and considering the Cauchy's equation $u_1 = r\,\varepsilon_{22}$ (see Eq. (3.2)), the boundary condition (4.10) is replaced by the strain boundary condition determined as

$$(\varepsilon_{22m})_{r=R_2} = (\varepsilon_{22e})_{r=R_2} = \frac{(u_{1e})_{r=R_2}}{R_2} = -p_1\,\varrho_{2e}^{me}, \tag{4.12}$$

where $(\varepsilon_{22q})_{r=R_2}$ is the tangential strain in the spherical envelope $(q=e)$ and in the cell matrix $(q=m)$ on the matrix-envelope boundary, thus for $r = R_2$. Consequently, the thermal stresses and the corresponding quantities for the spherical particle are function of the coefficient ϱ_{2e}^{me} (see Section 1.4).

As mentioned at the beginning of Chapter 4, the set $\{a_{1m}u_{1rm}, \ldots, a_{nm}u_{nrm}\}$, represented by the solutions $u_{1rm}, \ldots, u_{nrm}$ of a linear differential equation with non-zero or zero right sides derived by one of the presented different mathematical techniques, contains functions of the variable r as well as constant functions regarding the variable r. The absolute value $|u_{1m}|$ of the radial displacement u_{1m} of the infinitesimal spherical cap in the cell matrix is required to represent a decreasing function of the variable $r \in \langle R_1, r_c\rangle$ and $r \in \langle R_2, r_c\rangle$, exhibiting maximal values on the particle-matrix and matrix-envelope boundaries related to the multi-particle-matrix and multi-particle-envelope-matrix systems, thus for $r = R_1$ and $r = R_2$, respectively. Provided that $\beta_p \neq \beta_e = \beta_m$ resulting in the boundary condition (4.12), and considering the Cauchy's equation $u_1 = r\,\varepsilon_{22}$ (see Eq. (3.2)), the maximal value on the matrix-envelope boundary is $\left|(u_{1m})_{r=R_2}\right| = |p_1\,\varrho_{2e}^{me} R_2|$.

Finally, disregarding the conditions concerning the coefficients β_p, β_e, β_m, the dependence $u_{1m} - r$ is required to represent a monotonous function of the variable $r \in \langle R_1, r_c\rangle$, $r \in \langle R_1, r_c\rangle$ (see Section 12.4.3).

Accordingly, with regard to the boundary conditions (4.9), (4.11) or (4.9), (4.11), (4.14) for the multi-particle-matrix system, at the very most one of the terms u_{irm}, u_{jrm} or u_{irm}, u_{jrm}, u_{krm} $(i \neq j \neq k;\ i, j, k = 1, \ldots, n)$ from the set $\{u_{1rm}, \ldots, u_{nrm}\}$ can be constant, and the decreasing course of the $|u_{1m}| - r$ dependence is ensured by the integration constants C_{im}, C_{jm} or C_{im}, C_{jm}, C_{km}, respectively. In case of the multi-particle-envelope-matrix system, the condition (4.9) is replaced by (4.10) or (4.12) provided that $\beta_p \neq \beta_e \neq \beta_m$ or $\beta_p \neq \beta_e = \beta_m$, respectively. If the terms u_{irm}, u_{jrm} are constant functions and the term u_{krm} is r-dependent, thus $\partial u_{irm}/\partial r = 0$, $\partial u_{jrm}/\partial r = 0$, consequently the boundary condition (4.14), regarding Eq. (4.1), is transformed to the form

$$a_{km}C_{km}\left(\frac{\partial u_{krm}}{\partial r}\right)_{r=r_c} = 0, \tag{4.13}$$

and then the coefficient $C_{km} = 0$. Accordingly, the radial displacement $u_{1m} = a_{im}C_{im}u_{irm} + a_{jm}C_{jm}u_{jrm}$ is a constant function regarding the variable $r \in \langle R_1, r_c\rangle$ and $r \in \langle R_2, r_c\rangle$, related to the multi-particle-matrix and multi-particle-envelope-matrix systems, respectively, what is physically unacceptable.

Finally, the boundary conditions for the cell matrix define a course of the dependence $u_{1m} - r$ as a monotonous function to exhibit non-zero and zero values on the particle-matrix, matrix-envelope boundaries and on the cell surface, respectively, the former derived by Eqs. (4.9), (4.10), (4.12), (4.14) considering Eqs. (3.31)–(3.34). On the one hand, the dependencies $\sigma_{iim} - r$, $\sigma_{12+jm} - r$ $(i=1,2,3;\ j=0,1)$ are assumed to represent monotonous functions of the variable $r \in \langle R_1, r_c \rangle$ and $r \in \langle R_2, r_c \rangle$ related to the multi-particle-matrix and multi-particle-envelope-matrix systems, respectively. On the other hand, except for Eqs. (4.9), (4.10), no additional boundary conditions exist for the stresses σ_{iim}, σ_{12+jm} $(i=1,2,3;\ j=0,1)$ on the cell surface, thus for $r = r_c = |OC|$ (see Figs. 2.2, 2.3), and consequently, the dependencies $\sigma_{iim} - r$, $\sigma_{12+jm} - r$ result from the dependence $u_{1m} - r$ only. Additionally, the monotonousness of the dependence $u_{1m} - r$ needs not to result in the monotonousness of the dependencies $\sigma_{iim} - r$, $\sigma_{12+jm} - r$.

4.1.3.2 Additional boundary conditions

The point C on the cell surface (see Figs. 2.2, 2.3) is related to two neighbouring identical cells with the Cartesian systems $(O_1x_1'x_2'x_3')$, $(O_2x_1'x_2'x_3')$ and with the distance $r_c = |O_1C| = |O_2C|$ as length of the abscissae O_1C, O_2C in a radial direction determined by the angles φ, ν and represented by the axis x_1'. With regard to the imaginary division of the infinite matrix, the surface of the cell to represent the part of the multi-particle-(envelope)-matrix system related to one spherical particle is not a physical boundary, and consequently a smooth course of the r-dependent functions $u_{1m} = u_{1m}(r)$ and $w_m = w_m(r)$ is assumed to be exhibited on the cell surface. Consequently, the dependencies $u_{1m}^{(1)} - r$ and $u_{1m}^{(2)} - r$ related to the Cartesian systems $(O_1x_1'x_2'x_3')$ and $(O_2x_1'x_2'x_3')$ are required not to mutually create a singular point in the point C, as also required for the dependencies $w_m^{(1)} - r$ and $w_m^{(2)} - r$, respectively.

To fulfil the non-singularity assumption, the functions $u_{1m}^{(i)}$, $w_m^{(i)}$ $(i=1,2)$ are required to be extremal on the cell surface, thus for $r = r_c$, and consequently the functions $f_u = u_{1m}^{(1)} + u_{2m}^{(1)}$, $f_w = w_m^{(1)} + w_m^{(1)}$ are smooth regarding the variable r and for $r = r_c$. Finally, resulting from decreasing courses of the r-dependent functions $|u_{1m}|$, w_m, the dependencies $|u_{1m}| - r$, $w_m - r$ exhibit minima on the cell surface. Considering the boundary condition (4.11), provided that $u_{1m} > 0$ or $u_{1m} < 0$, the required monotonous dependence $u_{1m} - r$ accordingly exhibit the minimum or the maximum on the cell surface (see Section 12.4.3), thus for $r = r_c$, respectively.

Accordingly, in addition to the standard boundary conditions given by Eqs. (4.9)–(4.12) to be mandatory for the cell matrix, zero radial strain ε_{11m} on the cell surface is considered in the form

$$(\varepsilon_{11m})_{r=r_c} = \left(\frac{\partial u_{1m}}{\partial r}\right)_{r=r_c} = 0. \tag{4.14}$$

representing a zero strain boundary condition as the third boundary condition to the pairs (4.9), (4.11) or (4.10), (4.11) or (4.11), (4.12). The boundary condition (4.14) is considered in connection with Eqs. (6.4), (7.3) for the radial displacement u_{1q} of the infinitesimal spherical cap, containing three integration constants C_{1q}, C_{2q}, C_{3q}.

As presented in Section 3.1.8, the thermal-stress induced elastic energy density $w_q \propto \sigma_q^2 > 0$, as a decreasing function of the variable r, includes the shear strains ε_{12q} and ε_{13q} as functions of $\partial u_{1q}/\partial\varphi$ and $\partial u_{1q}/\partial\nu$, respectively, depending on $\partial C_{iq}/\partial\varphi$ and $\partial C_{iq}/\partial\nu$. On the one hand, a condition of the minimum of the $w_m - r$ dependence, exhibited in the cell matrix on the cell surface, thus for $r = r_c$, might be considered to be a reasonable assumption. On the other hand, representing a differential equation to include the terms C_{iq}, $\partial C_{iq}/\partial\varphi$, $\partial C_{iq}/\partial\nu$ for which no conditions exist, the assumed boundary condition derived in the form

$$\left(\frac{\partial w_m}{\partial r}\right)_{r=r_c} = 0, \tag{4.15}$$

can not be accordingly considered as the fourth boundary condition for the cell with an *isotropic* matrix.

4.2 One-particle-(envelope)-matrix system

Boundary conditions for the spherical particle, for the spherical envelope and for the infinite matrix of the one-particle-(envelope)-matrix system are identical to those presented in Sections 4.1.1–4.1.3, except for Eq. (4.11) replacing by Eqs. (4.16) or (4.17) derived as

$$(u_{1m})_{r\to\infty} = 0, \tag{4.16}$$

$$(\sigma_{iim})_{r\to\infty} = 0, \quad (\sigma_{12+jm})_{r\to\infty} = 0, \quad i = 1,2,3; \quad j = 0,1, \tag{4.17}$$

and accordingly, as presented in Section 4.1.3.1 as well, the absolute values $|u_{1m}|$, $|\sigma_{iim}|$, $|\sigma_{12+jm}|$ ($i = 1,2,3$; $j = 0,1$) are required to be decreasing functions of the variable $r \in \langle R_1, \infty)$ and $r \in \langle R_2, \infty)$, related to the one-particle-matrix and one-particle-envelope-matrix systems, respectively. Accordingly disregarding the conditions concerning the coefficients β_p, β_e, β_m, the dependencies $u_{1m} - r$, $\sigma_{iim} - r$, $\sigma_{12+jm} - r$ are required to represent monotonous functions of the variable $r \in \langle R_1, r_c\rangle$, $r \in \langle R_1, r_c\rangle$ (see Section 12.4.3).

The boundary conditions (4.16) and (4.17) define courses of the dependencies $u_{1m} - r$ and $\sigma_{11m} - r$, and consequently Eqs. (4.9) or (4.10) or (4.12) determine one integration constant only, and additionally Equations (4.16) and (4.17) are both considered for the determination of the thermal stresses and the corresponding quantities in the one-particle-(envelope)-matrix system. Finally, with regard to

(4.16) or (4.17), the assumed boundary condition (4.14) for $r \to \infty$ is accordingly fulfilled, and another integration constant is not required to be considered.

4.3 Supplement

As presented in Sections 4.1.1 and 4.1.3, the boundary conditions for the spherical particle and for the cell matrix given by Eqs. (4.6) and (4.12), provided that $\beta_p = \beta_e \neq \beta_m$ and $\beta_p \neq \beta_e = \beta_m$, respectively, are related to the radial displacement u_{1e} in the spherical envelope determined from the boundary conditions given by Eqs. (4.7) and (4.8), respectively. A detailed analysis of reasons for the determination of the boundary conditions (4.6) and (4.12) by the radial displacement u_{1e} is presented in Sections 4.3.1 and 4.3.2 concerning the conditions $\beta_p = \beta_e \neq \beta_m$ and $\beta_p \neq \beta_e = \beta_m$, respectively.

With regard to the one-particle-(envelope)-matrix system, the boundary condition (4.11) mentioned in Sections 4.3.1, 4.3.2 is meant to be replaced by (4.16), and additionally the boundary condition (4.14) is not considered.

4.3.1 Condition $\beta_p = \beta_e \neq \beta_m$

With regard to Eq. (4.6), the integration constant C_{ip} included in the solution $u_{1p} = C_{ip} a_{ip} u_{irp}$ $(i = 1, \ldots, \mathrm{n})$ (see Eq. (4.1)) for the spherical particle is derived by the coefficient ϱ_{1e}^{pe} related to the spherical envelope. Conversely, assuming two boundary conditions for the spherical envelope given by Eq. (4.8) and determined as

$$(\varepsilon_{22e})_{r=R_1} = (\varepsilon_{22p})_{r=R_1} = \frac{(u_{1p})_{r=R_1}}{R_1} = -p_2\, \varrho_{1p}^{pe}, \tag{4.18}$$

the integration constants C_{je}, C_{ke} $(j, k = 1, \ldots, \mathrm{n};\ j \neq k)$ included in the solution $u_{1e} = C_{je} a_{je} u_{jre} + C_{ke} a_{ke} u_{kre}$ (see Eq. (4.1)) are derived by the coefficient ϱ_{1m}^{pe} related to the spherical particle. Consequently, the radial displacement u_{1p} in the spherical particle could be determined using the boundary condition (4.3), transforming to the form

$$C_{lp} a_{lp} \left(u_{lrp}\right)_{r=0} = 0, \tag{4.19}$$

and accordingly, neither the integration constant C_{lp} $(l = 1, \ldots, \mathrm{n})$ related to the term u_{lrp} representing an increasing function of the variable $r \in \langle 0, R_1 \rangle$, and consequently nor the coefficient ϱ_{1p}^{pe} can be determined. Additionally, assuming the absolute value $|u_{1p}|$ to represent an increasing function of the variable $r \in \langle 0, R_1 \rangle$, and provided that $C_{lp} = 0$, we get $u_{1p} = C_{lp} a_{lp} u_{lrp} = 0$ (see Eq. (4.1)) what is physically unacceptable.

4.3.2 Condition $\beta_p \neq \beta_e = \beta_m$

With regard to Eqs. (4.11), (4.12), the integration constants C_{im}, C_{jm} included in the solution $u_{1m} = C_{im}a_{im}u_{irm} + C_{jm}a_{jm}u_{jrm}$ $(i, j = 1, \ldots, \text{n};\ i \neq j)$ (see Eq. (4.1)) for the cell matrix are derived by the coefficient ϱ_{2e}^{me} related to the spherical envelope. Conversely, assuming two boundary conditions for the spherical envelope given by Eq. (4.7) and determined as

$$(\varepsilon_{22e})_{r=R_2} = (\varepsilon_{22m})_{r=R_2} = \frac{(u_{1m})_{r=R_2}}{R_2} = -p_1\,\varrho_{2m}^{me}, \tag{4.20}$$

the integration constants C_{ke}, C_{le} included in the solution $u_{1e} = C_{ke}a_{ke}u_{kre} + C_{le}a_{le}u_{lre}$ $(k, l = 1, \ldots, \text{n};\ k \neq l)$ (see Eq. (4.1)) for the spherical envelope are derived by the coefficient ϱ_{2m}^{me} related to the cell matrix. Consequently, the radial displacement u_{1m} in the cell matrix could be determined using one boundary condition only, given by Eq. (4.11). However, the boundary boundary condition (4.11) is transformed to the form

$$C_{sm}a_{sm}\,(u_{srm})_{r=r_c} = 0, \tag{4.21}$$

and accordingly, neither the integration constant C_{sm} $(s = 1, \ldots, \text{n})$ related to the term u_{srm} representing a decreasing function of the variable $r \in \langle R_2, r_c\rangle$, and consequently nor the coefficient ϱ_{2m}^{me} can be determined. Additionally, assuming the absolute value $|u_{1m}|$ to represent a decreasing function of the variable $r \in \langle R_2, r_c\rangle$, and provided that $C_{sm} = 0$, we get $u_{1m} = C_{sm}a_{sm}u_{srm} = 0$ (see Eq. (4.1)) what is physically unacceptable.

Similarly, provided that the radial displacement u_{1m} could be determined using two boundary conditions given by Eqs. (4.11), (4.14), and considering at least one of the terms u_{xrm}, u_{yrm} to represent a decreasing function of the variable $r \in \langle R_2, r_c\rangle$ $(x, y = 1, \ldots, \text{n};\ x \neq y)$, we get

$$C_{xm}a_{xm}\,(u_{xrm})_{r=r_c} + C_{ym}a_{ym}\,(u_{yrm})_{r=r_c} = 0, \tag{4.22}$$

$$C_{xm}a_{xm}\left(\frac{u_{xrm}}{\partial r}\right)_{r=r_c} + C_{ym}a_{ym}\left(\frac{u_{yrm}}{\partial r}\right)_{r=r_c} = 0. \tag{4.23}$$

Accordingly, the integration constants C_{xm}, C_{ym}, determined from the system of the linear algebraic equations (4.22), (4.23), are equal to zero, and finally $u_{1m} = C_{xm}a_{xm}u_{xrm} + C_{ym}a_{ym}u_{yrm} = 0$ (see Eq. (4.1)) what is physically unacceptable.

Chapter 5

Isotropic multi- and one-particle-(envelope)-matrix systems. Solutions 1, 2, 3

Chapter 5 presents three different mathematical techniques (see Sections 5.1–5.3) applied to a system of the differential equations (3.35), (3.36) and resulting in three different linear differential equations with non-zero right sides (see Eqs. (5.5), (5.13), (5.18)) with three different solutions for the radial displacement u_{1q} of the infinitesimal spherical cap given by Eqs. (5.7), (5.15), (5.20). With regard to the analysis presented in Section 5.4 to consider the boundary conditions (see Section 4.1) defined for the spherical particle ($q = p$), for the spherical envelope ($q = e$) and for the cell matrix ($q = m$) as components of composite materials, the three different solutions are suitable for the determination of the thermal stresses and the corresponding quantities, the latter represented by w_q, W_q, σ_{iq}, w_{iq} ($i = 1,2,3$) (see Eqs. (3.41), (3.46), (3.47)), in each of the components.

As a consequence of the reduction of the coefficients $a_q^{(1)}$, $a_q^{(2)}$, $a_q^{(3)}$ (see Eqs. (5.6), (5.14), (5.19)) within the determination of the integration constants C_{1q}, C_{2q} using the boundary conditions, formulae for the thermal stresses and for the corresponding quantities, related to the solutions (5.7), (5.15), (5.20), exhibit the same form, except for the coefficient $\Lambda_{1+iq} = \Lambda_{1+iq}^{(j)}$ ($i = 0,1$; $j = 1,2,3$; $q = p,e,m$) substituted by $\Lambda_{1+iq}^{(1)}$, $\Lambda_{1+iq}^{(2)}$, $\Lambda_{1+iq}^{(3)}$ (see Eqs. (5.8), (5.16), (5.21)) related to the mathematical techniques 1, 2, 3 presented in Sections 5.1, 5.2, 5.3, respectively, resulting in the terms u_{1rq}, u_{2rq} to represent increasing and decreasing functions of the variable r due to the exponents $\lambda_{1q} > 1$ and $\lambda_{2q} < 0$ (see Eqs. (5.9), (5.10), (5.4)), respectively. Consequently, with regard to the boundary conditions presented in Section 4.1, one or two of the terms u_{1rq}, u_{2rq} are considered for the determination of the integration constants C_{1q}, C_{2q} included in the radial displacement u_{1q} as a total solution representing a linear combination of the particular solutions $C_{1q}a_{1q}u_{1rq}$, $C_{2q}a_{2q}u_{2rq}$, regarding Eq. (4.1) and accordingly $a_{1+iq} = \Lambda_{1+iq}$ ($i = 0,1$).

Although the terms u_{1rq}, u_{2rq} related to the three different mathematical techniques are the same, the radial displacement u_{1q} given by Eqs. (5.7), (5.15), (5.20), resulting from the linear differential equations (5.5), (5.13), (5.18) with non-zero right sides, is different due to the coefficients $a_{1q} = \Lambda_{1q}^{(i)}$, $a_{2q} = \Lambda_{2q}^{(i)}$ ($i = 1,2,3$) (see Eqs. (5.8), (5.16), (5.21)), regarding Eq. (4.1).

The radial displacement u_{1q}, the thermal stresses and the corresponding quantities are functions of the angles φ, ν included in the φ, ν-dependent distance r_c as a function of the experimentally determined parameters v, R_{1ex}, d_{1ex}, d_{2ex}, d_{3ex} (see Sections 2.2.1, 2.3). As a consequence of symmetry of the multi-particle-(envelope)-matrix system resulting from the matrix infinity and from the periodical distribution of the spherical particles shown in Fig. 2.1a,b, the angles φ, ν are both from the interval $\langle 0, \pi/2 \rangle$.

With regard to completeness of the presented topic, Section 5.7 presents the analytical model of the thermal stresses and the corresponding quantities in the one-particle-(envelope)-matrix system containing, in contrast to the multi-particle-(envelope)-matrix system, one spherical particle only. Considering the boundary conditions defined in Section 4.2, the analytical model related to the one-particle-(envelope)-matrix system is required to be derived regarding the analysis presented in Section 2.2.2 and concerning the condition $v \to 0$.

To derive the elastic energy density W induced by the thermal stresses, formulae for the integration of the r-dependent integrand wr^2 (see Eq. (3.42)) are presented in Section 12.4.4. Consequently, the integral in the formula $W = \int_0^{\pi/2} \int_0^{\pi/2} \Omega \, d\varphi \, d\nu$ (see Eqs. (5.33), (5.51), (5.70), (5.84), (5.103), (5.121), (5.136), (5.154)), (5.156), (5.175), (5.192)), related to the integrand represented by the φ, ν-dependent coefficient Ω and to the variables $\varphi, \nu \in \langle 0, \pi/2 \rangle$, can be derived numerically for a real composite material, considering Eq. (1.5).

Formulae for the thermal stresses and for the corresponding quantities in the spherical particle and envelope related to the multi-particle-envelope-matrix system provided $\beta_p \neq \beta_e = \beta_m$ and $\varepsilon_{11te} - \varepsilon_{11tp} = f(\varphi, \nu)$ are presented in Sections 9.4.1, 9.4.2, considering the analysis and the mathematical techniques to result in the different linear differential equations (9.2), (9.6), (9.10), (9.13), (9.17), (9.21), (9.24), (9.28), (9.32), (9.36) with zero right sides presented in Sections 9.1 and 9.2, respectively.

5.1 Mathematical techniques 1

Performing $r\,[\partial \text{Eq.}(3.35)/\partial r]$, the differential equation (3.35) is transformed to the form [1]

[1]The mathematical techniques to result in Eqs. (5.1)–(5.7) are presented in Section 12.4.5 in detail, including a detailed explanation of the Wronskian's method illustratively applied to the differential equation (5.5).

$$r^2 \frac{\partial^2 U_{1q}}{\partial r^2} + (1 - c_{3q})\, r \, \frac{\partial U_{1q}}{\partial r} = 0, \tag{5.1}$$

where the function U_{1q} and the coefficient c_{3q} are given by Eqs. (3.37) and (3.40), respectively. Substituting Eq. (3.35) to Eq. (5.1), we get

$$r^2 \frac{\partial^2 U_{1q}}{\partial r^2} + c_{3q}\,(1 - c_{3q})\, U_{1q} = 0. \tag{5.2}$$

and assuming $U_{1q} = r^{\lambda_q}$ (see Eq. (3.37)), we get

$$U_{1q} = \frac{\partial^2 u_{1q}}{\partial \varphi^2} + \frac{\partial^2 u_{1q}}{\partial \nu^2} = \sum_{i=0}^{1} C_{1+iq}\, r^{\lambda_{1+iq}}, \tag{5.3}$$

where C_{1q}, C_{2q} are integration constants and the exponent λ_{1+iq} ($i = 0{,}1$) is derived as

$$\begin{aligned} \lambda_{1+iq} &= \frac{1}{2}\left[1 + (\delta_{0i} - \delta_{1i})\sqrt{1 - 4c_{3q}\,(1 - c_{3q})}\right] \\ &= \frac{1}{2}\left\{1 + (\delta_{0i} - \delta_{1i})\sqrt{1 + 16\,(1 - \mu_m)\,[1 + 4\,(1 - \mu_m)]}\right\}, \quad i = 0, 1. \end{aligned} \tag{5.4}$$

With regard to the Poisson's number of an elastic solid continuum and real materials, $\mu_q = 1/4$ and $\mu_q < 0.5$ [1, p 38], respectively, the coefficient $c_{3q} < 0$ (see Eq. (3.40)), the discriminant $D = 1 - 4c_{3q}\,(1 - c_{3q}) > 0$ (see Eqs. (12.74), (12.51)), and accordingly the real exponents $\lambda_{1q} > 3$, $\lambda_{2q} < -2$ with respect to $\mu_q < 0.5$.

Considering Eq. (5.3), the differential equation (3.36) is transformed to the form

$$\frac{\partial^2 u_{1q}}{\partial r^2} + \frac{2}{r}\frac{\partial u_{1q}}{\partial r} - \frac{2u_{1q}}{r^2} = -3a_q^{(1)} \sum_{i=0}^{1} C_{1+iq}\, r^{\lambda_{1+iq}-2}, \tag{5.5}$$

representing a linear differential equation of the second rank regarding the variable r with a non-zero right side, and the constant $a_q^{(1)}$ is derived as

$$a_q^{(1)} = \frac{1}{3s_{44q}\,(c_{1q} + c_{2q})}, \tag{5.6}$$

where the elastic modulus s_{44q} and the coefficients c_{1q}, c_{2q} are given by Eqs. (3.25) and (3.38), (3.39), respectively.

Using the Wronskian's method [5] presented in Section 12.4.5, representing a short introduction to problems of linear differential equations with non-zero right sides, the solution of the differential equation (5.5) for the radial displacement u_{1q} of the infinitesimal spherical cap (see Figs. 3.2, 3.3) has the form

$$u_{1q} = a_q^{(1)} \sum_{i=0}^{1} C_{1+iq}\, \Lambda_{1+iq}^{(1)}\, r^{\lambda_{1+iq}}, \tag{5.7}$$

where the coefficient $\Lambda_{1+iq}^{(1)}$ $(i=0,1)$ is derived as

$$\Lambda_{1+iq}^{(1)} = \frac{1}{\lambda_{1+iq}+2} - \frac{1}{\lambda_{1+iq}-1}, \quad i=0,1. \tag{5.8}$$

The integration constants C_{1q}, C_{2q} are derived from boundary conditions (see Chapter 4), considering one or two of the terms u_{1rq} and u_{2rq}, regarding Eq. (4.1), in the forms

$$u_{1rq} = r^{\lambda_{1q}}, \tag{5.9}$$

$$u_{2rq} = r^{\lambda_{2q}}, \tag{5.10}$$

representing increasing and decreasing functions of the variable r, resulting from the exponents $\lambda_{1q} > 1$ and $\lambda_{2q} < 0$, respectively.

With regard to the analysis, presented in Section 5.4 and concerning the integration constants C_{1q}, C_{2q}, the subscript q in formulae in Sections 5.1–5.3 is related to material parameters of the spherical particle $(q=p)$, the spherical envelope $(q=e)$ and the cell matrix $(q=m)$.

5.2 Mathematical techniques 2

Substituting Eq. (5.3) to Eq. (3.35), the differential equation (3.35) is transformed to the form

$$r\,\frac{\partial U_{1q}}{\partial r} = c_{3q}\sum_{i=0}^{1} C_{1+iq}\, r^{\lambda_{1+iq}}, \tag{5.11}$$

and performing $r\,[\partial\text{Eq.}(3.36)/\partial r]$, we get

$$r^3\,\frac{\partial^3 u_{1q}}{\partial r^3} + 4\,r^2\,\frac{\partial^2 u_{1q}}{\partial r^2} + \frac{r}{s_{44q}\,(c_{1q}+c_{2q})}\,\frac{\partial U_{1q}}{\partial r} = 0, \tag{5.12}$$

where the elastic modulus s_{44q}, the exponents $\lambda_{1q} > 1$, $\lambda_{2q} < 0$ $(q=p,e,m)$ and the coefficients c_{1q}, c_{2q}, c_{3q} are given by Eqs. (3.25), (5.4) and (3.38)–(3.40), respectively.

Considering Eq. (5.11), the differential equation (5.12) is transformed to the form

$$\frac{\partial^3 u_{1q}}{\partial r^3} + \frac{4}{r}\,\frac{\partial^2 u_{1q}}{\partial r^2} = -a_q^{(2)}\sum_{i=0}^{1} C_{1+iq}\, r^{\lambda_{1+iq}-3}, \tag{5.13}$$

representing a linear differential equation of the third rank regarding the variable r with a non-zero right side, and the constant $a_q^{(2)}$, as a function of $a_q^{(1)}$ given by Eq. (5.6), is derived as

$$a_q^{(2)} = 3c_{3q}a_q^{(1)}. \tag{5.14}$$

Using the Wronskian's method [5] presented in Section 12.4.5, representing a short introduction to problems of linear differential equations with non-zero right sides, the solution of the differential equation (5.13) for the radial displacement u_{1q} of the infinitesimal spherical cap (see Figs. 3.2, 3.3) has the form

$$u_{1q} = a_q^{(2)} \sum_{i=0}^{1} C_{1+iq} \, \Lambda_{1+iq}^{(2)} \, r^{\lambda_{1+iq}}, \tag{5.15}$$

where the coefficient $\Lambda_{1+iq}^{(2)}$ $(i=0,1)$ is derived as

$$\Lambda_{1+iq}^{(2)} = \frac{1}{2\lambda_{1+iq}} - \frac{1}{3\left(\lambda_{1+iq} - 1\right)} - \frac{1}{6\left(\lambda_{1+iq} + 2\right)}, \quad i = 0, 1. \tag{5.16}$$

5.3 Mathematical techniques 3

Performing $r\left[\partial \mathrm{Eq.}(5.3)/\partial r\right]$, we get

$$r\,\frac{\partial U_{1q}}{\partial r} = \sum_{i=0}^{1} C_{1+iq}\lambda_{1+iq}\, r^{\lambda_{1+iq}}, \tag{5.17}$$

and consequently substituting Eq. (5.17) to Eq. (5.12), the differential equation (5.12) is transformed to the form

$$\frac{\partial^3 u_{1q}}{\partial r^3} + \frac{4}{r}\frac{\partial^2 u_{1q}}{\partial r^2} = -a_q^{(3)} \sum_{i=0}^{1} C_{1+iq}\lambda_{1+iq}\, r^{\lambda_{1+iq}-3}, \tag{5.18}$$

representing a linear differential equation of the third rank regarding the variable r with a non-zero right side. The exponents $\lambda_{1q} > 1$, $\lambda_{2q} < 0$ $(q = p, e, m)$ are given by Eq (5.4) and the constant $a_q^{(3)}$, as a function of $a_q^{(1)}$ given by Eq. (5.6), is derived as

$$a_q^{(3)} = 3a_q^{(1)}. \tag{5.19}$$

Using the Wronskian's method [5] presented in Section 12.4.5, representing a short introduction to problems of linear differential equations with non-zero right sides, the solution of the differential equation (5.18) for the radial displacement u_{1q} of the infinitesimal spherical cap (see Figs. 3.2, 3.3) has the form

$$u_{1q} = a_q^{(3)} \sum_{i=0}^{1} C_{1+iq} \, \Lambda_{1+iq}^{(3)} \, r^{\lambda_{1+iq}}, \tag{5.20}$$

where the coefficient $\Lambda_{1+iq}^{(3)}$ $(i=0,1)$, as a function of $\Lambda_{1+iq}^{(2)}$ given by Eq. (5.16), is derived as

$$\Lambda_{1+iq}^{(3)} = \lambda_{1+iq}\,\Lambda_{1+iq}^{(2)}, \quad i = 0, 1. \tag{5.21}$$

5.4 Analysis of solution

5.4.1 Spherical particle

With regard to the geometric boundary condition $(u_{1p})_{r=0} = 0$ given by Eq. (4.3) related to the multi-particle-matrix and multi-particle-envelope-matrix systems provided that $\beta_p \neq \beta_m$, and $\beta_p \neq \beta_e \neq \beta_m$ or $\beta_p = \beta_e \neq \beta_m$ (see Eqs. (3.48)–(3.50)), the absolute value $|u_{1p}|$ of the radial displacement u_{1p} of the infinitesimal spherical cap in the spherical particle is required to represent an increasing function of the variable $r \in \langle 0, R_1 \rangle$, exhibiting maximal values on the particle-matrix and particle-envelope boundaries, respectively.

Accordingly, the term u_{1rp} is considered, and thus the integration constants $C_{1p} \neq 0$, $C_{2p} = 0$ (see Eqs. (5.7), (5.9), (5.15), (5.20)). Additionally, the exponent λ_{1p} is required to be $\lambda_{1p} \geq 1$, as fulfilled regarding Eq. (5.4), otherwise we get $(\sigma_{iip})_{r\to 0} \propto \left(r^{\lambda_{1p}-1}\right)_{r\to 0} \to \infty$, $(\sigma_{12+jp})_{r\to 0} \propto \left(r^{\lambda_{1p}-1}\right)_{r\to 0} \to \infty$ $(i = 1,2,3;\ j = 0,1)$ (see Eqs. (3.31)–(3.34)) what is physically unacceptable. Provided that $\beta_p = \beta_e \neq \beta_m$ the maximal value on the particle-envelope boundary is $\left|(u_{1p})_{r=R_1}\right| = |p_2\, \varrho_{1e}^{pe} R_1|$.

Explaining in detail, resulting from $\lambda_{2p} < 0$ (see Eq. (5.4)), the radial displacement $(u_{1p})_{r\to 0} \propto (u_{2rp})_{r\to 0} = \left(r^{\lambda_{2p}}\right)_{r\to 0} \to \infty$ what is physically unacceptable. Similarly, the stress $(\sigma_{ijp})_{r\to 0} \propto \left(r^{\lambda_{2p}-1}\right)_{r\to 0} \to \infty$ $(i, j = 1,2,3)$ (see Eqs. (3.31)–(3.34)) what is physically unacceptable, and thus the integration constant $C_{2p} = 0$.

The analysis presented in Section 5.4.1 is also considered for the spherical particle in the one-particle-(envelope)-matrix system (see Section 5.7.1).

5.4.2 Spherical envelope

The analysis presented in Section 5.4.2 is considered for the spherical envelope in the multi- and one-particle-envelope-matrix systems (see Sections 5.6.2, 5.7.1).

With regard to the stress boundary conditions for the spherical envelope given by Eqs. (4.7), (4.8) provided that $\beta_p \neq \beta_e \neq \beta_m$ (see Eqs. (3.48)–(3.50)), the terms u_{1re}, u_{2re} (see Eqs. (5.9), (5.10)) both as functions of the variable $r \in \langle R_1, R_2 \rangle$ are considered for the determination of the radial displacement u_{1e} of the infinitesimal spherical cap in the spherical envelope as a function of the variable $r \in \langle R_1, R_2 \rangle$, and thus the integration constants $C_{1e} \neq 0$, $C_{2e} \neq 0$ (see Eqs. (5.7), (5.15), (5.20)).

Similarly, with regard to the stress boundary condition given by Eq. (4.10) provided that $\beta_p \neq \beta_e = \beta_m$, the absolute value $|u_{1e}|$ exhibits a maximal value on the particle-envelope boundary, thus for $r = R_1$. Accordingly, $|u_{1e}|$ is required to represent a decreasing function of the variable $r \in \langle R_1, R_2 \rangle$ or, at the very most, a constant function regarding the variable r. Consequently, the term u_{2re} is considered, and thus the integration constant $C_{1e} = 0$, $C_{2e} \neq 0$.

Finally, with regard to the stress boundary condition given by Eq. (4.8) provided that $\beta_p = \beta_e \neq \beta_m$, the absolute value $|u_{1e}|$ exhibits a maximal value on the matrix-envelope boundary, thus for $r = R_2$. Accordingly, $|u_{1e}|$ is required to represent an increasing function of the variable $r \in \langle R_1, R_2\rangle$ or, at the very most, a constant function regarding the variable r. Consequently, the term u_{1re} is considered, and thus the integration constants $C_{1e} \neq 0$, $C_{2e} = 0$.

5.4.3 Cell matrix

With regard to the stress, geometric boundary conditions for the cell matrix given by Eqs. (4.9), (4.11) and (4.10), (4.11), related to the multi-particle-matrix and multi-particle-envelope-matrix systems provided that $\beta_p \neq \beta_m$, and $\beta_p \neq \beta_e \neq \beta_m$ or $\beta_p = \beta_e \neq \beta_m$ (see Eqs. (3.48)–(3.50)), the absolute value $|u_{1m}|$ of the radial displacement u_{1m} of the infinitesimal spherical cap in the cell matrix, exhibiting maximal values on the particle-matrix and matrix-envelope boundaries, thus for $r = R_1$ and $r = R_2$, is required to represent a decreasing function of the variable $r \in \langle R_1, r_c\rangle$ and $r \in \langle R_2, r_c\rangle$, respectively. Accordingly, the terms u_{1rm}, u_{2rm} are considered, and thus the integration constants $C_{1m} \neq 0$, $C_{2m} \neq 0$ (see Eqs. (5.7), (5.9), (5.10), (5.15), (5.20)).

With regard to the multi-particle-envelope-matrix system related to the condition $\beta_p \neq \beta_e = \beta_m$, resulting in the geometric and zero strain boundary conditions given by Eqs. (4.11) and (4.12), respectively, the absolute value $|u_{1m}|$ is also required to represent a decreasing function of the variable $r \in \langle R_2, r_c\rangle$, exhibiting the maximal value $\left|(u_{1m})_{r=R_2}\right| = |p_1\, \varrho_{2e}^{me} R_2|$ on the matrix-envelope boundary, thus for $r = R_2$. Accordingly, the terms u_{1rm}, u_{2rm} are also considered, and thus the integration constants $C_{1m} \neq 0$, $C_{2m} \neq 0$.

5.4.4 Infinite matrix

As presented in Section 4.2, one integration constant is considered only, regarding $|u_{1m}|$, $|\sigma_{11m}|$ to represent decreasing functions of the variable $r \in \langle R_1, \infty)$ and $r \in \langle R_2, \infty)$ related to the one-particle-matrix and one-particle-envelope-matrix systems, respectively. Accordingly, the term u_{2rm} is considered, and thus the integration constants $C_{1m} = 0$, $C_{2m} \neq 0$ (see Eqs. (5.7), (5.10), (5.15), (5.20)).

5.5 Multi-particle-matrix system

The variable r related to the thermal stresses and to the corresponding quantities in the spherical particle with the radius R_1 and in the cell matrix is from the intervals $\langle 0, R_1\rangle$ and $\langle R_1, r_c\rangle$, respectively, where the φ, ν-dependent distance r_c as a function of the experimentally determined parameters v, R_{1ex}, d_{1ex}, d_{2ex}, d_{3ex} (see Section 2.2.1) is derived in Section 2.3. With regard to Eq. (3.58), the compressive or tensile radial stress $p_1 > 0$ or $p_1 < 0$ acting on the particle-matrix boundary, thus for $r = R_1$, respectively, is given by Eq. (10.1). The determination of the radial stress p_1, as a consequence of the condition $\beta_p \neq \beta_m$ (see Eqs. (3.48)–(3.50)), considers the tangential strains $(\varepsilon_{22p})_{r=R_1}$ and $(\varepsilon_{22m})_{r=R_1}$ in the spherical particle and in the cell matrix on the particle-matrix boundary (see Eq. (3.58)), resulting in the coefficients ϱ_{1p} and ϱ_{1m} (see Section 10.1.1), respectively.

5.5.1 Thermal stresses in spherical particle

Formulae for the thermal stresses and for the corresponding quantities related to the spherical particle presented in Section 5.5.1 are derived for the multi-particle-matrix and multi-particle-envelope-matrix systems with regard to the conditions $\beta_p \neq \beta_m$ and $\beta_p \neq \beta_e \neq \beta_m$, $\beta_p \neq \beta_e = \beta_m$ (see Eqs. (3.48)–(3.50)), respectively, where the condition $\beta_p \neq \beta_e = \beta_m$ is connected with the φ, ν-dependent difference $\varepsilon_{11te} - \varepsilon_{11tp} = f(\varphi, \nu)$ (see Eq. (12.23)).

With regard to the boundary conditions (4.3)–(4.5) to determine one integration constant only, and to the analysis presented in Section 5.4.1 considering the term u_{1rp} (see Eq. (5.9)) to represent an increasing function of the variable $r \in \langle 0, R_1 \rangle$, and thus the integration constant $C_{2p} = 0$, the radial displacement u_{1p} of the infinitesimal spherical cap in the spherical particle, regarding Eqs. (3.31), (5.7), has the form

$$u_{1p} = -p_1 \, \varrho_{1p} \, r \left(\frac{r}{R_1} \right)^{\lambda_{1p}-1}, \tag{5.22}$$

where the coefficient ϱ_{1p}, the exponent $\lambda_{1p} > 1$ as a function of $c_{3p} < 0$ (see Eq. (3.40)) and the radial stress p_1 acting on the particle-matrix boundary are given by Eqs. (5.36), (5.4) and (10.1) respectively.

Consequently, the radial stress σ_{11p}, the tangential and shear stresses σ_{22p}, σ_{33p} and σ_{12p}, σ_{13p} acting in the spherical particle, respectively, regarding Eqs. (3.31)–(3.34), (5.22), are derived as

$$\sigma_{11p} = -p_1 \left(\frac{r}{R_1} \right)^{\lambda_{1p}-1}, \tag{5.23}$$

$$\sigma_{22p} = \sigma_{33p} = -p_1 \, \varrho_{1p} \left(c_{1p} - \lambda_{1p} \, c_{2p} \right) \left(\frac{r}{R_1} \right)^{\lambda_{1p}-1}, \tag{5.24}$$

$$\sigma_{12p} = -\frac{\varrho_{1p}}{s_{44p}} \frac{\partial p_1}{\partial \varphi} \left(\frac{r}{R_1} \right)^{\lambda_{1p}-1}, \tag{5.25}$$

$$\sigma_{13p} = -\frac{\varrho_{1p}}{s_{44p}} \frac{\partial p_1}{\partial \nu} \left(\frac{r}{R_1} \right)^{\lambda_{1p}-1}, \tag{5.26}$$

where the elastic modulus s_{44p} is given by Eq. (3.25).

Finally, the radial strain ε_{11p}, the tangential and shear strains ε_{22p}, ε_{33p} and ε_{12p}, ε_{13p} in the spherical particle, respectively, regarding Eqs. (3.1)–(3.5), (5.22), have the forms

$$\varepsilon_{11p} = -p_1 \varrho_{1p} \lambda_{1p} \left(\frac{r}{R_1} \right)^{\lambda_{1p}-1}, \tag{5.27}$$

$$\varepsilon_{22p} = \varepsilon_{33p} = -p_1 \varrho_{1p} \left(\frac{r}{R_1} \right)^{\lambda_{1p}-1}, \tag{5.28}$$

$$\varepsilon_{12p} = -\varrho_{1p} \frac{\partial p_1}{\partial \varphi} \left(\frac{r}{R_1} \right)^{\lambda_{1p}-1}, \tag{5.29}$$

$$\varepsilon_{13p} = -\varrho_{1p} \frac{\partial p_1}{\partial \nu} \left(\frac{r}{R_1} \right)^{\lambda_{1p}-1}. \tag{5.30}$$

With regard to the determination of the compressive or tensile radial stress $p_1 > 0$ or $p_1 < 0$, respectively, acting on the particle-matrix boundary, thus for $r = R_1$, considering Eq. (3.58), the tangential strain on the particle-matrix boundary, $(\varepsilon_{22p})_{r=R_1}$, is required to be derived in the form

$$(\varepsilon_{22p})_{r=R_1} = -p_1 \varrho_{1p}. \tag{5.31}$$

Considering Eqs. (3.41), (3.42), (5.23)–(5.30), the thermal-stress induced elastic energy density w_p and the thermal-stress induced elastic energy W_p, accumulated in an arbitrary point of the spherical particle and in the volume of the spherical particle, respectively, have the forms

$$w_p = \frac{\omega_{1p} \varrho_{1p}^2}{2} \left(\frac{r}{R_1} \right)^{2(\lambda_{1p}-1)}, \tag{5.32}$$

$$W_p = 8 \int_0^{\pi/2} \int_0^{\pi/2} \int_0^{R_1} w_p r^2 \, dr \, d\varphi \, d\nu = \int_0^{\pi/2} \int_0^{\pi/2} \Omega_{1p} \, d\varphi \, d\nu, \tag{5.33}$$

where the coefficient Ω_{1p} given by Eq. (5.38) is a function of ω_{1p} given by Eq. (5.37).

With regard to the determination of the thermal-stress induced strengthening presented in Volume III, the stress σ_{ip} acting in a direction of the axis x_i and the elastic energy density w_{ip} induced by the stress σ_{ip} ($i = 1,2,3$) (see Eqs. (3.46), (3.47), (5.23)–(5.26)), both related to an arbitrary point of the spherical particle, are derived as

$$\sigma_{ip} = -\xi_{i1p} \varrho_{1p} \left(\frac{r}{R_1} \right)^{\lambda_{1p}-1}, \quad i = 1, 2, 3, \tag{5.34}$$

$$w_{ip} = \frac{s_{11p} \left(\xi_{i1p} \varrho_{1p} \right)^2}{2} \left(\frac{r}{R_1} \right)^{2(\lambda_{1p}-1)}, \quad i = 1, 2, 3, \tag{5.35}$$

where the elastic modulus s_{11p} and the coefficient ξ_{i1p} ($i = 1,2,3$) are given by Eqs. (3.23) and (5.39), respectively.

Appendix. Considering the elastic modulus s_{44p}, the coefficients c_{1p}, c_{2p}, a'_{1i}, a'_{2i}, a'_{3i} $(i=1,2,3)$, and the exponent $\lambda_{1p} > 1$ as a function of the coefficient c_{3p} (see Eqs. (3.25), (3.38)–(3.40), (12.26)–(12.34), (5.4)), the coefficients ϱ_{1p}, ω_{1+ip}, Ω_{1+ip} $(i=0,1)$, $\xi_{i\,1+jp}$ $(i=1,2,3;\ j=0,1)$ have the forms

$$\varrho_{1p} = \frac{1}{\lambda_{1p}\left(c_{1p}+c_{2p}\right) - 2c_{2p}}, \tag{5.36}$$

$$\omega_{1+ip} = p_{1+i}^2\left[\lambda_{1p}^2\left(c_{1p}+c_{2p}\right) - 4\lambda_{1p}c_{2p} + 2c_{1p}\right] + \frac{2}{s_{44p}}\left[\left(\frac{\partial p_{1+i}}{\partial\varphi}\right)^2 + \left(\frac{\partial p_{1+i}}{\partial\nu}\right)^2\right],$$
$$i = 0, 1, \tag{5.37}$$

$$\Omega_{1+ip} = \frac{4\,\omega_{1+ip}\,R_1^3\left(\varrho_{1p}^2\,\delta_{0i} + \delta_{1i}\right)}{2\lambda_{1p}+1}, \quad i = 0, 1, \tag{5.38}$$

$$\xi_{i\,1+jp} = p_{1+j}\left[\frac{a'_{1i}}{\varrho_{1p}} + \left(a'_{2i}+a'_{3i}\right)\left(c_{1p}-\lambda_{1p}c_{2p}\right)\right]$$
$$+ \frac{1}{s_{44p}}\left[\left(a'_{1i}+a'_{2i}\right)\frac{\partial p_{1+j}}{\partial\varphi} + \left(a'_{1i}+a'_{3i}\right)\frac{\partial p_{1+j}}{\partial\nu}\right],$$
$$i = 1, 2, 3; \quad j = 0, 1. \tag{5.39}$$

5.5.2 Thermal stresses in cell matrix

With regard to the boundary conditions (4.9), (4.11) and the analysis presented in Section 5.4.3, considering the terms u_{1rm} and u_{2rm} (see Eqs. (5.9), (5.10)) to represent increasing and decreasing functions of the variable $r \in \langle R_1, r_c\rangle$, respectively, the radial displacement u_{1m} of the infinitesimal spherical cap in the cell matrix, regarding Eqs. (3.31), (5.7), has the form

$$u_{1m} = -p_1 \sum_{i=0}^{1} \kappa_{1+2im}^{(1)}\,\Lambda_{1+im}\,r^{\lambda_{1+im}}, \tag{5.40}$$

where the exponents $\lambda_{1m} > 1$, $\lambda_{2m} < 0$ as functions of $c_{3m} < 0$ (see Eq. (3.40)), the coefficient $\Lambda_{1+im} = \Lambda_{1+im}^{(j)}$ $(i=0,1;\ j=1,2,3)$ and the radial stress p_1 acting on the particle-matrix boundary are given by Eqs. (5.4), (5.8), (5.16), (5.21) and (10.1), respectively. The φ,ν-dependent distance r_c, as a function of the experimentally determined parameters v, R_{1ex}, d_{1ex}, d_{2ex}, d_{3ex} (see Section 2.2.1), included in the coefficient $\kappa_{1+2im}^{(1)}$ $(i=0,1)$ is derived in Section 2.3.

With regard to Section 5.5.2, the coefficients $\psi_{11\eta m}^{(1)}$, $\psi_{12\eta m}^{(1)}$ $(\eta=\varphi,\nu)$, ϱ_{1m} and $\omega_{11+i\,1+jm}$ $(i,j=0,1)$, ξ_{i11+jm} $(i=1,2,3;\ j=0,1)$ mentioned below (see Eqs. (5.43),

(5.44), (5.47)–(5.50), (5.52), (5.53)) depend on $\kappa_{1m}^{(1)}$, $\kappa_{3m}^{(1)}$ and $\kappa_{1m}^{(1)}$, $\kappa_{3m}^{(1)}$, $\psi_{11\eta m}^{(1)}$, $\psi_{12\eta m}^{(1)}$, respectively, where the coefficients $\kappa_{1m}^{(1)}$, $\kappa_{3m}^{(1)}$ (see Eq. (12.35)) are functions of χ_{11+im}, χ_{21+im} $(i=0,1)$ given by Eqs. (5.54), (5.55), respectively.

Consequently, the radial stress σ_{11m}, the tangential and shear stresses σ_{22m}, σ_{33m} and σ_{12m}, σ_{13m} acting in the cell matrix, respectively, regarding Eqs. (3.31)–(3.34), (5.40), are derived as

$$\sigma_{11m} = -p_1 \sum_{i=0}^{1} \kappa_{1+2im}^{(1)} \Lambda_{1+im} \left[\lambda_{1+im} \left(c_{1m} + c_{2m}\right) - 2c_{2m}\right] r^{\lambda_{1+im}-1}, \tag{5.41}$$

$$\sigma_{22m} = \sigma_{33m} = -p_1 \sum_{i=0}^{1} \kappa_{1+2im}^{(1)} \Lambda_{1+im} \left(c_{1m} - \lambda_{1+im}\, c_{2m}\right) r^{\lambda_{1+im}-1}, \tag{5.42}$$

$$\sigma_{12m} = -\frac{1}{s_{44m}} \sum_{i=0}^{1} \psi_{11+i\varphi m}^{(1)} \Lambda_{1+im}\, r^{\lambda_{1+im}-1}, \tag{5.43}$$

$$\sigma_{13m} = -\frac{1}{s_{44m}} \sum_{i=0}^{1} \psi_{11+i\nu m}^{(1)} \Lambda_{1+im}\, r^{\lambda_{1+im}-1}, \tag{5.44}$$

where the elastic modulus s_{44m} and the coefficients c_{1m}, c_{2m}, $\psi_{11+i\eta m}^{(1)}$ $(i=0,1;$ $\eta=\varphi,\nu)$ are given by Eqs. (3.25) and (3.38), (3.39), (12.36), respectively.

Finally, the radial strain ε_{11m}, the tangential and shear strains ε_{22m}, ε_{33m} and ε_{12m}, ε_{13m} in the cell matrix, respectively, regarding Eqs. (3.1)–(3.5), (5.40), have the forms

$$\varepsilon_{11m} = -p_1 \sum_{i=0}^{1} \kappa_{1+2im}^{(1)} \lambda_{1+im} \Lambda_{1+im}\, r^{\lambda_{1+im}-1}, \tag{5.45}$$

$$\varepsilon_{22m} = \varepsilon_{33m} = -p_1 \sum_{i=0}^{1} \kappa_{1+2im}^{(1)} \Lambda_{1+im}\, r^{\lambda_{1+im}-1}, \tag{5.46}$$

$$\varepsilon_{12m} = -\sum_{i=0}^{1} \psi_{11+i\varphi m}^{(1)} \Lambda_{1+im}\, r^{\lambda_{1+im}-1}, \tag{5.47}$$

$$\varepsilon_{13m} = -\sum_{i=0}^{1} \psi_{11+i\nu m}^{(1)} \Lambda_{1+im}\, r^{\lambda_{1+im}-1}. \tag{5.48}$$

With regard to the determination of the compressive or tensile radial stress $p_1 > 0$ or $p_1 < 0$ acting on the particle-matrix boundary, thus for $r = R_1$, respectively, considering Eq. (3.58), the tangential strain on the particle-matrix boundary, $(\varepsilon_{22m})_{r=R_1}$, is required to be derived in the form

$$(\varepsilon_{22m})_{r=R_1} = -p_1\, \varrho_{1m}, \tag{5.49}$$

where the coefficient ϱ_{1m} is given by Eq. (5.56).

Considering Eqs. (3.41), (3.42), (5.41)–(5.48), the thermal-stress induced elastic energy density w_m and the thermal-stress induced elastic energy W_m, accumulated in an arbitrary point of the cell matrix and in the volume of the cell matrix, respectively, have the forms

$$w_m = \frac{1}{2} \sum_{i,j=0}^{1} \omega_{11+i\,1+jm}\, r^{\lambda_{1+im}+\lambda_{1+jm}-2}, \tag{5.50}$$

$$W_m = 8 \int_0^{\pi/2} \int_0^{\pi/2} \int_{R_1}^{r_c} w_m\, r^2\, dr\, d\varphi\, d\nu = \int_0^{\pi/2} \int_0^{\pi/2} \Omega_{11m}\, d\varphi\, d\nu, \tag{5.51}$$

where the coefficient Ω_{11m} given by Eq. (5.58) is a function of $\omega_{11+i\,1+jm}$ $(i,j=0,1)$ given by Eq. (5.57).

With regard to the determination of the thermal-stress induced strengthening presented in Volume III, the stress σ_{im} acting in a direction of the axis x_i and the elastic energy density w_{im} induced by the stress σ_{im} $(i=1,2,3)$ (see Eqs. (3.46), (3.47), (5.41)–(5.44)), both related to an arbitrary point of the cell matrix, are derived as

$$\sigma_{im} = -\left(\xi_{i11m}\, r^{\lambda_{1m}-1} + \xi_{i12m}\, r^{\lambda_{2m}-1}\right), \quad i = 1, 2, 3, \tag{5.52}$$

$$w_{im} = \frac{s_{11m}}{2}\left[\xi_{i11m}^2\, r^{2(\lambda_{1m}-1)} + \xi_{i12m}^2\, r^{2(\lambda_{2m}-1)} + 2\,\xi_{i11m}\,\xi_{i12m}\, r^{\lambda_{1m}+\lambda_{2m}-2}\right], \quad i = 1, 2, 3, \tag{5.53}$$

where the elastic modulus s_{11m} and the coefficients ξ_{i11m}, ξ_{i12m} $(i=1,2,3)$ are given by Eqs. (3.23) and (5.59), respectively.

Appendix. Considering the elastic modulus s_{44m}, the coefficients c_{1m}, c_{2m}, $\Lambda_{1+im} = \Lambda_{1+im}^{(j)}$ $(i=0,1;\ j=1,2,3)$, a'_{1i}, a'_{2i}, a'_{3i} $(i=1,2,3)$, and the exponents $\lambda_{1m} > 1$, $\lambda_{2m} < 0$ as functions of the coefficient c_{3m} (see Eqs. (3.25), (3.38)–(3.40), (5.8), (5.16), (5.21), (12.26)–(12.34), (5.4)), the coefficients χ_{11+im}, χ_{21+im}, ϱ_{1+im}, $\omega_{1+i\,1+j\,1+km}$, $\Omega_{1+i\,1+jm}$[2] $(i,j,k=0,1)$, $\xi_{i1+j\,1+km}$ $(i=1,2,3;\ j,k=0,1)$ have the forms

$$\chi_{11+im} = \Lambda_{1+im} R_1^{\lambda_{1+im}-1}\left[\lambda_{1+im}\left(c_{1m}+c_{2m}\right) - 2c_{2m}\right], \quad i = 0, 1, \tag{5.54}$$

$$\chi_{21+im} = \Lambda_{1+im} r_c^{\lambda_{1+im}}, \quad i = 0, 1, \tag{5.55}$$

[2]The coefficient Ω_{21m} is determined but not used in formulae for the elastic energy W_m of the cell matrix.

$$\varrho_{1+im} = \sum_{j=0}^{1} \kappa_{1+2jm}^{(1)} \Lambda_{1+jm} R_{1+i}^{\lambda_{1+jm}-1}, \quad i = 0, 1, \tag{5.56}$$

$$\begin{aligned}\omega_{1+i\ 1+j\ 1+km} &= \Lambda_{1+jm} \Lambda_{1+km} \\ &\times \Big\{ p_{1+i}^2 \kappa_{1+2jm}^{(1)} \kappa_{1+2km}^{(1)} \left[\lambda_{1+jm}\lambda_{1+km}\left(c_{1m}+c_{2m}\right)-4\lambda_{1+jm}c_{2m}+2c_{1m}\right] \\ &+ \frac{2}{s_{44m}} \left[\psi_{1+i\ 1+j\varphi m}^{(1)} \psi_{1+i\ 1+k\varphi m}^{(1)} + \psi_{1+i\ 1+j\nu m}^{(1)} \psi_{1+i\ 1+k\nu m}^{(1)}\right] \Big\}, \\ & i, j, k = 0, 1, \end{aligned} \tag{5.57}$$

$$\begin{aligned}\Omega_{1+i\ 1+jm} &= 4 \sum_{k,l=0}^{1} \frac{\omega_{1+i\ 1+k\ 1+lm}}{\lambda_{1+km} + \lambda_{1+lm} + 1} \left(r_c^{\lambda_{1+km}+\lambda_{1+lm}+1} - R_{1+j}^{\lambda_{1+km}+\lambda_{1+lm}+1} \right), \\ & i, j = 0, 1, \end{aligned} \tag{5.58}$$

$$\begin{aligned}\xi_{i\,1+j\ 1+km} &= \Lambda_{1+km} \\ \times \Big(p_{1+j} \kappa_{1+2km}^{(1)} & \left\{ a'_{1i} \left[\lambda_{1+km}\left(c_{1m}+c_{2m}\right) - 2c_{2m}\right] + \left(a'_{2i}+a'_{3i}\right)\left(c_{1m} - \lambda_{1+km}c_{2m}\right) \right\} \\ + \frac{1}{s_{44m}} & \left[\left(a'_{1i}+a'_{2i}\right) \psi_{1+j\ 1+k\varphi m}^{(1)} + \left(a'_{1i}+a'_{3i}\right) \psi_{1+j\ 1+k\nu m}^{(1)} \right] \Big), \\ i = 1, 2, 3; & \quad j, k = 0, 1. \end{aligned} \tag{5.59}$$

5.6 Multi-particle-envelope-matrix system

The variable r related to the thermal stresses and to the corresponding quantities in the spherical particle with the radius R_1, in the spherical envelope with the radii $R_1 < R_2$ and in the cell matrix is from the intervals $\langle 0, R_1 \rangle$, $\langle R_1, R_2 \rangle$ and $\langle R_2, r_c \rangle$, respectively, where the φ, ν-dependent distance r_c as a function of the experimentally determined parameters v, R_{1ex}, d_{1ex}, d_{2ex}, d_{3ex} (see Section 2.2.1) is derived in Section 2.3.

The compressive or tensile radial stresses $p_1 > 0$ or $p_1 < 0$ and $p_2 > 0$ or $p_2 < 0$, as consequences of the conditions $\beta_p \neq \beta_m$ and $\beta_m \neq \beta_e$, acting on the particle-envelope and matrix-envelope boundaries, thus for $r = R_1$ and $r = R_2$, respectively, are given by Eqs. (10.2)–(10.5) as dependent on the following conditions concerning the coefficients β_p, β_e, β_m (see Eqs. (3.48)–(3.50)).

With regard to the condition $\beta_p \neq \beta_e \neq \beta_m$, the radial stresses p_1, p_2 are determined by the tangential strains $(\varepsilon_{22p})_{r=R_1}$, $(\varepsilon_{22e})_{r=R_{1+i}}$ $(i = 0,1)$ and $(\varepsilon_{22m})_{r=R_2}$ in the spherical particle, in the spherical envelope and in the cell matrix, resulting in the coefficients ϱ_{1p}, ϱ_{1+ie} and ϱ_{2m}, respectively (see Eqs. (3.63), (3.64), (10.2), (10.3)).

Similarly, provided that $\beta_p \neq \beta_e = \beta_m$, the radial stress p_1 is determined by the tangential strains $(\varepsilon_{22p})_{r=R_1}$ and $(\varepsilon_{22e})_{r=R_1}$, resulting in the coefficients ϱ_{1p} and ϱ_{1e}^{me}, respectively (see Eqs. (3.63), (10.4)).

Finally, provided that $\beta_p = \beta_e \neq \beta_m$, the radial stress p_2 is determined by the tangential strains $(\varepsilon_{22m})_{r=R_2}$ and $(\varepsilon_{22e})_{r=R_2}$, resulting in the coefficients ϱ_{2m} and ϱ_{2e}^{pe}, respectively (see Eqs. (3.64), (10.5)).

5.6.1 Thermal stresses in spherical particle

5.6.1.1 Conditions $\beta_p \neq \beta_e = \beta_m$ and $\varepsilon_{11te} - \varepsilon_{11tp} = f(\varphi, \nu)$

Formulae for the thermal stresses and for the corresponding quantities related to the spherical particle in the multi-particle-envelope-matrix system provided that $\beta_p \neq \beta_e \neq \beta_m$ and $\beta_p \neq \beta_e = \beta_m$, the latter considering the φ, ν-dependent difference $\varepsilon_{11te} - \varepsilon_{11tp} = f(\varphi, \nu)$ (see Eq. (12.23)), are identical with those presented in Section 5.5.1 to result from the mathematical techniques in Sections 5.1–5.3, considering the radial stress p_1 (see Eq. (10.2)) acting on the particle-envelope boundary.

5.6.1.2 Condition $\beta_p = \beta_e \neq \beta_m$

With regard to the boundary conditions (4.3), (4.4), (4.6) to determine one integration constant only, and to the analysis presented in Section 5.4.1 considering the term u_{1rp} (see Eq. (5.9)) to represent an increasing function of the variable $r \in \langle 0, R_1 \rangle$, and thus the integration constant $C_{2p} = 0$, the radial displacement u_{1p} of the infinitesimal spherical cap in the spherical particle, regarding Eqs. (3.2), (5.7), has the form

$$u_{1p} = -p_2\, \varrho_{1e}^{pe}\, r \left(\frac{r}{R_1}\right)^{\lambda_{1p}-1}, \tag{5.60}$$

where the exponent $\lambda_{1p} > 1$ as a function of $c_{3p} < 0$ (see Eq. (3.40)), the radial stress p_2 acting on the matrix-envelope boundary and the coefficient ϱ_{1e}^{pe} are given by Eqs. (5.4), (10.5) and (5.124), (6.264), (6.279), (7.251), (7.266), respectively.

Consequently, the radial stress σ_{11p}, the tangential and shear stresses σ_{22p}, σ_{33p} and σ_{12p}, σ_{13p} acting in the spherical particle, respectively, regarding Eqs. (3.31)–(3.34), (5.60), are derived as

$$\sigma_{11p} = -\frac{p_2\, \varrho_{1e}^{pe}}{\varrho_{1p}} \left(\frac{r}{R_1}\right)^{\lambda_{1p}-1}, \tag{5.61}$$

$$\sigma_{22p} = \sigma_{33p} = -p_2\, \varrho_{1e}^{pe} \left(c_{1p} - \lambda_{1p}\, c_{2p}\right) \left(\frac{r}{R_1}\right)^{\lambda_{1p}-1}, \tag{5.62}$$

$$\sigma_{12p} = -\frac{\varrho_{1e}^{pe}}{s_{44p}} \frac{\partial p_2}{\partial \varphi} \left(\frac{r}{R_1}\right)^{\lambda_{1p}-1}, \tag{5.63}$$

$$\sigma_{13p} = -\frac{\varrho_{1e}^{pe}}{s_{44p}} \frac{\partial p_2}{\partial \nu} \left(\frac{r}{R_1} \right)^{\lambda_{1p}-1}, \tag{5.64}$$

where the elastic modulus s_{44p} and the coefficients c_{1p}, c_{2p}, ϱ_{1p} are given by Eqs. (3.25) and (3.38), (3.39), (5.36), respectively.

Finally, the radial strain ε_{11p}, the tangential and shear strains ε_{22p}, ε_{33p} and ε_{12p}, ε_{13p} in the spherical particle, respectively, regarding Eqs. (3.1)–(3.5), (5.60), have the forms

$$\varepsilon_{11p} = -p_2 \, \varrho_{1e}^{pe} \lambda_{1p} \left(\frac{r}{R_1} \right)^{\lambda_{1p}-1}, \tag{5.65}$$

$$\varepsilon_{22p} = \varepsilon_{33p} = -p_2 \, \varrho_{1e}^{pe} \left(\frac{r}{R_1} \right)^{\lambda_{1p}-1}, \tag{5.66}$$

$$\varepsilon_{12p} = -\varrho_{1e}^{pe} \frac{\partial p_2}{\partial \varphi} \left(\frac{r}{R_1} \right)^{\lambda_{1p}-1}, \tag{5.67}$$

$$\varepsilon_{13p} = -\varrho_{1e}^{pe} \frac{\partial p_2}{\partial \nu} \left(\frac{r}{R_1} \right)^{\lambda_{1p}-1}. \tag{5.68}$$

Considering Eqs. (3.41), (3.42), (5.61)–(5.68), the thermal-stress induced elastic energy density w_p and the thermal-stress induced elastic energy W_p, accumulated in an arbitrary point of the spherical particle and in the volume of the spherical particle, respectively, are derived as

$$w_p = \frac{\left(\varrho_{1e}^{pe}\right)^2 \omega_{2p}}{2} \left(\frac{r}{R_1} \right)^{2(\lambda_{1p}-1)}, \tag{5.69}$$

$$W_p = 8 \int_0^{\pi/2} \int_0^{\pi/2} \int_0^{R_1} w_p \, r^2 \, dr \, d\varphi \, d\nu = \left(\varrho_{1e}^{pe}\right)^2 \int_0^{\pi/2} \int_0^{\pi/2} \Omega_{2p} \, d\varphi \, d\nu, \tag{5.70}$$

where the coefficient Ω_{2p} given by Eq. (5.38) is a function of ω_{2p} given by Eq. (5.37).

With regard to the determination of the thermal-stress induced strengthening presented in Volume III, the stress σ_{ip} acting in a direction of the axis x_i and the elastic energy density w_{ip} induced by the stress σ_{ip} ($i = 1,2,3$) (see Eqs. (3.46), (3.47), (5.61)–(5.64)), both related to an arbitrary point of the spherical particle, have the forms

$$\sigma_{ip} = -\xi_{i2p} \, \varrho_{1p} \left(\frac{r}{R_1} \right)^{\lambda_{1p}-1}, \quad i = 1, 2, 3, \tag{5.71}$$

$$w_{ip} = \frac{s_{11p} \left(\xi_{i2p} \, \varrho_{1p} \right)^2}{2} \left(\frac{r}{R_1} \right)^{2(\lambda_{1p}-1)}, \quad i = 1, 2, 3, \tag{5.72}$$

where the elastic modulus s_{11p} and the coefficient ξ_{i2p} $(i=1,2,3)$ are given by Eqs. (3.23) and (5.39), respectively.

5.6.2 Thermal stresses in spherical envelope

5.6.2.1 Condition $\beta_p \neq \beta_e \neq \beta_m$

With regard to the boundary conditions (4.7), (4.8) and the analysis presented in Section 5.4.2, considering the terms u_{1re} and u_{2re} (see Eqs. (5.9), (5.10)) to represent increasing and decreasing functions of the variable $r \in \langle R_1, R_2 \rangle$, respectively, the radial displacement u_{1e} of the infinitesimal spherical cap in the spherical envelope, regarding Eqs. (3.31), (5.7), has the form

$$u_{1e} = -\sum_{i,j=0}^{1} p_{1+i}\, \kappa^{(1)}_{1+i+2je}\, \Lambda_{1+je}\, (\delta_{0i} - \delta_{1i})\, r^{\lambda_{1+je}}, \tag{5.73}$$

where the exponents $\lambda_{1e} > 1$, $\lambda_{2e} < 0$ as functions of $c_{3e} < 0$ (see Eq. (3.40)), the coefficient $\Lambda_{1+ie} = \Lambda^{(j)}_{1+ie}$ $(i=0,1;\ j=1,2,3)$ and the radial stresses p_1, p_2 acting on the particle-envelope, matrix-envelope boundaries are given by Eqs. (5.4), (5.8), (5.16), (5.21) and (10.2), (10.3), respectively.

With regard to Section 5.6.2.1, the coefficients $\varrho_{1+i\,1+je}$, $\omega_{1+i\,1+je}$ $(i,j=0,1)$, $\xi_{i\,1+je}$ $(i=1,2,3;\ j=0,1)$ mentioned below (see Eqs. (5.82), (5.83), (5.85), (5.86)) depend on $\kappa^{(1)}_{1e}, \ldots, \kappa^{(1)}_{4e}$ (see Eq. (12.35)) representing functions of $\chi_{1+j\,1+ie}$ $(i,j=0,1)$ given by Eq. (5.87).

Consequently, the radial stress σ_{11e}, the tangential and shear stresses σ_{22e}, σ_{33e} and σ_{12e}, σ_{13e} acting in the spherical envelope, respectively, regarding Eqs. (3.31)–(3.34), (5.73), are derived as

$$\sigma_{11e} = -\sum_{i,j=0}^{1} p_{1+i}\, \kappa^{(1)}_{1+i+2je}\, \Lambda_{1+je}\, (\delta_{0i} - \delta_{1i})\, [\lambda_{1+je}\, (c_{1e} + c_{2e}) - 2c_{2e}]\, r^{\lambda_{1+je}-1}, \tag{5.74}$$

$$\sigma_{22e} = \sigma_{33e} = -\sum_{i,j=0}^{1} p_{1+i}\, \kappa^{(1)}_{1+i+2je}\, \Lambda_{1+je}\, (\delta_{0i} - \delta_{1i})\, (c_{1e} - \lambda_{1+je} c_{2e})\, r^{\lambda_{1+je}-1}, \tag{5.75}$$

$$\sigma_{12e} = -\frac{1}{s_{44e}} \sum_{i,j=0}^{1} \frac{\partial p_{1+i}}{\partial \varphi}\, \kappa^{(1)}_{1+i+2je}\, \Lambda_{1+je}\, (\delta_{0i} - \delta_{1i})\, r^{\lambda_{1+je}-1}, \tag{5.76}$$

$$\sigma_{13e} = -\frac{1}{s_{44e}} \sum_{i,j=0}^{1} \frac{\partial p_{1+i}}{\partial \nu}\, \kappa^{(1)}_{1+i+2je}\, \Lambda_{1+je}\, (\delta_{0i} - \delta_{1i})\, r^{\lambda_{1+je}-1}, \tag{5.77}$$

where the elastic modulus s_{44e} and the coefficients c_{1e}, c_{2e} are given by Eqs. (3.25) and (3.38), (3.39), respectively.

Finally, the radial strain ε_{11e}, the tangential and shear strains ε_{22e}, ε_{33e} and ε_{12e}, ε_{13e} in the spherical envelope, respectively, regarding Eqs. (3.1)–(3.5), (5.73), have the forms

$$\varepsilon_{11e} = -\sum_{i,j=0}^{1} p_{1+i}\,\kappa_{1+i+2je}^{(1)}\lambda_{1+je}\,\Lambda_{1+je}\left(\delta_{0i}-\delta_{1i}\right) r^{\lambda_{1+je}-1}, \tag{5.78}$$

$$\varepsilon_{22e} = \varepsilon_{33e} = -\sum_{i,j=0}^{1} p_{1+i}\,\kappa_{1+i+2je}^{(1)}\,\Lambda_{1+je}\left(\delta_{0i}-\delta_{1i}\right) r^{\lambda_{1+je}-1}, \tag{5.79}$$

$$\varepsilon_{12e} = -\sum_{i,j=0}^{1} \frac{\partial p_{1+i}}{\partial\varphi}\,\kappa_{1+i+2je}^{(1)}\,\Lambda_{1+je}\left(\delta_{0i}-\delta_{1i}\right) r^{\lambda_{1+je}-1}, \tag{5.80}$$

$$\varepsilon_{13e} = -\sum_{i,j=0}^{1} \frac{\partial p_{1+i}}{\partial\nu}\,\kappa_{1+i+2je}^{(1)}\,\Lambda_{1+je}\left(\delta_{0i}-\delta_{1i}\right) r^{\lambda_{1+je}-1}. \tag{5.81}$$

With regard to the determination of the compressive or tensile radial stresses $p_1 > 0$ or $p_1 < 0$ and $p_2 > 0$ or $p_2 < 0$ acting on the particle-envelope and matrix-envelope boundaries, thus for $r = R_1$ and $r = R_2$, respectively, considering Eqs. (3.63), (3.64), the tangential strains on the particle-envelope and matrix-envelope boundaries, $(\varepsilon_{22e})_{r=R_1}$ and $(\varepsilon_{22e})_{r=R_2}$, respectively, are required to be derived in the form

$$(\varepsilon_{22e})_{r=R_{1+i}} = -\sum_{j=0}^{1} p_{1+j}\,\varrho_{1+i\;1+je}, \quad i=0,1, \tag{5.82}$$

where the coefficient $\varrho_{1+i\;1+je}$ $(i,j=0,1)$ is given by Eq. (5.88).

Considering Eqs. (3.41), (3.42), (5.74)–(5.81), the thermal-stress induced elastic energy density w_e and the thermal-stress induced elastic energy W_e, accumulated in an arbitrary point of the spherical envelope and in the volume of the spherical envelope, respectively, have the forms

$$w_e = \frac{1}{2}\sum_{i,j=0}^{1} \omega_{1+i\;1+je}\, r^{\lambda_{1+ie}+\lambda_{1+je}-2}, \tag{5.83}$$

$$W_e = 8\int_0^{\pi/2}\int_0^{\pi/2}\int_{R_1}^{R_2} w_e\, r^2\, dr\, d\varphi\, d\nu = \int_0^{\pi/2}\int_0^{\pi/2} \Omega_e\, d\varphi\, d\nu, \tag{5.84}$$

where the coefficient Ω_e given by Eq. (5.90) is a function of $\omega_{1+i\;1+je}$ $(i,j=0,1)$ given by Eqs. (5.89).

With regard to the determination of the thermal-stress induced strengthening presented in Volume III, the stress σ_{ie} acting in a direction of the axis x_i and the elastic energy density w_{ie} induced by the stress σ_{ie} ($i=1,2,3$) (see Eqs. (3.46), (3.47), (5.74)–(5.77)), both related to an arbitrary point of the spherical envelope, are derived as

$$\sigma_{ie} = -\left(\xi_{i\,1e}\, r^{\lambda_{1e}-1} + \xi_{i\,2e}\, r^{\lambda_{2e}-1}\right), \quad i = 1, 2, 3, \tag{5.85}$$

$$w_{ie} = \frac{s_{11e}}{2}\left[\xi_{i\,1e}^2\, r^{2(\lambda_{1e}-1)} + \xi_{i\,2e}^2\, r^{2(\lambda_{2e}-1)} + 2\,\xi_{i\,1e}\,\xi_{i\,2e}\, r^{\lambda_{1e}+\lambda_{2e}-2}\right], \quad i = 1, 2, 3, \tag{5.86}$$

where the elastic modulus s_{11e} and the coefficient $\xi_{i\,1+je}$ ($i=1,2,3$; $j=0,1$) are given by Eqs. (3.23) and (5.91), respectively.

Appendix. Considering the elastic modulus s_{44e}, the coefficients c_{1e}, c_{2e}, $\Lambda_{1+ie} = \Lambda_{1+ie}^{(j)}$ ($i=0,1$; $j=1,2,3$), a'_{1i}, a'_{2i}, a'_{3i} ($i=1,2,3$), and the exponents $\lambda_{1e} > 1$, $\lambda_{2e} < 0$ as functions of the coefficient c_{3e} (see Eqs. (3.25), (3.38)–(3.40), (5.8), (5.16), (5.21), (12.26)–(12.34), (5.4)), the coefficients $\chi_{1+j\;1+ie}$, $\varrho_{1+i\;1+je}$, $\omega_{1+i\;1+je}$ ($i,j=0,1$), Ω_e, $\xi_{i\,1+je}$ ($i=1,2,3$; $j=0,1$) have the forms

$$\chi_{1+j\;1+ie} = \Lambda_{1+ie} R_{1+j}^{\lambda_{1+ie}-1}\left[\lambda_{1+ie}\left(c_{1e}+c_{2e}\right) - 2c_{2e}\right], \quad i,j = 0, 1, \tag{5.87}$$

$$\varrho_{1+i\;1+je} = \sum_{k=0}^{1} \kappa_{1+j+2ke}^{(1)}\, \Lambda_{1+ke}\left(\delta_{0j}-\delta_{1j}\right) R_{1+i}^{\lambda_{1+ke}-1}, \quad i,j = 0, 1, \tag{5.88}$$

$$\begin{aligned}\omega_{1+i\;1+je} = {} & \sum_{k,\,l=0}^{1} \kappa_{1+2i+ke}^{(1)}\, \kappa_{1+2j+le}^{(1)}\, \Lambda_{1+ie}\, \Lambda_{1+je}\left(\delta_{0k}-\delta_{1k}\right)\left(\delta_{0l}-\delta_{1l}\right) \\ & \times \left\{p_{1+k}\, p_{1+l}\left[\lambda_{1+ie}\lambda_{1+je}\left(c_{1e}+c_{2e}\right) - 4\lambda_{1+ie}c_{2e} + 2c_{1e}\right]\right. \\ & \left. + \frac{2}{s_{44e}}\left(\frac{\partial p_{1+k}}{\partial \varphi}\frac{\partial p_{1+l}}{\partial \varphi} + \frac{\partial p_{1+k}}{\partial \nu}\frac{\partial p_{1+l}}{\partial \nu}\right)\right\}, \quad i,j = 0, 1,\end{aligned} \tag{5.89}$$

$$\Omega_e = 4\sum_{i,j=0}^{1} \frac{\omega_{1+i\;1+je}}{\lambda_{1+ie}+\lambda_{1+je}+1}\left(R_2^{\lambda_{1+ie}+\lambda_{1+je}+1} - R_1^{\lambda_{1+ie}+\lambda_{1+je}+1}\right), \tag{5.90}$$

$$\begin{aligned}\xi_{i\,1+je} = {} & \Lambda_{1+je}\sum_{k=0}^{1} \kappa_{1+2j+ke}^{(1)}\left(\delta_{0k}-\delta_{1k}\right) \\ & \times \left(p_{1+k}\left\{a'_{1i}\left[\lambda_{1+je}\left(c_{1e}+c_{2e}\right) - 2c_{2e}\right] + \left(a'_{2i}+a'_{3i}\right)\left(c_{1e}-\lambda_{1+je}c_{2e}\right)\right\}\right. \\ & \left. + \frac{1}{s_{44e}}\left[\left(a'_{1i}+a'_{2i}\right)\frac{\partial p_{1+k}}{\partial \varphi} + \left(a'_{1i}+a'_{3i}\right)\frac{\partial p_{1+k}}{\partial \nu}\right]\right), \\ & i = 1, 2, 3; \quad j = 0, 1.\end{aligned} \tag{5.91}$$

5.6.2.2 Conditions $\beta_p \neq \beta_e = \beta_m$ and $\varepsilon_{11te} - \varepsilon_{11tp} = f(\varphi, \nu)$

The thermal stresses and the corresponding quantities presented in Section 5.6.2.2 and related to Eq. (5.4) are derived on the condition $\beta_p \neq \beta_e = \beta_m$, considering the φ, ν-dependent difference $\varepsilon_{11te} - \varepsilon_{11tp} = f(\varphi, \nu)$ (see Eq. (12.23)), resulting in the φ, ν-dependent difference $\beta_e - \beta_p$ (see Eqs. (3.48)–(3.50)) and consequently in the φ, ν-dependent radial stress $p_1 = (\beta_e - \beta_p) / (\varrho_{1p} - \varrho_{1e}^{me})$ (see Eq. (10.4)) acting on the particle-envelope boundary, derived from Eq. (3.63) and included in the formulae in Section 5.6.2.2, where the φ, ν-independent coefficients ϱ_{1p} and ϱ_{1e}^{me} are given by Eqs. (5.36) and (5.106), (6.207), (6.228), (6.246), (7.230), respectively, considering the Castigliano's theorem (see Section 3.1.9).

With regard to the boundary condition (4.7) and the analysis presented in Section 5.4.2, considering the term u_{2re} (see Eq. (5.10)) to represent a decreasing function of the variable $r \in \langle R_1, R_2 \rangle$, and thus the integration constant $C_{1e} = 0$, the radial displacement u_{1e} of the infinitesimal spherical cap in the spherical envelope, regarding Eqs. (3.31), (5.7), has the form

$$u_{1e} = -p_1 \, \varrho_{1e}^{me} \, r \left(\frac{r}{R_1} \right)^{\lambda_{2e}-1}, \tag{5.92}$$

where the coefficient ϱ_{1e}^{me} and the exponent $\lambda_{2e} < 0$ as a function of $c_{3e} < 0$ (see Eq. (3.40)) are given by Eqs. (5.106) and (5.4), respectively.

Consequently, the radial stress σ_{11e}, the tangential and shear stresses σ_{22e}, σ_{33e} and σ_{12e}, σ_{13e} acting in the spherical envelope, respectively, regarding Eqs. (3.31)–(3.34), (5.92), are derived as

$$\sigma_{11e} = -p_1 \left(\frac{r}{R_1} \right)^{\lambda_{2e}-1}, \tag{5.93}$$

$$\sigma_{22e} = \sigma_{33e} = -p_1 \, \varrho_{1e}^{me} \left(c_{1e} - \lambda_{2e} c_{2e} \right) \left(\frac{r}{R_1} \right)^{\lambda_{2e}-1}, \tag{5.94}$$

$$\sigma_{12e} = -\frac{\varrho_{1e}^{me}}{s_{44e}} \frac{\partial p_1}{\partial \varphi} \left(\frac{r}{R_1} \right)^{\lambda_{2e}-1}, \tag{5.95}$$

$$\sigma_{13e} = -\frac{\varrho_{1e}^{me}}{s_{44e}} \frac{\partial p_1}{\partial \nu} \left(\frac{r}{R_1} \right)^{\lambda_{2e}-1}, \tag{5.96}$$

where the elastic modulus s_{44e} is given by Eq. (3.25).

Finally, the radial strain ε_{11e}, the tangential and shear strains ε_{22e}, ε_{33e} and ε_{12e}, ε_{13e} in the spherical envelope, respectively, regarding Eqs. (3.1)–(3.5), (5.92), have the forms

$$\varepsilon_{11e} = -p_1 \, \varrho_{1e}^{me} \, \lambda_{2e} \left(\frac{r}{R_1} \right)^{\lambda_{2e}-1}, \tag{5.97}$$

$$\varepsilon_{22e} = \varepsilon_{33e} = -p_1\,\varrho_{1e}^{me}\left(\frac{r}{R_1}\right)^{\lambda_{2e}-1}, \tag{5.98}$$

$$\varepsilon_{12e} = -\varrho_{1e}^{me}\,\frac{\partial p_1}{\partial \varphi}\left(\frac{r}{R_1}\right)^{\lambda_{2e}-1}, \tag{5.99}$$

$$\varepsilon_{13e} = -\varrho_{1e}^{me}\,\frac{\partial p_1}{\partial \nu}\left(\frac{r}{R_1}\right)^{\lambda_{2e}-1}. \tag{5.100}$$

With regard to the determination of the compressive or tensile radial stress $p_1 > 0$ or $p_1 < 0$ acting on the particle-matrix boundary, thus for $r = R_1$, respectively, considering Eq. (3.59), and to the boundary condition for the cell matrix provided that $\beta_p \neq \beta_e = \beta_m$ (see Eq. (4.12)), considering the condition $\varepsilon_{11te} - \varepsilon_{11tp} = f(\varphi, \nu)$ (see Eq. (12.23)), the tangential strains on the particle-envelope and matrix-envelope boundaries, $(\varepsilon_{22e})_{r=R_1}$ and $(\varepsilon_{22e})_{r=R_2}$, respectively, are required to be derived in the form

$$(\varepsilon_{22e})_{r=R_{1+i}} = -p_1\,\varrho_{1+ie}^{me}, \quad i = 0, 1, \tag{5.101}$$

where the coefficient ϱ_{1+ie}^{me} $(i=0,1)$ is given by Eq. (5.106).

Considering Eqs. (3.41), (3.42), (5.93)–(5.100), the thermal-stress induced elastic energy density w_e and the thermal-stress induced elastic energy W_e, accumulated in an arbitrary point of the spherical envelope and in the volume of the spherical envelope, respectively, have the forms

$$w_e = \frac{\omega_{1e}\,(\varrho_{1e}^{me})^2}{2}\left(\frac{r}{R_1}\right)^{2(\lambda_{2e}-1)}, \tag{5.102}$$

$$W_e = 8\int_0^{\pi/2}\int_0^{\pi/2}\int_{R_1}^{R_2} w_e\, r^2 dr d\varphi d\nu = (\varrho_{1e}^{me})^2\int_0^{\pi/2}\int_0^{\pi/2}\Omega_{1e} d\varphi d\nu, \tag{5.103}$$

where the coefficient Ω_{1e} given by Eq. (5.108) is a function of ω_{1e} given by Eq. (5.107).

With regard to the determination of the thermal-stress induced strengthening presented in Volume III, the stress σ_{ie} acting in a direction of the axis x_i and the elastic energy density w_{ie} induced by the stress σ_{ie} $(i=1,2,3)$ (see Eqs. (3.46), (3.47), (5.93)–(5.96)), both related to an arbitrary point of the spherical envelope, are derived as

$$\sigma_{ie} = -\xi_{i1e}\,\varrho_{1e}^{me}\left(\frac{r}{R_1}\right)^{\lambda_{2e}-1}, \quad i = 1, 2, 3, \tag{5.104}$$

$$w_{ie} = \frac{s_{11e}\,(\xi_{i1e}\,\varrho_{1e}^{me})^2}{2}\left(\frac{r}{R_1}\right)^{2(\lambda_{2e}-1)}, \quad i = 1, 2, 3, \tag{5.105}$$

where the elastic modulus s_{11e} and the coefficient ξ_{i1e} ($i=1,2,3$) are given by Eqs. (3.23) and (5.109), respectively.

Appendix. Considering the elastic modulus s_{44e}, the coefficients c_{1e}, c_{2e}, a'_{1i}, a'_{2i}, a'_{3i} ($i=1,2,3$), and the exponent $\lambda_{2e} < 0$ as a function of the coefficient c_{3e} (see Eqs. (3.25), (3.38)–(3.40), (12.26)–(12.34), (5.4)), the coefficients ϱ^{me}_{1+ie}, ω_{1+ie}, Ω_{1+ie} ($i=0,1$), ξ_{i1+je} ($i=1,2,3$; $j=0,1$) have the forms

$$\varrho^{me}_{1+ie} = \frac{1}{\lambda_{2e}\left(c_{1e}+c_{2e}\right)-2c_{2e}}\left(\frac{R_{1+i}}{R_1}\right)^{\lambda_{2e}-1}, \quad i=0,1, \tag{5.106}$$

$$\omega_{1+ie} = p^2_{1+i}\left[\lambda^2_{2-ie}\left(c_{1e}+c_{2e}\right)-4\lambda_{2-ie}c_{2e}+2c_{1e}\right]+\frac{2}{s_{44e}}\left[\left(\frac{\partial p_{1+i}}{\partial\varphi}\right)^2+\left(\frac{\partial p_{1+i}}{\partial\nu}\right)^2\right],$$
$$i=0,1, \tag{5.107}$$

$$\Omega_{1+ie} = \frac{4\,\omega_{1+ie}R^3_{1+i}\left(\delta_{0i}-\delta_{1i}\right)}{2\lambda_{2-ie}+1}\left[\left(\frac{R_{2-i}}{R_{1+i}}\right)^{2\lambda_{2-ie}+1}-1\right], \quad i=0,1, \tag{5.108}$$

$$\begin{aligned}\xi_{i1+je} &= p_{1+j}\left\{a'_{1i}\left[\lambda_{2-je}\left(c_{1e}+c_{2e}\right)-2c_{2e}\right]+\left(a'_{2i}+a'_{3i}\right)\left(c_{1e}-\lambda_{2-je}c_{2e}\right)\right\}\\ &+\frac{1}{s_{44e}}\left[\left(a'_{1i}+a'_{2i}\right)\frac{\partial p_{1+j}}{\partial\varphi}+\left(a'_{1i}+a'_{3i}\right)\frac{\partial p_{1+j}}{\partial\nu}\right],\\ & i=1,2,3; \quad j=0,1.\end{aligned} \tag{5.109}$$

5.6.2.3 Condition $\beta_p = \beta_e \neq \beta_m$

With regard to the boundary condition (4.8) and the analysis presented in Section 5.4.2, considering the term u_{1re} (see Eqs. (5.9)) to represent an increasing function of the variable $r \in \langle R_1, R_2\rangle$, and thus the integration constant $C_{2e} = 0$, the radial displacement u_{1e} of the infinitesimal spherical cap in the spherical envelope, regarding Eqs. (3.31), (5.7), has the form

$$u_{1e} = -p_2\,\varrho^{pe}_{2e}\,r\left(\frac{r}{R_2}\right)^{\lambda_{1e}-1}, \tag{5.110}$$

where the coefficient ϱ^{pe}_{2e}, the exponent $\lambda_{1e} > 1$ as a function of $c_{3e} < 0$ (see Eq. (3.40)), and the radial stress p_2 acting on the matrix-envelope boundary are given by Eqs. (5.124), (5.4) and (10.5), respectively.

Consequently, the radial stress σ_{11e}, the tangential and shear stresses σ_{22e}, σ_{33e} and σ_{12e}, σ_{13e} acting in the spherical envelope, respectively, regarding Eqs. (3.31)–(3.34), (5.110), are derived as

$$\sigma_{11e} = -p_2 \left(\frac{r}{R_2} \right)^{\lambda_{1e}-1}, \tag{5.111}$$

$$\sigma_{22e} = \sigma_{33e} = -p_2 \, \varrho_{2e}^{pe} \left(c_{1e} - \lambda_{1e} c_{2e} \right) \left(\frac{r}{R_2} \right)^{\lambda_{1e}-1}, \tag{5.112}$$

$$\sigma_{12e} = -\frac{\varrho_{2e}^{pe}}{s_{44e}} \frac{\partial p_2}{\partial \varphi} \left(\frac{r}{R_2} \right)^{\lambda_{1e}-1}, \tag{5.113}$$

$$\sigma_{13e} = -\frac{\varrho_{2e}^{pe}}{s_{44e}} \frac{\partial p_2}{\partial \nu} \left(\frac{r}{R_2} \right)^{\lambda_{1e}-1}, \tag{5.114}$$

where the elastic modulus s_{44e} is given by Eq. (3.25).

Finally, the radial strain ε_{11e}, the tangential and shear strains ε_{22e}, ε_{33e} and ε_{12e}, ε_{13e} in the spherical envelope, respectively, regarding Eqs. (3.1)–(3.5), (5.110), have the forms

$$\varepsilon_{11e} = -p_2 \, \varrho_{2e}^{pe} \, \lambda_{1e} \left(\frac{r}{R_2} \right)^{\lambda_{1e}-1}, \tag{5.115}$$

$$\varepsilon_{22e} = \varepsilon_{33e} = -p_2 \, \varrho_{2e}^{pe} \left(\frac{r}{R_2} \right)^{\lambda_{1e}-1}, \tag{5.116}$$

$$\varepsilon_{12e} = -\varrho_{2e}^{pe} \frac{\partial p_2}{\partial \varphi} \left(\frac{r}{R_2} \right)^{\lambda_{1e}-1}, \tag{5.117}$$

$$\varepsilon_{13e} = -\varrho_{2e}^{pe} \frac{\partial p_2}{\partial \nu} \left(\frac{r}{R_2} \right)^{\lambda_{1e}-1}. \tag{5.118}$$

With regard to the determination of the compressive or tensile radial stress $p_2 > 0$ or $p_2 < 0$ acting on the matrix-envelope boundary, thus for $r = R_2$, respectively, considering Eq. (3.64), and to the boundary condition for the spherical particle provided that $\beta_p = \beta_e \neq \beta_m$ (see Eq. (4.6)), the tangential strains on the particle-envelope and matrix-envelope boundaries, $(\varepsilon_{22e})_{r=R_1}$ and $(\varepsilon_{22e})_{r=R_2}$, respectively, are required to be derived in the form

$$(\varepsilon_{22e})_{r=R_{1+i}} = -p_2 \, \varrho_{1+ie}^{pe}, \quad i = 0, 1, \tag{5.119}$$

where the coefficient ϱ_{1+ie}^{pe} ($i = 0{,}1$) is given by Eq. (5.124).

Considering Eqs. (3.41), (3.42), (5.111)–(5.118), the thermal-stress induced elastic energy density w_e and the thermal-stress induced elastic energy W_e, accumulated in an arbitrary point of the spherical envelope and in the volume of the spherical envelope, respectively, have the forms

$$w_e = \frac{\omega_{2e}\left(\varrho_{2e}^{pe}\right)^2}{2}\left(\frac{r}{R_2}\right)^{2(\lambda_{1e}-1)}, \tag{5.120}$$

$$W_e = 8\int_0^{\pi/2}\int_0^{\pi/2}\int_{R_1}^{R_2} w_e\, r^2\, dr\, d\varphi\, d\nu = \left(\varrho_{2e}^{pe}\right)^2\int_0^{\pi/2}\int_0^{\pi/2}\Omega_{2e}\, d\varphi\, d\nu, \tag{5.121}$$

where the coefficient Ω_{2e} given by Eq. (5.108) is a function of ω_{2e} given by Eq. (5.107).

With regard to the determination of the thermal-stress induced strengthening presented in Volume III, the stress σ_{ie} acting in a direction of the axis x_i and the elastic energy density w_{ie} induced by the stress σ_{ie} $(i = 1,2,3)$ (see Eqs. (3.46), (3.47), (5.111)–(5.114)), both related to an arbitrary point of the spherical envelope, are derived as

$$\sigma_{ie} = -\xi_{i2e}\,\varrho_{2e}^{pe}\left(\frac{r}{R_2}\right)^{\lambda_{1e}-1}, \quad i = 1, 2, 3, \tag{5.122}$$

$$w_{ie} = \frac{s_{11e}\left(\xi_{i2e}\,\varrho_{2e}^{pe}\right)^2}{2}\left(\frac{r}{R_2}\right)^{2(\lambda_{1e}-1)}, \quad i = 1, 2, 3, \tag{5.123}$$

where the elastic modulus s_{11e} and the coefficient ξ_{i2e} $(i = 1,2,3)$ are given by Eqs. (3.23) and (5.109), respectively.

Appendix. Considering the coefficients c_{1e}, c_{2e}, and the exponent $\lambda_{1e} > 1$ as a function of the coefficient c_{3e} (see Eqs. (3.38)–(3.40), (5.4)), the coefficient ϱ_{1+ie}^{pe} $(i = 0,1)$ has the form

$$\varrho_{1+ie}^{pe} = \frac{1}{\lambda_{1e}\left(c_{1e} + c_{2e}\right) - 2c_{2e}}\left(\frac{R_{1+i}}{R_2}\right)^{\lambda_{1e}-1}, \quad i = 0, 1. \tag{5.124}$$

5.6.3 Thermal stresses in cell matrix

5.6.3.1 Conditions $\beta_p \neq \beta_e \neq \beta_m$ or $\beta_p = \beta_e \neq \beta_m$

The thermal stresses and the corresponding quantities for the cell matrix presented in Section 5.6.3.1 are derived with regard to the conditions $\beta_p \neq \beta_e \neq \beta_m$ or $\beta_p = \beta_e \neq \beta_m$ (see Eqs. (3.48)–(3.50)).

With regard to the boundary conditions (4.10), (4.11) and the analysis presented in Section 5.4.3, considering the terms u_{1rm} and u_{2rm} (see Eqs. (5.9), (5.10)) to represent increasing and decreasing functions of the variable $r \in \langle R_2, r_c\rangle$, respectively, the radial displacement u_{1m} of the infinitesimal spherical cap in the cell matrix, regarding Eqs. (3.31), (5.7), has the form

$$u_{1m} = -p_2 \sum_{i=0}^{1} \kappa_{1+2im}^{(1)} \Lambda_{1+im} r^{\lambda_{1+im}}, \tag{5.125}$$

where the exponents $\lambda_{1m} > 1$, $\lambda_{2m} < 0$ as functions of $c_{3m} < 0$ (see Eq. (3.40)) and the coefficient $\Lambda_{1+im} = \Lambda_{1+im}^{(j)}$ $(i=0,1;\ j=1,2,3)$ are given by Eqs. (5.4) and (5.8), (5.16), (5.21), respectively. Provided that $\beta_p \neq \beta_e \neq \beta_m$ or $\beta_p = \beta_e \neq \beta_m$, the radial stress p_2 acting on the matrix-envelope boundary is given by Eqs. (10.3) or (10.5), respectively. The φ, ν-dependent distance r_c, as a function of the experimentally determined parameters v, R_{1ex}, d_{1ex}, d_{2ex}, d_{3ex} (see Section 2.2.1), included in the coefficient $\kappa_{1+2im}^{(1)}$ $(i=0,1)$ is derived in Section 2.3.

With regard to Section 5.6.3.1, the coefficients $\psi_{21\eta m}^{(1)}$, $\psi_{22\eta m}^{(1)}$ $(\eta=\varphi,\nu)$, ϱ_{2m} and $\omega_{21+i\ 1+jm}$ $(i,j=0,1)$, $\xi_{i\,21+jm}$ $(i=1,2,3;\ j=0,1)$ mentioned below (see Eqs. (5.128), (5.129), (5.132)–(5.135), (5.137), (5.138)) depend on $\kappa_{1m}^{(1)}$, $\kappa_{3m}^{(1)}$ and $\kappa_{1m}^{(1)}$, $\kappa_{3m}^{(1)}$, $\psi_{21\eta m}^{(1)}$, $\psi_{22\eta m}^{(1)}$, respectively, where the coefficients $\kappa_{1m}^{(1)}$, $\kappa_{3m}^{(1)}$ (see Eq. (12.35)) are functions of χ_{11+im}, χ_{21+im} $(i=0,1)$ given by Eqs. (5.139), (5.55), respectively.

Consequently, the radial stress σ_{11m}, the tangential and shear stresses σ_{22m}, σ_{33m} and σ_{12m}, σ_{13m} acting in the cell matrix, respectively, regarding Eqs. (3.31)–(3.34), (5.125), are derived as

$$\sigma_{11m} = -p_2 \sum_{i=0}^{1} \kappa_{1+2im}^{(1)} \Lambda_{1+im} \left[\lambda_{1+im} \left(c_{1m} + c_{2m}\right) - 2c_{2m}\right] r^{\lambda_{1+im}-1}, \tag{5.126}$$

$$\sigma_{22m} = \sigma_{33m} = -p_2 \sum_{i=0}^{1} \kappa_{1+2im}^{(1)} \Lambda_{1+im} \left(c_{1m} - \lambda_{1+im} c_{2m}\right) r^{\lambda_{1+im}-1}, \tag{5.127}$$

$$\sigma_{12m} = -\frac{1}{s_{44m}} \sum_{i=0}^{1} \psi_{21+i\varphi m}^{(1)} \Lambda_{1+im} r^{\lambda_{1+im}-1}, \tag{5.128}$$

$$\sigma_{13m} = -\frac{1}{s_{44m}} \sum_{i=0}^{1} \psi_{21+i\nu m}^{(1)} \Lambda_{1+im} r^{\lambda_{1+im}-1}, \tag{5.129}$$

where the elastic modulus s_{44m} and the coefficients c_{1m}, c_{2m}, $\psi_{21+i\eta m}^{(1)}$ $(i=0,1;\ \eta=\varphi,\nu)$ are given by Eqs. (3.25) and (3.38), (3.39), (12.36), respectively.

Finally, the radial strain ε_{11m}, the tangential and shear strains ε_{22m}, ε_{33m} and ε_{12m}, ε_{13m} in the cell matrix, respectively, regarding Eqs. (3.1)–(3.5), (5.125), have the forms

$$\varepsilon_{11m} = -p_2 \sum_{i=0}^{1} \kappa_{1+2im}^{(1)} \lambda_{1+im} \Lambda_{1+im} r^{\lambda_{1+im}-1}, \tag{5.130}$$

$$\varepsilon_{22m} = \varepsilon_{33m} = -p_2 \sum_{i=0}^{1} \kappa_{1+2im}^{(1)} \Lambda_{1+im} r^{\lambda_{1+im}-1}, \tag{5.131}$$

$$\varepsilon_{12m} = -\sum_{i=0}^{1} \psi_{21+i\varphi m}^{(1)} \Lambda_{1+im} r^{\lambda_{1+im}-1}, \tag{5.132}$$

$$\varepsilon_{13m} = -\sum_{i=0}^{1} \psi_{21+i\nu m}^{(1)} \Lambda_{1+im} r^{\lambda_{1+im}-1}. \tag{5.133}$$

With regard to the determination of the compressive or tensile radial stress $p_2 > 0$ or $p_2 < 0$ acting on the matrix-envelope boundary, thus for $r = R_2$, respectively, considering Eq. (3.64), the tangential strain on the matrix-envelope boundary, $(\varepsilon_{22m})_{r=R_2}$, is required to be derived in the form

$$(\varepsilon_{22m})_{r=R_2} = -p_2 \, \varrho_{2m}, \tag{5.134}$$

where the coefficient ϱ_{2m} is given by Eq. (5.56).

Considering Eqs. (3.41), (3.42), (5.126)–(5.133), the thermal-stress induced elastic energy density w_m and the thermal-stress induced elastic energy W_m, accumulated in an arbitrary point of the cell matrix and in the volume of the cell matrix, respectively, have the forms

$$w_m = \frac{1}{2} \sum_{i,j=0}^{1} \omega_{21+i\;1+jm} r^{\lambda_{1+im}+\lambda_{1+jm}-2}, \tag{5.135}$$

$$W_m = 8 \int_0^{\pi/2} \int_0^{\pi/2} \int_{R_2}^{r_c} w_m r^2 \, dr \, d\varphi \, d\nu = \int_0^{\pi/2} \int_0^{\pi/2} \Omega_{22m} \, d\varphi \, d\nu, \tag{5.136}$$

where the coefficient Ω_{22m} given by Eq. (5.58) is a function of $\omega_{21+i\;1+jm}$ $(i, j = 0,1)$ given by Eq. (5.57).

With regard to the determination of the thermal-stress induced strengthening presented in Volume III, the stress σ_{im} acting in a direction of the axis x_i and the elastic energy density w_{im} induced by the stress σ_{im} $(i = 1,2,3)$ (see Eqs. (3.46), (3.47), (5.126)–(5.129)), both related to an arbitrary point of the cell matrix, are derived as

$$\sigma_{im} = -\left(\xi_{i21m} r^{\lambda_{1m}-1} + \xi_{i22m} r^{\lambda_{2m}-1} \right), \quad i = 1, 2, 3, \tag{5.137}$$

$$w_{im} = \frac{s_{11m}}{2} \left[\xi_{i21m}^2 r^{2(\lambda_{1m}-1)} + \xi_{i22m}^2 r^{2(\lambda_{2m}-1)} + 2\, \xi_{i21m} \, \xi_{i22m} r^{\lambda_{1m}+\lambda_{2m}-2} \right], \quad i = 1, 2, 3, \tag{5.138}$$

where the elastic modulus s_{11m} and the coefficients ξ_{i21m}, ξ_{i22m} $(i = 1,2,3)$ are given by Eqs. (3.23) and (5.59), respectively.

Appendix. Considering the coefficients c_{1m}, c_{2m}, $\Lambda_{1+im} = \Lambda^{(j)}_{1+im}$ $(i=0,1;$ $j=1,2,3)$, and the exponents $\lambda_{1m} > 1$, $\lambda_{2m} < 0$ as functions of the coefficient c_{3m} (see Eqs. (3.38)–(3.40), (5.8), (5.16), (5.21), (5.4)), the coefficient χ_{11+im} $(i=0,1)$ has the form

$$\chi_{11+im} = \Lambda_{1+im} R_2^{\lambda_{1+im}-1} \left[\lambda_{1+im} \left(c_{1m} + c_{2m}\right) - 2c_{2m}\right], \quad i = 0, 1. \tag{5.139}$$

5.6.3.2 Conditions $\beta_p \neq \beta_e = \beta_m$, and $\varepsilon_{11te} - \varepsilon_{11tp} = f(\varphi, \nu)$ or $\varepsilon_{11te} - \varepsilon_{11tp} \neq f(\varphi, \nu)$

The formulae in Section 5.6.3.2 consider the φ, ν-dependent or φ, ν-independent radial stress $p_1 = (\beta_e - \beta_p) / (\varrho_{1p} - \varrho^{me}_{1e})$ (see Eq. (10.4)) acting on the particle-envelope boundary and derived from Eq. (3.63), where the φ, ν-independent coefficients ϱ_{1p} and ϱ^{me}_{1e}, along with ϱ^{me}_{2e} included in the formulae in Section 5.6.3.2, are given by Eqs. (5.36) or (9.50), and (5.106), (6.207), (6.228), (6.246), (7.230) or (9.63), respectively.

The φ, ν-dependence of the radial stress p_1 results from the φ, ν-dependent difference $\varepsilon_{11te} - \varepsilon_{11tp} = f(\varphi, \nu)$ (see Eq. (12.23)), resulting in the φ, ν-dependent difference $\beta_e - \beta_p$ (see Eqs. (3.48)–(3.50)). The φ, ν-independence of the radial stress p_1 results from the condition $\varepsilon_{11te} - \varepsilon_{11tp} \neq f(\varphi, \nu)$ (see Eqs. (12.23), (12.24)), regarding simplification, equivalent to those defined in Items 1, 2 in Section 9.1 and resulting in the φ, ν-independent difference $\beta_e - \beta_p$.

With regard to the boundary conditions (4.11), (4.12) and the analysis presented in Section 5.4.3, considering the terms u_{1rm} and u_{2rm} (see Eqs. (5.9), (5.10)) to represent increasing and decreasing functions of the variable $r \in \langle R_2, r_c \rangle$, respectively, the radial displacement u_{1m} of the infinitesimal spherical cap in the cell matrix for the φ, ν-dependent or φ, ν-independent radial stress p_1, regarding Eqs. (3.31), (5.7), has the form

$$u_{1m} = -p_1\, \varrho^{me}_{2e} \sum_{i=0}^{1} \kappa^{(1)}_{1+2im}\, \Lambda_{1+im}\, r^{\lambda_{1+im}}, \tag{5.140}$$

where the exponents $\lambda_{1m} > 1$, $\lambda_{2m} < 0$ as functions of $c_{3m} < 0$ (see Eq. (3.40)), and the coefficient $\Lambda_{1+im} = \Lambda^{(j)}_{1+im}$ $(i=0,1;\ j=1,2,3)$ are given by Eqs. (5.4) and (5.8), (5.16), (5.21), respectively. The φ, ν-dependent distance r_c, as a function of the experimentally determined parameters v, R_{1ex}, d_{1ex}, d_{2ex}, d_{3ex} (see Section 2.2.1), included in the coefficient $\kappa^{(1)}_{1+2im}$ $(i=0,1)$ is derived in Section 2.3.

With regard to Section 5.6.3.2, the coefficients $\psi^{(1)}_{11\eta m}$, $\psi^{(1)}_{12\eta m}$ $(\eta = \varphi, \nu)$, $\omega_{21+i\ 1+jm}$ $(i,j=0,1)$, ξ_{i1+jm} $(i=1,2,3;\ j=0,1)$ and $\omega_{11+i\ 1+jm}$ $(i,j=0,1)$, ξ_{i11+jm}, $(i=1,2,3;$ $j=0,1)$ mentioned below (see Eqs. (5.143), (5.144), (5.149), (5.150), (5.155), (5.159), (5.160) and (5.153), (5.157), (5.158)) depend on $\kappa^{(1)}_{1m}$, $\kappa^{(1)}_{3m}$ and $\kappa^{(1)}_{1m}$, $\kappa^{(1)}_{3m}$, $\psi^{(1)}_{11\eta m}$,

$\psi^{(1)}_{12\eta m}$, respectively, where the coefficients $\kappa^{(1)}_{1m}$, $\kappa^{(1)}_{3m}$ (see Eq. (12.35)) are functions of χ_{11+im}, χ_{21+im} $(i=0,1)$ given by Eqs. (5.161), (5.55), respectively.

Consequently, considering the φ,ν-dependent or φ,ν-independent radial stress p_1, the radial and tangential stresses, σ_{11m} and σ_{22m}, σ_{33m}, respectively, acting in the cell matrix, regarding Eqs. (3.31), (3.32), (5.140), are derived as

$$\sigma_{11m} = -p_1\,\varrho^{me}_{2e}\sum_{i=0}^{1}\kappa^{(1)}_{1+2im}\,\Lambda_{1+im}\left[\lambda_{1+im}\left(c_{1m}+c_{2m}\right)-2c_{2m}\right]r^{\lambda_{1+im}-1}, \tag{5.141}$$

$$\sigma_{22m} = \sigma_{33m} = -p_1\,\varrho^{me}_{2e}\sum_{i=0}^{1}\kappa^{(1)}_{1+2im}\,\Lambda_{1+im}\left(c_{1m}-\lambda_{1+im}\,c_{2m}\right)r^{\lambda_{1+im}-1}, \tag{5.142}$$

where the coefficients c_{1m}, c_{2m} are given by Eqs. (3.38), (3.39), respectively.

With regard to Eqs. (3.33), (3.34), (5.140), the shear stresses σ_{12m}, σ_{13m} acting in the cell matrix, considering the φ,ν-dependent radial stress p_1, are derived as

$$\sigma_{12m} = -\frac{\varrho^{me}_{2e}}{s_{44m}}\sum_{i=0}^{1}\psi^{(1)}_{11+i\varphi m}\,\Lambda_{1+im}\,r^{\lambda_{1+im}-1}, \tag{5.143}$$

$$\sigma_{13m} = -\frac{\varrho^{me}_{2e}}{s_{44m}}\sum_{i=0}^{1}\psi^{(1)}_{11+i\nu m}\,\Lambda_{1+im}\,r^{\lambda_{1+im}-1}, \tag{5.144}$$

where the elastic modulus s_{44m} and the coefficient $\psi^{(1)}_{11+i\eta m}$ $(i=0,1;\ \eta=\varphi,\nu)$ are given by Eqs. (3.25) and (12.36), respectively. Considering the φ,ν-independent radial stress p_1, we get

$$\sigma_{12m} = -\frac{p_1\,\varrho^{me}_{2e}}{s_{44m}}\sum_{i=0}^{1}\frac{\partial\kappa^{(1)}_{1+2im}}{\partial\varphi}\,\Lambda_{1+im}\,r^{\lambda_{1+im}-1}, \tag{5.145}$$

$$\sigma_{13m} = -\frac{p_1\,\varrho^{me}_{2e}}{s_{44m}}\sum_{i=0}^{1}\frac{\partial\kappa^{(1)}_{1+2im}}{\partial\nu}\,\Lambda_{1+im}\,r^{\lambda_{1+im}-1}. \tag{5.146}$$

Finally, considering the φ,ν-dependent or φ,ν-independent radial stress p_1, the radial and tangential strains, ε_{11m} and ε_{22m}, ε_{33m}, respectively, acting in the cell matrix, regarding Eqs. (3.1)–(3.3), (5.140), have the forms

$$\varepsilon_{11m} = -p_1\,\varrho^{me}_{2e}\sum_{i=0}^{1}\kappa^{(1)}_{1+2im}\lambda_{1+im}\,\Lambda_{1+im}\,r^{\lambda_{1+im}-1}, \tag{5.147}$$

$$\varepsilon_{22m} = \varepsilon_{33m} = -p_1\,\varrho^{me}_{2e}\sum_{i=0}^{1}\kappa^{(1)}_{1+2im}\,\Lambda_{1+im}\,r^{\lambda_{1+im}-1}. \tag{5.148}$$

With regard to Eqs. (3.4), (3.5), (5.140), the shear strains σ_{12m}, σ_{13m} acting in the cell matrix, considering the φ, ν-dependent radial stress p_1, are derived as

$$\varepsilon_{12m} = -\varrho_{2e}^{me} \sum_{i=0}^{1} \psi_{11+i\varphi m}^{(1)} \Lambda_{1+im} r^{\lambda_{1+im}-1}, \tag{5.149}$$

$$\varepsilon_{13m} = -\varrho_{2e}^{me} \sum_{i=0}^{1} \psi_{11+i\nu m}^{(1)} \Lambda_{1+im} r^{\lambda_{1+im}-1}, \tag{5.150}$$

and considering the φ, ν-independent radial stress p_1, we get

$$\varepsilon_{12m} = -p_1 \varrho_{2e}^{me} \sum_{i=0}^{1} \frac{\partial \kappa_{1+2im}^{(1)}}{\partial \varphi} \Lambda_{1+im} r^{\lambda_{1+im}-1}, \tag{5.151}$$

$$\varepsilon_{13m} = -p_1 \varrho_{2e}^{me} \sum_{i=0}^{1} \frac{\partial \kappa_{1+2im}^{(1)}}{\partial \nu} \Lambda_{1+im} r^{\lambda_{1+im}-1}. \tag{5.152}$$

With regard to the φ, ν-dependent radial stress p_1, considering Eqs. (3.41), (3.42), (5.141)–(5.144), (5.147)–(5.150), the thermal-stress induced elastic energy density w_m and the thermal-stress induced elastic energy W_m, accumulated in an arbitrary point of the cell matrix and in the volume of the cell matrix, respectively, are derived as

$$w_m = \frac{\left(\varrho_{2e}^{me}\right)^2}{2} \sum_{i,j=0}^{1} \omega_{11+i\ 1+jm} r^{\lambda_{1+im}+\lambda_{1+jm}-2}, \tag{5.153}$$

$$W_m = 8 \int_0^{\pi/2} \int_0^{\pi/2} \int_{R_2}^{r_c} w_m r^2 dr d\varphi d\nu = \left(\varrho_{2e}^{me}\right)^2 \int_0^{\pi/2} \int_0^{\pi/2} \Omega_{12m} d\varphi d\nu, \tag{5.154}$$

where the coefficient Ω_{12m} given by Eq. (5.58) is a function of $\omega_{11+i\ 1+jm}$ $(i, j = 0{,}1)$ given by Eq. (5.57). With regard to the φ, ν-independent radial stress p_1, considering Eqs. (3.41), (3.42), (5.141), (5.142), (5.145)–(5.148), (5.151), (5.152), we get

$$w_m = \frac{\left(p_1 \varrho_{2e}^{me}\right)^2}{2} \sum_{i,j=0}^{1} \omega_{21+i\ 1+jm} r^{\lambda_{1+im}+\lambda_{1+jm}-2}, \tag{5.155}$$

$$W_m = 8 \int_0^{\pi/2} \int_0^{\pi/2} \int_{R_2}^{r_c} w_m r^2\, dr\, d\varphi\, d\nu = \left(p_1 \varrho_{2e}^{me}\right)^2 \int_0^{\pi/2} \int_0^{\pi/2} \Omega_{22m}\, d\varphi\, d\nu, \tag{5.156}$$

where the coefficient Ω_{22m} given by Eq. (5.58) is a function of $\omega_{21+i\ 1+jm}$ $(i, j = 0{,}1)$ given by Eq. (5.162).

With regard to the determination of the thermal-stress induced strengthening presented in Volume III, the stress σ_{im} acting in a direction of the axis x_i and the elastic energy density w_{im} induced by the stress σ_{im} ($i=1,2,3$) (see Eqs. (3.46), (3.47), (5.141)–(5.144)), both related to an arbitrary point of the cell matrix, considering the φ, ν-dependent radial stress p_1, have the forms

$$\sigma_{im} = -\varrho_{2e}^{me} \left(\xi_{i11m} r^{\lambda_{1m}-1} + \xi_{i12m} r^{\lambda_{2m}-1} \right), \quad i = 1, 2, 3, \tag{5.157}$$

$$w_{im} = \frac{s_{11m} \left(\varrho_{2e}^{me}\right)^2}{2} \left[\xi_{i11m}^2 r^{2(\lambda_{1m}-1)} + \xi_{i12m}^2 r^{2(\lambda_{2m}-1)} + 2\, \xi_{i11m} \xi_{i12m} r^{\lambda_{1m}+\lambda_{2m}-2} \right], \quad i = 1, 2, 3, \tag{5.158}$$

where the elastic modulus s_{11m} and the coefficients ξ_{i11m}, ξ_{i12m} ($i=1,2,3$) are given by Eqs. (3.23) and (5.59), respectively. With regard to the φ, ν-independent radial stress p_1, considering Eqs. (3.46), (3.47), (5.141), (5.142), (5.145), (5.146), we get

$$\sigma_{im} = -p_1 \varrho_{2e}^{me} \left(\xi_{i1m} r^{\lambda_{1m}-1} + \xi_{i2m} r^{\lambda_{2m}-1} \right), \quad i = 1, 2, 3, \tag{5.159}$$

$$w_{im} = \frac{s_{11m} \left(p_1 \varrho_{2e}^{me}\right)^2}{2} \left[\xi_{i1m}^2 r^{2(\lambda_{1m}-1)} + \xi_{i2m}^2 r^{2(\lambda_{2m}-1)} + 2\, \xi_{i1m} \xi_{i2m} r^{\lambda_{1m}+\lambda_{2m}-2} \right], \quad i = 1, 2, 3, \tag{5.160}$$

where the elastic modulus s_{11m} and the coefficients ξ_{i1m}, ξ_{i2m} ($i=1,2,3$) are given by Eqs. (3.23) and (5.163), respectively.

Appendix. Considering the elastic modulus s_{44m}, the coefficients c_{1m}, c_{2m}, $\Lambda_{1+im} = \Lambda_{1+im}^{(j)}$ ($i=0,1$; $j=1,2,3$), a'_{1i}, a'_{2i}, a'_{3i} ($i=1,2,3$), and the exponents $\lambda_{1m} > 1$, $\lambda_{2m} < 0$ as functions of the coefficient c_{3m} (see Eqs. (3.25), (3.38)–(3.40), (5.8), (5.16), (5.21), (12.26)–(12.34), (5.4)), the coefficients χ_{11+im}, $\omega_{21+i\,1+jm}$ ($i,j=0,1$), $\xi_{i\,1+jm}$ ($i=1,2,3$; $j=0,1$) have the forms

$$\chi_{11+im} = \Lambda_{1+im} R_2^{\lambda_{1+im}-1}, \quad i = 0, 1, \tag{5.161}$$

$$\begin{aligned} \omega_{21+i\,1+jm} &= \Lambda_{1+im} \Lambda_{1+jm} \\ &\times \left\{ \kappa_{1+2im}^{(1)} \kappa_{1+2jm}^{(1)} \left[\lambda_{1+im} \lambda_{1+jm} \left(c_{1m} + c_{2m} \right) - 4 \lambda_{1+im} c_{2m} + 2 c_{1m} \right] \right. \\ &\left. + \frac{2}{s_{44m}} \left[\frac{\partial \kappa_{1+2im}^{(1)}}{\partial \varphi} \frac{\partial \kappa_{1+2im}^{(1)}}{\partial \varphi} + \frac{\partial \kappa_{1+2jm}^{(1)}}{\partial \nu} \frac{\partial \kappa_{1+2jm}^{(1)}}{\partial \nu} \right] \right\}, \\ & i, j = 0, 1, \end{aligned} \tag{5.162}$$

$$
\begin{aligned}
\xi_{i\,1+jm} &= \Lambda_{1+jm} \\
&\times \left(\kappa_{1+2jm}^{(1)} \left\{ a'_{1i} \left[\lambda_{1+jm} \left(c_{1m} + c_{2m} \right) - 2c_{2m} \right] + \left(a'_{2i} + a'_{3i} \right) \left(c_{1m} - \lambda_{1+jm} c_{2m} \right) \right\} \right. \\
&\left. + \frac{1}{s_{44m}} \left[\left(a'_{1i} + a'_{2i} \right) \frac{\partial \kappa_{1+2jm}^{(1)}}{\partial \varphi} + \left(a'_{1i} + a'_{3i} \right) \frac{\partial \kappa_{1+2jm}^{(1)}}{\partial \nu} \right] \right), \\
& i = 1, 2, 3; \quad j = 0, 1.
\end{aligned} \tag{5.163}
$$

5.7 One-particle-(envelope)-matrix system. Conditions $\varepsilon_{11tm} - \varepsilon_{11tp} = f(\varphi, \nu)$, $\varepsilon_{11te} - \varepsilon_{11tp} = f(\varphi, \nu)$, $\varepsilon_{11tm} - \varepsilon_{11te} = f(\varphi, \nu)$

In contrast to Section 9.5, the analytical model of the thermal stresses and the corresponding quantities presented in Section 5.7, the latter represented by w_q, σ_{iq}, w_{iq} ($i = 1,2,3$) (see Eqs. (3.41), (3.46), (3.47)), consider the φ, ν-dependent differences in the phase-transformation induced radial strains, $\varepsilon_{11tm} - \varepsilon_{11tp} = f(\varphi, \nu)$, and $\varepsilon_{11te} - \varepsilon_{11tp} = f(\varphi, \nu)$, $\varepsilon_{11tm} - \varepsilon_{11te} = f(\varphi, \nu)$, (see Eqs. (12.23)), resulting in the φ, ν-dependent differences $\beta_m - \beta_p$, and $\beta_e - \beta_p$, $\beta_m - \beta_e$ (see Eqs. (3.48)–(3.50)), respectively, and consequently resulting in the φ, ν-dependent radial stresses p_1 and p_2 (see Eqs. (10.1)–(10.5)) acting on the particle-matrix, particle-envelope and matrix-envelope boundaries related to the one-particle-matrix and one-particle-envelope-matrix systems, respectively. Accordingly, the analytical model results from a system of the differential equations (3.35), (3.36) with the solutions (5.7), (5.15), (5.20).

As presented in Sections 4.2, 5.4.4, considering one integration constant only, the solutions (5.7), (5.15), (5.20) exhibit identical formulae for the infinite matrix due to the reduction of the coefficients $a_q^{(1)}$, $a_q^{(2)}$, $a_q^{(3)}$, $\Lambda_{1+iq} = \Lambda_{1+iq}^{(j)}$ ($i = 0,1$; $j = 1,2,3$; $q = p,e,m$) (see Eqs. (5.6), (5.8), (5.14), (5.16), (5.19), (5.21)) within the determination of the integration constants $C_{2m} \neq 0$ related to the term u_{2rm} as a decreasing function of the variable r regarding Eqs. (4.16), (4.17), using one of the boundary conditions (4.9), (4.10), (4.12).

5.7.1 Thermal stresses in spherical particle and envelope

Formulae for the spherical particle and for the spherical envelope in the one-particle-(envelope)-matrix system are identical with those presented in Sections 5.5.1, 5.6.1.2 and 5.6.2, 6.4.1, 7.5.1, respectively, where the coefficient ϱ_{1+im} ($i = 0,1$) in formulae (10.1)–(10.3), (10.5) for the radial stresses p_1, p_2 is given by Eqs. (5.178), (6.476), considering the Castigliano's theorem (see Section 3.1.9). Section 5.5.1 is related to the one-particle-matrix and one-particle-envelope-matrix systems with regard to the conditions $\beta_p \neq \beta_m$ and $\beta_p \neq \beta_e \neq \beta_m$, $\beta_p \neq \beta_e = \beta_m$ (see Eqs. (3.48)–(3.50)), respectively.

5.7.2 Thermal stresses in infinite matrix

5.7.2.1 Conditions $\beta_p \neq \beta_m$, $\beta_p \neq \beta_e \neq \beta_m$, $\beta_p = \beta_e \neq \beta_m$

Considering Eqs. (5.164)–(5.181) containing the subscript $i = 0,1$, the thermal stresses and the corresponding quantities for the infinite matrix presented in Section 5.7.2.1 are derived

1. with regard to the condition $\beta_p \neq \beta_m$ related to the one-particle-matrix system, and consequently the subscript $i = 0$,

2. with regard to the conditions $\beta_p \neq \beta_e \neq \beta_m$, $\beta_p = \beta_e \neq \beta_m$ related to the one-particle-envelope-matrix system, and consequently the subscript $i = 1$.

With regard to the boundary conditions (4.9), (4.16) to determine one integration constant only, considering the analysis presented in Section 5.4.4 to define u_{1m} as a decreasing function of the variable $r \in \langle R_{1+i}, \infty)$ $(i = 0,1)$, and thus the integration constant $C_{1m} = 0$, the radial displacement u_{1m} of the infinitesimal spherical cap in the infinite matrix, regarding Eqs. (3.31), (5.7), has the form

$$u_{1m} = -p_{1+i}\, \varrho_{1+im}\, r \left(\frac{r}{R_{1+i}} \right)^{\lambda_{2m}-1}, \quad i = 0, 1, \tag{5.164}$$

where the coefficient ϱ_{1+im} $(i = 0,1)$ and the exponent $\lambda_{2m} < 0$ as a function of $c_{3m} < 0$ (see Eq. (3.40)) are given by Eqs. (5.178) and (5.4), respectively. Provided that $\beta_p \neq \beta_m$ and $\beta_p \neq \beta_e \neq \beta_m$ or $\beta_p = \beta_e \neq \beta_m$, the radial stresses p_1, p_2 are given by Eqs. (10.1) and (10.2), (10.3) or (10.5), respectively.

Consequently, the radial stress σ_{11m}, the tangential and shear stresses σ_{22m}, σ_{33m} and σ_{12m}, σ_{13m} acting in the infinite matrix, respectively, regarding Eqs. (3.31)–(3.34), (5.164), are derived as

$$\sigma_{11m} = -p_{1+i} \left(\frac{r}{R_{1+i}} \right)^{\lambda_{2m}-1}, \quad i = 0, 1, \tag{5.165}$$

$$\sigma_{22m} = \sigma_{33m} = -p_{1+i}\, \varrho_{1+im} \left(c_{1m} - \lambda_{2m}\, c_{2m} \right) \left(\frac{r}{R_{1+i}} \right)^{\lambda_{2m}-1}, \quad i = 0, 1, \tag{5.166}$$

$$\sigma_{12m} = -\frac{\varrho_{1+im}}{s_{44m}} \frac{\partial p_{1+i}}{\partial \varphi} \left(\frac{r}{R_{1+i}} \right)^{\lambda_{2m}-1}, \quad i = 0, 1, \tag{5.167}$$

$$\sigma_{13m} = -\frac{\varrho_{1+im}}{s_{44m}} \frac{\partial p_{1+i}}{\partial \nu} \left(\frac{r}{R_{1+i}} \right)^{\lambda_{2m}-1}, \quad i = 0, 1, \tag{5.168}$$

where the elastic modulus s_{44m} is given by Eq. (3.25).

Finally, the radial strain ε_{11m}, the tangential and shear strains ε_{22m}, ε_{33m} and ε_{12m}, ε_{13m} in the infinite matrix, respectively, regarding Eqs. (3.1)–(3.5), (5.164), have the forms

$$\varepsilon_{11m} = -p_{1+i}\,\varrho_{1+im}\,\lambda_{2m}\left(\frac{r}{R_{1+i}}\right)^{\lambda_{2m}-1}, \quad i=0,1, \tag{5.169}$$

$$\varepsilon_{22m} = \varepsilon_{33m} = -p_{1+i}\varrho_{1+im}\left(\frac{r}{R_{1+i}}\right)^{\lambda_{2m}-1}, \quad i=0,1, \tag{5.170}$$

$$\varepsilon_{12m} = -\varrho_{1+im}\frac{\partial p_{1+i}}{\partial\varphi}\left(\frac{r}{R_{1+i}}\right)^{\lambda_{2m}-1}, \quad i=0,1, \tag{5.171}$$

$$\varepsilon_{13m} = -\varrho_{1+im}\frac{\partial p_{1+i}}{\partial\nu}\left(\frac{r}{R_{1+i}}\right)^{\lambda_{2m}-1}, \quad i=0,1. \tag{5.172}$$

With regard to the determination of the compressive or tensile radial stress $p_{1+i} > 0$ or $p_{1+i} < 0$ acting on the particle-matrix, matrix-envelope boundary, thus for $r = R_1$, $r = R_2$, respectively, considering Eqs. (3.58), (3.63), (3.64), the tangential strains on the particle-matrix and matrix-envelope boundaries, $(\varepsilon_{22m})_{r=R_1}$ and $(\varepsilon_{22m})_{r=R_2}$, respectively, are required to be derived in the form

$$(\varepsilon_{22m})_{r=R_{1+i}} = -p_{1+i}\,\varrho_{1+im}, \quad i=0,1, \tag{5.173}$$

where the coefficient ϱ_{1+im} ($i=0,1$) is given by Eq. (5.178).

Considering Eqs. (3.41), (3.42), (5.165)–(5.172), the thermal-stress induced elastic energy density w_m and the thermal-stress induced elastic energy $W_m^{(1)}$[3], accumulated in an arbitrary point of the infinite matrix and in the volume of the infinite matrix, respectively, have the forms

$$w_m = \frac{\omega_{1+im}\,\varrho_{1+im}^2}{2}\left(\frac{r}{R_{1+i}}\right)^{2(\lambda_{2m}-1)}, \quad i=0,1, \tag{5.174}$$

$$W_m^{(1)} = 8\int_0^{\pi/2}\int_0^{\pi/2}\int_{R_{1+i}}^{\infty} w_m\, r^2 dr d\varphi d\nu = \int_0^{\pi/2}\int_0^{\pi/2}\Omega_{1+i\,1+im}d\varphi d\nu, \quad i=0,1, \tag{5.175}$$

where the coefficient $\Omega_{1+i\,1+im}$ ($i=0,1$) given by Eq. (5.180) is a function of ω_{1+im} given by Eq. (5.179).

With regard to the determination of the thermal-stress induced strengthening presented in Volume III, the stress σ_{jm} acting in a direction of the axis x_j and the elastic energy density w_{jm} induced by the stress σ_{jm} ($j=1,2,3$) (see Eqs. (3.46),

[3]The superscript (1) is related to the *one*-particle-(envelope)-matrix system.

(3.47), (5.165)–(5.168)), both related to an arbitrary point of the infinite matrix, are derived as

$$\sigma_{jm} = -\xi_{j\,1+im}\,\varrho_{\,1+im}\left(\frac{r}{R_{1+i}}\right)^{\lambda_{2m}-1}, \quad i=0,1; \quad j=1,2,3, \tag{5.176}$$

$$w_{jm} = \frac{s_{11m}\left(\xi_{j\,1+im}\,\varrho_{\,1+im}\right)^2}{2}\left(\frac{r}{R_{1+i}}\right)^{2(\lambda_{2m}-1)}, \quad i=0,1; \quad j=1,2,3, \tag{5.177}$$

where the elastic modulus s_{11m} and the coefficient $\xi_{j\,1+im}$ $(i=0,1;\ j=1,2,3)$ are given by Eqs. (3.23) and (5.181), respectively.

Appendix. Considering the elastic modulus s_{44m}, the coefficients c_{1m}, c_{2m}, a'_{1j}, a'_{2j}, a'_{3j} $(j=1,2,3)$, and the exponent $\lambda_{2m} < 0$ as a function of the coefficient c_{3m} (see Eqs. (3.25), (3.38)–(3.40), (12.26)–(12.34), (5.4)), the coefficients $\varrho_{\,1+im}$ [4], ω_{1+im}, $\Omega_{1+i\,1+jm}$ [5] $(i,j=0,1)$, $\xi_{j\,1+im}$ $(i=0,1;\ j=1,2,3)$ have the forms

$$\varrho_{\,1+im} = \frac{1}{\lambda_{2m}\left(c_{1m}+c_{2m}\right)-2c_{2m}}, \quad i=0,1, \tag{5.178}$$

$$\begin{aligned}\omega_{1+im} &= p_{1+i}^2\left[\lambda_{2m}^2\left(c_{1m}+c_{2m}\right)+2\left(c_{1m}-2\lambda_{2m}\,c_{2m}\right)\right]\\ &+\frac{2}{s_{44m}}\left[\left(\frac{\partial p_{1+i}}{\partial\varphi}\right)^2+\left(\frac{\partial p_{1+i}}{\partial\nu}\right)^2\right], \quad i=0,1,\end{aligned} \tag{5.179}$$

$$\Omega_{1+i\,1+jm} = \frac{4\,\omega_{1+im}\,\varrho_{1+im}^2 R_{1+j}^3}{|2\lambda_{2m}+1|}, \quad i,j=0,1, \tag{5.180}$$

$$\begin{aligned}\xi_{j\,1+im} &= p_{1+i}\left[\frac{a'_{1j}}{\varrho_{\,1+im}}+\left(a'_{2j}+a'_{3j}\right)\left(c_{1m}-\lambda_{2m}\,c_{2m}\right)\right]\\ &+\frac{1}{s_{44m}}\left[\left(a'_{1j}+a'_{2j}\right)\frac{\partial p_{1+i}}{\partial\varphi}+\left(a'_{1j}+a'_{3j}\right)\frac{\partial p_{1+i}}{\partial\nu}\right],\\ & i=0,1; \quad j=1,2,3.\end{aligned} \tag{5.181}$$

[4] The subscript $1+i$ in Eq. (5.178) is used due to uniform notations applied in formulae for the tangential strain $(\varepsilon_{22q})_{r=R_{1+i}}$ $(i=0,1;\ q=p,e,m)$, e.g. as presented in Eq. (5.173).

[5] With regard to $\lambda_{2m} < -2$ (see Eq. (5.4)) considering the Poisson's number $\mu_m < 0.5$ for real materials [1, p 38], the term $-(2\lambda_{2m}+1) > 0$ resulting in the integration (see Eq. (5.175)) is replaced by $|2\lambda_{2m}+1|$. Additionally, the coefficient Ω_{21m} is determined but not used in formulae for the elastic energy W_m of the infinite matrix.

5.7.2.2 Condition $\beta_p \neq \beta_e = \beta_m$

The formulae in Section 5.7.2.2 consider the φ, ν-dependent radial stress $p_1 = (\beta_e - \beta_p) / (\varrho_{1p} - \varrho_{1e}^{me})$ (see Eq. (10.4)) acting on the particle-envelope boundary and derived from Eq. (3.63), where the φ, ν-independent coefficients ϱ_{1p} and ϱ_{1e}^{me}, along with ϱ_{2e}^{me} included in the formulae in Section 5.7.2.2, are given by Eqs. (5.36) or (9.50), and (5.106), (6.207), (6.228), (6.246), (7.230) or (9.63), respectively. The φ, ν-dependence of the radial stress p_1 results from the φ, ν-dependent difference $\varepsilon_{11te} - \varepsilon_{11tp} = f(\varphi, \nu)$ (see Eq. (12.23)), resulting in the φ, ν-dependent difference $\beta_e - \beta_p$ (see Eqs. (3.48)–(3.50)).

With regard to the boundary conditions (4.12), (4.16) to determine one integration constant only, considering the analysis presented in Section 5.4.4 to define u_{1m} as a decreasing function of the variable $r \in \langle R_2, \infty)$, and thus the integration constant $C_{1m} = 0$, the radial displacement u_{1m} of the infinitesimal spherical cap in the infinite matrix, regarding Eqs. (3.31), (5.7), has the form

$$u_{1m} = -p_1 \, \varrho_{2e}^{me} \, \varrho_{1m} \, r \left(\frac{r}{R_2} \right)^{\lambda_{2m} - 1}, \tag{5.182}$$

where the coefficient ϱ_{1m} and the exponent $\lambda_{2m} < 0$ as a function of $c_{3m} < 0$ (see Eq. (3.40)) are given by Eqs. (5.178) and (5.4), respectively.

Consequently, the radial stress σ_{11m}, the tangential and shear stresses σ_{22m}, σ_{33m} and σ_{12m}, σ_{13m} acting in the infinite matrix, respectively, regarding Eqs. (3.31)–(3.34), (5.182), are derived as

$$\sigma_{11m} = -p_1 \, \varrho_{2e}^{me} \left(\frac{r}{R_2} \right)^{\lambda_{2m} - 1}, \tag{5.183}$$

$$\sigma_{22m} = \sigma_{33m} = -p_1 \, \varrho_{2e}^{me} \, \varrho_{1m} \, (c_{1m} - \lambda_{2m} \, c_{2m}) \left(\frac{r}{R_2} \right)^{\lambda_{2m} - 1}, \tag{5.184}$$

$$\sigma_{12m} = - \frac{\varrho_{2e}^{me} \, \varrho_{1m}}{s_{44m}} \frac{\partial p_1}{\partial \varphi} \left(\frac{r}{R_2} \right)^{\lambda_{2m} - 1}, \tag{5.185}$$

$$\sigma_{13m} = - \frac{\varrho_{2e}^{me} \, \varrho_{1m}}{s_{44m}} \frac{\partial p_1}{\partial \nu} \left(\frac{r}{R_2} \right)^{\lambda_{2m} - 1}, \tag{5.186}$$

where the elastic modulus s_{44m} is given by Eq. (3.25).

Finally, the radial strain ε_{11m}, the tangential and shear strains ε_{22m}, ε_{33m} and ε_{12m}, ε_{13m} in the infinite matrix, respectively, regarding Eqs. (3.1)–(3.5), (5.182), have the forms

$$\varepsilon_{11m} = -p_1 \, \varrho_{2e}^{me} \, \varrho_{1m} \, \lambda_{2m} \left(\frac{r}{R_2} \right)^{\lambda_{2m} - 1}, \tag{5.187}$$

$$\varepsilon_{22m} = \varepsilon_{33m} = -p_1\, \varrho_{2e}^{me}\, \varrho_{1m} \left(\frac{r}{R_2}\right)^{\lambda_{2m}-1}, \tag{5.188}$$

$$\varepsilon_{12m} = -\varrho_{2e}^{me}\, \varrho_{1m} \frac{\partial p_1}{\partial \varphi} \left(\frac{r}{R_2}\right)^{\lambda_{2m}-1}, \tag{5.189}$$

$$\varepsilon_{13m} = -\varrho_{2e}^{me}\, \varrho_{1m} \frac{\partial p_1}{\partial \nu} \left(\frac{r}{R_2}\right)^{\lambda_{2m}-1}. \tag{5.190}$$

Considering Eqs. (3.41), (3.42), (5.183)–(5.190), the thermal-stress induced elastic energy density w_m and the thermal-stress induced elastic energy $W_m^{(1)}$[6], accumulated in an arbitrary point of the infinite matrix and in the volume of the infinite matrix, respectively, have the forms

$$w_m = \frac{\omega_{1m} \left(\varrho_{2e}^{me}\, \varrho_{1m}\right)^2}{2} \left(\frac{r}{R_2}\right)^{2(\lambda_{2m}-1)}, \tag{5.191}$$

$$W_m^{(1)} = 8 \int_0^{\pi/2} \int_0^{\pi/2} \int_{R_2}^{\infty} w_m\, r^2 dr d\varphi d\nu = \left(\varrho_{2e}^{me}\right)^2 \int_0^{\pi/2} \int_0^{\pi/2} \Omega_{12m} d\varphi d\nu, \tag{5.192}$$

where the coefficient Ω_{12m} given by Eq. (5.180) is a function of ω_{1m} given by Eq. (5.179).

With regard to the determination of the thermal-stress induced strengthening presented in Volume III, the stress σ_{im} acting in a direction of the axis x_i and the elastic energy density w_{im} induced by the stress σ_{im} ($i=1,2,3$) (see Eqs. (3.46), (3.47), (5.183)–(5.186)), both related to an arbitrary point of the infinite matrix, are derived as

$$\sigma_{im} = -\xi_{i1m}\, \varrho_{2e}^{me}\, \varrho_{1m} \left(\frac{r}{R_2}\right)^{\lambda_{2m}-1}, \quad i = 1, 2, 3, \tag{5.193}$$

$$w_{im} = \frac{s_{11m} \left(\xi_{i1m}\, \varrho_{2e}^{me}\, \varrho_{1m}\right)^2}{2} \left(\frac{r}{R_2}\right)^{2(\lambda_{2m}-1)}, \quad i = 1, 2, 3, \tag{5.194}$$

where the elastic modulus s_{11m} and the coefficient ξ_{i1m} ($i=1,2,3$) are given by Eqs. (3.23) and (5.181), respectively.

[6]The superscript (1) is related to the *one*-particle-(envelope)-matrix system.

Chapter 6

Isotropic multi- and one-particle-(envelope)-matrix systems. Solution 4

With regard to the analysis presented in Section 6.2 to consider the boundary conditions defined in Section 4.1, the total solution for the radial displacement u_{1q} given by Eq. (6.4), representing a linear combination of the particular solutions $C_{1q}a_{1q}u_{1rq}$, $C_{2q}a_{2q}u_{2rq}$, $C_{2q}a_{2q}u_{3rq}$ (see Eqs. (4.1), (6.6)–(6.8)), and resulting from the mathematical techniques presented in Section 6.1 and applied to a system of the linear differential equations (3.35), (3.36), is suitable for the determination of the thermal stresses and the corresponding quantities, the latter represented by w_q, W_q, σ_{iq}, w_{iq} $(i=1,2,3)$ (see Eqs. (3.41), (3.46), (3.47)), in the spherical envelope $(q=e)$ and the cell matrix $(q=m)$ as two of components of composite materials, where $a_{1q}=1$, $a_{2q}=C_q^{(1)}$, $a_{3q}=3/2$ (see Eqs. (6.4), (6.5)). Consequently, one, two or three of the terms u_{1rq}, u_{2rq}, u_{3rq} are considered for the determination of the integration constants C_{1q}, C_{2q}, C_{3q} using the boundary conditions.

The radial displacement u_{1q}, the thermal stresses and the corresponding quantities are functions of the angles φ, ν included in the φ, ν-dependent distance r_c as a function of the experimentally determined parameters v, R_{1ex}, d_{1ex}, d_{2ex}, d_{3ex} (see Sections 2.2.1, 2.3). As a consequence of symmetry of the multi-particle-(envelope)-matrix system resulting from the matrix infinity and from the periodical distribution of the spherical particles shown in Fig. 2.1a,b, the angles φ, ν are both from the interval $\langle 0, \pi/2 \rangle$.

With regard to completeness of the presented topic, Section 6.5 presents the analytical model of the thermal stresses and the corresponding quantities in the infinite matrix of the one-particle-(envelope)-matrix system containing, in contrast to the multi-particle-(envelope)-matrix system, one spherical particle only. Considering the boundary conditions defined in Section 4.2, the analytical model related to the one-particle-(envelope)-matrix system is required to be derived regarding the analysis presented in Section 2.2.2 and concerning the condition $v \to 0$.

To derive the elastic energy density W induced by the thermal stresses, formulae for the integration of the r-dependent integrand wr^2 (see Eq. (3.42)) are presented in Section 12.4.4. Consequently, the integral in the formula $W = \int_0^{\pi/2}\int_0^{\pi/2} \Omega \, d\varphi \, d\nu$ (see Eqs. (6.25), (6.54), (6.76), (6.99), (6.134), (6.161), (6.182), (6.204), (6.225), (6.243), (6.261), (6.276), (6.291), (6.307), (6.322), (6.337), (6.355), (6.357), (6.387), (6.389), (6.413), (6.415), (6.440), (6.442), (6.473), (6.490)) related to the integrand represented by the φ, ν-dependent coefficient Ω and to the variables $\varphi, \nu \in \langle 0, \pi/2 \rangle$, can be derived numerically for a real composite material, considering Eq. (1.5).

6.1 Mathematical techniques 4

Performing ∂^2Eq.(3.35)$/\partial r^2$, the differential equation (3.35) is transformed to the form

$$r \, \frac{\partial^3 U_{1q}}{\partial r^3} + (2 - c_{3q}) \, \frac{\partial^2 U_{1q}}{\partial r^2} = 0, \tag{6.1}$$

and assuming $U_{1q} = r^{\lambda_q}$ (see Eq. (3.37)), we get

$$U_{1q} = \frac{\partial^2 u_{1q}}{\partial \varphi^2} + \frac{\partial^2 u_{1q}}{\partial \nu^2} = C_{1q} \, r + C_{2q} \, r^{c_{3q}} + C_{3q}, \tag{6.2}$$

where the coefficient c_{3q} is given by Eq. (3.40) and C_{1q}, C_{2q}, C_{3q} are integration constants.

Considering Eq. (6.2), the differential equation (3.36) is transformed to the form

$$\frac{\partial^2 u_{1q}}{\partial r^2} + \frac{2}{r} \, \frac{\partial u_{1q}}{\partial r} - \frac{2 u_{1q}}{r^2} = -3 a_q^{(1)} \left(\frac{C_{1q}}{r} + C_{2q} \, r^{c_{3q} - 2} + \frac{C_{3q}}{r^2} \right), \tag{6.3}$$

representing a linear differential equation of the second rank regarding the variable r with a non-zero right side, and the coefficient $a_q^{(1)}$ is given by Eq. (5.6).

Using the Wronskian's method [5] presented in Section 12.4.5, representing a short introduction to problems of linear differential equations with non-zero right sides, the solution of the differential equation (6.3) for the radial displacement u_{1q} of the infinitesimal spherical cap (see Figs. 3.2, 3.3) has the form

$$u_{1q} = a_q^{(1)} \left[C_{1q} \, r \left(\frac{1}{3} - \ln r \right) + C_{2q} \, C_q^{(1)} \, r^{c_{3q}} + \frac{3 C_{3q}}{2} \right], \tag{6.4}$$

where the coefficient $C_q^{(1)}$ is derived as

$$C_q^{(1)} = \frac{1}{c_{3q} + 2} - \frac{1}{c_{3q} - 1}. \tag{6.5}$$

The integration constants C_{1q}, C_{2q}, C_{3q} are derived from boundary conditions (see Sections 4.1.1–4.1.3, 4.2), considering one, two or three of the terms u_{1rq}, u_{2rq}, u_{3rq}, regarding Eq. (4.1), in the forms

$$u_{1rq} = r\left(\frac{1}{3} - \ln r\right), \tag{6.6}$$

$$u_{2rq} = r^{c_{3q}}, \tag{6.7}$$

$$u_{3rq} = 1, \tag{6.8}$$

where the term u_{2rq} is a decreasing function of the variable r due to $c_{3q} < 0$ (see Eq. (3.40)), the term u_{3rq} is a constant function regarding the variable r. The term u_{1rq} represents a decreasing function of the variable r on the condition [6, p 368]

$$\frac{\partial u_{1rq}}{\partial r} = -\left(\frac{2}{3} + \ln r\right) < 0, \tag{6.9}$$

resulting in the condition

$$r > R = e^{-2/3} \approx 0.51342, \tag{6.10}$$

where $e \approx 2.71828$ is the Euler's number [6, p 247]. Similarly, on the condition

$$r < R = e^{-2/3} \approx 0.51342, \tag{6.11}$$

the term u_{1rq} represents an increasing function of the variable r.

With regard to the analysis, presented in Section 6.2 and concerning the integration constants C_{1q}, C_{2q}, C_{3q}, the subscript q in formulae in Section 6.1 is related to material parameters of the spherical envelope ($q = e$) and the cell matrix ($q = m$).

6.2 Analysis of solution

6.2.1 Spherical particle

With regard to the geometric boundary condition $(u_{1p})_{r=0} = 0$ given by Eq. (4.3) related to the multi-particle-matrix and multi-particle-envelope-matrix systems provided that $\beta_p \neq \beta_m$, and $\beta_p \neq \beta_e \neq \beta_m$ or $\beta_p \neq \beta_e = \beta_m$ or $\beta_p = \beta_e \neq \beta_m$ (see Eqs. (3.48)–(3.50)), the absolute value $|u_{1p}|$ of the radial displacement u_{1p} of the infinitesimal spherical cap in the spherical particle is required to represent an increasing function of the variable $r \in \langle 0, R_1 \rangle$, exhibiting maximal values on the particle-matrix and particle-envelope boundaries, respectively. With regard to $(\sigma_{ijp})_{r\to 0} \propto (\partial u_{1rp}/\partial r,\ u_{1rp}/r)_{r\to 0} \propto (\ln r)_{r\to 0} \to -\infty$ $(i, j = 1,2,3)$ (see Eqs. (3.31)–(3.34)) and to $(u_{1p})_{r\to 0} \propto (u_{1rp})_{r\to 0} = (r^{c_{3q}})_{r\to 0} \to \infty$ resulting from $c_{3q} < 0$ (see Eq. (3.40)), the integration constants $C_{1p} = C_{2p} = 0$. Considering

$C_{1p} = C_{2p} = 0$ and $C_{3p} \neq 0$, the radial displacement u_{1p} of the infinitesimal spherical cap in the spherical particle does not fulfil the boundary condition given by Eq. (4.3) due to $(u_{1p})_{r=0} = C_{3p} / [2 s_{44p} (c_{1p} + c_{2p})] \neq 0$, and accordingly the coefficients $C_{1p} = C_{2p} = C_{3p} = 0$.

6.2.2 Spherical envelope

The analysis presented in Section 6.2.2 is considered for the spherical envelope in the multi- and one-particle-envelope-matrix systems (see Sections 6.4.1, 6.5.1).

With regard to the boundary conditions (4.7), (4.8) for the spherical envelope provided that $\beta_p \neq \beta_e \neq \beta_m$ (see Eqs. (3.48)–(3.50)), the combinations of the terms

1. u_{1re} and u_{2re}, corresponding to the integration constants $C_{1e} \neq 0$, $C_{2e} \neq 0$, $C_{3e} = 0$,
2. u_{1re} and u_{3re}, corresponding to the integration constants $C_{1e} \neq 0$, $C_{2e} = 0$, $C_{3e} \neq 0$,
3. u_{2re} and u_{3re}, corresponding to the integration constants $C_{1e} = 0$, $C_{2e} \neq 0$, $C_{3e} \neq 0$,

are considered for the determination of the radial displacement u_{1e} of the infinitesimal spherical cap in the spherical envelope as a function of the variable $r \in \langle R_1, R_2 \rangle$ (see Eqs. (6.4), (6.6)–(6.8)).

Similarly, with regard to the boundary condition given by Eq. (4.7) provided that $\beta_p \neq \beta_e = \beta_m$, the absolute value $|u_{1e}|$ exhibits a maximal value on the particle-envelope boundary, thus for $r = R_1$. Accordingly, $|u_{1e}|$ is required to represent a decreasing function of the variable $r \in \langle R_1, R_2 \rangle$ or, at the very most, a constant function regarding the variable r. Consequently, the term

4. u_{1re}, corresponding to the integration constants $C_{1e} \neq 0$, $C_{2e} = C_{3e} = 0$, is considered, related to the condition $r \geq R$ (see Eq. (6.10)) resulting, with regard to the variable $r \in \langle R_1, R_2 \rangle$, in the condition
$$R_1 \geq R \approx 0.51342, \tag{6.12}$$
5. u_{2re}, corresponding to the integration constants $C_{2e} \neq 0$, $C_{1e} = C_{3e} = 0$, is considered,
6. u_{3re}, corresponding to the integration constants $C_{3e} \neq 0$, $C_{1e} = C_{2e} = 0$, is considered.

Against the extreme value of the radius $R_1 \geq e^{-2/3} \approx 0.51342$ (see Eq. (6.12)) in comparison with radii of particles in real composite materials [1], regarding completeness of the presented topic, the thermal stresses and the corresponding quantities in the spherical envelope related to the term u_{1re} are presented in the part A of Section 6.4.1.2 as a curiosity only. The thermal stresses and the corresponding quantities presented in Section 6.4.1.2 are derived on the condition $\beta_p \neq \beta_e = \beta_m$, considering the φ, ν-dependent difference $\varepsilon_{11te} - \varepsilon_{11tp} = f(\varphi, \nu)$ (see Eq. (12.23)), resulting in the φ, ν-dependent difference $\beta_e - \beta_p$ (see Eqs. (3.48)–(3.50)) and consequently in the φ, ν-dependent radial stress $p_1 = (\beta_e - \beta_p) / (\varrho_{1p} - \varrho_{1e}^{me})$ (see Eq. (10.4)) acting on the particle-envelope boundary, derived from Eq. (3.63) and included in the formulae in Section 6.4.1.2, where the coefficients ϱ_{1p}, ϱ_{1e}^{me} given by Eqs. (5.36), (5.106), respectively, are not functions of the variables φ, ν.

Finally, with regard to the boundary condition given by Eq. (4.8) provided that $\beta_p = \beta_e \neq \beta_m$, the absolute value $|u_{1e}|$ exhibits a maximal value on the matrix-envelope boundary, thus for $r = R_2$. Accordingly, $|u_{1e}|$ is required to represent an increasing function of the variable $r \in \langle R_1, R_2 \rangle$ or, at the very most, a constant function regarding the variable r. Consequently, the term

7. u_{1re}, corresponding to the integration constants $C_{1e} \neq 0$, $C_{2e} = C_{3e} = 0$, is considered, related to the condition $r \leq R$ (see Eq. (6.11)) resulting, with regard to the variable $r \in \langle R_1, R_2 \rangle$, in the condition

$$R_2 \leq R \approx 0.51342, \tag{6.13}$$

8. u_{3re}, corresponding to the integration constants $C_{3e} \neq 0$, $C_{1e} = C_{2e} = 0$, is considered.

Resulting from the conditions $C_{3e} \neq 0$, $C_{1e} = C_{2e} = 0$, and as presented in Section 3.1.2.3 Item 5), the assumption for the radial displacement $u_{1e} \propto u_{3re}$ (see Eq. (6.8)) as a constant function regarding the variable r may be improbable, nevertheless it is considered, with regard to completeness of the presented topic, in Sections 6.4.1.2.C, 6.4.1.3.B.

6.2.3 Cell matrix

With regard to the boundary conditions (4.9), (4.11) and (4.10), (4.11) for the cell matrix related to the multi-particle-matrix and multi-particle-envelope-matrix systems provided that $\beta_p \neq \beta_m$ and $\beta_p \neq \beta_e \neq \beta_m$ or $\beta_p = \beta_e \neq \beta_m$ (see Eqs. (3.48)–(3.50)), the absolute value $|u_{1m}|$ of the radial displacement u_{1m} of the infinitesimal spherical cap in the cell matrix, exhibiting maximal values on the particle-matrix and matrix-envelope boundaries, thus for $r = R_1$ and $r = R_2$, is required to represent a decreasing function of the variable $r \in \langle R_1, r_c \rangle$ and $r \in \langle R_2, r_c \rangle$, respectively. Accordingly, the combinations of the terms

1. u_{1rm} and u_{2rm}, corresponding to the integration constants $C_{1m} \neq 0$, $C_{2m} \neq 0$, $C_{3m} = 0$,

2. u_{1rm} and u_{3rm}, corresponding to the integration constants $C_{1m} \neq 0$, $C_{2m} = 0$, $C_{3m} \neq 0$,

3. u_{2rm} and u_{3rm}, corresponding to the integration constants $C_{1m} = 0$, $C_{2m} \neq 0$, $C_{3m} \neq 0$,

are considered for the determination of the radial displacement u_{1m} in the cell matrix (see Eqs. (6.4), (6.6)–(6.8)).

With regard to the multi-particle-envelope-matrix system related to the condition $\beta_p \neq \beta_e = \beta_m$, resulting in the geometric and zero strain boundary conditions given by Eqs. (4.11) and (4.12), respectively, the absolute value $|u_{1m}|$ is also required to represent a decreasing function of the variable $r \in \langle R_2, r_c\rangle$, exhibiting the maximal value $\left|(u_{1m})_{r=R_2}\right| = |p_1\, \varrho_{2e}^{me} R_2|$ on the matrix-envelope boundary, thus for $r = R_2$. Accordingly, the combinations of the terms u_{1rm}, u_{2rm} or u_{1rm}, u_{3rm} or u_{2rm}, u_{3rm} are considered as in case of the conditions $\beta_p \neq \beta_m$ and $\beta_p \neq \beta_e \neq \beta_m$ or $\beta_p = \beta_e \neq \beta_m$, related to the multi-particle-matrix and multi-particle-envelope-matrix systems, respectively.

In addition to the standard boundary conditions for the cell matrix given by Eqs. (4.9), (4.11) and (4.10), (4.11) related to the multi-particle-matrix and multi-particle-envelope-matrix systems provided that $\beta_p \neq \beta_m$ and $\beta_p \neq \beta_e \neq \beta_m$ or $\beta_p = \beta_e \neq \beta_m$, respectively, and given by Eqs. (4.11), (4.12) related to the multi-particle-envelope-matrix system provided that $\beta_p \neq \beta_e = \beta_m$, zero radial strain ε_{11m} on the cell surface, thus for $r = r_c$, is considered as the third boundary condition (see Eq. (4.14)). The third boundary condition results from a possible assumption that the radial displacement u_{1m} as a function of the variable r is extremal on the cell surface. Accordingly, with regard to the variable $r \in \langle R_1, r_c\rangle$ and $r \in \langle R_2, r_c\rangle$ related to the multi-particle-matrix and multi-particle-envelope-matrix system, respectively, the minimum or the maximum are assumed to be exhibited for $r = r_c$ provided that the radial displacement u_{1m} represents a decreasing function or an increasing function of the variable r, thus for $u_{1m} > 0$ or $u_{1m} < 0$, respectively. The assumed boundary condition of the minimum or the maximum of the $u_{1m} - r$ dependence on the cell surface for $u_{1m} > 0$ or $u_{1m} < 0$, respectively, can be perhaps considered as a possible state of deformation of a solid continuum on cell boundaries. Finally, the terms

4. u_{1rm}, u_{2rm} and u_{3rm} corresponding to the integration constants $C_{1m} \neq 0$, $C_{2m} \neq 0$, $C_{3m} \neq 0$,

are considered for the determination of the radial displacement u_{1m} in the cell matrix.

6.2.4 Infinite matrix

As presented in Section 4.2, one integration constant is considered only, regarding $|u_{1m}|$, $|\sigma_{11m}|$ to represent decreasing functions of the variable $r \in \langle R_1, \infty)$ and $r \in \langle R_2, \infty)$ related to the one-particle-matrix and one-particle-envelope-matrix systems, respectively. Accordingly, the term u_{2rm} is considered, and thus the integration constants $C_{1m} = C_{3m} = 0$, $C_{2m} \neq 0$ (see Eqs. (6.4), (6.7)).

Considering the term u_{1rm} to represent a decreasing function of the variable $r \in \langle R_1, \infty\rangle$ and $r \in \langle R_2, \infty\rangle$ for $R_1 \geq e^{-2/3} \approx 0.51342$ and $R_2 \geq e^{-2/3} \approx 0.51342$, respectively, we get $(\sigma_{iim})_{r\to\infty} \propto (\ln r)_{r\to\infty} \to \infty$, $(\sigma_{12+jm})_{r\to\infty} \propto (\ln r)_{r\to\infty} \to \infty$ $(i=1,2,3;\ j=0,1)$ (see Eqs. (3.31)–(3.34)) what is physically unacceptable, and additionally the boundary condition (4.17) is not fulfilled.

6.3 Multi-particle-matrix system

The variable r related to the thermal stresses and to the corresponding quantities in the cell matrix is from the interval $\langle R_1, r_c\rangle$, where the φ, ν-dependent distance r_c as a function of the experimentally determined parameters v, R_{1ex}, d_{1ex}, d_{2ex}, d_{3ex} (see Section 2.2.1) is derived in Section 2.3. With regard to Eq. (3.58), the compressive or tensile radial stress $p_1 > 0$ or $p_1 < 0$ acting on the particle-matrix boundary, thus for $r = R_1$, respectively, is given by Eq. (10.1). The determination of the radial stress p_1, as a consequence of the condition $\beta_p \neq \beta_m$ (see Eqs. (3.48)–(3.50)), considers the tangential strains $(\varepsilon_{22p})_{r=R_1}$ and $(\varepsilon_{22m})_{r=R_1}$ in the spherical particle and in the cell matrix on the particle-matrix boundary (see Eq. (3.58)), resulting in the coefficients ϱ_{1p} and ϱ_{1m} (see Chapter 10), respectively.

6.3.1 Thermal stresses in cell matrix

A. Integration constants $C_{1m} \neq 0$, $C_{2m} \neq 0$, $C_{3m} = 0$

With regard to the boundary conditions (4.9), (4.11) and the analysis presented in Section 6.2.3, considering the terms u_{1rm} and u_{2rm} (see Eqs. (6.6), (6.7)) to represent functions of the variable $r \in \langle R_1, r_c\rangle$, the radial displacement u_{1m} of the infinitesimal spherical cap in the cell matrix, regarding Eqs. (3.31), (6.4), has the form

$$u_{1m} = -p_1 \left[\kappa_{1m}^{(1)} r \left(\frac{1}{3} - \ln r\right) + \kappa_{3m}^{(1)} C_m^{(1)} r^{c_{3m}}\right], \tag{6.14}$$

where the coefficients $c_{3m} < 0$, $C_m^{(1)}$ and the radial stress p_1 acting on the particle-matrix boundary are given by Eqs. (3.40), (6.5) and (10.1), respectively. The φ, ν-dependent distance r_c, as a function of the experimentally determined parameters v, R_{1ex}, d_{1ex}, d_{2ex}, d_{3ex} (see Section 2.2.1), included in the coefficients $\kappa_{1m}^{(1)}$, $\kappa_{3m}^{(1)}$ is derived in Section 2.3.

With regard to Section 6.3.1.A, the coefficients $\psi_{11\eta m}^{(1)}$, $\psi_{12\eta m}^{(1)}$ $(\eta = \varphi, \nu)$, ϱ_{1m} and $\omega_{11m}, \ldots, \omega_{16m}$, ξ_{i11m}, ξ_{i12m}, ξ_{i13m} $(i=1,2,3)$ mentioned below (see Eqs. (6.17), (6.18), (6.21)–(6.24), (6.26), (6.27)) depend on $\kappa_{1m}^{(1)}$, $\kappa_{3m}^{(1)}$ and $\kappa_{1m}^{(1)}$, $\kappa_{3m}^{(1)}$, $\psi_{11\eta m}^{(1)}$,

$\psi^{(1)}_{12\eta m}$, respectively, where the coefficients $\kappa^{(1)}_{1m}$, $\kappa^{(1)}_{3m}$ (see Eq. (12.35)) are functions of χ_{11m}, χ_{12m}, χ_{21m}, χ_{22m} given by Eqs. (6.28)–(6.31), respectively.

Consequently, the radial stress σ_{11m}, the tangential and shear stresses σ_{22m}, σ_{33m} and σ_{12m}, σ_{13m} acting in the cell matrix, respectively, regarding Eqs. (3.31)–(3.34), (6.14), are derived as

$$\sigma_{11m} = -p_1 \left\{ -\kappa^{(1)}_{1m} \left[\frac{2}{3} \left(c_{1m} + 2c_{2m} \right) + \left(c_{1m} - c_{2m} \right) \ln r \right] + \kappa^{(1)}_{3m} C^{(1)}_{m} \left[c_{3m} \left(c_{1m} + c_{2m} \right) - 2c_{2m} \right] r^{c_{3m}-1} \right\}, \tag{6.15}$$

$$\sigma_{22m} = \sigma_{33m} = -p_1 \left\{ \kappa^{(1)}_{1m} \left[\frac{1}{3} \left(c_{1m} + 2c_{2m} \right) - \left(c_{1m} - c_{2m} \right) \ln r \right] + \kappa^{(1)}_{3m} C^{(1)}_{m} \left(c_{1m} - c_{2m} c_{3m} \right) r^{c_{3m}-1} \right\}, \tag{6.16}$$

$$\sigma_{12m} = -\frac{1}{s_{44m}} \left[\psi^{(1)}_{11\varphi m} \left(\frac{1}{3} - \ln r \right) + \psi^{(1)}_{12\varphi m} C^{(1)}_{m} r^{c_{3m}-1} \right], \tag{6.17}$$

$$\sigma_{13m} = -\frac{1}{s_{44m}} \left[\psi^{(1)}_{11\nu m} \left(\frac{1}{3} - \ln r \right) + \psi^{(1)}_{12\nu m} C^{(1)}_{m} r^{c_{3m}-1} \right], \tag{6.18}$$

where the elastic modulus s_{44m} and the coefficients c_{1m}, c_{2m}, $\psi^{(1)}_{11\eta m}$, $\psi^{(1)}_{12\eta m}$ $(\eta = \varphi, \nu)$ are given by Eqs. (3.25) and (3.38), (3.39), (12.36), respectively.

Finally, the radial strain ε_{11m}, the tangential and shear strains ε_{22m}, ε_{33m} and ε_{12m}, ε_{13m} in the cell matrix, respectively, regarding Eqs. (3.1)–(3.5), (6.14), have the forms

$$\varepsilon_{11m} = -p_1 \left[-\kappa^{(1)}_{1m} \left(\frac{2}{3} + \ln r \right) + \kappa^{(1)}_{3m} C^{(1)}_{m} c_{3m} r^{c_{3m}-1} \right], \tag{6.19}$$

$$\varepsilon_{22m} = \varepsilon_{33m} = -p_1 \left[\kappa^{(1)}_{1m} \left(\frac{1}{3} - \ln r \right) + \kappa^{(1)}_{3m} C^{(1)}_{m} r^{c_{3m}-1} \right], \tag{6.20}$$

$$\varepsilon_{12m} = -\left[\psi^{(1)}_{11\varphi m} \left(\frac{1}{3} - \ln r \right) + \psi^{(1)}_{12\varphi m} C^{(1)}_{m} r^{c_{3m}-1} \right], \tag{6.21}$$

$$\varepsilon_{13m} = -\left[\psi^{(1)}_{11\nu m} \left(\frac{1}{3} - \ln r \right) + \psi^{(1)}_{12\nu m} C^{(1)}_{m} r^{c_{3m}-1} \right]. \tag{6.22}$$

With regard to the determination of the compressive or tensile radial stress $p_1 > 0$ or $p_1 < 0$ acting on the particle-matrix boundary, thus for $r = R_1$, respectively, considering Eq. (3.58), the tangential strain on the particle-matrix boundary, $(\varepsilon_{22m})_{r=R_1}$, is required to be derived in the form

$$(\varepsilon_{22m})_{r=R_1} = -p_1 \varrho_{1m}, \tag{6.23}$$

where the coefficient ϱ_{1m} is given by Eq. (6.32).

Considering Eqs. (3.41), (3.42), (6.15)–(6.22), the thermal-stress induced elastic energy density w_m and the thermal-stress induced elastic energy W_m, accumulated in an arbitrary point of the cell matrix and in the volume of the cell matrix, respectively, have the forms

$$w_m = \frac{1}{2}\left[\omega_{11m} + \omega_{12m}\, r^{c_{3m}-1} + \omega_{13m}\, r^{2(c_{3m}-1)} + \omega_{14m} \ln r + \omega_{15m} \ln^2 r + \omega_{16m}\, r^{c_{3m}-1} \ln r\right], \tag{6.24}$$

$$W_m = 8 \int_0^{\pi/2}\int_0^{\pi/2}\int_{R_1}^{r_c} w_m\, r^2 dr d\varphi d\nu = \int_0^{\pi/2}\int_0^{\pi/2} \Omega_{11m} d\varphi d\nu, \tag{6.25}$$

where the coefficient Ω_{11m} given by Eq. (6.39) is a function of $\omega_{11m}, \ldots, \omega_{16m}$ given by Eqs. (6.33)–(6.38), respectively.

With regard to the determination of the thermal-stress induced strengthening presented in Volume III, the stress σ_{im} acting in a direction of the axis x_i and the elastic energy density w_{im} induced by the stress σ_{im} ($i = 1,2,3$) (see Eqs. (3.46), (3.47), (6.15)–(6.18)), both related to an arbitrary point of the cell matrix, are derived as

$$\sigma_{im} = -\left(\xi_{i11m} + \xi_{i12m}\, r^{c_{3m}-1} - \xi_{i13m} \ln r\right), \quad i = 1, 2, 3, \tag{6.26}$$

$$w_{im} = \frac{s_{11m}}{2}\left[\xi_{i11m}^2 + \xi_{i12m}^2\, r^{2(c_{3m}-1)} + \xi_{i13m}^2 \ln^2 r + 2\,\xi_{i11m}\,\xi_{i12m}\, r^{c_{3m}-1} - 2\,\xi_{i11m}\,\xi_{i13m} \ln r - 2\,\xi_{i12m}\,\xi_{i13m}\, r^{c_{3m}-1} \ln r\right], \quad i = 1, 2, 3, \tag{6.27}$$

where the elastic modulus s_{11m} and the coefficients ξ_{i11m}, ξ_{i12m}, ξ_{i13m} ($i = 1,2,3$) are given by Eqs. (3.23) and (6.40)–(6.42), respectively.

Appendix. Considering the elastic modulus s_{44m} and the coefficients c_{1m}, c_{2m}, c_{3m}, $C_m^{(1)}$, a'_{1i}, a'_{2i}, a'_{3i} ($i = 1,2,3$), (see Eqs. (3.25), (3.38)–(3.40), (6.5), (12.26)–(12.34)), the coefficients χ_{11m}, χ_{12m}, χ_{21m}, χ_{22m}, ϱ_{1+im}, $\omega_{1+i1m}, \ldots, \omega_{1+i6m}$, $\Omega_{1+i\,1+jm}$[1] ($i, j = 0,1$), $\xi_{i1+j\,1m}$, $\xi_{i1+j\,2m}$, $\xi_{i1+j\,3m}$ ($i = 1,2,3$; $j = 0,1$) have the forms

[1]The coefficient Ω_{21m} is determined but not used in formulae for the elastic energy W_m of the cell matrix.

$$\chi_{11m} = -\left[\frac{2}{3}(c_{1m} + 2c_{2m}) + (c_{1m} - c_{2m}) \ln R_1\right], \tag{6.28}$$

$$\chi_{12m} = C_m^{(1)} \left[c_{3m}(c_{1m} + c_{2m}) - 2c_{2m}\right] R_1^{c_{3m}-1}, \tag{6.29}$$

$$\chi_{21m} = r_c \left(\frac{1}{3} - \ln r_c\right), \tag{6.30}$$

$$\chi_{22m} = C_m^{(1)} r_c^{c_{3m}}, \tag{6.31}$$

$$\varrho_{1+im} = \kappa_{1m}^{(1)} \left(\frac{1}{3} - \ln R_{1+i}\right) + \kappa_{3m}^{(1)} C_m^{(1)} R_{1+i}^{c_{3m}-1}, \quad i = 0, 1, \tag{6.32}$$

$$\omega_{1+i\,1m} = \frac{2}{9}\left(3p_{1+i}^2 \left[\kappa_{1m}^{(1)}\right]^2 (c_{1m} + 2c_{2m}) + \frac{1}{s_{44m}}\left\{\left[\psi_{1+i\,1\varphi m}^{(1)}\right]^2 + \left[\psi_{1+i\,1\nu m}^{(1)}\right]^2\right\}\right), \quad i = 0, 1, \tag{6.33}$$

$$\omega_{1+i\,2m} = \frac{4C_m^{(1)}}{3}\left\{p_{1+i}^2 \kappa_{1m}^{(1)} \kappa_{3m}^{(1)} (c_{1m} + 2c_{2m})(1 - c_{3m}) + \frac{1}{s_{44m}}\left[\psi_{1+i\,1\varphi m}^{(1)} \psi_{1+i\,2\varphi m}^{(1)} + \psi_{1+i\,1\nu m}^{(1)} \psi_{1+i\,2\nu m}^{(1)}\right]\right\}, \quad i = 0, 1, \tag{6.34}$$

$$\omega_{1+i\,3m} = \left[C_m^{(1)}\right]^2 \left(p_{1+i}^2 \left[\kappa_{3m}^{(1)}\right]^2 \left[c_{3m}^2 (c_{1m} + c_{2m}) + 2(c_{1m} - 2c_{2m}c_{3m})\right] + \frac{2}{s_{44m}}\left\{\left[\psi_{1+i\,2\varphi m}^{(1)}\right]^2 + \left[\psi_{1+i\,2\nu m}^{(1)}\right]^2\right\}\right), \quad i = 0, 1, \tag{6.35}$$

$$\omega_{1+i\,4m} = -\frac{4}{3s_{44m}}\left\{\left[\psi_{1+i\,1\varphi m}^{(1)}\right]^2 + \left[\psi_{1+i\,1\nu m}^{(1)}\right]^2\right\}, \quad i = 0, 1, \tag{6.36}$$

$$\omega_{1+i\,5m} = 3p_{1+i}^2 \left[\kappa_{1m}^{(1)}\right]^2 (c_{1m} - c_{2m}) + \frac{2}{s_{44m}}\left\{\left[\psi_{1+i\,1\varphi m}^{(1)}\right]^2 + \left[\psi_{1+i\,1\nu m}^{(1)}\right]^2\right\}, \quad i = 0, 1, \tag{6.37}$$

$$\omega_{1+i\,6m} = -2C_m^{(1)}\left\{p_{1+i}^2 \kappa_{1m}^{(1)} \kappa_{3m}^{(1)} (c_{1m} - c_{2m})(c_{3m} + 2) + \frac{2}{s_{44m}}\left[\psi_{1+i\,1\varphi m}^{(1)} \psi_{1+i\,2\varphi m}^{(1)} + \psi_{1+i\,1\nu m}^{(1)} \psi_{1+i\,2\nu m}^{(1)}\right]\right\}, \quad i = 0, 1, \tag{6.38}$$

$$\Omega_{1+i\,1+jm} = 4\left(\frac{\omega_{1+i\,1m}}{3}\left(r_c^3 - R_{1+j}^3\right)\right.$$
$$+ \frac{\omega_{1+i\,2m}}{c_{3m}+2}\left(r_c^{c_{3m}+2} - R_{1+j}^{c_{3m}+2}\right) + \frac{\omega_{1+i\,3m}}{2c_{3m}+1}\left(r_c^{2c_{3m}+1} - R_{1+j}^{2c_{3m}+1}\right)$$
$$+ \frac{\omega_{1+i\,4m}}{3}\left[r_c^3\left(\ln r_c - \frac{1}{3}\right) - R_{1+j}^3\left(\ln R_{1+j} - \frac{1}{3}\right)\right]$$
$$+ \frac{\omega_{1+i\,5m}}{3}\left\{r_c^3\left[\left(\ln r_c - \frac{1}{3}\right)^2 + \frac{1}{9}\right] - R_{1+j}^3\left[\left(\ln R_{1+j} - \frac{1}{3}\right)^2 + \frac{1}{9}\right]\right\}$$
$$\left.+ \frac{\omega_{1+i\,6m}}{c_{3m}+2}\left[r_c^{c_{3m}+2}\left(\ln r_c - \frac{1}{c_{3m}+2}\right) - R_{1+j}^{c_{3m}+2}\left(\ln R_{1+j} - \frac{1}{c_{3m}+2}\right)\right]\right),$$
$$i, j = 0, 1, \tag{6.39}$$

$$\xi_{i\,1+j\;1m} = \frac{1}{3}\left\{p_{1+j}\kappa_{1m}^{(1)}\left(a'_{2i} + a'_{3i} - 2a'_{1i}\right)\left(c_{1m} + 2c_{2m}\right)\right.$$
$$\left.+ \frac{1}{s_{44m}}\left[\left(a'_{1i} + a'_{2i}\right)\psi_{1+j\,1\varphi m}^{(1)} + \left(a'_{1i} + a'_{3i}\right)\psi_{1+j\,1\nu m}^{(1)}\right]\right\},$$
$$i = 1, 2, 3; \quad j = 0, 1, \tag{6.40}$$

$$\xi_{i\,1+j\;2m} = C_m^{(1)}\left(p_{1+j}\kappa_{3m}^{(1)}\left\{a'_{1i}\left[c_{3m}\left(c_{1m} + c_{2m}\right) - 2c_{2m}\right] + \left(a'_{2i} + a'_{3i}\right)\left(c_{1m} - c_{2m}c_{3m}\right)\right\}\right.$$
$$\left.+ \frac{1}{s_{44m}}\left[\left(a'_{1i} + a'_{2i}\right)\psi_{1+j\,2\varphi m}^{(1)} + \left(a'_{1i} + a'_{3i}\right)\psi_{1+j\,2\nu m}^{(1)}\right]\right),$$
$$i = 1, 2, 3; \quad j = 0, 1, \tag{6.41}$$

$$\xi_{i\,1+j\;3m} = p_{1+j}\kappa_{1m}^{(1)}\left(a'_{1i} + a'_{2i} + a'_{3i}\right)\left(c_{1m} - c_{2m}\right)$$
$$+ \frac{1}{s_{44m}}\left[\left(a'_{1i} + a'_{2i}\right)\psi_{1+j\,1\varphi m}^{(1)} + \left(a'_{1i} + a'_{3i}\right)\psi_{1+j\,1\nu m}^{(1)}\right],$$
$$i = 1, 2, 3; \quad j = 0, 1. \tag{6.42}$$

B. Integration constants $C_{1m} \neq 0$, $C_{2m} = 0$, $C_{3m} \neq 0$

With regard to the boundary conditions (4.9), (4.11) and the analysis presented in Section 6.2.3, considering the terms u_{1rm} and u_{3rm} (see Eqs. (6.6), (6.8)), the former as a function of the variable $r \in \langle R_1, r_c\rangle$ and the latter as a constant function regarding the variable r, respectively, the radial displacement u_{1m} of the infinitesimal spherical cap in the cell matrix, regarding Eqs. (3.31), (6.4), has the form

$$u_{1m} = -p_1\left[\kappa_{1m}^{(1)}r\left(\frac{1}{3} - \ln r\right) + \frac{3\kappa_{3m}^{(1)}}{2}\right], \tag{6.43}$$

where the radial stress p_1 acting on the particle-matrix boundary is given by Eq. (10.1). The φ,ν-dependent distance r_c, as a function of the experimentally determined parameters v, R_{1ex}, d_{1ex}, d_{2ex}, d_{3ex} (see Section 2.2.1), included in the coefficients $\kappa_{1m}^{(1)}$, $\kappa_{3m}^{(1)}$ is derived in Section 2.3.

With regard to Section 6.3.1.B, the coefficients $\psi_{11\eta m}^{(1)}$, $\psi_{12\eta m}^{(1)}$ $(\eta = \varphi, \nu)$, ϱ_{1m} and $\omega_{11m}, \ldots, \omega_{16m}$, ξ_{i11m}, ξ_{i12m}, ξ_{i13m} $(i = 1,2,3)$ mentioned below (see Eqs. (6.46), (6.47), (6.50)–(6.53), (6.55), (6.56)) depend on $\kappa_{1m}^{(1)}$, $\kappa_{3m}^{(1)}$ and $\kappa_{1m}^{(1)}$, $\kappa_{3m}^{(1)}$, $\psi_{11\eta m}^{(1)}$, $\psi_{12\eta m}^{(1)}$, respectively, where the coefficients $\kappa_{1m}^{(1)}$, $\kappa_{3m}^{(1)}$ (see Eq. (12.35)) are functions of χ_{11m}, χ_{12m}, χ_{21m}, χ_{22m} given by Eqs. (6.28), (6.57), (6.30), (6.58), respectively.

Consequently, the radial stress σ_{11m}, the tangential and shear stresses σ_{22m}, σ_{33m} and σ_{12m}, σ_{13m} acting in the cell matrix, respectively, regarding Eqs. (3.31)–(3.34), (6.43), are derived as

$$\sigma_{11m} = -p_1 \left\{ -\kappa_{1m}^{(1)} \left[\frac{2}{3} \left(c_{1m} + 2c_{2m} \right) + \left(c_{1m} - c_{2m} \right) \ln r \right] - \frac{3\kappa_{3m}^{(1)} c_{2m}}{r} \right\}, \tag{6.44}$$

$$\sigma_{22m} = \sigma_{33m} = -p_1 \left\{ \kappa_{1m}^{(1)} \left[\frac{1}{3} \left(c_{1m} + 2c_{2m} \right) - \left(c_{1m} - c_{2m} \right) \ln r \right] + \frac{3\kappa_{3m}^{(1)} c_{1m}}{2r} \right\}, \tag{6.45}$$

$$\sigma_{12m} = -\frac{1}{s_{44m}} \left[\psi_{11\varphi m}^{(1)} \left(\frac{1}{3} - \ln r \right) + \frac{3\psi_{12\varphi m}^{(1)}}{2r} \right], \tag{6.46}$$

$$\sigma_{13m} = -\frac{1}{s_{44m}} \left[\psi_{11\nu m}^{(1)} \left(\frac{1}{3} - \ln r \right) + \frac{3\psi_{12\nu m}^{(1)}}{2r} \right], \tag{6.47}$$

where the elastic modulus s_{44m} and the coefficients c_{1m}, c_{2m}, $\psi_{11\eta m}^{(1)}$, $\psi_{12\eta m}^{(1)}$ $(\eta = \varphi, \nu)$ are given by Eqs. (3.25) and (3.38), (3.39), (12.36), respectively.

Finally, the radial strain ε_{11m}, the tangential and shear strains ε_{22m}, ε_{33m} and ε_{12m}, ε_{13m} in the cell matrix, respectively, regarding Eqs. (3.1)–(3.5), (6.43), have the forms[2]

$$\varepsilon_{11m} = p_1 \kappa_{1m}^{(1)} \left(\frac{2}{3} + \ln r \right), \tag{6.48}$$

$$\varepsilon_{22m} = \varepsilon_{33m} = -p_1 \left[\kappa_{1m}^{(1)} \left(\frac{1}{3} - \ln r \right) + \frac{3\kappa_{3m}^{(1)}}{2r} \right], \tag{6.49}$$

$$\varepsilon_{12m} = -\left[\psi_{11\varphi m}^{(1)} \left(\frac{1}{3} - \ln r \right) + \frac{3\psi_{12\varphi m}^{(1)}}{2r} \right], \tag{6.50}$$

[2]To avoid misunderstanding, a sign on the right side in Eq. (6.48) is really plus, in contrast to minus sign on the right sides in Eqs. (6.49)–(6.51).

$$\varepsilon_{13m} = -\left[\psi_{11\nu m}^{(1)}\left(\frac{1}{3} - \ln r\right) + \frac{3\psi_{12\nu m}^{(1)}}{2r}\right]. \tag{6.51}$$

With regard to the determination of the compressive or tensile radial stress $p_1 > 0$ or $p_1 < 0$ acting on the particle-matrix boundary, thus for $r = R_1$, respectively, considering Eq. (3.58), the tangential strain on the particle-matrix boundary, $(\varepsilon_{22m})_{r=R_1}$, is required to be derived in the form

$$(\varepsilon_{22m})_{r=R_1} = -p_1\,\varrho_{1m}, \tag{6.52}$$

where the coefficient ϱ_{1m} is given by Eq. (6.59).

Considering Eqs. (3.41), (3.42), (6.44)–(6.51), the thermal-stress induced elastic energy density w_m and the thermal-stress induced elastic energy W_m, accumulated in an arbitrary point of the cell matrix and in the volume of the cell matrix, respectively, have the forms

$$w_m = \frac{1}{2}\left(\omega_{11m} + \frac{\omega_{12m}}{r} + \frac{\omega_{13m}}{r^2} + \omega_{14m}\ln r + \omega_{15m}\ln^2 r + \omega_{16m}\frac{\ln r}{r}\right), \tag{6.53}$$

$$W_m = 8\int_0^{\pi/2}\int_0^{\pi/2}\int_{R_1}^{r_c} w_m\, r^2 dr d\varphi d\nu = \int_0^{\pi/2}\int_0^{\pi/2} \Omega_{11m} d\varphi d\nu, \tag{6.54}$$

where the coefficient Ω_{11m} given by Eq. (6.63) is a function of $\omega_{11m}, \ldots, \omega_{16m}$ given by Eqs. (6.33), (6.60), (6.61), (6.36), (6.37), (6.62), respectively.

With regard to the determination of the thermal-stress induced strengthening presented in Volume III, the stress σ_{im} acting in a direction of the axis x_i and the elastic energy density w_{im} induced by the stress σ_{im} ($i=1,2,3$) (see Eqs. (3.46), (3.47), (6.44)–(6.47)), both related to an arbitrary point of the cell matrix, are derived as

$$\sigma_{im} = -\left(\xi_{i11m} + \frac{\xi_{i12m}}{r} - \xi_{i13m}\ln r\right), \quad i = 1, 2, 3, \tag{6.55}$$

$$w_{im} = \frac{s_{11m}}{2}\left(\xi_{i11m}^2 + \frac{\xi_{i12m}^2}{r^2} + \xi_{i13m}^2\ln^2 r + \frac{2\,\xi_{i11m}\,\xi_{i12m}}{r} - 2\,\xi_{i11m}\,\xi_{i13m}\ln r - 2\,\xi_{i12m}\,\xi_{i13m}\frac{\ln r}{r}\right), \quad i = 1, 2, 3, \tag{6.56}$$

where the elastic modulus s_{11m} and the coefficients ξ_{i11m}, ξ_{i12m}, ξ_{i13m} ($i=1,2,3$) are given by Eqs. (3.23) and (6.40), (6.64), (6.42), respectively.

Appendix. Considering the elastic modulus s_{44m} and the coefficients c_{1m}, c_{2m}, a'_{1i}, a'_{2i}, a'_{3i} ($i=1,2,3$) (see Eqs. (3.25), (3.38), (3.39), (12.26)–(12.34)), the coefficients χ_{12m}, χ_{22m}, ϱ_{1+im}, $\omega_{1+i\,2m}$, $\omega_{1+i\,3m}$, $\omega_{1+i\,6m}$, $\Omega_{1+i\;1+jm}$[3] ($i,j=0,1$), $\xi_{i\,1+j\;2m}$ ($i=1,2,3$; $j=0,1$) have the forms

$$\chi_{12m} = -\frac{3c_{2m}}{R_1}, \tag{6.57}$$

$$\chi_{22m} = \frac{3}{2}, \tag{6.58}$$

$$\varrho_{1+im} = \kappa_{1m}^{(1)}\left(\frac{1}{3} - \ln R_{1+i}\right) + \frac{3\kappa_{3m}^{(1)}}{2R_{1+i}}, \quad i=0,1, \tag{6.59}$$

$$\begin{aligned}\omega_{1+i\,2m} = 2\Big\{&p_{1+i}^2\kappa_{1m}^{(1)}\kappa_{3m}^{(1)}\left(c_{1m}+2c_{2m}\right)\\ &+\frac{1}{s_{44m}}\left[\psi_{1+i\,1\varphi m}^{(1)}\psi_{1+i\,2\varphi m}^{(1)} + \psi_{1+i\,1\nu m}^{(1)}\psi_{1+i\,2\nu m}^{(1)}\right]\Big\}, \quad i=0,1,\end{aligned} \tag{6.60}$$

$$\begin{aligned}\omega_{1+i\,3m} = &\frac{9}{2}\left(p_{1+i}^2\left[\kappa_{3m}^{(1)}\right]^2 c_{1m} + \frac{1}{s_{44m}}\left\{\left[\psi_{1+i\,2\varphi m}^{(1)}\right]^2 + \left[\psi_{1+i\,2\nu m}^{(1)}\right]^2\right\}\right),\\ &i=0,1,\end{aligned} \tag{6.61}$$

$$\begin{aligned}\omega_{1+i\,6m} = -6\Big\{&p_{1+i}^2\kappa_{1m}^{(1)}\kappa_{3m}^{(1)}\left(c_{1m}-c_{2m}\right)\\ &+\frac{1}{s_{44m}}\left[\psi_{1+i\,1\varphi m}^{(1)}\psi_{1+i\,2\varphi m}^{(1)} + \psi_{1+i\,1\nu m}^{(1)}\psi_{1+i\,2\nu m}^{(1)}\right]\Big\}, \quad i=0,1,\end{aligned} \tag{6.62}$$

$$\begin{aligned}\Omega_{1+i\;1+jm} = 4\Bigg(&\frac{\omega_{1+i\,1m}}{3}\left(r_c^3 - R_{1+j}^3\right) + \frac{\omega_{1+i\,2m}}{2}\left(r_c^2 - R_{1+j}^2\right)\\ &+\omega_{1+i\,3m}\left(r_c - R_{1+j}\right) + \frac{\omega_{1+i\,4m}}{3}\left[r_c^3\left(\ln r_c - \frac{1}{3}\right) - R_{1+j}^3\left(\ln R_{1+j} - \frac{1}{3}\right)\right]\\ &+\frac{\omega_{1+i\,5m}}{3}\left\{r_c^3\left[\left(\ln r_c - \frac{1}{3}\right)^2 + \frac{1}{9}\right] - R_{1+j}^3\left[\left(\ln R_{1+j} - \frac{1}{3}\right)^2 + \frac{1}{9}\right]\right\}\\ &+\frac{\omega_{1+i\,6m}}{2}\left[r_c^2\left(\ln r_c - \frac{1}{2}\right) - R_{1+j}^2\left(\ln R_{1+j} - \frac{1}{2}\right)\right]\Bigg), \quad i,j=0,1,\end{aligned} \tag{6.63}$$

$$\begin{aligned}\xi_{i\,1+j\;2m} = &\frac{3}{2}\Big\{p_{1+j}\kappa_{3m}^{(1)}\left[c_{1m}\left(a'_{2i}+a'_{3i}\right) - 2a'_{1i}\,c_{2m}\right]\\ &+\frac{1}{s_{44m}}\left[\left(a'_{1i}+a'_{2i}\right)\psi_{1+j\,2\varphi m}^{(1)} + \left(a'_{1i}+a'_{3i}\right)\psi_{1+j\,2\nu m}^{(1)}\right]\Big\},\\ &i=1,2,3; \quad j=0,1.\end{aligned} \tag{6.64}$$

[3]The coefficient Ω_{21m} is determined but not used in formulae for the elastic energy W_m of the cell matrix.

C. Integration constants $C_{1m} = 0$, $C_{2m} \neq 0$, $C_{3m} \neq 0$

With regard to the boundary conditions (4.9), (4.11) and the analysis presented in Section 6.2.3, considering the terms u_{2rm} and u_{3rm} (see Eqs. (6.7), (6.8)), the former as a function of the variable $r \in \langle R_1, r_c \rangle$ and the latter as a constant function regarding the variable r, respectively, the radial displacement u_{1m} of the infinitesimal spherical cap in the cell matrix, regarding Eqs. (3.31), (6.4), has the form

$$u_{1m} = -p_1 \left[\kappa_{1m}^{(1)} C_m^{(1)} r^{c_{3m}} + \frac{3\kappa_{3m}^{(1)}}{2} \right], \tag{6.65}$$

where the coefficients $c_{3m} < 0$, $C_m^{(1)}$ and the radial stress p_1 acting on the particle-matrix boundary are given by Eqs. (3.40), (6.5) and (10.1), respectively. The φ, ν-dependent distance r_c, as a function of the experimentally determined parameters v, R_{1ex}, d_{1ex}, d_{2ex}, d_{3ex} (see Section 2.2.1), included in the coefficients $\kappa_{1m}^{(1)}$, $\kappa_{3m}^{(1)}$ is derived in Section 2.3.

With regard to Section 6.3.1.C, the coefficients $\psi_{11\eta m}^{(1)}$, $\psi_{12\eta m}^{(1)}$ $(\eta = \varphi, \nu)$, ϱ_{1m} and ω_{11m}, ω_{12m}, ω_{13m}, ξ_{i11m}, ξ_{i12m} $(i = 1,2,3)$ mentioned below (see Eqs. (6.68), (6.69), (6.72)–(6.75), (6.77), (6.78)) depend on $\kappa_{1m}^{(1)}$, $\kappa_{3m}^{(1)}$ and $\kappa_{1m}^{(1)}$, $\kappa_{3m}^{(1)}$, $\psi_{11\eta m}^{(1)}$, $\psi_{12\eta m}^{(1)}$, respectively, where the coefficients $\kappa_{1m}^{(1)}$, $\kappa_{3m}^{(1)}$ (see Eq. (12.35)) are functions of χ_{11m}, χ_{12m}, χ_{21m}, χ_{22m} given by Eqs. (6.79), (6.57), (6.80), (6.58), respectively.

Consequently, the radial stress σ_{11m}, the tangential and shear stresses σ_{22m}, σ_{33m} and σ_{12m}, σ_{13m} acting in the cell matrix, respectively, regarding Eqs. (3.31)–(3.34), (6.65), are derived as

$$\sigma_{11m} = -p_1 \left\{ \kappa_{1m}^{(1)} C_m^{(1)} \left[c_{3m} \left(c_{1m} + c_{2m} \right) - 2c_{2m} \right] r^{c_{3m}-1} - \frac{3\kappa_{3m}^{(1)} c_{2m}}{r} \right\}, \tag{6.66}$$

$$\sigma_{22m} = \sigma_{33m} = -p_1 \left[\kappa_{1m}^{(1)} C_m^{(1)} \left(c_{1m} - c_{2m} c_{3m} \right) r^{c_{3m}-1} + \frac{3\kappa_{3m}^{(1)} c_{1m}}{2r} \right], \tag{6.67}$$

$$\sigma_{12m} = -\frac{1}{s_{44m}} \left[\psi_{11\varphi m}^{(1)} C_m^{(1)} r^{c_{3m}-1} + \frac{3\psi_{12\varphi m}^{(1)}}{2r} \right], \tag{6.68}$$

$$\sigma_{13m} = -\frac{1}{s_{44m}} \left[\psi_{11\nu m}^{(1)} C_m^{(1)} r^{c_{3m}-1} + \frac{3\psi_{12\nu m}^{(1)}}{2r} \right], \tag{6.69}$$

where the elastic modulus s_{44m} and the coefficients c_{1m}, c_{2m}, $\psi_{11\eta m}^{(1)}$, $\psi_{12\eta m}^{(1)}$ $(\eta = \varphi, \nu)$ are given by Eqs. (3.25) and (3.38), (3.39), (12.36), respectively.

Finally, the radial strain ε_{11m}, the tangential and shear strains ε_{22m}, ε_{33m} and ε_{12m}, ε_{13m} in the cell matrix, respectively, regarding Eqs. (3.1)–(3.5), (6.65), have the forms

$$\varepsilon_{11m} = -p_1 \kappa_{1m}^{(1)} C_m^{(1)} c_{3m}\, r^{c_{3m}-1}, \tag{6.70}$$

$$\varepsilon_{22m} = \varepsilon_{33m} = -p_1 \left[\kappa_{1m}^{(1)} C_m^{(1)}\, r^{c_{3m}-1} + \frac{3\kappa_{3m}^{(1)}}{2r} \right], \tag{6.71}$$

$$\varepsilon_{12m} = -\left[\psi_{11\varphi m}^{(1)} C_m^{(1)}\, r^{c_{3m}-1} + \frac{3\psi_{12\varphi m}^{(1)}}{2r} \right], \tag{6.72}$$

$$\varepsilon_{13m} = -\left[\psi_{11\nu m}^{(1)} C_m^{(1)}\, r^{c_{3m}-1} + \frac{3\psi_{12\nu m}^{(1)}}{2r} \right]. \tag{6.73}$$

With regard to the determination of the compressive or tensile radial stress $p_1 > 0$ or $p_1 < 0$ acting on the particle-matrix boundary, thus for $r = R_1$, respectively, considering Eq. (3.58), the tangential strain on the particle-matrix boundary, $(\varepsilon_{22m})_{r=R_1}$, is required to be derived in the form

$$(\varepsilon_{22m})_{r=R_1} = -p_1\, \varrho_{1m}, \tag{6.74}$$

where the coefficient ϱ_{1m} is given by Eq. (6.81).

Considering Eqs. (3.41), (3.42), (6.66)–(6.73), the thermal-stress induced elastic energy density w_m and the thermal-stress induced elastic energy W_m, accumulated in an arbitrary point of the cell matrix and in the volume of the cell matrix, respectively, have the forms

$$w_m = \frac{1}{2} \left[\frac{\omega_{11m}}{r^2} + \omega_{12m}\, r^{c_{3m}-2} + \omega_{13m}\, r^{2(c_{3m}-1)} \right], \tag{6.75}$$

$$W_m = 8 \int_0^{\pi/2} \int_0^{\pi/2} \int_{R_1}^{r_c} w_m\, r^2 dr d\varphi d\nu = \int_0^{\pi/2} \int_0^{\pi/2} \Omega_{11m} d\varphi d\nu, \tag{6.76}$$

where the coefficient Ω_{11m} given by Eq. (6.85) is a function of ω_{11m}, ω_{12m}, ω_{13m} given by Eqs. (6.82)–(6.84), respectively.

With regard to the determination of the thermal-stress induced strengthening presented in Volume III, the stress σ_{im} acting in a direction of the axis x_i and the elastic energy density w_{im} induced by the stress σ_{im} ($i=1,2,3$) (see Eqs. (3.46), (3.47), (6.66)–(6.69)), both related to an arbitrary point of the cell matrix, are derived as

$$\sigma_{im} = -\left(\frac{\xi_{i\,11m}}{r} + \xi_{i\,12m}\, r^{c_{3m}-1}\right), \quad i = 1, 2, 3, \tag{6.77}$$

$$w_{im} = \frac{s_{11m}}{2}\left[\frac{\xi^2_{i\,11m}}{r^2} + \xi^2_{i\,12m}\, r^{2(c_{3m}-1)} + 2\,\xi_{i\,11m}\,\xi_{i\,12m}\, r^{c_{3m}-2}\right], \quad i = 1, 2, 3, \tag{6.78}$$

where the elastic modulus s_{11m} and the coefficients $\xi_{i\,11m}$, $\xi_{i\,12m}$ $(i=1,2,3)$ are given by Eqs. (3.23) and (6.86), (6.87), respectively.

Appendix. Considering the elastic modulus s_{44m} and the coefficients c_{1m}, c_{2m}, c_{3m}, $C^{(1)}_m$, a'_{1i}, a'_{2i}, a'_{3i} $(i=1,2,3)$ (see Eqs. (3.25), (3.38)–(3.40), (6.5), (12.26)–(12.34)), the coefficients χ_{11m}, χ_{21m}, ϱ_{1+im}, $\omega_{1+i\,1m}$, $\omega_{1+i\,2m}$, $\omega_{1+i\,3m}$, $\Omega_{1+i\;1+jm}$ [4] $(i,j=0,1)$, $\xi_{i\,1+j\;1m}$, $\xi_{i\,1+j\;2m}$ $(i=1,2,3;\ j=0,1)$ have the forms

$$\chi_{11m} = C^{(1)}_m\left[c_{3m}\left(c_{1m}+c_{2m}\right) - 2c_{2m}\right] R_1^{c_{3m}-1}, \tag{6.79}$$

$$\chi_{21m} = C^{(1)}_m r_c^{c_{3m}}, \tag{6.80}$$

$$\varrho_{1+im} = \kappa^{(1)}_{1m} C^{(1)}_m R_{1+i}^{c_{3m}-1} + \frac{3\kappa^{(1)}_{3m}}{2R_{1+i}}, \quad i = 0, 1, \tag{6.81}$$

$$\omega_{1+i\,1m} = \frac{9}{2}\left(p^2_{1+i}\left[\kappa^{(1)}_{3m}\right]^2 c_{1m} + \frac{1}{s_{44m}}\left\{\left[\psi^{(1)}_{1+i\,2\varphi m}\right]^2 + \left[\psi^{(1)}_{1+i\,2\nu m}\right]^2\right\}\right),$$
$$i = 0, 1, \tag{6.82}$$

$$\omega_{1+i\,2m} = 6C^{(1)}_m\left\{p^2_{1+i}\kappa^{(1)}_{1m}\kappa^{(1)}_{3m}\left(c_{1m} - c_{2m}c_{3m}\right)\right.$$
$$\left. + \frac{1}{s_{44m}}\left[\psi^{(1)}_{1+i\,1\varphi m}\psi^{(1)}_{1+i\,2\varphi m} + \psi^{(1)}_{1+i\,1\nu m}\psi^{(1)}_{1+i\,2\nu m}\right]\right\}, \quad i = 0, 1, \tag{6.83}$$

$$\omega_{1+i\,3m} = \left[C^{(1)}_m\right]^2\left(p^2_{1+i}\left[\kappa^{(1)}_{1m}\right]^2\left[c^2_{3m}\left(c_{1m}+c_{2m}\right) + 2\left(c_{1m} - 2c_{2m}c_{3m}\right)\right]\right.$$
$$\left. + \frac{2}{s_{44m}}\left\{\left[\psi^{(1)}_{1+i\,1\varphi m}\right]^2 + \left[\psi^{(1)}_{1+i\,1\nu m}\right]^2\right\}\right), \quad i = 0, 1, \tag{6.84}$$

[4] The coefficient Ω_{21m} is determined but not used in formulae for the elastic energy W_m of the cell matrix.

$$\Omega_{1+i\;1+jm} = 4\left[\omega_{1+i\,1m}\left(r_c - R_{1+j}\right) + \frac{\omega_{1+i\,2m}}{c_{3m}+1}\left(r_c^{c_{3m}+1} - R_{1+j}^{c_{3m}+1}\right) + \frac{\omega_{1+i\,3m}}{2c_{3m}+1}\left(r_c^{2c_{3m}+1} - R_{1+j}^{2c_{3m}+1}\right)\right], \quad i,j = 0,1, \tag{6.85}$$

$$\xi_{i\,1+j\;1m} = \frac{3}{2}\left\{p_{1+j}\kappa_{3m}^{(1)}\left[c_{1m}\left(a'_{2i} + a'_{3i}\right) - 2a'_{1i}\,c_{2m}\right] + \frac{1}{s_{44m}}\left[\left(a'_{1i} + a'_{2i}\right)\psi_{1+j\,2\varphi m}^{(1)} + \left(a'_{1i} + a'_{3i}\right)\psi_{1+j\,2\nu m}^{(1)}\right]\right\}, \quad i = 1,2,3; \quad j = 0,1, \tag{6.86}$$

$$\xi_{i\,1+j\;2m} = C_m^{(1)}\left(p_{1+j}\kappa_{1m}^{(1)}\left\{a'_{1i}\left[c_{3m}\left(c_{1m} + c_{2m}\right) - 2c_{2m}\right] + \left(a'_{2i} + a'_{3i}\right)\left(c_{1m} - c_{2m}c_{3m}\right)\right\} + \frac{1}{s_{44m}}\left[\left(a'_{1i} + a'_{2i}\right)\psi_{1+j\,1\varphi m}^{(1)} + \left(a'_{1i} + a'_{3i}\right)\psi_{1+j\,1\nu m}^{(1)}\right]\right), \quad i = 1,2,3; \quad j = 0,1. \tag{6.87}$$

D. Integration constants $C_{1m} \neq 0$, $C_{2m} \neq 0$, $C_{3m} \neq 0$

With regard to the boundary conditions (4.9), (4.11), (4.14) and the analysis presented in Section 6.2.3, considering the terms u_{1rm}, u_{2rm} and u_{3rm} (see Eqs. (6.6)–(6.8)) to represent functions of the variable $r \in \langle R_1, r_c\rangle$ and a constant function regarding the variable r, respectively, the radial displacement u_{1m} of the infinitesimal spherical cap in the cell matrix, regarding Eqs. (3.31), (6.4), has the form

$$u_{1m} = -p_1\left[\kappa_{4m}^{(2)} r\left(\frac{1}{3} - \ln r\right) + \kappa_{5m}^{(2)} C_m^{(1)}\, r^{c_{3m}} + \frac{3\kappa_{6m}^{(2)}}{2}\right], \tag{6.88}$$

where the coefficients $c_{3m} < 0$, $C_m^{(1)}$ and the radial stress p_1 acting on the particle-matrix boundary are given by Eqs. (3.40), (6.5) and (10.1), respectively. The φ, ν-dependent distance r_c, as a function of the experimentally determined parameters v, R_{1ex}, d_{1ex}, d_{2ex}, d_{3ex} (see Section 2.2.1), included in the coefficients $\kappa_{4m}^{(2)}$, $\kappa_{5m}^{(2)}$, $\kappa_{6m}^{(2)}$ is derived in Section 2.3.

With regard to Section 6.3.1.D, the coefficients $\psi_{11\eta m}^{(2)}$, $\psi_{12\eta m}^{(2)}$, $\psi_{13\eta m}^{(2)}$ $(\eta = \varphi, \nu)$, ϱ_{1m} and $\omega_{11m}, \ldots, \omega_{110m}$, $\xi_{i11m}, \ldots, \xi_{i14m}$ $(i = 1,2,3)$ mentioned below (see Eqs. (6.91), (6.92), (6.95)–(6.98), (6.100), (6.101)) depend on $\kappa_{4m}^{(2)}$, $\kappa_{5m}^{(2)}$, $\kappa_{6m}^{(2)}$ and $\kappa_{4m}^{(2)}$, $\kappa_{5m}^{(2)}$, $\kappa_{6m}^{(2)}$, $\psi_{11\eta m}^{(2)}$, $\psi_{12\eta m}^{(2)}$, $\psi_{13\eta m}^{(2)}$, respectively, where the coefficients $\kappa_{4m}^{(2)}$, $\kappa_{5m}^{(2)}$, $\kappa_{6m}^{(2)}$

(see Eq. (12.39)) are functions of χ_{11m}, χ_{12m}, χ_{13m}, χ_{21m}, χ_{22m}, χ_{23m}, χ_{31m}, χ_{32m}, χ_{33m} given by Eqs. (6.28), (6.29), (6.102), (6.30), (6.31), (6.103)–(6.106), respectively.

Consequently, the radial stress σ_{11m}, the tangential and shear stresses σ_{22m}, σ_{33m} and σ_{12m}, σ_{13m} acting in the cell matrix, respectively, regarding Eqs. (3.31)–(3.34), (6.88), are derived as

$$\sigma_{11m} = -p_1 \left\{ -\kappa_{4m}^{(2)} \left[\frac{2}{3} \left(c_{1m} + 2c_{2m} \right) + \left(c_{1m} - c_{2m} \right) \ln r \right] + \kappa_{5m}^{(2)} C_m^{(1)} \left[c_{3m} \left(c_{1m} + c_{2m} \right) - 2c_{2m} \right] r^{c_{3m}-1} - \frac{3\kappa_{6m}^{(2)} c_{2m}}{r} \right\}, \quad (6.89)$$

$$\sigma_{22m} = \sigma_{33m} = -p_1 \left\{ \kappa_{4m}^{(2)} \left[\frac{1}{3} \left(c_{1m} + 2c_{2m} \right) - \left(c_{1m} - c_{2m} \right) \ln r \right] + \kappa_{5m}^{(2)} C_m^{(1)} \left(c_{1m} - c_{2m} c_{3m} \right) r^{c_{3m}-1} + \frac{3\kappa_{6m}^{(2)} c_{1m}}{2r} \right\}, \quad (6.90)$$

$$\sigma_{12m} = -\frac{1}{s_{44m}} \left[\psi_{11\varphi m}^{(2)} \left(\frac{1}{3} - \ln r \right) + \psi_{12\varphi m}^{(2)} C_m^{(1)} r^{c_{3m}-1} + \frac{3\psi_{13\varphi m}^{(2)}}{2r} \right], \quad (6.91)$$

$$\sigma_{13m} = -\frac{1}{s_{44m}} \left[\psi_{11\nu m}^{(2)} \left(\frac{1}{3} - \ln r \right) + \psi_{12\nu m}^{(2)} C_m^{(1)} r^{c_{3m}-1} + \frac{3\psi_{13\nu m}^{(2)}}{2r} \right], \quad (6.92)$$

where the elastic modulus s_{44m} and the coefficients c_{1m}, c_{2m}, $\psi_{11\eta m}^{(2)}$, $\psi_{12\eta m}^{(2)}$, $\psi_{13\eta m}^{(2)}$ $(\eta = \varphi, \nu)$ are given by Eqs. (3.25) and (3.38), (3.39), (12.40), respectively.

Finally, the radial strain ε_{11m}, the tangential and shear strains ε_{22m}, ε_{33m} and ε_{12m}, ε_{13m} in the cell matrix, respectively, regarding Eqs. (3.1)–(3.5), (6.88), have the forms

$$\varepsilon_{11m} = -p_1 \left[-\kappa_{4m}^{(2)} \left(\frac{2}{3} + \ln r \right) + \kappa_{5m}^{(2)} C_m^{(1)} c_{3m} \, r^{c_{3m}-1} \right], \quad (6.93)$$

$$\varepsilon_{22m} = \varepsilon_{33m} = -p_1 \left[\kappa_{4m}^{(2)} \left(\frac{1}{3} - \ln r \right) + \kappa_{5m}^{(2)} C_m^{(1)} r^{c_{3m}-1} + \frac{3\kappa_{6m}^{(2)}}{2r} \right], \quad (6.94)$$

$$\varepsilon_{12m} = -\left[\psi_{11\varphi m}^{(2)} \left(\frac{1}{3} - \ln r \right) + \psi_{12\varphi m}^{(2)} C_m^{(1)} r^{c_{3m}-1} + \frac{3\psi_{13\varphi m}^{(2)}}{2r} \right], \quad (6.95)$$

$$\varepsilon_{13m} = -\left[\psi^{(2)}_{11\nu m}\left(\frac{1}{3} - \ln r\right) + \psi^{(2)}_{12\nu m} C^{(1)}_m r^{c_{3m}-1} + \frac{3\psi^{(2)}_{13\nu m}}{2r}\right]. \tag{6.96}$$

With regard to the determination of the compressive or tensile radial stress $p_1 > 0$ or $p_1 < 0$ acting on the particle-matrix boundary, thus for $r = R_1$, respectively, considering Eq. (3.58), the tangential strain on the particle-matrix boundary, $(\varepsilon_{22m})_{r=R_1}$, is required to be derived in the form

$$(\varepsilon_{22m})_{r=R_1} = -p_1\,\varrho_{1m}, \tag{6.97}$$

where the coefficient ϱ_{1m} is given by Eq. (6.107).

Considering Eqs. (3.41), (3.42), (6.89)–(6.96), the thermal-stress induced elastic energy density w_m and the thermal-stress induced elastic energy W_m, accumulated in an arbitrary point of the cell matrix and in the volume of the cell matrix, respectively, have the forms

$$w_m = \frac{1}{2}\left[\omega_{11m} + \frac{\omega_{12m}}{r} + \frac{\omega_{13m}}{r^2} + \omega_{14m}\,r^{c_{3m}-1} + \omega_{15m}\,r^{c_{3m}-2} + \omega_{16m}\,r^{2(c_{3m}-1)} + \omega_{17m}\ln r + \omega_{18m}\ln^2 r + \frac{\omega_{19m}\ln r}{r} + \omega_{110m}\,r^{c_{3m}-1}\ln r\right], \tag{6.98}$$

$$W_m = 8\int_0^{\pi/2}\int_0^{\pi/2}\int_{R_1}^{r_c} w_m\,r^2 dr d\varphi d\nu = \int_0^{\pi/2}\int_0^{\pi/2}\Omega_{11m} d\varphi d\nu, \tag{6.99}$$

where the coefficient Ω_{11m} given by Eq. (6.118) is a function of $\omega_{11m}, \ldots, \omega_{110m}$ Eqs. (6.108)–(6.117), respectively.

With regard to the determination of the thermal-stress induced strengthening presented in Volume III, the stress σ_{im} acting in a direction of the axis x_i and the elastic energy density w_{im} induced by the stress σ_{im} $(i = 1,2,3)$ (see Eqs. (3.46), (3.47), (6.89)–(6.92)), both related to an arbitrary point of the cell matrix, are derived as

$$\sigma_{im} = -\left(\xi_{i\,11m} + \frac{\xi_{i\,12m}}{r} + \xi_{i\,13m}\,r^{c_{3m}-1} - \xi_{i\,14m}\ln r\right), \quad i = 1,2,3, \tag{6.100}$$

$$w_{im} = \frac{s_{11m}}{2}\left[\xi^2_{i\,11m} + \frac{\xi^2_{i\,12m}}{r^2} + \xi^2_{i\,13m}\,r^{2(c_{3m}-1)} + \xi^2_{i\,14m}\ln^2 r + \frac{2\,\xi_{i\,11m}\,\xi_{i\,12m}}{r} + 2\,\xi_{i\,11m}\,\xi_{i\,13m}\,r^{c_{3m}-1} - 2\,\xi_{i\,11m}\,\xi_{i\,14m}\ln r + 2\,\xi_{i\,12m}\,\xi_{i\,13m}\,r^{c_{3m}-2} - 2\,\xi_{i\,12m}\,\xi_{i\,14m}\frac{\ln r}{r} - 2\,\xi_{i\,13m}\,\xi_{i\,14m}\,r^{c_{3m}-1}\ln r\right],$$

$$i = 1,2,3, \tag{6.101}$$

where the elastic modulus s_{11m} and the coefficients $\xi_{i\,11m}, \ldots, \xi_{i\,14m}$ $(i=1,2,3)$ are given by Eqs. (3.23) and (6.119)–(6.122), respectively.

Appendix. Considering the elastic modulus s_{44m} and the coefficients c_{1m}, c_{2m}, c_{3m}, $C_m^{(1)}$, a'_{1i}, a'_{2i}, a'_{3i} $(i=1,2,3)$ (see Eqs. (3.25), (3.38)–(3.40), (6.5), (12.26)–(12.34)), the coefficients χ_{13m}, χ_{23m}, χ_{31m}, χ_{32m}, χ_{33m}, ϱ_{1+im}, $\omega_{1+i\,1m}, \ldots, \omega_{1+i\,10m}$, $\Omega_{1+i\,1+jm}$[5] $(i,j=0,1)$, $\xi_{i\,1+j\,1m}, \ldots, \xi_{i\,1+j\,4m}$ $(i=1,2,3;\ j=0,1)$ have the forms

$$\chi_{13m} = -\frac{3c_{2m}}{R_1}, \tag{6.102}$$

$$\chi_{23m} = \frac{3}{2}, \tag{6.103}$$

$$\chi_{31m} = -\left(\frac{2}{3} + \ln r_c\right), \tag{6.104}$$

$$\chi_{32m} = C_m^{(1)} c_{3m}\, r_c^{c_{3m}-1}, \tag{6.105}$$

$$\chi_{33m} = 0, \tag{6.106}$$

$$\varrho_{1+im} = \kappa_{4m}^{(2)}\left(\frac{1}{3} - \ln R_{1+i}\right) + \kappa_{5m}^{(2)} C_m^{(1)} R_{1+i}^{c_{3m}-1} + \frac{3\kappa_{6m}^{(2)}}{2R_{1+i}}, \quad i=0,1, \tag{6.107}$$

$$\omega_{1+i\,1m} = \frac{2}{9}\left(3p_{1+i}^2\left[\kappa_{4m}^{(2)}\right]^2 (c_{1m}+2c_{2m}) + \frac{1}{s_{44m}}\left\{\left[\psi_{1+i\,1\varphi m}^{(2)}\right]^2 + \left[\psi_{1+i\,1\nu m}^{(2)}\right]^2\right\}\right),$$
$$i=0,1, \tag{6.108}$$

$$\omega_{1+i\,2m} = 2\left\{p_{1+i}^2 \kappa_{4m}^{(2)}\kappa_{6m}^{(2)} (c_{1m}+2c_{2m}) + \frac{1}{s_{44m}}\left[\psi_{1+i\,1\varphi m}^{(2)}\psi_{1+i\,3\varphi m}^{(2)} + \psi_{1+i\,1\nu m}^{(2)}\psi_{1+i\,3\nu m}^{(2)}\right]\right\}, \quad i=0,1, \tag{6.109}$$

$$\omega_{1+i\,3m} = \frac{9}{2}\left(p_{1+i}^2\left[\kappa_{6m}^{(2)}\right]^2 c_{1m} + \frac{1}{s_{44m}}\left\{\left[\psi_{1+i\,3\varphi m}^{(2)}\right]^2 + \left[\psi_{1+i\,3\nu m}^{(2)}\right]^2\right\}\right),$$
$$i=0,1, \tag{6.110}$$

[5]The coefficient Ω_{21m} is determined but not used in formulae for the elastic energy W_m of the cell matrix.

$$\omega_{1+i\,4m} = \frac{4C_m^{(1)}}{3} \left\{ p_{1+i}^2 \kappa_{4m}^{(2)} \kappa_{5m}^{(2)} \left[c_{1m} \left(c_{3m} + 1 \right) + 2c_{2m} \right] \right. \\ \left. + \frac{1}{s_{44m}} \left[\psi_{1+i\,1\varphi m}^{(2)} \psi_{1+i\,2\varphi m}^{(2)} + \psi_{1+i\,1\nu m}^{(2)} \psi_{1+i\,2\nu m}^{(2)} \right] \right\}, \\ i = 0, 1, \tag{6.111}$$

$$\omega_{1+i\,5m} = 6C_m^{(1)} \left\{ p_{1+i}^2 \kappa_{5m}^{(2)} \kappa_{6m}^{(2)} \left(c_{1m} - c_{2m} c_{3m} \right) \right. \\ \left. + \frac{1}{s_{44m}} \left[\psi_{1+i\,2\varphi m}^{(2)} \psi_{1+i\,3\varphi m}^{(2)} + \psi_{1+i\,2\nu m}^{(2)} \psi_{1+i\,3\nu m}^{(2)} \right] \right\}, \\ i = 0, 1, \tag{6.112}$$

$$\omega_{1+i\,6m} = \left[C_m^{(1)} \right]^2 \left(p_{1+i}^2 \left[\kappa_{5m}^{(2)} \right]^2 \left[c_{3m}^2 \left(c_{1m} + c_{2m} \right) + 2 \left(c_{1m} - 2c_{2m} c_{3m} \right) \right] \right. \\ \left. + \frac{2}{s_{44m}} \left\{ \left[\psi_{1+i\,2\varphi m}^{(2)} \right]^2 + \left[\psi_{1+i\,2\nu m}^{(2)} \right]^2 \right\} \right), \\ i = 0, 1, \tag{6.113}$$

$$\omega_{1+i\,7m} = -\frac{4}{3s_{44m}} \left\{ \left[\psi_{1+i\,1\varphi m}^{(2)} \right]^2 + \left[\psi_{1+i\,1\nu m}^{(2)} \right]^2 \right\}, \quad i = 0, 1, \tag{6.114}$$

$$\omega_{1+i\,8m} = 3p_{1+i}^2 \left[\kappa_{4m}^{(2)} \right]^2 \left(c_{1m} - c_{2m} \right) + \frac{2}{s_{44m}} \left\{ \left[\psi_{1+i\,1\varphi m}^{(2)} \right]^2 + \left[\psi_{1+i\,1\nu m}^{(2)} \right]^2 \right\}, \\ i = 0, 1, \tag{6.115}$$

$$\omega_{1+i\,9m} = -6 \left\{ p_{1+i}^2 \kappa_{4m}^{(2)} \kappa_{6m}^{(2)} \left(c_{1m} - c_{2m} \right) \right. \\ \left. + \frac{1}{s_{44m}} \left[\psi_{1+i\,1\varphi m}^{(2)} \psi_{1+i\,3\varphi m}^{(2)} + \psi_{1+i\,1\nu m}^{(2)} \psi_{1+i\,3\nu m}^{(2)} \right] \right\}, \quad i=0, 1, \tag{6.116}$$

$$\omega_{1+i\,10m} = -2C_m^{(1)} \left\{ p_{1+i}^2 \kappa_{4m}^{(2)} \kappa_{5m}^{(2)} \left[c_{1m} \left(2 - c_{3m} \right) - c_{2m} \left(3c_{3m} + 2 \right) \right] \right. \\ \left. + \frac{2}{s_{44m}} \left[\psi_{1+i\,1\varphi m}^{(2)} \psi_{1+i\,2\varphi m}^{(2)} + \psi_{1+i\,1\nu m}^{(2)} \psi_{1+i\,2\nu m}^{(2)} \right] \right\}, \\ i = 0, 1, \tag{6.117}$$

$$\Omega_{1+i\,1+jm} = 4\left(\frac{\omega_{1+i\,1m}}{3}\left(r_c^3 - R_{1+j}^3\right) + \frac{\omega_{1+i\,2m}}{2}\left(r_c^2 - R_{1+j}^2\right)\right.$$
$$+ \omega_{1+i\,3m}\left(r_c - R_{1+j}\right) + \frac{\omega_{1+i\,4m}}{c_{3m}+2}\left(r_c^{c_{3m}+2} - R_{1+j}^{c_{3m}+2}\right)$$
$$+ \frac{\omega_{1+i\,5m}}{c_{3m}+1}\left(r_c^{c_{3m}+1} - R_{1+j}^{c_{3m}+1}\right) + \frac{\omega_{1+i\,6m}}{2c_{3m}+1}\left(r_c^{2c_{3m}+1} - R_{1+j}^{2c_{3m}+1}\right)$$
$$+ \frac{\omega_{1+i\,7m}}{3}\left[r_c^3\left(\ln r_c - \frac{1}{3}\right) - R_{1+j}^3\left(\ln R_{1+j} - \frac{1}{3}\right)\right]$$
$$+ \frac{\omega_{1+i\,8m}}{3}\left\{r_c^3\left[\left(\ln r_c - \frac{1}{3}\right)^2 + \frac{1}{9}\right] - R_{1+j}^3\left[\left(\ln R_{1+j} - \frac{1}{3}\right)^2 + \frac{1}{9}\right]\right\}$$
$$+ \frac{\omega_{1+i\,9m}}{2}\left[r_c^2\left(\ln r_c - \frac{1}{2}\right) - R_{1+j}^2\left(\ln R_{1+j} - \frac{1}{2}\right)\right]$$
$$\left. + \frac{\omega_{1+i\,10m}}{c_{3m}+2}\left[r_c^{c_{3m}+2}\left(\ln r_c - \frac{1}{c_{3m}+2}\right) - R_{1+j}^{c_{3m}+2}\left(\ln R_{1+j} - \frac{1}{c_{3m}+2}\right)\right]\right),$$
$$i, j = 0, 1, \tag{6.118}$$

$$\xi_{i\,1+j\;1m} = \frac{1}{3}\left\{p_{1+j}\kappa_{4m}^{(2)}\left(a'_{2i} + a'_{3i} - 2a'_{1i}\right)\left(c_{1m} + 2c_{2m}\right)\right.$$
$$\left. + \frac{1}{s_{44m}}\left[\left(a'_{1i} + a'_{2i}\right)\psi_{1+j\,1\varphi m}^{(2)} + \left(a'_{1i} + a'_{3i}\right)\psi_{1+j\,1\nu m}^{(2)}\right]\right\},$$
$$i = 1, 2, 3; \quad j = 0, 1, \tag{6.119}$$

$$\xi_{i\,1+j\;2m} = \frac{3}{2}\left\{p_{1+j}\kappa_{6m}^{(2)}\left[c_{1m}\left(a'_{2i} + a'_{3i}\right) - 2a'_{1i}\,c_{2m}\right]\right.$$
$$\left. + \frac{1}{s_{44m}}\left[\left(a'_{1i} + a'_{2i}\right)\psi_{1+j\,3\varphi m}^{(2)} + \left(a'_{1i} + a'_{3i}\right)\psi_{1+j\,3\nu m}^{(2)}\right]\right\},$$
$$i = 1, 2, 3; \quad j = 0, 1, \tag{6.120}$$

$$\xi_{i\,1+j\;3m} = C_m^{(1)}\left(p_{1+j}\kappa_{5m}^{(2)}\left\{a'_{1i}\left[c_{3m}\left(c_{1m} + c_{2m}\right) - 2c_{2m}\right] + \left(a'_{2i} + a'_{3i}\right)\left(c_{1m} - c_{2m}c_{3m}\right)\right\}\right.$$
$$\left. + \frac{1}{s_{44m}}\left[\left(a'_{1i} + a'_{2i}\right)\psi_{1+j\,2\varphi m}^{(2)} + \left(a'_{1i} + a'_{3i}\right)\psi_{1+j\,2\nu m}^{(2)}\right]\right),$$
$$i = 1, 2, 3; \quad j = 0, 1, \tag{6.121}$$

$$\xi_{i\,1+j\;4m} = p_{1+j}\kappa_{4m}^{(2)}\left(a'_{1i} + a'_{2i} + a'_{3i}\right)\left(c_{1m} - c_{2m}\right)$$
$$+ \frac{1}{s_{44m}}\left[\left(a'_{1i} + a'_{2i}\right)\psi_{1+j\,1\varphi m}^{(2)} + \left(a'_{1i} + a'_{3i}\right)\psi_{1+j\,1\nu m}^{(2)}\right],$$
$$i = 1, 2, 3; \quad j = 0, 1. \tag{6.122}$$

6.4 Multi-particle-envelope-matrix system

The variable r related to the thermal stresses and to the corresponding quantities in the spherical envelope with the radii $R_1 < R_2$ and in the cell matrix is from the intervals $\langle R_1, R_2\rangle$ and $\langle R_2, r_c\rangle$, respectively, where the φ, ν-dependent distance r_c as a function of the experimentally determined parameters v, R_{1ex}, d_{1ex}, d_{2ex}, d_{3ex} (see Section 2.2.1) is derived in Section 2.3.

The compressive or tensile radial stresses $p_1 > 0$ or $p_1 < 0$ and $p_2 > 0$ or $p_2 < 0$, as consequences of the conditions $\beta_p \neq \beta_m$ and $\beta_m \neq \beta_e$, acting on the particle-envelope and matrix-envelope boundaries, thus for $r = R_1$ and $r = R_2$, respectively, are given by Eqs. (10.2)–(10.5) as dependent on the following conditions concerning the coefficients β_p, β_e, β_m (see Eqs. (3.48)–(3.50)).

With regard to the condition $\beta_p \neq \beta_e \neq \beta_m$, the radial stresses p_1, p_2 are determined by the tangential strains $(\varepsilon_{22p})_{r=R_1}$, $(\varepsilon_{22e})_{r=R_{1+i}}$ $(i=0,1)$ and $(\varepsilon_{22m})_{r=R_2}$ in the spherical particle, in the spherical envelope and in the cell matrix, resulting in the coefficients ϱ_{1p}, ϱ_{1+ie} and ϱ_{2m}, respectively (see Eqs. (3.59), (3.60), (10.2), (10.3)).

Similarly, provided that $\beta_p \neq \beta_e = \beta_m$ and regarding the φ, ν-dependent difference $\varepsilon_{11te} - \varepsilon_{11tp} = f(\varphi, \nu)$ (see Eq. (12.23)), the radial stress p_1 is determined by the tangential strains $(\varepsilon_{22p})_{r=R_1}$ and $(\varepsilon_{22e})_{r=R_1}$, resulting in the coefficients ϱ_{1p} and ϱ_{1e}^{me}, respectively (see Eqs. (3.59), (10.4)).

Finally, provided that $\beta_p = \beta_e \neq \beta_m$, the radial stress p_2 is determined by the tangential strains $(\varepsilon_{22m})_{r=R_2}$ and $(\varepsilon_{22e})_{r=R_2}$, resulting in the coefficients ϱ_{2m} and ϱ_{2e}^{pe}, respectively (see Eqs. (3.60), (10.5)).

6.4.1 Thermal stresses in spherical envelope

6.4.1.1 Condition $\beta_p \neq \beta_e \neq \beta_m$

A. Integration constants $C_{1e} \neq 0$, $C_{2e} \neq 0$, $C_{3e} = 0$

With regard to the boundary conditions (4.7), (4.8) and the analysis presented in Section 6.2.2, considering the terms u_{1re} and u_{2re} (see Eqs. (6.6), (6.7)) to represent functions of the variable $r \in \langle R_1, R_2\rangle$, the radial displacement u_{1e} of the infinitesimal spherical cap in the spherical envelope, regarding Eqs. (3.31), (6.4), has the form

$$u_{1e} = -\left\{\left[p_1\kappa_{1e}^{(1)} - p_2\kappa_{2e}^{(1)}\right] r\left(\frac{1}{3} - \ln r\right) + \left[p_1\kappa_{3e}^{(1)} - p_2\kappa_{4e}^{(1)}\right] C_e^{(1)}\, r^{c_{3e}}\right\}, \quad (6.123)$$

where the coefficients $c_{3e} < 0$, $C_e^{(1)}$ and the radial stresses p_1, p_2 acting on the particle-envelope, matrix-envelope boundaries are given by Eqs. (3.40), (6.5) and (10.2), (10.3), respectively.

With regard to Section 6.4.1.1.A, the coefficients $\varrho_{1+i\,1+je}$ $(i,j=0,1)$, $\omega_{1e},\ldots,\omega_{6e}$, ξ_{i1e}, ξ_{i2e}, ξ_{i3e} $(i=1,2,3)$ mentioned below (see Eqs. (6.132), (6.133),

(6.135), (6.136)) depend on $\kappa_{1e}^{(1)}, \ldots, \kappa_{4e}^{(1)}$ (see Eq. (12.35)) representing functions of χ_{11e}, χ_{12e}, χ_{21e}, χ_{22e} given by Eqs. (6.137), (6.138).

Consequently, the radial stress σ_{11e}, the tangential and shear stresses σ_{22e}, σ_{33e} and σ_{12e}, σ_{13e} acting in the spherical envelope, respectively, regarding Eqs. (3.31)–(3.34), (6.123), are derived as

$$\sigma_{11e} = -\left\{ - \left[p_1 \kappa_{1e}^{(1)} - p_2 \kappa_{2e}^{(1)} \right] \left[\frac{2}{3} \left(c_{1e} + 2c_{2e} \right) + \left(c_{1e} - c_{2e} \right) \ln r \right] + \left[p_1 \kappa_{3e}^{(1)} - p_2 \kappa_{4e}^{(1)} \right] C_e^{(1)} \left[c_{3e} \left(c_{1e} + c_{2e} \right) - 2c_{2e} \right] r^{c_{3e}-1} \right\}, \quad (6.124)$$

$$\sigma_{22e} = \sigma_{33e} = -\left\{ \left[p_1 \kappa_{1e}^{(1)} - p_2 \kappa_{2e}^{(1)} \right] \left[\frac{1}{3} \left(c_{1e} + 2c_{2e} \right) - \left(c_{1e} - c_{2e} \right) \ln r \right] + \left[p_1 \kappa_{3e}^{(1)} - p_2 \kappa_{4e}^{(1)} \right] C_e^{(1)} \left(c_{1e} - c_{2e} c_{3e} \right) r^{c_{3e}-1} \right\}, \quad (6.125)$$

$$\sigma_{12e} = -\frac{1}{s_{44e}} \left\{ \left[\kappa_{1e}^{(1)} \frac{\partial p_1}{\partial \varphi} - \kappa_{2e}^{(1)} \frac{\partial p_2}{\partial \varphi} \right] \left(\frac{1}{3} - \ln r \right) + \left[\kappa_{3e}^{(1)} \frac{\partial p_1}{\partial \varphi} - \kappa_{4e}^{(1)} \frac{\partial p_2}{\partial \varphi} \right] C_e^{(1)} r^{c_{3e}-1} \right\}, \quad (6.126)$$

$$\sigma_{13e} = -\frac{1}{s_{44e}} \left\{ \left[\kappa_{1e}^{(1)} \frac{\partial p_1}{\partial \nu} - \kappa_{2e}^{(1)} \frac{\partial p_2}{\partial \nu} \right] \left(\frac{1}{3} - \ln r \right) + \left[\kappa_{3e}^{(1)} \frac{\partial p_1}{\partial \nu} - \kappa_{4e}^{(1)} \frac{\partial p_2}{\partial \nu} \right] C_e^{(1)} r^{c_{3e}-1} \right\}, \quad (6.127)$$

where the elastic modulus s_{44e} and the coefficients c_{1e}, c_{2e} are given by Eqs. (3.25) and (3.38), (3.39), respectively.

Finally, the radial strain ε_{11e}, the tangential and shear strains ε_{22e}, ε_{33e} and ε_{12e}, ε_{13e} in the spherical envelope, respectively, regarding Eqs. (3.26)–(3.30), (6.123), have the forms

$$\varepsilon_{11e} = -\left\{ - \left[p_1 \kappa_{1e}^{(1)} - p_2 \kappa_{2e}^{(1)} \right] \left(\frac{2}{3} + \ln r \right) + \left[p_1 \kappa_{3e}^{(1)} - p_2 \kappa_{4e}^{(1)} \right] C_e^{(1)} c_{3e} r^{c_{3e}-1} \right\}, \quad (6.128)$$

$$\varepsilon_{22e} = \varepsilon_{33e} = -\left\{ \left[p_1 \kappa_{1e}^{(1)} - p_2 \kappa_{2e}^{(1)} \right] \left(\frac{1}{3} - \ln r \right) + \left[p_1 \kappa_{3e}^{(1)} - p_2 \kappa_{4e}^{(1)} \right] C_e^{(1)} r^{c_{3e}-1} \right\}, \quad (6.129)$$

$$\varepsilon_{12e} = -\left\{\left[\kappa_{1e}^{(1)}\frac{\partial p_1}{\partial\varphi} - \kappa_{2e}^{(1)}\frac{\partial p_2}{\partial\varphi}\right]\left(\frac{1}{3} - \ln r\right) + \left[\kappa_{3e}^{(1)}\frac{\partial p_1}{\partial\varphi} - \kappa_{4e}^{(1)}\frac{\partial p_2}{\partial\varphi}\right]C_e^{(1)}\, r^{c_{3e}-1}\right\}, \tag{6.130}$$

$$\varepsilon_{13e} = -\left\{\left[\kappa_{1e}^{(1)}\frac{\partial p_1}{\partial\nu} - \kappa_{2e}^{(1)}\frac{\partial p_2}{\partial\nu}\right]\left(\frac{1}{3} - \ln r\right) + \left[\kappa_{3e}^{(1)}\frac{\partial p_1}{\partial\nu} - \kappa_{4e}^{(1)}\frac{\partial p_2}{\partial\nu}\right]C_e^{(1)}\, r^{c_{3e}-1}\right\}. \tag{6.131}$$

With regard to the determination of the compressive or tensile radial stresses $p_1 > 0$ or $p_1 < 0$ and $p_2 > 0$ or $p_2 < 0$ acting on the particle-envelope and matrix-envelope boundaries, thus for $r = R_1$ and $r = R_2$, respectively, considering Eqs. (3.59), (3.60), the tangential strains on the particle-envelope and matrix-envelope boundaries, $(\varepsilon_{22e})_{r=R_1}$ and $(\varepsilon_{22e})_{r=R_2}$, respectively, are required to be derived in the form

$$(\varepsilon_{22e})_{r=R_{1+i}} = -\sum_{j=0}^{1} p_{1+j}\varrho_{1+i\;1+je}, \quad i = 0, 1, \tag{6.132}$$

where the coefficient $\varrho_{1+i\;1+je}$ $(i, j = 0,1)$ is given by Eq. (6.139).

Considering Eqs. (3.41), (3.42), (6.124)–(6.131), the thermal-stress induced elastic energy density w_e and the thermal-stress induced elastic energy W_e, accumulated in an arbitrary point of the spherical envelope and in the volume of the spherical envelope, respectively, have the forms

$$w_e = \frac{1}{2}\left[\omega_{1e} + \omega_{2e}\, r^{c_{3e}-1} + \omega_{3e}\, r^{2(c_{3e}-1)} + \omega_{4e}\ln r + \omega_{5e}\ln^2 r + \omega_{6e}\, r^{c_{3e}-1}\ln r\right], \tag{6.133}$$

$$W_e = 8\int_0^{\pi/2}\int_0^{\pi/2}\int_{R_1}^{R_2} w_e\, r^2 dr d\varphi d\nu = \int_0^{\pi/2}\int_0^{\pi/2}\Omega_e d\varphi d\nu, \tag{6.134}$$

where the coefficient Ω_e given by Eq. (6.146) is a function of $\omega_{1e}, \ldots, \omega_{6e}$ given by Eqs. (6.140)–(6.145), respectively.

With regard to the determination of the thermal-stress induced strengthening presented in Volume III, the stress σ_{ie} acting in a direction of the axis x_i and the elastic energy density w_{ie} induced by the stress σ_{ie} $(i = 1,2,3)$ (see Eqs. (3.46), (3.47), (6.124)–(6.127)), both related to an arbitrary point of the spherical envelope, are derived as

$$\sigma_{ie} = -\left(\xi_{i1e} + \xi_{i2e}\, r^{c_{3e}-1} - \xi_{i3e} \ln r\right), \quad i = 1, 2, 3, \tag{6.135}$$

$$w_{ie} = \frac{s_{11e}}{2}\left[\xi_{i1e}^2 + \xi_{i2e}^2\, r^{2(c_{3e}-1)} + \xi_{i3e}^2 \ln^2 r + 2\,\xi_{i1e}\,\xi_{i2e}\, r^{c_{3e}-1} - 2\,\xi_{i1e}\,\xi_{i3e} \ln r - 2\,\xi_{i2e}\,\xi_{i3e}\, r^{c_{3e}-1} \ln r\right], \quad i = 1, 2, 3, \tag{6.136}$$

where the elastic modulus s_{11e} and the coefficients ξ_{i1e}, ξ_{i2e}, ξ_{i3e} $(i=1,2,3)$ are given by Eqs. (3.23) and (6.147)–(6.149), respectively.

Appendix. Considering the elastic modulus s_{44e} and the coefficients c_{1e}, c_{2e}, c_{3e}, $C_e^{(1)}$, a'_{1i}, a'_{2i}, a'_{3i} $(i=1,2,3)$ (see Eqs. (3.25), (3.38)–(3.40), (6.5), (12.26)–(12.34)), the coefficients $\chi_{1+i\,1e}$, $\chi_{1+i\,2e}$, $\varrho_{1+i\;1+je}$ $(i,j=0,1)$, $\omega_{1e},\ldots,\omega_{6e}$, Ω_e, $\xi_{i\,1e}$, $\xi_{i\,2e}$, $\xi_{i\,3e}$ $(i=1,2,3)$ have the forms

$$\chi_{1+i\,1e} = -\left[\frac{2}{3}\left(c_{1e} + 2c_{2e}\right) + \left(c_{1e} - c_{2e}\right) \ln R_{1+i}\right], \quad i = 0, 1, \tag{6.137}$$

$$\chi_{1+i\,2e} = C_e^{(1)}\left[c_{3e}\left(c_{1e} + c_{2e}\right) - 2c_{2e}\right] R_{1+i}^{c_{3e}-1}, \quad i = 0, 1, \tag{6.138}$$

$$\varrho_{1+i\;1+je} = \left(\delta_{0j} - \delta_{1j}\right)\left[\kappa_{1+je}^{(1)}\left(\frac{1}{3} - \ln R_{1+i}\right) + \kappa_{3+je}^{(1)} C_e^{(1)} R_{1+i}^{c_{3e}-1}\right], \quad i,j = 0, 1, \tag{6.139}$$

$$\omega_{1e} = \frac{2}{9}\left(3\left[p_1\kappa_{1e}^{(1)} - p_2\kappa_{2e}^{(1)}\right]^2\left(c_{1e} + 2c_{2e}\right) + \frac{1}{s_{44e}}\left\{\left[\kappa_{1e}^{(1)}\frac{\partial p_1}{\partial\varphi} - \kappa_{2e}^{(1)}\frac{\partial p_2}{\partial\varphi}\right]^2 + \left[\kappa_{1e}^{(1)}\frac{\partial p_1}{\partial\nu} - \kappa_{2e}^{(1)}\frac{\partial p_2}{\partial\nu}\right]^2\right\}\right), \tag{6.140}$$

$$\omega_{2e} = \frac{4C_e^{(1)}}{3}\left(\left[p_1\kappa_{1e}^{(1)} - p_2\kappa_{2e}^{(1)}\right]\left[p_1\kappa_{3e}^{(1)} - p_2\kappa_{4e}^{(1)}\right]\left(c_{1e} + 2c_{2e}\right)\left(1 - c_{3e}\right) + \frac{1}{s_{44e}}\left\{\left[\kappa_{1e}^{(1)}\frac{\partial p_1}{\partial\varphi} - \kappa_{2e}^{(1)}\frac{\partial p_2}{\partial\varphi}\right]\left[\kappa_{3e}^{(1)}\frac{\partial p_1}{\partial\varphi} - \kappa_{4e}^{(1)}\frac{\partial p_2}{\partial\varphi}\right] + \left[\kappa_{1e}^{(1)}\frac{\partial p_1}{\partial\nu} - \kappa_{2e}^{(1)}\frac{\partial p_2}{\partial\nu}\right]\left[\kappa_{3e}^{(1)}\frac{\partial p_1}{\partial\nu} - \kappa_{4e}^{(1)}\frac{\partial p_2}{\partial\nu}\right]\right\}\right), \tag{6.141}$$

$$\omega_{3e} = \left[C_e^{(1)}\right]^2\left(\left[p_1\kappa_{3e}^{(1)} - p_2\kappa_{4e}^{(1)}\right]^2\left[c_{3e}^2\left(c_{1e} + c_{2e}\right) + 2\left(c_{1e} - 2c_{2e}c_{3e}\right)\right] + \frac{2}{s_{44e}}\left\{\left[\kappa_{3e}^{(1)}\frac{\partial p_1}{\partial\varphi} - \kappa_{4e}^{(1)}\frac{\partial p_2}{\partial\varphi}\right]^2 + \left[\kappa_{3e}^{(1)}\frac{\partial p_1}{\partial\nu} - \kappa_{4e}^{(1)}\frac{\partial p_2}{\partial\nu}\right]^2\right\}\right), \tag{6.142}$$

$$\omega_{4e} = -\frac{4}{3s_{44e}} \left\{ \left[\kappa_{1e}^{(1)} \frac{\partial p_1}{\partial \varphi} - \kappa_{2e}^{(1)} \frac{\partial p_2}{\partial \varphi} \right]^2 + \left[\kappa_{1e}^{(1)} \frac{\partial p_1}{\partial \nu} - \kappa_{2e}^{(1)} \frac{\partial p_2}{\partial \nu} \right]^2 \right\}, \tag{6.143}$$

$$\begin{aligned} \omega_{5e} = {} & 3 \left[p_1 \kappa_{1e}^{(1)} - p_2 \kappa_{2e}^{(1)} \right]^2 (c_{1e} - c_{2e}) \\ & + \frac{2}{s_{44e}} \left\{ \left[\kappa_{1e}^{(1)} \frac{\partial p_1}{\partial \varphi} - \kappa_{2e}^{(1)} \frac{\partial p_2}{\partial \varphi} \right]^2 + \left[\kappa_{1e}^{(1)} \frac{\partial p_1}{\partial \nu} - \kappa_{2e}^{(1)} \frac{\partial p_2}{\partial \nu} \right]^2 \right\}, \end{aligned} \tag{6.144}$$

$$\begin{aligned} \omega_{6e} = {} & -2C_e^{(1)} \left(\left[p_1 \kappa_{1e}^{(1)} - p_2 \kappa_{2e}^{(1)} \right] \left[p_1 \kappa_{3e}^{(1)} - p_2 \kappa_{4e}^{(1)} \right] (c_{1e} - c_{2e}) (c_{3e} + 2) \right. \\ & + \frac{2}{s_{44e}} \left\{ \left[\kappa_{1e}^{(1)} \frac{\partial p_1}{\partial \varphi} - \kappa_{2e}^{(1)} \frac{\partial p_2}{\partial \varphi} \right] \left[\kappa_{3e}^{(1)} \frac{\partial p_1}{\partial \varphi} - \kappa_{4e}^{(1)} \frac{\partial p_2}{\partial \varphi} \right] \right. \\ & \left. \left. + \left[\kappa_{1e}^{(1)} \frac{\partial p_1}{\partial \nu} - \kappa_{2e}^{(1)} \frac{\partial p_2}{\partial \nu} \right] \left[\kappa_{3e}^{(1)} \frac{\partial p_1}{\partial \nu} - \kappa_{4e}^{(1)} \frac{\partial p_2}{\partial \nu} \right] \right\} \right), \end{aligned} \tag{6.145}$$

$$\begin{aligned} \Omega_e = {} & 4 \left(\frac{\omega_{1e}}{3} \left(R_2^3 - R_1^3 \right) + \frac{\omega_{2e}}{c_{3e} + 2} \left(R_2^{c_{3e}+2} - R_1^{c_{3e}+2} \right) \right. \\ & + \frac{\omega_{3e}}{2c_{3e} + 1} \left(R_2^{2c_{3e}+1} - R_1^{2c_{3e}+1} \right) + \frac{\omega_{4e}}{3} \left[R_2^3 \left(\ln R_2 - \frac{1}{3} \right) - R_1^3 \left(\ln R_1 - \frac{1}{3} \right) \right] \\ & + \frac{\omega_{5e}}{3} \left\{ R_2^3 \left[\left(\ln R_2 - \frac{1}{3} \right)^2 + \frac{1}{9} \right] - R_1^3 \left[\left(\ln R_1 - \frac{1}{3} \right)^2 + \frac{1}{9} \right] \right\} \\ & \left. + \frac{\omega_{6e}}{c_{3e} + 2} \left[R_2^{c_{3e}+2} \left(\ln R_2 - \frac{1}{c_{3e} + 2} \right) - R_1^{c_{3e}+2} \left(\ln R_1 - \frac{1}{c_{3e} + 2} \right) \right] \right), \end{aligned} \tag{6.146}$$

$$\begin{aligned} \xi_{i1e} = {} & \frac{1}{3} \left(\left[p_1 \kappa_{1e}^{(1)} - p_2 \kappa_{2e}^{(1)} \right] \left(a'_{2i} + a'_{3i} - 2a'_{1i} \right) (c_{1e} + 2c_{2e}) \right. \\ & + \frac{1}{s_{44e}} \left\{ \left(a'_{1i} + a'_{2i} \right) \left[\kappa_{1e}^{(1)} \frac{\partial p_1}{\partial \varphi} - \kappa_{2e}^{(1)} \frac{\partial p_2}{\partial \varphi} \right] \right. \\ & \left. \left. + \left(a'_{1i} + a'_{3i} \right) \left[\kappa_{1e}^{(1)} \frac{\partial p_1}{\partial \nu} - \kappa_{2e}^{(1)} \frac{\partial p_2}{\partial \nu} \right] \right\} \right), \quad i = 1, 2, 3, \end{aligned} \tag{6.147}$$

$$\begin{aligned} \xi_{i2e} = {} & C_e^{(1)} \left(\left[p_1 \kappa_{3e}^{(1)} - p_2 \kappa_{4e}^{(1)} \right] \left\{ a'_{1i} \left[c_{3e} (c_{1e} + c_{2e}) - 2c_{2e} \right] + \left(a'_{2i} + a'_{3i} \right) (c_{1e} - c_{2e} c_{3e}) \right\} \right. \\ & + \frac{1}{s_{44e}} \left\{ \left(a'_{1i} + a'_{2i} \right) \left[\kappa_{3e}^{(1)} \frac{\partial p_1}{\partial \varphi} - \kappa_{4e}^{(1)} \frac{\partial p_2}{\partial \varphi} \right] \right. \\ & \left. \left. + \left(a'_{1i} + a'_{3i} \right) \left[\kappa_{3e}^{(1)} \frac{\partial p_1}{\partial \nu} - \kappa_{4e}^{(1)} \frac{\partial p_2}{\partial \nu} \right] \right\} \right), \quad i = 1, 2, 3, \end{aligned} \tag{6.148}$$

$$\xi_{i3e} = \left[p_1\kappa_{1e}^{(1)} - p_2\kappa_{2e}^{(1)}\right]\left(a'_{1i} + a'_{2i} + a'_{3i}\right)\left(c_{1e} - c_{2e}\right)$$
$$+ \frac{1}{s_{44e}}\left\{\left(a'_{1i} + a'_{2i}\right)\left[\kappa_{1e}^{(1)}\frac{\partial p_1}{\partial \varphi} - \kappa_{2e}^{(1)}\frac{\partial p_2}{\partial \varphi}\right] + \left(a'_{1i} + a'_{3i}\right)\left[\kappa_{1e}^{(1)}\frac{\partial p_1}{\partial \nu} - \kappa_{2e}^{(1)}\frac{\partial p_2}{\partial \nu}\right]\right\},$$
$$i = 1, 2, 3. \tag{6.149}$$

B. Integration constants $C_{1e} \neq 0$, $C_{2e} = 0$, $C_{3e} \neq 0$

With regard to the boundary conditions (4.7), (4.8) and the analysis presented in Section 6.2.2, considering the terms u_{1re} and u_{3re} (see Eqs. (6.6), (6.8)), the former as a function of the variable $r \in \langle R_1, R_2\rangle$ and the latter as a constant function regarding the variable r, respectively, the radial displacement u_{1e} of the infinitesimal spherical cap in the spherical envelope, regarding Eqs. (3.31), (6.4), has the form

$$u_{1e} = -\left\{\left[p_1\kappa_{1e}^{(1)} - p_2\kappa_{2e}^{(1)}\right] r\left(\frac{1}{3} - \ln r\right) + \frac{3}{2}\left[p_1\kappa_{3e}^{(1)} - p_2\kappa_{4e}^{(1)}\right]\right\}, \tag{6.150}$$

where the radial stresses p_1, p_2 acting on the particle-envelope, matrix-envelope boundaries are given by Eqs. (10.2), (10.3), respectively.

With regard to Section 6.4.1.1.B, the coefficients $\varrho_{1+i\,1+je}$ $(i, j=0,1)$, $\omega_{1e}, \ldots, \omega_{6e}$, ξ_{i1e}, ξ_{i2e}, ξ_{i3e} $(i=1,2,3)$ mentioned below (see Eqs. (6.159), (6.160), (6.162), (6.163)) depend on $\kappa_{1e}^{(1)}, \ldots, \kappa_{4e}^{(1)}$ (see Eq. (12.35)) representing functions of χ_{11e}, χ_{12e}, χ_{21e}, χ_{22e} given by Eqs. (6.137), (6.164).

Consequently, the radial stress σ_{11e}, the tangential and shear stresses σ_{22e}, σ_{33e} and σ_{12e}, σ_{13e} acting in the spherical envelope, respectively, regarding Eqs. (3.31)–(3.34), (6.150), are derived as

$$\sigma_{11e} = -\left\{-\left[p_1\kappa_{1e}^{(1)} - p_2\kappa_{2e}^{(1)}\right]\left[\frac{2}{3}\left(c_{1e} + 2c_{2e}\right) + \left(c_{1e} - c_{2e}\right)\ln r\right]\right.$$
$$\left. - \frac{3c_{2e}}{r}\left[p_1\kappa_{3e}^{(1)} - p_2\kappa_{4e}^{(1)}\right]\right\}, \tag{6.151}$$

$$\sigma_{22e} = \sigma_{33e} = -\left\{\left[p_1\kappa_{1e}^{(1)} - p_2\kappa_{2e}^{(1)}\right]\left[\frac{1}{3}\left(c_{1e} + 2c_{2e}\right) - \left(c_{1e} - c_{2e}\right)\ln r\right]\right.$$
$$\left. + \frac{3c_{1e}}{2r}\left[p_1\kappa_{3e}^{(1)} - p_2\kappa_{4e}^{(1)}\right]\right\}, \tag{6.152}$$

$$\sigma_{12e} = -\frac{1}{s_{44e}}\left\{\left[\kappa_{1e}^{(1)}\frac{\partial p_1}{\partial \varphi} - \kappa_{2e}^{(1)}\frac{\partial p_2}{\partial \varphi}\right]\left(\frac{1}{3} - \ln r\right)\right.$$
$$\left. + \frac{3}{2r}\left[\kappa_{3e}^{(1)}\frac{\partial p_1}{\partial \varphi} - \kappa_{4e}^{(1)}\frac{\partial p_2}{\partial \varphi}\right]\right\}, \tag{6.153}$$

$$\sigma_{13e} = -\frac{1}{s_{44e}}\left\{\left[\kappa_{1e}^{(1)}\frac{\partial p_1}{\partial \nu} - \kappa_{2e}^{(1)}\frac{\partial p_2}{\partial \nu}\right]\left(\frac{1}{3} - \ln r\right) + \frac{3}{2r}\left[\kappa_{3e}^{(1)}\frac{\partial p_1}{\partial \nu} - \kappa_{4e}^{(1)}\frac{\partial p_2}{\partial \nu}\right]\right\}, \qquad (6.154)$$

where the elastic modulus s_{44e} and the coefficients c_{1e}, c_{2e} are given by Eqs. (3.25) and (3.38), (3.39), respectively.

Finally, the radial strain ε_{11e}, the tangential and shear strains ε_{22e}, ε_{33e} and ε_{12e}, ε_{13e} in the spherical envelope, respectively, regarding Eqs. (3.26)–(3.30), (6.150), have the forms[6]

$$\varepsilon_{11e} = \left[p_1\kappa_{1e}^{(1)} - p_2\kappa_{2e}^{(1)}\right]\left(\frac{2}{3} + \ln r\right), \qquad (6.155)$$

$$\varepsilon_{22e} = \varepsilon_{33e} = -\left\{\left[p_1\kappa_{1e}^{(1)} - p_2\kappa_{2e}^{(1)}\right]\left(\frac{1}{3} - \ln r\right) + \frac{3}{2r}\left[p_1\kappa_{3e}^{(1)} - p_2\kappa_{4e}^{(1)}\right]\right\}, \qquad (6.156)$$

$$\varepsilon_{12e} = -\left\{\left[\kappa_{1e}^{(1)}\frac{\partial p_1}{\partial \varphi} - \kappa_{2e}^{(1)}\frac{\partial p_2}{\partial \varphi}\right]\left(\frac{1}{3} - \ln r\right) + \frac{3}{2r}\left[\kappa_{3e}^{(1)}\frac{\partial p_1}{\partial \varphi} - \kappa_{4e}^{(1)}\frac{\partial p_2}{\partial \varphi}\right]\right\}, \qquad (6.157)$$

$$\varepsilon_{13e} = -\left\{\left[\kappa_{1e}^{(1)}\frac{\partial p_1}{\partial \nu} - \kappa_{2e}^{(1)}\frac{\partial p_2}{\partial \nu}\right]\left(\frac{1}{3} - \ln r\right) + \frac{3}{2r}\left[\kappa_{3e}^{(1)}\frac{\partial p_1}{\partial \nu} - \kappa_{4e}^{(1)}\frac{\partial p_2}{\partial \nu}\right]\right\}. \qquad (6.158)$$

With regard to the determination of the compressive or tensile radial stresses $p_1 > 0$ or $p_1 < 0$ and $p_2 > 0$ or $p_2 < 0$ acting on the particle-envelope and matrix-envelope boundaries, thus for $r = R_1$ and $r = R_2$, respectively, considering Eqs. (3.59), (3.60), the tangential strains on the particle-envelope and matrix-envelope boundaries, $(\varepsilon_{22e})_{r=R_1}$ and $(\varepsilon_{22e})_{r=R_2}$, respectively, are required to be derived in the form

$$(\varepsilon_{22e})_{r=R_{1+i}} = -\sum_{j=0}^{1} p_{1+j}\varrho_{1+i\,1+je}, \quad i = 0, 1, \qquad (6.159)$$

where the coefficient $\varrho_{1+i\,1+je}$ $(i, j = 0,1)$ is given by Eq. (6.165).

Considering Eqs. (3.41), (3.42), (6.151)–(6.158), the thermal-stress induced elastic energy density w_e and the thermal-stress induced elastic energy W_e, accumulated in an arbitrary point of the spherical envelope and in the volume of the spherical envelope, respectively, have the forms

$$w_e = \frac{1}{2}\left(\omega_{1e} + \frac{\omega_{2e}}{r} + \frac{\omega_{3e}}{r^2} + \omega_{4e}\ln r + \omega_{5e}\ln^2 r + \omega_{6e}\frac{\ln r}{r}\right), \qquad (6.160)$$

[6]To avoid misunderstanding, a sign on the right side in Eq. (6.155) is really plus, in contrast to minus sign on the right sides in Eqs. (6.156)–(6.158).

$$W_e = 8 \int_0^{\pi/2} \int_0^{\pi/2} \int_{R_1}^{R_2} w_e \, r^2 dr d\varphi d\nu = \int_0^{\pi/2} \int_0^{\pi/2} \Omega_e d\varphi d\nu, \tag{6.161}$$

where the coefficient Ω_e given by Eq. (6.169) is a function of $\omega_{1e}, \ldots, \omega_{6e}$ given by Eqs. (6.140), (6.166), (6.167), (6.143), (6.144), (6.168), respectively.

With regard to the determination of the thermal-stress induced strengthening presented in Volume III, the stress σ_{ie} acting in a direction of the axis x_i and the elastic energy density w_{ie} induced by the stress σ_{ie} ($i = 1,2,3$) (see Eqs. (3.46), (3.47), (6.151)–(6.154)), both related to an arbitrary point of the spherical envelope, are derived as

$$\sigma_{ie} = -\left(\xi_{i1e} + \frac{\xi_{i2e}}{r} - \xi_{i3e} \ln r \right), \quad i = 1, 2, 3, \tag{6.162}$$

$$w_{ie} = \frac{s_{11e}}{2} \left(\xi_{i1e}^2 + \frac{\xi_{i2e}^2}{r^2} + \xi_{i3e}^2 \ln^2 r + \frac{2\,\xi_{i1e}\,\xi_{i2e}}{r} - 2\,\xi_{i1e}\,\xi_{i3e} \ln r - 2\,\xi_{i2e}\,\xi_{i3e} \frac{\ln r}{r} \right),$$
$$i = 1, 2, 3, \tag{6.163}$$

where the elastic modulus s_{11e} and the coefficients ξ_{i1e}, ξ_{i2e}, ξ_{i3e} ($i = 1,2,3$) are given by Eqs. (3.23) and (6.147), (6.170), (6.149), respectively.

Appendix. Considering the elastic modulus s_{44e} and the coefficients c_{1e}, c_{2e}, a'_{1i}, a'_{2i}, a'_{3i} ($i = 1,2,3$) (see Eqs. (3.25), (3.38), (3.39), (12.26)–(12.34)), the coefficients $\chi_{1+i\,2e}$, $\varrho_{1+i\;1+je}$ ($i, j = 0,1$), ω_{2e}, ω_{3e}, ω_{6e}, Ω_e, $\xi_{i\,2e}$ ($i = 1,2,3$) have the forms

$$\chi_{1+i\,2e} = -\frac{3c_{2e}}{R_{1+i}}, \quad i = 0, 1, \tag{6.164}$$

$$\varrho_{1+i\;1+je} = (\delta_{0j} - \delta_{1j}) \left[\kappa_{1+je}^{(1)} \left(\frac{1}{3} - \ln R_{1+i} \right) + \frac{3\kappa_{3+je}^{(1)}}{2R_{1+i}} \right], \quad i, j = 0, 1, \tag{6.165}$$

$$\begin{aligned} \omega_{2e} = 2 \Bigg(& \left[p_1 \kappa_{1e}^{(1)} - p_2 \kappa_{2e}^{(1)} \right] \left[p_1 \kappa_{3e}^{(1)} - p_2 \kappa_{4e}^{(1)} \right] (c_{1e} + 2c_{2e}) \\ & + \frac{1}{s_{44e}} \left\{ \left[\kappa_{1e}^{(1)} \frac{\partial p_1}{\partial \varphi} - \kappa_{2e}^{(1)} \frac{\partial p_2}{\partial \varphi} \right] \left[\kappa_{3e}^{(1)} \frac{\partial p_1}{\partial \varphi} - \kappa_{4e}^{(1)} \frac{\partial p_2}{\partial \varphi} \right] \right. \\ & \left. + \left[\kappa_{1e}^{(1)} \frac{\partial p_1}{\partial \nu} - \kappa_{2e}^{(1)} \frac{\partial p_2}{\partial \nu} \right] \left[\kappa_{3e}^{(1)} \frac{\partial p_1}{\partial \nu} - \kappa_{4e}^{(1)} \frac{\partial p_2}{\partial \nu} \right] \right\} \Bigg), \end{aligned} \tag{6.166}$$

$$\begin{aligned} \omega_{3e} = \frac{9}{2} \Bigg(& \left[p_1 \kappa_{3e}^{(1)} - p_2 \kappa_{4e}^{(1)} \right]^2 c_{1e} \\ & + \frac{1}{s_{44e}} \left\{ \left[\kappa_{3e}^{(1)} \frac{\partial p_1}{\partial \varphi} - \kappa_{4e}^{(1)} \frac{\partial p_2}{\partial \varphi} \right]^2 + \left[\kappa_{3e}^{(1)} \frac{\partial p_1}{\partial \nu} - \kappa_{4e}^{(1)} \frac{\partial p_2}{\partial \nu} \right]^2 \right\} \Bigg), \end{aligned} \tag{6.167}$$

$$\omega_{6e} = -6 \left(\left[p_1 \kappa_{1e}^{(1)} - p_2 \kappa_{2e}^{(1)} \right] \left[p_1 \kappa_{3e}^{(1)} - p_2 \kappa_{4e}^{(1)} \right] (c_{1e} - c_{2e}) + \frac{1}{s_{44e}} \left\{ \left[\kappa_{1e}^{(1)} \frac{\partial p_1}{\partial \varphi} - \kappa_{2e}^{(1)} \frac{\partial p_2}{\partial \varphi} \right] \left[\kappa_{3e}^{(1)} \frac{\partial p_1}{\partial \varphi} - \kappa_{4e}^{(1)} \frac{\partial p_2}{\partial \varphi} \right] + \left[\kappa_{1e}^{(1)} \frac{\partial p_1}{\partial \nu} - \kappa_{2e}^{(1)} \frac{\partial p_2}{\partial \nu} \right] \left[\kappa_{3e}^{(1)} \frac{\partial p_1}{\partial \nu} - \kappa_{4e}^{(1)} \frac{\partial p_2}{\partial \nu} \right] \right\} \right), \quad (6.168)$$

$$\Omega_e = 4 \left(\frac{\omega_{1e}}{3} \left(R_2^3 - R_1^3 \right) + \frac{\omega_{2e}}{2} \left(R_2^2 - R_1^2 \right) + \omega_{3e} \left(R_2 - R_1 \right) + \frac{\omega_{4e}}{3} \left[R_2^3 \left(\ln R_2 - \frac{1}{3} \right) - R_1^3 \left(\ln R_1 - \frac{1}{3} \right) \right] + \frac{\omega_{5e}}{3} \left\{ R_2^3 \left[\left(\ln R_2 - \frac{1}{3} \right)^2 + \frac{1}{9} \right] - R_1^3 \left[\left(\ln R_1 - \frac{1}{3} \right)^2 + \frac{1}{9} \right] \right\} + \frac{\omega_{6e}}{2} \left[R_2^2 \left(\ln R_2 - \frac{1}{2} \right) - R_1^2 \left(\ln R_1 - \frac{1}{2} \right) \right] \right), \quad (6.169)$$

$$\xi_{i2e} = \frac{3}{2} \left(\left[p_1 \kappa_{3e}^{(1)} - p_2 \kappa_{4e}^{(1)} \right] \left[c_{1e} \left(a'_{2i} + a'_{3i} \right) - 2 a'_{1i} c_{2e} \right] + \frac{1}{s_{44e}} \left\{ \left(a'_{1i} + a'_{2i} \right) \left[\kappa_{3e}^{(1)} \frac{\partial p_1}{\partial \varphi} - \kappa_{4e}^{(1)} \frac{\partial p_2}{\partial \varphi} \right] + \left(a'_{1i} + a'_{3i} \right) \left[\kappa_{3e}^{(1)} \frac{\partial p_1}{\partial \nu} - \kappa_{4e}^{(1)} \frac{\partial p_2}{\partial \nu} \right] \right\} \right), \quad i = 1, 2, 3. \quad (6.170)$$

C. Integration constants $C_{1e} = 0$, $C_{2e} \neq 0$, $C_{3e} \neq 0$

With regard to the boundary conditions (4.7), (4.8) and the analysis presented in Section 6.2.2, considering the terms u_{2re} and u_{3re} (see Eqs. (6.7), (6.8)), the former as a decreasing function of the variable $r \in \langle R_1, R_2 \rangle$ and the latter as a constant function regarding the variable r, respectively, the radial displacement u_{1e} of the infinitesimal spherical cap in the spherical envelope, regarding Eqs. (3.31), (6.4), has the form

$$u_{1e} = - \left\{ \left[p_1 \kappa_{1e}^{(1)} - p_2 \kappa_{2e}^{(1)} \right] C_e^{(1)} r^{c_{3e}} + \frac{3}{2} \left[p_1 \kappa_{3e}^{(1)} - p_2 \kappa_{4e}^{(1)} \right] \right\}, \quad (6.171)$$

where the coefficients $c_{3e} < 0$, $C_e^{(1)}$ and the radial stresses p_1, p_2 acting on the particle-envelope, matrix-envelope boundaries are given by Eqs. (3.40), (6.5) and (10.2), (10.3), respectively.

With regard to Section 6.4.1.1.C, the coefficients $\varrho_{1+i\ 1+je}$ $(i, j = 0,1)$, ω_{1e}, ω_{2e}, ω_{3e}, ξ_{i1e}, ξ_{i2e} $(i = 1,2,3)$ mentioned below (see Eqs. (6.180), (6.181), (6.183), (6.184)) depend on $\kappa_{1e}^{(1)}, \ldots, \kappa_{4e}^{(1)}$ (see Eq. (12.35)) representing functions of χ_{11e}, χ_{12e}, χ_{21e}, χ_{22e} given by Eqs. (6.185), (6.164).

Consequently, the radial stress σ_{11e}, the tangential and shear stresses σ_{22e}, σ_{33e} and σ_{12e}, σ_{13e} acting in the spherical envelope, respectively, regarding Eqs. (3.31)–(3.34), (6.171), are derived as

$$\sigma_{11e} = -\left\{ \left[p_1 \kappa_{1e}^{(1)} - p_2 \kappa_{2e}^{(1)} \right] C_e^{(1)} \left[c_{3e} \left(c_{1e} + c_{2e} \right) - 2 c_{2e} \right] r^{c_{3e}-1} - \frac{3 c_{2e}}{r} \left[p_1 \kappa_{3e}^{(1)} - p_2 \kappa_{4e}^{(1)} \right] \right\}, \tag{6.172}$$

$$\sigma_{22e} = \sigma_{33e} = -\left\{ \left[p_1 \kappa_{1e}^{(1)} - p_2 \kappa_{2e}^{(1)} \right] C_e^{(1)} \left(c_{1e} - c_{2e} c_{3e} \right) r^{c_{3e}-1} + \frac{3 c_{1e}}{2r} \left[p_1 \kappa_{3e}^{(1)} - p_2 \kappa_{4e}^{(1)} \right] \right\}, \tag{6.173}$$

$$\sigma_{12e} = -\frac{1}{s_{44e}} \left\{ \left[\kappa_{1e}^{(1)} \frac{\partial p_1}{\partial \varphi} - \kappa_{2e}^{(1)} \frac{\partial p_2}{\partial \varphi} \right] C_e^{(1)} r^{c_{3e}-1} + \frac{3}{2r} \left[\kappa_{3e}^{(1)} \frac{\partial p_1}{\partial \varphi} - \kappa_{4e}^{(1)} \frac{\partial p_2}{\partial \varphi} \right] \right\}, \tag{6.174}$$

$$\sigma_{13e} = -\frac{1}{s_{44e}} \left\{ \left[\kappa_{1e}^{(1)} \frac{\partial p_1}{\partial \nu} - \kappa_{2e}^{(1)} \frac{\partial p_2}{\partial \nu} \right] C_e^{(1)} r^{c_{3e}-1} + \frac{3}{2r} \left[\kappa_{3e}^{(1)} \frac{\partial p_1}{\partial \nu} - \kappa_{4e}^{(1)} \frac{\partial p_2}{\partial \nu} \right] \right\}, \tag{6.175}$$

where the elastic modulus s_{44e} and the coefficients c_{1e}, c_{2e} are given by Eqs. (3.25) and (3.38), (3.39), respectively.

Finally, the radial strain ε_{11e}, the tangential and shear strains ε_{22e}, ε_{33e} and ε_{12e}, ε_{13e} in the spherical envelope, respectively, regarding Eqs. (3.26)–(3.30), (6.171), have the forms

$$\varepsilon_{11e} = -C_e^{(1)} c_{3e} \left[p_1 \kappa_{1e}^{(1)} - p_2 \kappa_{2e}^{(1)} \right] r^{c_{3e}-1}, \tag{6.176}$$

$$\varepsilon_{22e} = \varepsilon_{33e} = -\left\{ \left[p_1 \kappa_{1e}^{(1)} - p_2 \kappa_{2e}^{(1)} \right] C_e^{(1)} r^{c_{3e}-1} + \frac{3}{2r} \left[p_1 \kappa_{3e}^{(1)} - p_2 \kappa_{4e}^{(1)} \right] \right\}, \tag{6.177}$$

$$\varepsilon_{12e} = -\left\{ \left[\kappa_{1e}^{(1)} \frac{\partial p_1}{\partial \varphi} - \kappa_{2e}^{(1)} \frac{\partial p_2}{\partial \varphi} \right] C_e^{(1)} r^{c_{3e}-1} + \frac{3}{2r} \left[\kappa_{3e}^{(1)} \frac{\partial p_1}{\partial \varphi} - \kappa_{4e}^{(1)} \frac{\partial p_2}{\partial \varphi} \right] \right\}, \tag{6.178}$$

$$\varepsilon_{13e} = -\left\{ \left[\kappa_{1e}^{(1)} \frac{\partial p_1}{\partial \nu} - \kappa_{2e}^{(1)} \frac{\partial p_2}{\partial \nu} \right] C_e^{(1)} r^{c_{3e}-1} + \frac{3}{2r} \left[\kappa_{3e}^{(1)} \frac{\partial p_1}{\partial \nu} - \kappa_{4e}^{(1)} \frac{\partial p_2}{\partial \nu} \right] \right\}. \tag{6.179}$$

With regard to the determination of the compressive or tensile radial stresses $p_1 > 0$ or $p_1 < 0$ and $p_2 > 0$ or $p_2 < 0$ acting on the particle-envelope and matrix-envelope boundaries, thus for $r = R_1$ and $r = R_2$, respectively, considering Eqs. (3.59), (3.60), the tangential strains on the particle-envelope and matrix-envelope boundaries, $(\varepsilon_{22e})_{r=R_1}$ and $(\varepsilon_{22e})_{r=R_2}$, respectively, are required to be derived in the form

$$(\varepsilon_{22e})_{r=R_{1+i}} = -\sum_{j=0}^{1} p_{1+j}\varrho_{1+i\ 1+je}, \quad i = 0, 1, \tag{6.180}$$

where the coefficient $\varrho_{1+i\ 1+je}$ $(i, j = 0,1)$ is given by Eq. (6.186).

Considering Eqs. (3.41), (3.42), (6.172)–(6.179), the thermal-stress induced elastic energy density w_e and the thermal-stress induced elastic energy W_e, accumulated in an arbitrary point of the spherical envelope and in the volume of the spherical envelope, respectively, have the forms

$$w_e = \frac{1}{2}\left[\frac{\omega_{1e}}{r^2} + \omega_{2e}\, r^{c_{3e}-2} + \omega_{3e}\, r^{2(c_{3e}-1)}\right], \tag{6.181}$$

$$W_e = 8\int_0^{\pi/2}\int_0^{\pi/2}\int_{R_1}^{R_2} w_e\, r^2 dr d\varphi d\nu = \int_0^{\pi/2}\int_0^{\pi/2} \Omega_e d\varphi d\nu, \tag{6.182}$$

where the coefficient Ω_e given by Eq. (6.190) is a function of ω_{1e}, ω_{2e}, ω_{3e} given by Eqs. (6.187)–(6.189), respectively.

With regard to the determination of the thermal-stress induced strengthening presented in Volume III, the stress σ_{ie} acting in a direction of the axis x_i and the elastic energy density w_{ie} induced by the stress σ_{ie} $(i = 1,2,3)$ (see Eqs. (3.46), (3.47), (6.172)–(6.175)), both related to an arbitrary point of the spherical envelope, are derived as

$$\sigma_{ie} = -\left(\frac{\xi_{i1e}}{r} + \xi_{i2e}\, r^{c_{3e}-1}\right), \quad i = 1, 2, 3, \tag{6.183}$$

$$w_{ie} = \frac{s_{11e}}{2}\left[\frac{\xi_{i1e}^2}{r^2} + \xi_{i2e}^2\, r^{2(c_{3e}-1)} + 2\,\xi_{i1e}\,\xi_{i2e}\, r^{c_{3e}-2}\right], \quad i = 1, 2, 3, \tag{6.184}$$

where the elastic modulus s_{11e} and the coefficients ξ_{i1e}, ξ_{i2e} $(i = 1,2,3)$ are given by Eqs. (3.23) and (6.191), (6.192), respectively.

Appendix. Considering the elastic modulus s_{44e} and the coefficients c_{1e}, c_{2e}, c_{3e}, $C_e^{(1)}$, a'_{1i}, a'_{2i}, a'_{3i} $(i = 1,2,3)$ (see Eqs. (3.25), (3.38)–(3.40), (6.5), (12.26)–(12.34)), the coefficients $\chi_{1+i\,1e}$, $\varrho_{1+i\ 1+je}$ $(i, j = 0,1)$, ω_{1e}, ω_{2e}, ω_{3e}, Ω_e, $\xi_{i\,1e}$, $\xi_{i\,2e}$ $(i = 1,2,3)$ have the forms

$$\chi_{1+i\,1e} = C_e^{(1)}\left[c_{3e}\left(c_{1e}+c_{2e}\right)-2c_{2e}\right]R_{1+i}^{c_{3e}-1}, \quad i=0,1, \tag{6.185}$$

$$\varrho_{1+i\,1+je} = \left(\delta_{0j}-\delta_{1j}\right)\left[\kappa_{1+je}^{(1)}C_e^{(1)}R_{1+i}^{c_{3e}-1}+\frac{3\kappa_{3+je}^{(1)}}{2R_{1+i}}\right], \quad i,j=0,1, \tag{6.186}$$

$$\omega_{1e} = \frac{9}{2}\left(\left[p_1\kappa_{3e}^{(1)}-p_2\kappa_{4e}^{(1)}\right]^2 c_{1e} + \frac{1}{s_{44e}}\left\{\left[\kappa_{3e}^{(1)}\frac{\partial p_1}{\partial\varphi}-\kappa_{4e}^{(1)}\frac{\partial p_2}{\partial\varphi}\right]^2+\left[\kappa_{3e}^{(1)}\frac{\partial p_1}{\partial\nu}-\kappa_{4e}^{(1)}\frac{\partial p_2}{\partial\nu}\right]^2\right\}\right), \tag{6.187}$$

$$\omega_{2e} = 6C_e^{(1)}\left(\left[p_1\kappa_{1e}^{(1)}-p_2\kappa_{2e}^{(1)}\right]\left[p_1\kappa_{3e}^{(1)}-p_2\kappa_{4e}^{(1)}\right]\left(c_{1e}-c_{2e}c_{3e}\right) + \frac{1}{s_{44e}}\left\{\left[\kappa_{1e}^{(1)}\frac{\partial p_1}{\partial\varphi}-\kappa_{2e}^{(1)}\frac{\partial p_2}{\partial\varphi}\right]\left[\kappa_{3e}^{(1)}\frac{\partial p_1}{\partial\varphi}-\kappa_{4e}^{(1)}\frac{\partial p_2}{\partial\varphi}\right] + \left[\kappa_{1e}^{(1)}\frac{\partial p_1}{\partial\nu}-\kappa_{2e}^{(1)}\frac{\partial p_2}{\partial\nu}\right]\left[\kappa_{3e}^{(1)}\frac{\partial p_1}{\partial\nu}-\kappa_{4e}^{(1)}\frac{\partial p_2}{\partial\nu}\right]\right\}\right), \tag{6.188}$$

$$\omega_{3e} = \left[C_e^{(1)}\right]^2\left(\left[p_1\kappa_{1e}^{(1)}-p_2\kappa_{2e}^{(1)}\right]^2\left[c_{3e}^2\left(c_{1e}+c_{2e}\right)+2\left(c_{1e}-2c_{2e}c_{3e}\right)\right] + \frac{2}{s_{44e}}\left\{\left[\kappa_{1e}^{(1)}\frac{\partial p_1}{\partial\varphi}-\kappa_{2e}^{(1)}\frac{\partial p_2}{\partial\varphi}\right]^2+\left[\kappa_{1e}^{(1)}\frac{\partial p_1}{\partial\nu}-\kappa_{2e}^{(1)}\frac{\partial p_2}{\partial\nu}\right]^2\right\}\right), \tag{6.189}$$

$$\Omega_e = 4\left[\omega_{1e}\left(R_2-R_1\right)+\frac{\omega_{2e}}{c_{3e}+1}\left(R_2^{c_{3e}+1}-R_1^{c_{3e}+1}\right) + \frac{\omega_{3e}}{2c_{3e}+1}\left(R_2^{2c_{3e}+1}-R_1^{2c_{3e}+1}\right)\right], \tag{6.190}$$

$$\xi_{i\,1e} = \frac{3}{2}\left(\left[p_1\kappa_{3e}^{(1)}-p_2\kappa_{4e}^{(1)}\right]\left[c_{1e}\left(a'_{2i}+a'_{3i}\right)-2a'_{1i}\,c_{2e}\right] + \frac{1}{s_{44e}}\left\{\left(a'_{1i}+a'_{2i}\right)\left[\kappa_{3e}^{(1)}\frac{\partial p_1}{\partial\varphi}-\kappa_{4e}^{(1)}\frac{\partial p_2}{\partial\varphi}\right] + \left(a'_{1i}+a'_{3i}\right)\left[\kappa_{3e}^{(1)}\frac{\partial p_1}{\partial\nu}-\kappa_{4e}^{(1)}\frac{\partial p_2}{\partial\nu}\right]\right\}\right), \quad i=1,2,3, \tag{6.191}$$

$$\xi_{i\,2e} = C_e^{(1)}\left(\left[p_1\kappa_{1e}^{(1)}-p_2\kappa_{2e}^{(1)}\right]\left\{a'_{1i}\left[c_{3e}\left(c_{1e}+c_{2e}\right)-2c_{2e}\right]+\left(a'_{2i}+a'_{3i}\right)\left(c_{1e}-c_{2e}c_{3e}\right)\right\} + \frac{1}{s_{44e}}\left\{\left(a'_{1i}+a'_{2i}\right)\left[\kappa_{1e}^{(1)}\frac{\partial p_1}{\partial\varphi}-\kappa_{2e}^{(1)}\frac{\partial p_2}{\partial\varphi}\right] + \left(a'_{1i}+a'_{3i}\right)\left[\kappa_{1e}^{(1)}\frac{\partial p_1}{\partial\nu}-\kappa_{2e}^{(1)}\frac{\partial p_2}{\partial\nu}\right]\right\}\right), \quad i=1,2,3. \tag{6.192}$$

6.4.1.2 Conditions $\beta_p \neq \beta_e = \beta_m$ and $\varepsilon_{11te} - \varepsilon_{11tp} = f(\varphi, \nu)$

The thermal stresses and the corresponding quantities presented in Section 6.4.1.2 and related to Eq. (6.4) are derived on the condition $\beta_p \neq \beta_e = \beta_m$, considering the φ, ν-dependent difference $\varepsilon_{11te} - \varepsilon_{11tp} = f(\varphi, \nu)$ (see Eq. (12.23)), resulting in the φ, ν-dependent difference $\beta_e - \beta_p$ (see Eqs. (3.48)–(3.50)) and consequently in the φ, ν-dependent radial stress $p_1 = (\beta_e - \beta_p) / (\varrho_{1p} - \varrho_{1e}^{me})$ (see Eq. (10.4)) acting on the particle-envelope boundary, derived from Eq. (3.63) and included in the formulae in Section 6.4.1.2, where the φ, ν-independent coefficients ϱ_{1p} and ϱ_{1e}^{me} are given by Eqs. (5.36) and (5.106), (6.207), (6.228), (6.246), (7.230), respectively, considering the Castigliano's theorem (see Section 3.1.9).

A. Integration constants $C_{1e} \neq 0$, $C_{2e} = C_{3e} = 0$

With regard to the boundary condition (4.7) and the analysis presented in Section 6.2.2, considering the term u_{1re} (see Eq. (6.6)) to represent a decreasing function of the variable $r \in \langle R_1, R_2 \rangle$ provided that $R_1 > R = e^{-2/3} \approx 0.51342$ (see Eqs. (6.10), (6.12)), the radial displacement u_{1e} of the infinitesimal spherical cap in the spherical envelope, regarding Eqs. (3.31), (6.4), has the form

$$u_{1e} = -\frac{p_1 r (3 \ln r - 1)}{2(c_{1e} + 2c_{2e}) + 3(c_{1e} - c_{2e}) \ln R_1}, \tag{6.193}$$

where the coefficients c_{1e}, c_{2e} and the radial stress p_1 acting on the particle-envelope boundary are given by Eqs. (3.38), (3.39) and (10.4), respectively.

Against the extreme value of the radius $R_1 \geq e^{-2/3} \approx 0.51342$ (see Eq. (6.10)) in comparison with radii of particles in real composite materials [1], regarding completeness of the presented topic, the thermal stresses and the corresponding quantities in the spherical envelope related to the term u_{1re} are presented as a curiosity only.

Consequently, the radial stress σ_{11e}, the tangential and shear stresses σ_{22e}, σ_{33e} and σ_{12e}, σ_{13e} acting in the spherical envelope, respectively, regarding Eqs. (3.31)–(3.34), (6.193), are derived as

$$\sigma_{11e} = -\frac{p_1 [2(c_{1e} + 2c_{2e}) + 3(c_{1e} - c_{2e}) \ln r]}{2(c_{1e} + 2c_{2e}) + 3(c_{1e} - c_{2e}) \ln R_1}, \tag{6.194}$$

$$\sigma_{22e} = \sigma_{33e} = -\frac{p_1 [3(c_{1e} - c_{2e}) \ln r - (c_{1e} + 2c_{2e})]}{2(c_{1e} + 2c_{2e}) + 3(c_{1e} - c_{2e}) \ln R_1}, \tag{6.195}$$

$$\sigma_{12e} = -\frac{3 \ln r - 1}{s_{44e} [2(c_{1e} + 2c_{2e}) + 3(c_{1e} - c_{2e}) \ln R_1]} \frac{\partial p_1}{\partial \varphi}, \tag{6.196}$$

$$\sigma_{13e} = -\frac{3 \ln r - 1}{s_{44e} [2(c_{1e} + 2c_{2e}) + 3(c_{1e} - c_{2e}) \ln R_1]} \frac{\partial p_1}{\partial \nu}, \tag{6.197}$$

where the elastic modulus s_{44e} is given by Eq. (3.25).

Finally, the radial strain ε_{11e}, the tangential and shear strains ε_{22e}, ε_{33e} and ε_{12e}, ε_{13e} in the spherical envelope, respectively, regarding Eqs. (3.26)–(3.30), (6.193), have the forms

$$\varepsilon_{11e} = -\frac{p_1\left(3\ln r + 2\right)}{2\left(c_{1e} + 2c_{2e}\right) + 3\left(c_{1e} - c_{2e}\right)\ln R_1}, \tag{6.198}$$

$$\varepsilon_{22e} = \varepsilon_{33e} = -\frac{p_1\left(3\ln r - 1\right)}{2\left(c_{1e} + 2c_{2e}\right) + 3\left(c_{1e} - c_{2e}\right)\ln R_1}, \tag{6.199}$$

$$\varepsilon_{12e} = -\frac{3\ln r - 1}{2\left(c_{1e} + 2c_{2e}\right) + 3\left(c_{1e} - c_{2e}\right)\ln R_1}\frac{\partial p_1}{\partial \varphi}, \tag{6.200}$$

$$\varepsilon_{13e} = -\frac{3\ln r - 1}{2\left(c_{1e} + 2c_{2e}\right) + 3\left(c_{1e} - c_{2e}\right)\ln R_1}\frac{\partial p_1}{\partial \nu}. \tag{6.201}$$

With regard to the determination of the compressive or tensile radial stress $p_1 > 0$ or $p_1 < 0$ acting on the particle-matrix boundary, thus for $r = R_1$, respectively, considering Eq. (3.59), and to the boundary condition for the cell matrix provided that $\beta_p \neq \beta_e = \beta_m$, the tangential strains on the particle-envelope and matrix-envelope boundaries, $(\varepsilon_{22e})_{r=R_1}$ and $(\varepsilon_{22e})_{r=R_2}$, respectively, are required to be derived in the form

$$(\varepsilon_{22e})_{r=R_{1+i}} = -p_1\,\varrho^{me}_{1+ie}, \quad i = 0, 1, \tag{6.202}$$

where the coefficient ϱ^{me}_{1+ie} $(i=0,1)$ is given by Eq. (6.207).

Considering Eqs. (3.41), (3.42), (6.194)–(6.201), the thermal-stress induced elastic energy density w_e and the thermal-stress induced elastic energy W_e, accumulated in an arbitrary point of the spherical envelope and in the volume of the spherical envelope, respectively, have the forms

$$w_e = \frac{\omega_{1e} + \omega_{2e}\ln r + \omega_{3e}\ln^2 r}{2\left[2\left(c_{1e} + 2c_{2e}\right) + 3\left(c_{1e} - c_{2e}\right)\ln R_1\right]^2}, \tag{6.203}$$

$$W_e = 8\int_0^{\pi/2}\int_0^{\pi/2}\int_{R_1}^{R_2} w_e\, r^2 dr d\varphi d\nu = \int_0^{\pi/2}\int_0^{\pi/2} \Omega_{1e} d\varphi d\nu, \tag{6.204}$$

where the coefficient Ω_{1e} given by Eq. (6.211) is a function of ω_{1e}, ω_{2e}, ω_{3e} given by Eqs. (6.208)–(6.210), respectively.

With regard to the determination of the thermal-stress induced strengthening presented in Volume III, the stress σ_{ie} acting in a direction of the axis x_i and the elastic energy density w_{ie} induced by the stress σ_{ie} $(i=1,2,3)$ (see Eqs. (3.46), (3.47),

(6.194)–(6.197)), both related to an arbitrary point of the spherical envelope, are derived as[7]

$$\sigma_{ie} = \frac{\xi_{i1e} - \xi_{i2e} \ln r}{2\,(c_{1e} + 2c_{2e}) + 3\,(c_{1e} - c_{2e}) \ln R_1}, \quad i = 1, 2, 3, \tag{6.205}$$

$$w_{ie} = \frac{s_{11e} \left(\xi_{i1e}^2 + \xi_{i2e}^2 \ln^2 r - 2\,\xi_{i1e} \xi_{i2e} \ln r\right)}{2\,[2\,(c_{1e} + 2c_{2e}) + 3\,(c_{1e} - c_{2e}) \ln R_1]^2}, \quad i = 1, 2, 3, \tag{6.206}$$

where the elastic modulus s_{11e} and the coefficients ξ_{i1e}, ξ_{i2e} $(i = 1,2,3)$ are given by Eqs. (3.23) and (6.212), (6.213), respectively.

Appendix. Considering the elastic modulus s_{44e} and the coefficients c_{1e}, c_{2e}, a'_{1i}, a'_{2i}, a'_{3i} $(i = 1,2,3)$ (see Eqs. (3.25), (3.38), (3.39), (12.26)–(12.34)), the coefficients ϱ_{1+ie}^{me}, ω_{1+3ie}, ω_{2+3ie}, ω_{3+3ie}, Ω_{1+ie} $(i = 0,1)$, ξ_{i1+2je}, ξ_{i2+2je}, ξ_{i3+2je} $(i = 1,2,3;\ j = 0,1)$ have the forms

$$\varrho_{1+ie}^{me} = \frac{3 \ln R_{1+i} - 1}{2\,(c_{1e} + 2c_{2e}) + 3\,(c_{1e} - c_{2e}) \ln R_1}, \quad i = 0, 1, \tag{6.207}$$

$$\omega_{1+3ie} = 2 \left\{ 3p_{1+i}^2 \,(c_{1e} + 2c_{2e}) + \frac{1}{s_{44e}} \left[\left(\frac{\partial p_{1+i}}{\partial \varphi} \right)^2 + \left(\frac{\partial p_{1+i}}{\partial \nu} \right)^2 \right] \right\}, \quad i = 0, 1, \tag{6.208}$$

$$\omega_{2+3ie} = -\frac{12}{s_{44e}} \left[\left(\frac{\partial p_{1+i}}{\partial \varphi} \right)^2 + \left(\frac{\partial p_{1+i}}{\partial \nu} \right)^2 \right], \quad i = 0, 1, \tag{6.209}$$

$$\omega_{3+3ie} = 9 \left\{ 3p_{1+i}^2 \,(c_{1e} - c_{2e}) + \frac{2}{s_{44e}} \left[\left(\frac{\partial p_{1+i}}{\partial \varphi} \right)^2 + \left(\frac{\partial p_{1+i}}{\partial \nu} \right)^2 \right] \right\}, \quad i = 0, 1, \tag{6.210}$$

$$\begin{aligned} \Omega_{1+ie} &= \left[\frac{2}{2\,(c_{1e} + 2c_{2e}) + 3\,(c_{1e} - c_{2e}) \ln R_{1+i}} \right]^2 \\ &\times \left(\frac{\omega_{1+3ie}}{3} \left(R_2^3 - R_1^3\right) + \frac{\omega_{2+3ie}}{3} \left[R_2^3 \left(\ln R_2 - \frac{1}{3} \right) - R_1^3 \left(\ln R_1 - \frac{1}{3} \right) \right] \right. \\ &\left. + \frac{\omega_{3+3ie}}{3} \left\{ R_2^3 \left[\left(\ln R_2 - \frac{1}{3} \right)^2 + \frac{1}{9} \right] - R_1^3 \left[\left(\ln R_1 - \frac{1}{3} \right)^2 + \frac{1}{9} \right] \right\} \right), \\ & i = 0, 1, \end{aligned} \tag{6.211}$$

[7]To avoid misunderstanding, the fraction in Eq. (6.205) really contains the plus sign, in contrast to the minus sign in Eq. (6.226)).

$$\begin{aligned} \xi_{i\,1+2je} &= p_{1+j}\left(a'_{2i} + a'_{3i} - 2a'_{1i}\right)\left(c_{1e} + 2c_{2e}\right) \\ &+ \frac{1}{s_{44e}}\left[\left(a'_{1i} + a'_{2i}\right)\frac{\partial p_{1+j}}{\partial \varphi} + \left(a'_{1i} + a'_{3i}\right)\frac{\partial p_{1+j}}{\partial \nu}\right], \\ & i = 1, 2, 3; \quad j = 0, 1, \end{aligned} \tag{6.212}$$

$$\begin{aligned} \xi_{i\,2+2je} &= 3\left\{p_{1+j}\left(a'_{1i} + a'_{2i} + a'_{3i}\right)\left(c_{1e} - c_{2e}\right)\right. \\ &+ \left.\frac{1}{s_{44e}}\left[\left(a'_{1i} + a'_{2i}\right)\frac{\partial p_{1+j}}{\partial \varphi} + \left(a'_{1i} + a'_{3i}\right)\frac{\partial p_{1+j}}{\partial \nu}\right]\right\}, \\ & i = 1, 2, 3; \quad j = 0, 1. \end{aligned} \tag{6.213}$$

B. Integration constants $C_{2e} \neq 0$, $C_{1e} = C_{3e} = 0$

With regard to the boundary condition (4.7) and the analysis presented in Section 6.2.2, considering the term u_{2re} (see Eq. (6.7)) to represent a decreasing function of the variable $r \in \langle R_1, R_2 \rangle$, the radial displacement u_{1e} of the infinitesimal spherical cap in the spherical envelope, regarding Eqs. (3.31), (6.4), has the form

$$u_{1e} = -\frac{p_1 r}{c_{3e}\left(c_{1e} + c_{2e}\right) - 2c_{2e}}\left(\frac{r}{R_1}\right)^{c_{3e}-1}, \tag{6.214}$$

where the coefficients c_{1e}, c_{2e}, $c_{3e} < 0$ and the radial stress p_1 acting on the particle-envelope boundary are given by Eqs. (3.38)–(3.40) and (10.4), respectively.

Consequently, the radial stress σ_{11e}, the tangential and shear stresses σ_{22e}, σ_{33e} and σ_{12e}, σ_{13e} acting in the spherical envelope, respectively, regarding Eqs. (3.31)–(3.34), (6.214), are derived as

$$\sigma_{11e} = -p_1\left(\frac{r}{R_1}\right)^{c_{3e}-1}, \tag{6.215}$$

$$\sigma_{22e} = \sigma_{33e} = -\frac{p_1\left(c_{1e} - c_{2e}c_{3e}\right)}{c_{3e}\left(c_{1e} + c_{2e}\right) - 2c_{2e}}\left(\frac{r}{R_1}\right)^{c_{3e}-1}, \tag{6.216}$$

$$\sigma_{12e} = -\frac{1}{s_{44e}\left[c_{3e}\left(c_{1e} + c_{2e}\right) - 2c_{2e}\right]}\frac{\partial p_1}{\partial \varphi}\left(\frac{r}{R_1}\right)^{c_{3e}-1}, \tag{6.217}$$

$$\sigma_{13e} = -\frac{1}{s_{44e}\left[c_{3e}\left(c_{1e} + c_{2e}\right) - 2c_{2e}\right]}\frac{\partial p_1}{\partial \nu}\left(\frac{r}{R_1}\right)^{c_{3e}-1}, \tag{6.218}$$

where the elastic modulus s_{44e} is given by Eq. (3.25).

Finally, the radial strain ε_{11e}, the tangential and shear strains ε_{22e}, ε_{33e} and ε_{12e}, ε_{13e} in the spherical envelope, respectively, regarding Eqs. (3.26)–(3.30), (6.214), have the forms

$$\varepsilon_{11e} = -\frac{p_1 c_{3e}}{c_{3e}(c_{1e} + c_{2e}) - 2c_{2e}} \left(\frac{r}{R_1}\right)^{c_{3e}-1}, \tag{6.219}$$

$$\varepsilon_{22e} = \varepsilon_{33e} = -\frac{p_1}{c_{3e}(c_{1e} + c_{2e}) - 2c_{2e}} \left(\frac{r}{R_1}\right)^{c_{3e}-1}, \tag{6.220}$$

$$\varepsilon_{12e} = -\frac{1}{c_{3e}(c_{1e} + c_{2e}) - 2c_{2e}} \frac{\partial p_1}{\partial \varphi} \left(\frac{r}{R_1}\right)^{c_{3e}-1}, \tag{6.221}$$

$$\varepsilon_{13e} = -\frac{1}{c_{3e}(c_{1e} + c_{2e}) - 2c_{2e}} \frac{\partial p_1}{\partial \nu} \left(\frac{r}{R_1}\right)^{c_{3e}-1}. \tag{6.222}$$

With regard to the determination of the compressive or tensile radial stress $p_1 > 0$ or $p_1 < 0$ acting on the particle-matrix boundary, thus for $r = R_1$, respectively, considering Eq. (3.59), and to the boundary condition for the cell matrix provided that $\beta_p \neq \beta_e = \beta_m$, the tangential strains on the particle-envelope and matrix-envelope boundaries, $(\varepsilon_{22e})_{r=R_1}$ and $(\varepsilon_{22e})_{r=R_2}$, respectively, are required to be derived in the form

$$(\varepsilon_{22e})_{r=R_{1+i}} = -p_1\, \varrho^{me}_{1+ie}, \quad i = 0, 1, \tag{6.223}$$

where the coefficient ϱ^{me}_{1+ie} $(i=0,1)$ is given by Eq. (6.228).

Considering Eqs. (3.41), (3.42), (6.215)–(6.222), the thermal-stress induced elastic energy density w_e and the thermal-stress induced elastic energy W_e, accumulated in an arbitrary point of the spherical envelope and in the volume of the spherical envelope, respectively, have the forms

$$w_e = \frac{\omega_e}{2\left[c_{3e}(c_{1e} + c_{2e}) - 2c_{2e}\right]^2} \left(\frac{r}{R_1}\right)^{2(c_{3e}-1)}, \tag{6.224}$$

$$W_e = 8 \int\limits_0^{\pi/2} \int\limits_0^{\pi/2} \int\limits_{R_1}^{R_2} w_e\, r^2 dr d\varphi d\nu = \int\limits_0^{\pi/2} \int\limits_0^{\pi/2} \Omega_e d\varphi d\nu, \tag{6.225}$$

where the coefficient Ω_e given by Eq. (6.230) is a function of ω_e given by Eq. (6.229).

With regard to the determination of the thermal-stress induced strengthening presented in Volume III, the stress σ_{ie} acting in a direction of the axis x_i and the elastic energy density w_{ie} induced by the stress σ_{ie} $(i=1,2,3)$ (see Eqs. (3.46), (3.47), (6.215)–(6.218)), both related to an arbitrary point of the spherical envelope, are derived as

$$\sigma_{ie} = -\frac{\xi_{ie}}{c_{3e}(c_{1e} + c_{2e}) - 2c_{2e}} \left(\frac{r}{R_1}\right)^{c_{3e}-1}, \quad i = 1, 2, 3, \tag{6.226}$$

$$w_{ie} = \frac{s_{11e}}{2} \left[\frac{\xi_{ie}}{c_{3e} \left(c_{1e} + c_{2e} \right) - 2c_{2e}} \right]^2 \left(\frac{r}{R_1} \right)^{2(c_{3e}-1)}, \quad i = 1, 2, 3, \tag{6.227}$$

where the elastic modulus s_{11e} and the coefficient ξ_{ie} $(i = 1,2,3)$ is given by Eqs. (3.23) and (6.231), respectively.

Appendix. Considering the elastic modulus s_{44e} and the coefficients c_{1e}, c_{2e}, c_{3e}, a'_{1i}, a'_{2i}, a'_{3i} $(i = 1,2,3)$ (see Eqs. (3.25), (3.38)–(3.40), (12.26)–(12.34)), the coefficients ϱ^{me}_{1+ie} $(i = 0,1)$, ω_e, Ω_e, ξ_{ie} $(i = 1,2,3)$ have the forms

$$\varrho^{me}_{1+ie} = \frac{1}{c_{3e} \left(c_{1e} + c_{2e} \right) - 2c_{2e}} \left(\frac{R_{1+i}}{R_1} \right)^{c_{3e}-1}, \quad i = 0, 1, \tag{6.228}$$

$$\omega_e = p_1^2 \left[c_{3e}^2 \left(c_{1e} + c_{2e} \right) - 4c_{2e}c_{3e} + 2c_{1e} \right] + \frac{2}{s_{44e}} \left[\left(\frac{\partial p_1}{\partial \varphi} \right)^2 + \left(\frac{\partial p_1}{\partial \nu} \right)^2 \right], \tag{6.229}$$

$$\Omega_e = \frac{\omega_e R_1^3}{2c_{3e} + 1} \left[\frac{2}{c_{3e} \left(c_{1e} + c_{2e} \right) - 2c_{2e}} \right]^2 \left[\left(\frac{R_2}{R_1} \right)^{2c_{3e}+1} - 1 \right], \tag{6.230}$$

$$\begin{aligned} \xi_{ie} &= p_1 \left\{ a'_{1i} \left[c_{3e} \left(c_{1e} + c_{2e} \right) - 2c_{2e} \right] + \left(a'_{2i} + a'_{3i} \right) \left(c_{1e} - c_{2e}c_{3e} \right) \right\} \\ &+ \frac{1}{s_{44e}} \left[\left(a'_{1i} + a'_{2i} \right) \frac{\partial p_1}{\partial \varphi} + \left(a'_{1i} + a'_{3i} \right) \frac{\partial p_1}{\partial \nu} \right], \quad i = 1, 2, 3. \end{aligned} \tag{6.231}$$

C. Integration constants $C_{3e} \neq 0$, $C_{1e} = C_{2e} = 0$

With regard to the boundary condition (4.7) and the analysis presented in Section 6.2.2, considering the term u_{3re} (see Eq. (6.8)) to represent a constant function regarding the variable $r \in \langle R_1, R_2 \rangle$, the radial displacement u_{1e} of the infinitesimal spherical cap in the spherical envelope, regarding Eqs. (3.31), (6.4), has the form

$$u_{1e} = \frac{p_1 R_1}{2c_{2e}}, \tag{6.232}$$

where the coefficient c_{2e} and the radial stress p_1 acting on the particle-envelope boundary are given by Eq. (3.39) and (10.4), respectively.

Consequently, the radial stress σ_{11e}, the tangential and shear stresses σ_{22e}, σ_{33e} and σ_{12e}, σ_{13e} acting in the spherical envelope, respectively, regarding Eqs. (3.31)–(3.34), (6.232), are derived as

$$\sigma_{11e} = -\frac{p_1 R_1}{r}, \tag{6.233}$$

$$\sigma_{22e} = \sigma_{33e} = \frac{c_{1e} p_1 R_1}{2c_{2e} r}, \tag{6.234}$$

$$\sigma_{12e} = \frac{R_1}{2s_{44e}c_{2e}r}\frac{\partial p_1}{\partial \varphi}, \tag{6.235}$$

$$\sigma_{13e} = \frac{R_1}{2s_{44e}c_{2e}r}\frac{\partial p_1}{\partial \nu}, \tag{6.236}$$

where the elastic modulus s_{44e} and the coefficient c_{1e} are given by Eqs. (3.25) and (3.38), respectively.

Finally, the radial strain ε_{11e}, the tangential and shear strains ε_{22e}, ε_{33e} and ε_{12e}, ε_{13e} in the spherical envelope, respectively, regarding Eqs. (3.26)–(3.30), (6.232), have the forms

$$\varepsilon_{11e} = 0, \tag{6.237}$$

$$\varepsilon_{22e} = \varepsilon_{33e} = \frac{p_1 R_1}{2c_{2e}r}, \tag{6.238}$$

$$\varepsilon_{12e} = \frac{R_1}{2c_{2e}r}\frac{\partial p_1}{\partial \varphi}, \tag{6.239}$$

$$\varepsilon_{13e} = \frac{R_1}{2c_{2e}r}\frac{\partial p_1}{\partial \nu}. \tag{6.240}$$

With regard to the determination of the compressive or tensile radial stress $p_1 > 0$ or $p_1 < 0$ acting on the particle-matrix boundary, thus for $r = R_1$, respectively, considering Eq. (3.59), and to the boundary condition for the cell matrix provided that $\beta_p \neq \beta_e = \beta_m$, the tangential strains on the particle-envelope and matrix-envelope boundaries, $(\varepsilon_{22e})_{r=R_1}$ and $(\varepsilon_{22e})_{r=R_2}$, respectively, are required to be derived in the form

$$(\varepsilon_{22e})_{r=R_{1+i}} = -p_1\, \varrho^{me}_{1+ie}, \quad i = 0, 1, \tag{6.241}$$

where the coefficient ϱ^{me}_{1+ie} $(i=0,1)$ is given by Eq. (6.246).

Considering Eqs. (3.41), (3.42), (6.215)–(6.222), the thermal-stress induced elastic energy density w_e and the thermal-stress induced elastic energy W_e, accumulated in an arbitrary point of the spherical envelope and in the volume of the spherical envelope, respectively, have the forms

$$w_e = \frac{\omega_{1e} R_1^2}{4c_{2e}^2 r^2}, \tag{6.242}$$

$$W_e = 8\int_0^{\pi/2}\int_0^{\pi/2}\int_{R_1}^{R_2} w_e\, r^2 dr d\varphi d\nu = \int_0^{\pi/2}\int_0^{\pi/2} \Omega_{1e} d\varphi d\nu, \tag{6.243}$$

where the coefficient Ω_{1e} given by Eq. (6.248) is a function of ω_{1e} given by Eq. (6.247).

With regard to the determination of the thermal-stress induced strengthening presented in Volume III, the stress σ_{ie} acting in a direction of the axis x_i and the elastic energy density w_{ie} induced by the stress σ_{ie} ($i = 1,2,3$) (see Eqs. (3.46), (3.47), (6.215)–(6.218)), both related to an arbitrary point of the spherical envelope, are derived as[8]

$$\sigma_{ie} = \frac{\xi_{i1e} R_1}{2 c_{2e} r}, \quad i = 1, 2, 3, \tag{6.244}$$

$$w_{ie} = \frac{s_{11e}\, \xi_{i1e}^2 R_1^2}{8 c_{2e}^2 r^2}, \quad i = 1, 2, 3, \tag{6.245}$$

where the elastic modulus s_{11e} and the coefficient ξ_{i1e} ($i = 1,2,3$) is given by Eqs. (3.23) and (6.249), respectively.

Appendix. Considering the elastic modulus s_{44e} and the coefficients c_{1e}, c_{2e}, a'_{1i}, a'_{2i}, a'_{3i} ($i = 1,2,3$) (see Eqs. (3.25), (3.38), (3.39), (12.26)–(12.34)), the coefficients ϱ^{me}_{1+ie}, ω_{1+ie}, Ω_{1+ie} ($i = 0,1$), ξ_{i1+je} ($i = 1,2,3$; $j = 0,1$) have the forms

$$\varrho^{me}_{1+ie} = -\frac{R_1}{2 c_{2e} R_{1+i}}, \quad i = 0, 1, \tag{6.246}$$

$$\omega_{1+ie} = p_{1+i}^2 c_{1e} + \frac{1}{s_{44e}} \left[\left(\frac{\partial p_{1+i}}{\partial \varphi} \right)^2 + \left(\frac{\partial p_{1+i}}{\partial \nu} \right)^2 \right], \quad i = 0, 1, \tag{6.247}$$

$$\Omega_{1+ie} = 2\, \omega_{1+ie} \left(R_2 - R_1 \right) \left(\frac{R_{1+i}}{c_{2e}} \right)^2, \quad i = 0, 1, \tag{6.248}$$

$$\begin{aligned} \xi_{i1+je} = {} & p_{1+j} \left[c_{1e} \left(a'_{2i} + a'_{3i} \right) - 2 a'_{1i} c_{2e} \right] \\ & + \frac{1}{s_{44e}} \left[\left(a'_{1i} + a'_{2i} \right) \frac{\partial p_{1+j}}{\partial \varphi} + \left(a'_{1i} + a'_{3i} \right) \frac{\partial p_{1+j}}{\partial \nu} \right], \\ & i = 1, 2, 3; \quad j = 0, 1. \end{aligned} \tag{6.249}$$

6.4.1.3 Condition $\beta_p = \beta_e \neq \beta_m$

A. Integration constants $C_{1e} \neq 0$, $C_{2e} = C_{3e} = 0$

With regard to the boundary condition (4.8) and the analysis presented in Section 6.2.2, considering the term u_{1re} (see Eq. (6.6)) to represent an increasing function of the variable $r \in \langle R_1, R_2 \rangle$ provided that $R_2 < R = e^{-2/3} \approx 0.51342$ (see

[8]To avoid misunderstanding, Equations (6.244) and (6.246) really contain the plus and minus signs, respectively.

Eq. (6.13)), as fulfilled for the radii $R_1 < R_2$ of envelopes on the surface of particles in real composite materials [1], the radial displacement u_{1e} of the infinitesimal spherical cap in the spherical envelope, regarding Eqs. (3.31), (6.4), has the form

$$u_{1e} = -\frac{p_2\, r\,(3\ln r - 1)}{2\,(c_{1e} + 2c_{2e}) + 3\,(c_{1e} - c_{2e})\ln R_2}, \tag{6.250}$$

where the coefficients c_{1e}, c_{2e} and the radial stress p_2 acting on the matrix-envelope boundary are given by Eqs. (3.38), (3.39) and (10.5), respectively.

Consequently, the radial stress σ_{11e}, the tangential and shear stresses σ_{22e}, σ_{33e} and σ_{12e}, σ_{13e} acting in the spherical envelope, respectively, regarding Eqs. (3.31)–(3.34), (6.250), are derived as

$$\sigma_{11e} = -\frac{p_2\,[2\,(c_{1e} + 2c_{2e}) + 3\,(c_{1e} - c_{2e})\ln r]}{2\,(c_{1e} + 2c_{2e}) + 3\,(c_{1e} - c_{2e})\ln R_2}, \tag{6.251}$$

$$\sigma_{22e} = \sigma_{33e} = -\frac{p_2\,[3\,(c_{1e} - c_{2e})\ln r - (c_{1e} + 2c_{2e})]}{2\,(c_{1e} + 2c_{2e}) + 3\,(c_{1e} - c_{2e})\ln R_2}, \tag{6.252}$$

$$\sigma_{12e} = -\frac{3\ln r - 1}{s_{44e}\,[2\,(c_{1e} + 2c_{2e}) + 3\,(c_{1e} - c_{2e})\ln R_2]}\frac{\partial p_2}{\partial \varphi}, \tag{6.253}$$

$$\sigma_{13e} = -\frac{3\ln r - 1}{s_{44e}\,[2\,(c_{1e} + 2c_{2e}) + 3\,(c_{1e} - c_{2e})\ln R_2]}\frac{\partial p_2}{\partial \nu}, \tag{6.254}$$

where the elastic modulus s_{44e} is given by Eq. (3.25).

Finally, the radial strain ε_{11e}, the tangential and shear strains ε_{22e}, ε_{33e} and ε_{12e}, ε_{13e} in the spherical envelope, respectively, regarding Eqs. (3.26)–(3.30), (6.250), have the forms

$$\varepsilon_{11e} = -\frac{p_2\,(3\ln r + 2)}{2\,(c_{1e} + 2c_{2e}) + 3\,(c_{1e} - c_{2e})\ln R_2}, \tag{6.255}$$

$$\varepsilon_{22e} = \varepsilon_{33e} = -\frac{p_2\,(3\ln r - 1)}{2\,(c_{1e} + 2c_{2e}) + 3\,(c_{1e} - c_{2e})\ln R_2}, \tag{6.256}$$

$$\varepsilon_{12e} = -\frac{3\ln r - 1}{2\,(c_{1e} + 2c_{2e}) + 3\,(c_{1e} - c_{2e})\ln R_2}\frac{\partial p_2}{\partial \varphi}, \tag{6.257}$$

$$\varepsilon_{13e} = -\frac{3\ln r - 1}{2\,(c_{1e} + 2c_{2e}) + 3\,(c_{1e} - c_{2e})\ln R_2}\frac{\partial p_2}{\partial \nu}. \tag{6.258}$$

With regard to the determination of the compressive or tensile radial stress $p_2 > 0$ or $p_2 < 0$ acting on the matrix-envelope boundary, thus for $r = R_2$, respectively, considering Eq. (3.60), and to the boundary condition for the spherical particle provided that $\beta_p = \beta_e \neq \beta_m$, the tangential strains on the particle-envelope and matrix-envelope boundaries, $(\varepsilon_{22e})_{r=R_1}$ and $(\varepsilon_{22e})_{r=R_2}$, respectively, are required to be derived in the form

$$(\varepsilon_{22e})_{r=R_{1+i}} = -p_2\, \varrho^{pe}_{1+ie}, \quad i = 0, 1, \tag{6.259}$$

where the coefficient ϱ^{pe}_{1+ie} $(i=0,1)$ is given by Eq. (6.264).

Considering Eqs. (3.41), (3.42), (6.251)–(6.258), the thermal-stress induced elastic energy density w_e and the thermal-stress induced elastic energy W_e, accumulated in an arbitrary point of the spherical envelope and in the volume of the spherical envelope, respectively, have the forms

$$w_e = \frac{\omega_{4e} + \omega_{5e} \ln r + \omega_{6e} \ln^2 r}{2\left[2\left(c_{1e} + 2c_{2e}\right) + 3\left(c_{1e} - c_{2e}\right) \ln R_1\right]^2}, \tag{6.260}$$

$$W_e = 8 \int_0^{\pi/2} \int_0^{\pi/2} \int_{R_1}^{R_2} w_e\, r^2 dr d\varphi d\nu = \int_0^{\pi/2} \int_0^{\pi/2} \Omega_{2e} d\varphi d\nu, \tag{6.261}$$

where the coefficient Ω_{2e} given by Eq. (6.211) is a function of ω_{4e}, ω_{5e}, ω_{6e} given by Eqs. (6.208)–(6.210), respectively.

With regard to the determination of the thermal-stress induced strengthening presented in Volume III, the stress σ_{ie} acting in a direction of the axis x_i and the elastic energy density w_{ie} induced by the stress σ_{ie} $(i=1,2,3)$ (see Eqs. (3.46), (3.47), (6.251)–(6.254)), both related to an arbitrary point of the spherical envelope, are derived as

$$\sigma_{ie} = \frac{\xi_{i3e} - \xi_{i4e} \ln r}{2\left(c_{1e} + 2c_{2e}\right) + 3\left(c_{1e} - c_{2e}\right) \ln R_1}, \quad i = 1, 2, 3, \tag{6.262}$$

$$w_{ie} = \frac{s_{11e}\left(\xi^2_{i3e} + \xi^2_{i4e} \ln^2 r - 2\,\xi_{i3e}\xi_{i4e} \ln r\right)}{2\left[2\left(c_{1e} + 2c_{2e}\right) + 3\left(c_{1e} - c_{2e}\right) \ln R_1\right]^2}, \quad i = 1, 2, 3, \tag{6.263}$$

where the elastic modulus s_{11e} and the coefficients ξ_{i3e}, ξ_{i4e} $(i=1,2,3)$ are given by Eqs. (3.23) and (6.212), (6.213), respectively.

Appendix. Considering the coefficients c_{1e}, c_{2e} (see Eq. (3.38), (3.39)), the coefficient ϱ^{pe}_{1+ie} $(i=0,1)$ has the form

$$\varrho^{pe}_{1+ie} = \frac{3 \ln R_{1+i} - 1}{2\left(c_{1e} + 2c_{2e}\right) + 3\left(c_{1e} - c_{2e}\right) \ln R_2}, \quad i = 0, 1. \tag{6.264}$$

B. Integration constants $C_{3e} \neq 0$, $C_{1e} = C_{2e} = 0$

With regard to the boundary condition (4.8) and the analysis presented in Section 6.2.2, considering the term u_{3re} (see Eq. (6.8)) to represent a constant function regarding the variable $r \in \langle R_1, R_2 \rangle$, the radial displacement u_{1e} of the infinitesimal spherical cap in the spherical envelope, regarding Eqs. (3.31), (6.4), has the form

$$u_{1e} = \frac{p_2 R_2}{2c_{2e}}, \tag{6.265}$$

where the coefficient c_{2e} and the radial stress p_2 acting on the matrix-envelope boundary are given by Eqs. (3.39) and (10.5), respectively.

Consequently, the radial stress σ_{11e}, the tangential and shear stresses σ_{22e}, σ_{33e} and σ_{12e}, σ_{13e} acting in the spherical envelope, respectively, regarding Eqs. (3.31)–(3.34), (6.265), are derived as[9]

$$\sigma_{11e} = -\frac{p_2 R_2}{r}, \tag{6.266}$$

$$\sigma_{22e} = \sigma_{33e} = \frac{c_{1e} p_2 R_2}{2c_{2e} r}, \tag{6.267}$$

$$\sigma_{12e} = \frac{R_2}{2s_{44e} c_{2e} r} \frac{\partial p_2}{\partial \varphi}, \tag{6.268}$$

$$\sigma_{13e} = \frac{R_2}{2s_{44e} c_{2e} r} \frac{\partial p_2}{\partial \nu}, \tag{6.269}$$

where the elastic modulus s_{44e} and the coefficient c_{1e} are given by Eqs. (3.25) and (3.38), respectively.

Finally, the radial strain ε_{11e}, the tangential and shear strains ε_{22e}, ε_{33e} and ε_{12e}, ε_{13e} in the spherical envelope, respectively, regarding Eqs. (3.26)–(3.30), (6.265), have the forms

$$\varepsilon_{11e} = 0, \tag{6.270}$$

$$\varepsilon_{22e} = \varepsilon_{33e} = \frac{p_2 R_2}{2c_{2e} r}, \tag{6.271}$$

$$\varepsilon_{12e} = \frac{R_2}{2c_{2e} r} \frac{\partial p_2}{\partial \varphi}, \tag{6.272}$$

$$\varepsilon_{13e} = \frac{R_2}{2c_{2e} r} \frac{\partial p_2}{\partial \nu}. \tag{6.273}$$

[9]To avoid misunderstanding, Equation (6.266) really contains the minus sign, in contrast to the plus sign in Eqs. (6.267)–(6.269).

With regard to the determination of the compressive or tensile radial stress $p_2 > 0$ or $p_2 < 0$ acting on the matrix-envelope boundary, thus for $r = R_2$, respectively, considering Eq. (3.60), and to the boundary condition for the spherical particle provided that $\beta_p = \beta_e \neq \beta_m$, the tangential strains on the particle-envelope and matrix-envelope boundaries, $(\varepsilon_{22e})_{r=R_1}$ and $(\varepsilon_{22e})_{r=R_2}$, respectively, are required to be derived in the form

$$(\varepsilon_{22e})_{r=R_{1+i}} = -p_2\, \varrho^{pe}_{1+ie}, \quad i = 0, 1, \tag{6.274}$$

where the coefficient ϱ^{pe}_{1+ie} ($i = 0,1$) is given by Eq. (6.279).

Considering Eqs. (3.41), (3.42), (6.266)–(6.273), the thermal-stress induced elastic energy density w_e and the thermal-stress induced elastic energy W_e, accumulated in an arbitrary point of the spherical envelope and in the volume of the spherical envelope, respectively, have the forms

$$w_e = \frac{\omega_{2e} R_2^2}{4c_{2e}^2 r^2}, \tag{6.275}$$

$$W_e = 8 \int_0^{\pi/2} \int_0^{\pi/2} \int_{R_1}^{R_2} w_e\, r^2 dr d\varphi d\nu = \int_0^{\pi/2} \int_0^{\pi/2} \Omega_{2e} d\varphi d\nu, \tag{6.276}$$

where the coefficient Ω_{2e} given by Eq. (6.248) is a function of ω_{2e} given by Eq. (6.247).

With regard to the determination of the thermal-stress induced strengthening presented in Volume III, the stress σ_{ie} acting in a direction of the axis x_i and the elastic energy density w_{ie} induced by the stress σ_{ie} ($i = 1,2,3$) (see Eqs. (3.46), (3.47), (6.266)–(6.269)), both related to an arbitrary point of the spherical envelope, are derived as

$$\sigma_{ie} = \frac{\xi_{i2e} R_2}{2c_{2e} r}, \quad i = 1, 2, 3, \tag{6.277}$$

$$w_{ie} = \frac{s_{11e}\, \xi_{i2e}^2 R_2^2}{8c_{2e}^2 r^2}, \quad i = 1, 2, 3, \tag{6.278}$$

where the elastic modulus s_{11e} and the coefficient ξ_{i2e} ($i = 1,2,3$) is given by Eqs. (3.23) and (6.249), respectively.

Appendix. Considering the coefficient c_{2e} (see Eq. (3.39)), the coefficient ϱ^{pe}_{1+ie} ($i = 0,1$) has the form [10]

$$\varrho^{pe}_{1+ie} = -\frac{R_2}{2c_{2e} R_{1+i}}, \quad i = 0, 1. \tag{6.279}$$

[10] To avoid misunderstanding, Equation (6.279) really contains minus sign.

6.4.2 Thermal stresses in cell matrix

6.4.2.1 Conditions $\beta_p \neq \beta_e \neq \beta_m$ or $\beta_p = \beta_e \neq \beta_m$

The thermal stresses and the corresponding quantities in the cell matrix presented in Section 6.4.2.1 are derived regarding the conditions $\beta_p \neq \beta_e \neq \beta_m$ or $\beta_p = \beta_e \neq \beta_m$ (see Eqs. (3.48)–(3.50)).

A. Integration constants $C_{1m} \neq 0$, $C_{2m} \neq 0$, $C_{3m} = 0$

With regard to the boundary conditions (4.10), (4.11) and the analysis presented in Section 6.2.3, considering the terms u_{1rm} and u_{2rm} (see Eqs. (6.6), (6.7)) to represent functions of the variable $r \in \langle R_2, r_c\rangle$, the radial displacement u_{1m} of the infinitesimal spherical cap in the cell matrix, regarding Eqs. (3.31), (6.4), has the form

$$u_{1m} = -p_2 \left[\kappa_{1m}^{(1)} r \left(\frac{1}{3} - \ln r\right) + \kappa_{3m}^{(1)} C_m^{(1)} r^{c_{3m}}\right], \tag{6.280}$$

where the coefficients $c_{3m} < 0$, $C_m^{(1)}$ are given by Eqs. (3.40), (6.5), respectively. Provided that $\beta_p \neq \beta_e \neq \beta_m$ or $\beta_p = \beta_e \neq \beta_m$, the radial stress p_2 acting on the matrix-envelope boundary is given by Eqs. (10.3) or (10.5), respectively. The φ, ν-dependent distance r_c, as a function of the experimentally determined parameters v, R_{1ex}, d_{1ex}, d_{2ex}, d_{3ex} (see Section 2.2.1), included in the coefficients $\kappa_{1m}^{(1)}$, $\kappa_{3m}^{(1)}$ is derived in Section 2.3.

With regard to Section 6.4.2.1.A, the coefficients $\psi_{21\eta m}^{(1)}$, $\psi_{22\eta m}^{(1)}$ $(\eta = \varphi, \nu)$, ϱ_{2m} and $\omega_{21m}, \ldots, \omega_{26m}, \xi_{i21m}, \xi_{i22m}, \xi_{i23m}$ $(i = 1,2,3)$ mentioned below (see Eqs. (6.283), (6.284), (6.287)–(6.290), (6.292), (6.293)) depend on $\kappa_{1m}^{(1)}$, $\kappa_{3m}^{(1)}$ and $\kappa_{1m}^{(1)}$, $\kappa_{3m}^{(1)}$, $\psi_{21\eta m}^{(1)}$, $\psi_{22\eta m}^{(1)}$, respectively, where the coefficients $\kappa_{1m}^{(1)}$, $\kappa_{3m}^{(1)}$ (see Eq. (12.35)) are functions of χ_{11m}, χ_{12m}, χ_{21m}, χ_{22m} given by Eqs. (6.294), (6.295), (6.30), (6.31), respectively.

Consequently, the radial stress σ_{11m}, the tangential and shear stresses σ_{22m}, σ_{33m} and σ_{12m}, σ_{13m} acting in the cell matrix, respectively, regarding Eqs. (3.31)–(3.34), (6.280), are derived as

$$\sigma_{11m} = -p_2 \left\{-\kappa_{1m}^{(1)} \left[\frac{2}{3}(c_{1m} + 2c_{2m}) + (c_{1m} - c_{2m}) \ln r\right] + \kappa_{3m}^{(1)} C_m^{(1)} \left[c_{3m}(c_{1m} + c_{2m}) - 2c_{2m}\right] r^{c_{3m}-1}\right\}, \tag{6.281}$$

$$\sigma_{22m} = \sigma_{33m} = -p_2 \left\{\kappa_{1m}^{(1)} \left[\frac{1}{3}(c_{1m} + 2c_{2m}) - (c_{1m} - c_{2m}) \ln r\right] + \kappa_{3m}^{(1)} C_m^{(1)} (c_{1m} - c_{2m} c_{3m}) r^{c_{3m}-1}\right\}, \tag{6.282}$$

$$\sigma_{12m} = -\frac{1}{s_{44m}}\left[\psi^{(1)}_{21\varphi m}\left(\frac{1}{3} - \ln r\right) + \psi^{(1)}_{22\varphi m} C^{(1)}_m \, r^{c_{3m}-1}\right], \tag{6.283}$$

$$\sigma_{13m} = -\frac{1}{s_{44m}}\left[\psi^{(1)}_{21\nu m}\left(\frac{1}{3} - \ln r\right) + \psi^{(1)}_{22\nu m} C^{(1)}_m \, r^{c_{3m}-1}\right], \tag{6.284}$$

where the elastic modulus s_{44m} and the coefficients c_{1m}, c_{2m}, $\psi^{(1)}_{21\eta m}$, $\psi^{(1)}_{22\eta m}$ $(\eta = \varphi, \nu)$ are given by Eqs. (3.25) and (3.38), (3.39), (12.36), respectively.

Finally, the radial strain ε_{11m}, the tangential and shear strains ε_{22m}, ε_{33m} and ε_{12m}, ε_{13m} in the cell matrix, respectively, regarding Eqs. (3.1)–(3.5), (6.280), have the forms

$$\varepsilon_{11m} = -p_2\left[-\kappa^{(1)}_{1m}\left(\frac{2}{3} + \ln r\right) + \kappa^{(1)}_{3m} C^{(1)}_m c_{3m} \, r^{c_{3m}-1}\right], \tag{6.285}$$

$$\varepsilon_{22m} = \varepsilon_{33m} = -p_2\left[\kappa^{(1)}_{1m}\left(\frac{1}{3} - \ln r\right) + \kappa^{(1)}_{3m} C^{(1)}_m \, r^{c_{3m}-1}\right], \tag{6.286}$$

$$\varepsilon_{12m} = -\left[\psi^{(1)}_{21\varphi m}\left(\frac{1}{3} - \ln r\right) + \psi^{(1)}_{22\varphi m} C^{(1)}_m \, r^{c_{3m}-1}\right], \tag{6.287}$$

$$\varepsilon_{13m} = -\left[\psi^{(1)}_{21\nu m}\left(\frac{1}{3} - \ln r\right) + \psi^{(1)}_{22\nu m} C^{(1)}_m \, r^{c_{3m}-1}\right]. \tag{6.288}$$

With regard to the determination of the compressive or tensile radial stress $p_2 > 0$ or $p_2 < 0$ acting on the matrix-envelope boundary, thus for $r = R_2$, respectively, considering Eq. (3.58), the tangential strain on the matrix-envelope boundary, $(\varepsilon_{22m})_{r=R_2}$, is required to be derived in the form

$$(\varepsilon_{22m})_{r=R_2} = -p_2 \, \varrho_{2m}, \tag{6.289}$$

where the coefficient ϱ_{2m} is given by Eq. (6.32).

Considering Eqs. (3.41), (3.42), (6.281)–(6.288), the thermal-stress induced elastic energy density w_m and the thermal-stress induced elastic energy W_m, accumulated in an arbitrary point of the cell matrix and in the volume of the cell matrix, respectively, have the forms

$$w_m = \frac{1}{2}\left[\omega_{21m} + \omega_{22m} \, r^{c_{3m}-1} + \omega_{23m} \, r^{2(c_{3m}-1)} + \omega_{24m} \ln r \right.$$
$$\left. + \, \omega_{25m} \ln^2 r + \omega_{26m} \, r^{c_{3m}-1} \ln r\right], \tag{6.290}$$

$$W_m = 8 \int_0^{\pi/2} \int_0^{\pi/2} \int_{R_2}^{r_c} w_m \, r^2 dr d\varphi d\nu = \int_0^{\pi/2} \int_0^{\pi/2} \Omega_{22m} d\varphi d\nu, \tag{6.291}$$

where the coefficient Ω_{22m} given by Eq. (6.39) is a function of $\omega_{21m}, \ldots, \omega_{26m}$ given by Eqs. (6.33)–(6.38), respectively.

With regard to the determination of the thermal-stress induced strengthening presented in Volume III, the stress σ_{im} acting in a direction of the axis x_i and the elastic energy density w_{im} induced by the stress σ_{im} ($i=1,2,3$) (see Eqs. (3.46), (3.47), (6.281)–(6.284)), both related to an arbitrary point of the cell matrix, are derived as

$$\sigma_{im} = -\left(\xi_{i21m} + \xi_{i22m}\, r^{c_{3m}-1} - \xi_{i23m} \ln r\right), \quad i = 1, 2, 3, \tag{6.292}$$

$$\begin{aligned} w_{im} = \frac{s_{11m}}{2} \Big[& \xi_{i21m}^2 + \xi_{i22m}^2\, r^{2(c_{3m}-1)} + \xi_{i23m}^2 \ln^2 r + 2\, \xi_{i21m}\, \xi_{i22m}\, r^{c_{3m}-1} \\ & - 2\, \xi_{i21m}\, \xi_{i23m} \ln r - 2\, \xi_{i22m}\, \xi_{i23m}\, r^{c_{3m}-1} \ln r \Big], \\ i = 1, 2, 3, & \end{aligned} \tag{6.293}$$

where the elastic modulus s_{11m} and the coefficients ξ_{i21m}, ξ_{i22m}, ξ_{i23m} ($i=1,2,3$) are given by Eqs. (3.23) and (6.40)–(6.42), respectively.

Appendix. Considering the elastic modulus s_{44m} and the coefficients c_{1m}, c_{2m}, c_{3m}, $C_m^{(1)}$ (see Eqs. (3.25), (3.38)–(3.40), (6.5)), the coefficients χ_{11m}, χ_{12m} have the forms

$$\chi_{11m} = -\left[\frac{2}{3}\left(c_{1m} + 2c_{2m}\right) + \left(c_{1m} - c_{2m}\right) \ln R_2\right], \tag{6.294}$$

$$\chi_{12m} = C_m^{(1)} \left[c_{3m}\left(c_{1m} + c_{2m}\right) - 2c_{2m}\right] R_2^{c_{3m}-1}. \tag{6.295}$$

B. Integration constants $C_{1m} \neq 0$, $C_{2m} = 0$, $C_{3m} \neq 0$

With regard to the boundary conditions (4.10), (4.11) and the analysis presented in Section 6.2.3, considering the terms u_{1rm} and u_{3rm} (see Eqs. (6.6), (6.8)), the former as a function of the variable $r \in \langle R_2, r_c \rangle$ and the latter as a constant function regarding the variable r, respectively, the radial displacement u_{1m} of the infinitesimal spherical cap in the cell matrix, regarding Eqs. (3.31), (6.4), has the form

$$u_{1m} = -p_2 \left[\kappa_{1m}^{(1)}\, r \left(\frac{1}{3} - \ln r\right) + \frac{3\kappa_{3m}^{(1)}}{2}\right]. \tag{6.296}$$

Provided that $\beta_p \neq \beta_e \neq \beta_m$ or $\beta_p = \beta_e \neq \beta_m$, the radial stress p_2 acting on the matrix-envelope boundary is given by Eqs. (10.3) or (10.5), respectively. The φ, ν-dependent distance r_c, as a function of the experimentally determined parameters v, R_{1ex}, d_{1ex}, d_{2ex}, d_{3ex} (see Section 2.2.1), included in the coefficients $\kappa_{1m}^{(1)}$, $\kappa_{3m}^{(1)}$ is derived in Section 2.3.

With regard to Section 6.4.2.1.B, the coefficients $\psi_{21\eta m}^{(1)}$, $\psi_{22\eta m}^{(1)}$ $(\eta = \varphi, \nu)$, ϱ_{2m} and $\omega_{21m}, \ldots, \omega_{26m}$, ξ_{i21m}, ξ_{i22m}, ξ_{i23m} $(i = 1,2,3)$ mentioned below (see Eqs. (6.299), (6.300), (6.303)–(6.306), (6.308), (6.309)) depend on $\kappa_{1m}^{(1)}$, $\kappa_{3m}^{(1)}$ and $\kappa_{1m}^{(1)}$, $\kappa_{3m}^{(1)}$, $\psi_{21\eta m}^{(1)}$, $\psi_{22\eta m}^{(1)}$, respectively, where the coefficients $\kappa_{1m}^{(1)}$, $\kappa_{3m}^{(1)}$ (see Eq. (12.35)) are functions of χ_{11m}, χ_{12m}, χ_{21m}, χ_{22m} given by Eqs. (6.294), (6.310), (6.30), (6.58), respectively.

Consequently, the radial stress σ_{11m}, the tangential and shear stresses σ_{22m}, σ_{33m} and σ_{12m}, σ_{13m} acting in the cell matrix, respectively, regarding Eqs. (3.31)–(3.34), (6.296), are derived as

$$\sigma_{11m} = -p_2 \left\{ -\kappa_{1m}^{(1)} \left[\frac{2}{3} (c_{1m} + 2c_{2m}) + (c_{1m} - c_{2m}) \ln r \right] - \frac{3\kappa_{3m}^{(1)} c_{2m}}{r} \right\}, \tag{6.297}$$

$$\begin{aligned} \sigma_{22m} &= \sigma_{33m} \\ &= -p_2 \left\{ \kappa_{1m}^{(1)} \left[\frac{1}{3} (c_{1m} + 2c_{2m}) - (c_{1m} - c_{2m}) \ln r \right] + \frac{3\kappa_{3m}^{(1)} c_{1m}}{2r} \right\}, \end{aligned} \tag{6.298}$$

$$\sigma_{12m} = -\frac{1}{s_{44m}} \left[\psi_{21\varphi m}^{(1)} \left(\frac{1}{3} - \ln r \right) + \frac{3\psi_{22\varphi m}^{(1)}}{2r} \right], \tag{6.299}$$

$$\sigma_{13m} = -\frac{1}{s_{44m}} \left[\psi_{21\nu m}^{(1)} \left(\frac{1}{3} - \ln r \right) + \frac{3\psi_{22\nu m}^{(1)}}{2r} \right], \tag{6.300}$$

where the elastic modulus s_{44m} and the coefficients c_{1m}, c_{2m}, $\psi_{21\eta m}^{(1)}$, $\psi_{22\eta m}^{(1)}$ $(\eta = \varphi, \nu)$ are given by Eqs. (3.25) and (3.38), (3.39), (12.36), respectively.

Finally, the radial strain ε_{11m}, the tangential and shear strains ε_{22m}, ε_{33m} and ε_{12m}, ε_{13m} in the cell matrix, respectively, regarding Eqs. (3.1)–(3.5), (6.296), have the forms[11]

$$\varepsilon_{11m} = p_2 \kappa_{1m}^{(1)} \left(\frac{2}{3} + \ln r \right), \tag{6.301}$$

$$\varepsilon_{22m} = \varepsilon_{33m} = -p_2 \left[\kappa_{1m}^{(1)} \left(\frac{1}{3} - \ln r \right) + \frac{3\kappa_{3m}^{(1)}}{2r} \right], \tag{6.302}$$

$$\varepsilon_{12m} = -\left[\psi_{21\varphi m}^{(1)} \left(\frac{1}{3} - \ln r \right) + \frac{3\psi_{22\varphi m}^{(1)}}{2r} \right], \tag{6.303}$$

$$\varepsilon_{13m} = -\left[\psi_{21\nu m}^{(1)} \left(\frac{1}{3} - \ln r \right) + \frac{3\psi_{22\nu m}^{(1)}}{2r} \right]. \tag{6.304}$$

[11] To avoid misunderstanding, a sign on the right side in Eq. (6.301) is really plus, in contrast to minus sign on the right sides in Eqs. (6.302)–(6.304).

With regard to the determination of the compressive or tensile radial stress $p_2 > 0$ or $p_2 < 0$ acting on the matrix-envelope boundary, thus for $r = R_2$, respectively, considering Eq. (3.58), the tangential strain on the matrix-envelope boundary, $(\varepsilon_{22m})_{r=R_2}$, is required to be derived in the form

$$(\varepsilon_{22m})_{r=R_2} = -p_2\,\varrho_{2m}, \tag{6.305}$$

where the coefficient ϱ_{2m} is given by Eq. (6.59).

Considering Eqs. (3.41), (3.42), (6.297)–(6.304), the thermal-stress induced elastic energy density w_m and the thermal-stress induced elastic energy W_m, accumulated in an arbitrary point of the cell matrix and in the volume of the cell matrix, respectively, have the forms

$$w_m = \frac{1}{2}\left[\omega_{21m} + \frac{\omega_{22m}}{r} + \frac{\omega_{23m}}{r^2} + \omega_{24m}\ln r + \omega_{25m}\ln^2 r + \omega_{26m}\frac{\ln r}{r}\right], \tag{6.306}$$

$$W_m = 8\int_0^{\pi/2}\int_0^{\pi/2}\int_{R_2}^{r_c} w_m\, r^2 dr d\varphi d\nu = \int_0^{\pi/2}\int_0^{\pi/2}\Omega_{22m} d\varphi d\nu, \tag{6.307}$$

where the coefficient Ω_{22m} given by Eq. (6.63) is a function of $\omega_{21m}, \ldots, \omega_{26m}$ given by Eqs. (6.33), (6.60), (6.61), (6.36), (6.37), (6.62), respectively.

With regard to the determination of the thermal-stress induced strengthening presented in Volume III, the stress σ_{im} acting in a direction of the axis x_i and the elastic energy density w_{im} induced by the stress σ_{im} $(i=1,2,3)$ (see Eqs. (3.46), (3.47), (6.297)–(6.300)), both related to an arbitrary point of the cell matrix, are derived as

$$\sigma_{im} = -\left(\xi_{i21m} + \frac{\xi_{i22m}}{r} - \xi_{i23m}\ln r\right), \quad i = 1, 2, 3, \tag{6.308}$$

$$w_{im} = \frac{s_{11m}}{2}\left(\xi^2_{i21m} + \frac{\xi^2_{i22m}}{r^2} + \xi^2_{i23m}\ln^2 r + \frac{2\,\xi_{i21m}\,\xi_{i22m}}{r} - 2\,\xi_{i21m}\,\xi_{i23m}\ln r - 2\,\xi_{i22m}\,\xi_{i23m}\frac{\ln r}{r}\right), \quad i = 1, 2, 3, \tag{6.309}$$

where the elastic modulus s_{11m} and the coefficients ξ_{i21m}, ξ_{i22m} ξ_{i23m} $(i=1,2,3)$ are given by Eqs. (3.23) and (6.40), (6.64), (6.42), respectively.

Appendix. Considering the elastic modulus s_{44m} and the coefficients c_{1m}, c_{2m} (see Eqs. (3.25), (3.38), (3.39)), the coefficient χ_{12m} has the form

$$\chi_{12m} = -\frac{3c_{2m}}{R_2}. \tag{6.310}$$

C. Integration constants $C_{1m} = 0$, $C_{2m} \neq 0$, $C_{3m} \neq 0$

With regard to the boundary conditions (4.10), (4.11) and the analysis presented in Section 6.2.3, considering the terms u_{2rm} and u_{3rm} (see Eqs. (6.7), (6.8)), the former as a function of the variable $r \in \langle R_2, r_c \rangle$ and the latter as a constant function regarding the variable r, respectively, the radial displacement u_{1m} of the infinitesimal spherical cap in the cell matrix, regarding Eqs. (3.31), (6.4), has the form

$$u_{1m} = -p_2 \left[\kappa_{1m}^{(1)} C_m^{(1)} r^{c_{3m}} + \frac{3\kappa_{3m}^{(1)}}{2} \right], \tag{6.311}$$

where the coefficients $c_{3m} < 0$, $C_m^{(1)}$ are given by Eqs. (3.40), (6.5), respectively. Provided that $\beta_p \neq \beta_e \neq \beta_m$ or $\beta_p = \beta_e \neq \beta_m$, the radial stress p_2 acting on the matrix-envelope boundary is given by Eqs. (10.3) or (10.5), respectively. The φ, ν-dependent distance r_c, as a function of the experimentally determined parameters v, R_{1ex}, d_{1ex}, d_{2ex}, d_{3ex} (see Section 2.2.1), included in the coefficients $\kappa_{1m}^{(1)}$, $\kappa_{3m}^{(1)}$ is derived in Section 2.3.

With regard to Section 6.4.2.1.C, the coefficients $\psi_{21\eta m}^{(1)}$, $\psi_{22\eta m}^{(1)}$ $(\eta = \varphi, \nu)$, ϱ_{2m} and ω_{21m}, ω_{22m}, ω_{23m}, ξ_{i21m}, ξ_{i22m} $(i = 1,2,3)$ mentioned below (see Eqs. (6.314), (6.315), (6.318)–(6.321), (6.323), (6.324)) depend on $\kappa_{1m}^{(1)}$, $\kappa_{3m}^{(1)}$ and $\kappa_{1m}^{(1)}$, $\kappa_{3m}^{(1)}$, $\psi_{21\eta m}^{(1)}$, $\psi_{22\eta m}^{(1)}$, respectively, where the coefficients $\kappa_{1m}^{(1)}$, $\kappa_{3m}^{(1)}$ (see Eq. (12.35)) are functions of χ_{11m}, χ_{12m}, χ_{21m}, χ_{22m} given by Eqs. (6.325), (6.310), (6.80), (6.58), respectively.

Consequently, the radial stress σ_{11m}, the tangential and shear stresses σ_{22m}, σ_{33m} and σ_{12m}, σ_{13m} acting in the cell matrix, respectively, regarding Eqs. (3.31)–(3.34), (6.311), are derived as

$$\sigma_{11m} = -p_2 \left\{ \kappa_{1m}^{(1)} C_m^{(1)} \left[c_{3m} \left(c_{1m} + c_{2m} \right) - 2c_{2m} \right] r^{c_{3m}-1} - \frac{3\kappa_{3m}^{(1)} c_{2m}}{r} \right\}, \tag{6.312}$$

$$\sigma_{22m} = \sigma_{33m} = -p_2 \left[\kappa_{1m}^{(1)} C_m^{(1)} \left(c_{1m} - c_{2m} c_{3m} \right) r^{c_{3m}-1} + \frac{3\kappa_{3m}^{(1)} c_{1m}}{2r} \right], \tag{6.313}$$

$$\sigma_{12m} = -\frac{1}{s_{44m}} \left[\psi_{21\varphi m}^{(1)} C_m^{(1)} r^{c_{3m}-1} + \frac{3\psi_{22\varphi m}^{(1)}}{2r} \right], \tag{6.314}$$

$$\sigma_{13m} = -\frac{1}{s_{44m}} \left[\psi_{21\nu m}^{(1)} C_m^{(1)} r^{c_{3m}-1} + \frac{3\psi_{22\nu m}^{(1)}}{2r} \right], \tag{6.315}$$

where the elastic modulus s_{44m} and the coefficients c_{1m}, c_{2m}, $\psi_{21\eta m}^{(1)}$, $\psi_{22\eta m}^{(1)}$ $(\eta = \varphi, \nu)$ are given by Eqs. (3.25) and (3.38), (3.39), (12.36), respectively.

Finally, the radial strain ε_{11m}, the tangential and shear strains ε_{22m}, ε_{33m} and ε_{12m}, ε_{13m} in the cell matrix, respectively, regarding Eqs. (3.1)–(3.5), (6.311), have the forms

$$\varepsilon_{11m} = -p_2 \kappa_{1m}^{(1)} C_m^{(1)} c_{3m}\, r^{c_{3m}-1}, \tag{6.316}$$

$$\varepsilon_{22m} = \varepsilon_{33m} = -p_2 \left[\kappa_{1m}^{(1)} C_m^{(1)}\, r^{c_{3m}-1} + \frac{3\kappa_{3m}^{(1)}}{2r} \right], \tag{6.317}$$

$$\varepsilon_{12m} = -\left[\psi_{21\varphi m}^{(1)} C_m^{(1)}\, r^{c_{3m}-1} + \frac{3\psi_{22\varphi m}^{(1)}}{2r} \right], \tag{6.318}$$

$$\varepsilon_{13m} = -\left[\psi_{21\nu m}^{(1)} C_m^{(1)}\, r^{c_{3m}-1} + \frac{3\psi_{22\nu m}^{(1)}}{2r} \right]. \tag{6.319}$$

With regard to the determination of the compressive or tensile radial stress $p_2 > 0$ or $p_2 < 0$ acting on the matrix-envelope boundary, thus for $r = R_2$, respectively, considering Eq. (3.58), the tangential strain on the matrix-envelope boundary, $(\varepsilon_{22m})_{r=R_2}$, is required to be derived in the form

$$(\varepsilon_{22m})_{r=R_2} = -p_2\, \varrho_{2m}, \tag{6.320}$$

where the coefficient ϱ_{2m} is given by Eq. (6.81).

Considering Eqs. (3.41), (3.42), (6.312)–(6.319), the thermal-stress induced elastic energy density w_m and the thermal-stress induced elastic energy W_m, accumulated in an arbitrary point of the cell matrix and in the volume of the cell matrix, respectively, have the forms

$$w_m = \frac{1}{2} \left[\frac{\omega_{21m}}{r^2} + \omega_{22m}\, r^{c_{3m}-2} + \omega_{23m}\, r^{2(c_{3m}-1)} \right], \tag{6.321}$$

$$W_m = 8 \int_0^{\pi/2} \int_0^{\pi/2} \int_{R_2}^{r_c} w_m\, r^2 dr d\varphi d\nu = \int_0^{\pi/2} \int_0^{\pi/2} \Omega_{22m} d\varphi d\nu, \tag{6.322}$$

where the coefficient Ω_{22m} given by Eq. (6.85) is a function of ω_{21m}, ω_{22m}, ω_{23m} given by Eqs. (6.82)–(6.84), respectively.

With regard to the determination of the thermal-stress induced strengthening presented in Volume III, the stress σ_{im} acting in a direction of the axis x_i and the elastic energy density w_{im} induced by the stress σ_{im} $(i = 1,2,3)$ (see Eqs. (3.46), (3.47), (6.312)–(6.315)), both related to an arbitrary point of the cell matrix, are derived as

$$\sigma_{im} = -\left(\frac{\xi_{i\,21m}}{r} + \xi_{i\,22m}\, r^{c_{3m}-1}\right), \quad i = 1, 2, 3, \tag{6.323}$$

$$w_{im} = \frac{s_{11m}}{2}\left[\frac{\xi^2_{i\,21m}}{r^2} + \xi^2_{i\,22m}\, r^{2(c_{3m}-1)} + 2\,\xi_{i\,21m}\,\xi_{i\,22m}\, r^{c_{3m}-2}\right],$$
$$i = 1, 2, 3, \tag{6.324}$$

where the elastic modulus s_{11m} and the coefficients $\xi_{i\,21m}$, $\xi_{i\,22m}$ $(i=1,2,3)$ are given by Eqs. (3.23) and (6.86), (6.87), respectively.

Appendix. Considering the elastic modulus s_{44m} and the coefficients c_{1m}, c_{2m}, c_{3m}, $C_m^{(1)}$ (see Eqs. (3.25), (3.38)–(3.40), (6.5)), the coefficient χ_{11m} has the form

$$\chi_{11m} = C_m^{(1)}\left[c_{3m}\left(c_{1m} + c_{2m}\right) - 2c_{2m}\right] R_2^{c_{3m}-1}. \tag{6.325}$$

D. Integration constants $C_{1m} \neq 0$, $C_{2m} \neq 0$, $C_{3m} \neq 0$

With regard to the boundary conditions (4.10), (4.11), (4.14) and the analysis presented in Section 6.2.3, considering the terms u_{1rm}, u_{2rm} and u_{3rm} (see Eqs. (6.6)–(6.8)) to represent functions of the variable $r \in \langle R_2, r_c\rangle$ and a constant function regarding the variable r, respectively, the radial displacement u_{1m} of the infinitesimal spherical cap in the cell matrix, regarding Eqs. (3.31), (6.4), has the form

$$u_{1m} = -p_2\left[\kappa_{4m}^{(2)} r\left(\frac{1}{3} - \ln r\right) + \kappa_{5m}^{(2)} C_m^{(1)}\, r^{c_{3m}} + \frac{3\kappa_{6m}^{(2)}}{2}\right], \tag{6.326}$$

where the coefficients $c_{3m} < 0$, $C_m^{(1)}$ are given by Eqs. (3.40), (6.5), respectively. Provided that $\beta_p \neq \beta_e \neq \beta_m$ or $\beta_p = \beta_e \neq \beta_m$, the radial stress p_2 acting on the matrix-envelope boundary is given by Eqs. (10.3) or (10.5), respectively. The φ,ν-dependent distance r_c, as a function of the experimentally determined parameters v, R_{1ex}, d_{1ex}, d_{2ex}, d_{3ex} (see Section 2.2.1), included in the coefficients $\kappa_{4m}^{(2)}$, $\kappa_{5m}^{(2)}$, $\kappa_{6m}^{(2)}$ is derived in Section 2.3.

With regard to Section 6.4.2.1.D, the coefficients $\psi_{21\eta m}^{(2)}$, $\psi_{22\eta m}^{(2)}$, $\psi_{23\eta m}^{(2)}$ $(\eta=\varphi,\nu)$, ϱ_{2m} and $\omega_{21m},\ldots,\omega_{210m}$, $\xi_{i21m},\ldots,\xi_{i24m}$ $(i=1,2,3)$ mentioned below (see Eqs. (6.329), (6.330), (6.333)–(6.336), (6.338), (6.339)) depend on $\kappa_{4m}^{(2)}$, $\kappa_{5m}^{(2)}$, $\kappa_{6m}^{(2)}$ and $\kappa_{4m}^{(2)}$, $\kappa_{5m}^{(2)}$, $\kappa_{6m}^{(2)}$, $\psi_{21\eta m}^{(2)}$, $\psi_{22\eta m}^{(2)}$, $\psi_{23\eta m}^{(2)}$, respectively, where the coefficients $\kappa_{4m}^{(2)}$, $\kappa_{5m}^{(2)}$, $\kappa_{6m}^{(2)}$ (see Eq. (12.39)) are functions of χ_{11m}, χ_{12m}, χ_{13m}, χ_{21m}, χ_{22m}, χ_{23m}, χ_{31m}, χ_{32m}, χ_{33m} given by Eqs. (6.294), (6.295), (6.340), (6.30), (6.31), (6.103)–(6.106), respectively.

Consequently, the radial stress σ_{11m}, the tangential and shear stresses σ_{22m}, σ_{33m} and σ_{12m}, σ_{13m} acting in the cell matrix, respectively, regarding Eqs. (3.31)–(3.34), (6.326), are derived as

$$\sigma_{11m} = -p_2 \left\{ -\kappa_{4m}^{(2)} \left[\frac{2}{3} (c_{1m} + 2c_{2m}) + (c_{1m} - c_{2m}) \ln r \right] \right. \\ \left. + \kappa_{5m}^{(2)} C_m^{(1)} \left[c_{3m} (c_{1m} + c_{2m}) - 2c_{2m} \right] r^{c_{3m}-1} - \frac{3\kappa_{6m}^{(2)} c_{2m}}{r} \right\}, \quad (6.327)$$

$$\sigma_{22m} = \sigma_{33m} = -p_2 \left\{ \kappa_{4m}^{(2)} \left[\frac{1}{3} (c_{1m} + 2c_{2m}) - (c_{1m} - c_{2m}) \ln r \right] \right. \\ \left. + \kappa_{5m}^{(2)} C_m^{(1)} (c_{1m} - c_{2m} c_{3m}) r^{c_{3m}-1} + \frac{3\kappa_{6m}^{(2)} c_{1m}}{2r} \right\}, \quad (6.328)$$

$$\sigma_{12m} = -\frac{1}{s_{44m}} \left[\psi_{21\varphi m}^{(2)} \left(\frac{1}{3} - \ln r \right) + \psi_{22\varphi m}^{(2)} C_m^{(1)} r^{c_{3m}-1} + \frac{3\psi_{23\varphi m}^{(2)}}{2r} \right], \quad (6.329)$$

$$\sigma_{13m} = -\frac{1}{s_{44m}} \left[\psi_{21\nu m}^{(2)} \left(\frac{1}{3} - \ln r \right) + \psi_{22\nu m}^{(2)} C_m^{(1)} r^{c_{3m}-1} + \frac{3\psi_{23\nu m}^{(2)}}{2r} \right], \quad (6.330)$$

where the elastic modulus s_{44m} and the coefficients c_{1m}, c_{2m}, $\psi_{21\eta m}^{(2)}$, $\psi_{22\eta m}^{(2)}$, $\psi_{23\eta m}^{(2)}$ $(\eta = \varphi, \nu)$ are given by Eqs. (3.25) and (3.38), (3.39), (12.40), respectively.

Finally, the radial strain ε_{11m}, the tangential and shear strains ε_{22m}, ε_{33m} and ε_{12m}, ε_{13m} in the cell matrix, respectively, regarding Eqs. (3.1)–(3.5), (6.326), have the forms

$$\varepsilon_{11m} = -p_2 \left[-\kappa_{4m}^{(2)} \left(\frac{2}{3} + \ln r \right) + \kappa_{5m}^{(2)} C_m^{(1)} c_{3m} r^{c_{3m}-1} \right], \quad (6.331)$$

$$\varepsilon_{22m} = \varepsilon_{33m} = -p_2 \left[\kappa_{4m}^{(2)} \left(\frac{1}{3} - \ln r \right) + \kappa_{5m}^{(2)} C_m^{(1)} r^{c_{3m}-1} + \frac{3\kappa_{6m}^{(2)}}{2r} \right], \quad (6.332)$$

$$\varepsilon_{12m} = -\left[\psi_{21\varphi m}^{(2)} \left(\frac{1}{3} - \ln r \right) + \psi_{22\varphi m}^{(2)} C_m^{(1)} r^{c_{3m}-1} + \frac{3\psi_{23\varphi m}^{(2)}}{2r} \right], \quad (6.333)$$

$$\varepsilon_{13m} = -\left[\psi_{21\nu m}^{(2)} \left(\frac{1}{3} - \ln r \right) + \psi_{22\nu m}^{(2)} C_m^{(1)} r^{c_{3m}-1} + \frac{3\psi_{23\nu m}^{(2)}}{2r} \right]. \quad (6.334)$$

With regard to the determination of the compressive or tensile radial stress $p_2 > 0$ or $p_2 < 0$ acting on the matrix-envelope boundary, thus for $r = R_2$, respectively, considering Eq. (3.58), the tangential strain on the matrix-envelope boundary, $(\varepsilon_{22m})_{r=R_2}$, is required to be derived in the form

$$(\varepsilon_{22m})_{r=R_2} = -p_2\,\varrho_{2m}, \tag{6.335}$$

where the coefficient ϱ_{2m} is given by Eq. (6.107).

Considering Eqs. (3.41), (3.42), (6.327)–(6.334), the thermal-stress induced elastic energy density w_m and the thermal-stress induced elastic energy W_m, accumulated in an arbitrary point of the cell matrix and in the volume of the cell matrix, respectively, have the forms

$$\begin{aligned} w_m = \frac{1}{2}\Big[& \omega_{21m} + \frac{\omega_{22m}}{r} + \frac{\omega_{23m}}{r^2} + \omega_{24m}\, r^{c_{3m}-1} + \omega_{25m}\, r^{c_{3m}-2} \\ & + \omega_{26m}\, r^{2(c_{3m}-1)} + \omega_{27m} \ln r + \omega_{28m} \ln^2 r \\ & + \frac{\omega_{29m} \ln r}{r} + \omega_{210m}\, r^{c_{3m}-1} \ln r \Big], \end{aligned} \tag{6.336}$$

$$W_m = 8 \int_0^{\pi/2}\int_0^{\pi/2}\int_{R_2}^{r_c} w_m\, r^2 dr d\varphi d\nu = \int_0^{\pi/2}\int_0^{\pi/2} \Omega_{22m} d\varphi d\nu, \tag{6.337}$$

where the coefficient Ω_{22m} given by Eq. (6.118) is a function of $\omega_{21m},\ldots,\omega_{210m}$ Eqs. (6.108)–(6.117), respectively.

With regard to the determination of the thermal-stress induced strengthening presented in Volume III, the stress σ_{im} acting in a direction of the axis x_i and the elastic energy density w_{im} induced by the stress σ_{im} ($i = 1,2,3$) (see Eqs. (3.46), (3.47), (6.327)–(6.330)), both related to an arbitrary point of the cell matrix, are derived as

$$\sigma_{im} = -\left(\xi_{i21m} + \frac{\xi_{i22m}}{r} + \xi_{i23m}\, r^{c_{3m}-1} - \xi_{i24m} \ln r\right), \quad i = 1, 2, 3, \tag{6.338}$$

$$\begin{aligned} w_{im} = \frac{s_{11m}}{2}\Bigg[& \xi_{i21m}^2 + \frac{\xi_{i22m}^2}{r^2} + \xi_{i23m}^2\, r^{2(c_{3m}-1)} + \xi_{i24m}^2 \ln^2 r + \frac{2\,\xi_{i21m}\,\xi_{i22m}}{r} \\ & + 2\,\xi_{i21m}\,\xi_{i23m}\, r^{c_{3m}-1} - 2\,\xi_{i21m}\,\xi_{i24m} \ln r + 2\,\xi_{i22m}\,\xi_{i23m}\, r^{c_{3m}-2} \\ & - 2\,\xi_{i22m}\,\xi_{i24m} \frac{\ln r}{r} - 2\,\xi_{i23m}\,\xi_{i24m}\, r^{c_{3m}-1} \ln r \Bigg], \\ & i = 1, 2, 3, \end{aligned} \tag{6.339}$$

where the elastic modulus s_{11m} and the coefficients $\xi_{i21m},\ldots,\xi_{i24m}$ ($i = 1,2,3$) are given by Eqs. (3.23) and (6.119)–(6.122), respectively.

Appendix. Considering the elastic modulus s_{44m} and the coefficients c_{1m}, c_{2m} (see Eqs. (3.25), (3.38), (3.39)), the coefficient χ_{13m} has the form

$$\chi_{13m} = -\frac{3c_{2m}}{R_2}. \tag{6.340}$$

6.4.2.2 Conditions $\beta_p \neq \beta_e = \beta_m$, and $\varepsilon_{11te} - \varepsilon_{11tp} = f(\varphi, \nu)$ or $\varepsilon_{11te} - \varepsilon_{11tp} \neq f(\varphi, \nu)$

The formulae in Section 6.4.2.2 consider the φ, ν-dependent or φ, ν-independent radial stress $p_1 = (\beta_e - \beta_p) / (\varrho_{1p} - \varrho_{1e}^{me})$ (see Eq. (10.4)) acting on the particle-envelope boundary and derived from Eq. (3.63), where the φ, ν-independent coefficients ϱ_{1p} and ϱ_{1e}^{me}, along with ϱ_{2e}^{me} included in the formulae in Section 6.4.2.2, are given by Eqs. (5.36) or (9.50), and (5.106), (6.207), (6.228), (6.246), (7.230) or (9.63), respectively.

The φ, ν-dependence of the radial stress p_1 results from the condition $\varepsilon_{11te} - \varepsilon_{11tp} = f(\varphi, \nu)$, resulting in the φ, ν-dependent difference $\beta_e - \beta_p$ (see Eqs. (3.48)–(3.50)), where at least one of the phase-transformation induced radial strains ε_{11tp}, ε_{11te} is φ, ν-dependent (see Eq. (12.23)). The φ, ν-independence of the radial stress p_1 results from the condition $\varepsilon_{11te} - \varepsilon_{11tp} \neq f(\varphi, \nu)$ (see Eqs. (12.23), (12.24)), regarding simplification, equivalent to those defined in Items 1, 2 in Section 9.1 and resulting in the φ, ν-independent difference $\beta_e - \beta_p$.

A. Integration constants $C_{1m} \neq 0$, $C_{2m} \neq 0$, $C_{3m} = 0$

With regard to the boundary conditions (4.11), (4.12) and the analysis presented in Section 6.2.3, considering the terms u_{1rm} and u_{2rm} (see Eqs. (6.6), (6.7)) to represent functions of the variable $r \in \langle R_2, r_c \rangle$, the radial displacement u_{1m} of the infinitesimal spherical cap in the cell matrix for the φ, ν-dependent or φ, ν-independent radial stress p_1, regarding Eqs. (3.31), (6.4), has the form

$$u_{1m} = -p_1 \, \varrho_{2e}^{me} \left[\kappa_{1m}^{(1)} r \left(\frac{1}{3} - \ln r \right) + \kappa_{3m}^{(1)} C_m^{(1)} \, r^{c_{3m}} \right], \tag{6.341}$$

where the coefficients $c_{3m} < 0$, $C_m^{(1)}$ are given by Eqs. (3.40), (6.5), respectively. The radial stress p_1 acting on the particle-envelope boundary and the coefficient ϱ_{2e}^{me} are analysed at the beginning of Section 6.4.2.2. The φ, ν-dependent distance r_c, as a function of the experimentally determined parameters v, R_{1ex}, d_{1ex}, d_{2ex}, d_{3ex} (see Section 2.2.1), included in the coefficients $\kappa_{1m}^{(1)}$, $\kappa_{3m}^{(1)}$ is derived in Section 2.3.

With regard to Section 6.4.2.2.A, the coefficients $\psi_{11\eta m}^{(1)}$, $\psi_{12\eta m}^{(1)}$ $(\eta = \varphi, \nu)$, $\omega_{21m}, \ldots, \omega_{26m}$, ξ_{i1m}, ξ_{i2m}, ξ_{i3m} $(i = 1,2,3)$ and $\omega_{11m}, \ldots, \omega_{16m}$, ξ_{i11m}, ξ_{i12m}, ξ_{i13m} $(i = 1,2,3)$ mentioned below (see Eqs. (6.344), (6.345), (6.350), (6.351), (6.356), (6.360), (6.361) and (6.354), (6.358), (6.359)) depend on $\kappa_{1m}^{(1)}$, $\kappa_{3m}^{(1)}$ and $\kappa_{1m}^{(1)}$, $\kappa_{3m}^{(1)}$, $\psi_{11\eta m}^{(1)}$, $\psi_{12\eta m}^{(1)}$, respectively, where the coefficients $\kappa_{1m}^{(1)}$, $\kappa_{3m}^{(1)}$ (see Eq. (12.35)) are functions of χ_{11m}, χ_{12m}, χ_{21m}, χ_{22m} given by Eqs. (6.362), (6.363), (6.30), (6.31), respectively.

Consequently, considering the φ, ν-dependent or φ, ν-independent radial stress p_1, the radial and tangential stresses, σ_{11m} and σ_{22m}, σ_{33m}, respectively, acting in the cell matrix, regarding Eqs. (3.31), (3.32), (6.341), are derived as

$$\sigma_{11m} = -p_1\, \varrho_{2e}^{me} \left\{ -\kappa_{1m}^{(1)} \left[\frac{2}{3} \left(c_{1m} + 2c_{2m} \right) + \left(c_{1m} - c_{2m} \right) \ln r \right] + \kappa_{3m}^{(1)} C_m^{(1)} \left[c_{3m} \left(c_{1m} + c_{2m} \right) - 2c_{2m} \right] r^{c_{3m}-1} \right\}, \tag{6.342}$$

$$\sigma_{22m} = \sigma_{33m} = -p_1\, \varrho_{2e}^{me} \left\{ \kappa_{1m}^{(1)} \left[\frac{1}{3} \left(c_{1m} + 2c_{2m} \right) - \left(c_{1m} - c_{2m} \right) \ln r \right] + \kappa_{3m}^{(1)} C_m^{(1)} \left(c_{1m} - c_{2m} c_{3m} \right) r^{c_{3m}-1} \right\}, \tag{6.343}$$

where the coefficients c_{1m}, c_{2m} are given by Eqs. (3.38), (3.39), respectively.

With regard to Eqs. (3.33), (3.34), (6.341), the shear stresses σ_{12m}, σ_{13m} acting in the cell matrix, considering the φ, ν-dependent radial stress p_1, are derived as

$$\sigma_{12m} = -\frac{\varrho_{2e}^{me}}{s_{44m}} \left[\psi_{11\varphi m}^{(1)} \left(\frac{1}{3} - \ln r \right) + \psi_{12\varphi m}^{(1)} C_m^{(1)} r^{c_{3m}-1} \right], \tag{6.344}$$

$$\sigma_{13m} = -\frac{\varrho_{2e}^{me}}{s_{44m}} \left[\psi_{11\nu m}^{(1)} \left(\frac{1}{3} - \ln r \right) + \psi_{12\nu m}^{(1)} C_m^{(1)} r^{c_{3m}-1} \right], \tag{6.345}$$

where the elastic modulus s_{44m} and the coefficients $\psi_{11\eta m}^{(1)}$, $\psi_{12\eta m}^{(1)}$ $(\eta = \varphi, \nu)$ are given by Eqs. (3.25) and (12.36), respectively. Considering the φ, ν-independent radial stress p_1, we get

$$\sigma_{12m} = -\frac{p_1\, \varrho_{2e}^{me}}{s_{44m}} \left[\frac{\partial \kappa_{1m}^{(1)}}{\partial \varphi} \left(\frac{1}{3} - \ln r \right) + \frac{\partial \kappa_{3m}^{(1)}}{\partial \varphi} C_m^{(1)} r^{c_{3m}-1} \right], \tag{6.346}$$

$$\sigma_{13m} = -\frac{p_1\, \varrho_{2e}^{me}}{s_{44m}} \left[\frac{\partial \kappa_{1m}^{(1)}}{\partial \nu} \left(\frac{1}{3} - \ln r \right) + \frac{\partial \kappa_{3m}^{(1)}}{\partial \nu} C_m^{(1)} r^{c_{3m}-1} \right]. \tag{6.347}$$

Finally, considering the φ, ν-dependent or φ, ν-independent radial stress p_1, the radial and tangential strains, ε_{11m} and ε_{22m}, ε_{33m}, respectively, acting in the cell matrix, regarding Eqs. (3.1)–(3.3), (6.341), have the forms

$$\varepsilon_{11m} = -p_1\, \varrho_{2e}^{me} \left[-\kappa_{1m}^{(1)} \left(\frac{2}{3} + \ln r \right) + \kappa_{3m}^{(1)} C_m^{(1)} c_{3m}\, r^{c_{3m}-1} \right], \tag{6.348}$$

$$\varepsilon_{22m} = \varepsilon_{33m} = -p_1\, \varrho_{2e}^{me} \left[\kappa_{1m}^{(1)} \left(\frac{1}{3} - \ln r \right) + \kappa_{3m}^{(1)} C_m^{(1)} r^{c_{3m}-1} \right]. \tag{6.349}$$

With regard to Eqs. (3.4), (3.5), (6.341), the shear strains σ_{12m}, σ_{13m} acting in the cell matrix, considering the φ, ν-dependent radial stress p_1, are derived as

$$\varepsilon_{12m} = -\varrho_{2e}^{me} \left[\psi_{11\varphi m}^{(1)} \left(\frac{1}{3} - \ln r \right) + \psi_{12\varphi m}^{(1)} C_m^{(1)} \, r^{c_{3m}-1} \right], \tag{6.350}$$

$$\varepsilon_{13m} = -\varrho_{2e}^{me} \left[\psi_{11\nu m}^{(1)} \left(\frac{1}{3} - \ln r \right) + \psi_{12\nu m}^{(1)} C_m^{(1)} \, r^{c_{3m}-1} \right], \tag{6.351}$$

and considering the φ, ν-independent radial stress p_1, we get

$$\varepsilon_{12m} = -p_1 \, \varrho_{2e}^{me} \left[\frac{\partial \kappa_{1m}^{(1)}}{\partial \varphi} \left(\frac{1}{3} - \ln r \right) + \frac{\partial \kappa_{3m}^{(1)}}{\partial \varphi} \, C_m^{(1)} \, r^{c_{3m}-1} \right], \tag{6.352}$$

$$\varepsilon_{13m} = -p_1 \, \varrho_{2e}^{me} \left[\frac{\partial \kappa_{1m}^{(1)}}{\partial \nu} \left(\frac{1}{3} - \ln r \right) + \frac{\partial \kappa_{3m}^{(1)}}{\partial \nu} \, C_m^{(1)} \, r^{c_{3m}-1} \right]. \tag{6.353}$$

With regard to the φ, ν-dependent radial stress p_1, considering Eqs. (3.41), (3.42), (6.342)–(6.345), (6.348)–(6.351), the thermal-stress induced elastic energy density w_m and the thermal-stress induced elastic energy W_m, accumulated in an arbitrary point of the cell matrix and in the volume of the cell matrix, respectively, are derived as

$$w_m = \frac{(\varrho_{2e}^{me})^2}{2} \left[\omega_{11m} + \omega_{12m} \, r^{c_{3m}-1} + \omega_{13m} \, r^{2(c_{3m}-1)} + \omega_{14m} \ln r \right.$$
$$\left. + \, \omega_{15m} \ln^2 r + \omega_{16m} \, r^{c_{3m}-1} \ln r \right], \tag{6.354}$$

$$W_m = 8 \int_0^{\pi/2} \int_0^{\pi/2} \int_{R_2}^{r_c} w_m \, r^2 dr d\varphi d\nu = (\varrho_{2e}^{me})^2 \int_0^{\pi/2} \int_0^{\pi/2} \Omega_{12m} d\varphi d\nu, \tag{6.355}$$

where the coefficient Ω_{12m} given by Eq. (6.39) is a function of $\omega_{11m}, \ldots, \omega_{16m}$ given by Eqs. (6.33)–(6.38), respectively.

With regard to the φ, ν-independent radial stress p_1, considering Eqs. (3.41), (3.42), (6.342), (6.343), (6.346)–(6.349), (6.352), (6.353), we get

$$w_m = \frac{(p_1 \, \varrho_{2e}^{me})^2}{2} \left[\omega_{21m} + \omega_{22m} \, r^{c_{3m}-1} + \omega_{23m} \, r^{2(c_{3m}-1)} + \omega_{24m} \ln r \right.$$
$$\left. + \, \omega_{25m} \ln^2 r + \omega_{26m} \, r^{c_{3m}-1} \ln r \right], \tag{6.356}$$

$$W_m = 8 \int_0^{\pi/2} \int_0^{\pi/2} \int_{R_2}^{r_c} w_m \, r^2 \, dr \, d\varphi \, d\nu = (p_1 \, \varrho_{2e}^{me})^2 \int_0^{\pi/2} \int_0^{\pi/2} \Omega_{22m} \, d\varphi \, d\nu, \tag{6.357}$$

where the coefficient Ω_{22m} given by Eq. (6.39) is a function of $\omega_{21m}, \ldots, \omega_{26m}$ given by Eqs. (6.364)–(6.369), respectively.

With regard to the determination of the thermal-stress induced strengthening presented in Volume III, the stress σ_{im} acting in a direction of the axis x_i and the elastic energy density w_{im} induced by the stress σ_{im} $(i=1,2,3)$ (see Eqs. (3.46), (3.47), (6.342)–(6.345)), both related to an arbitrary point of the cell matrix, considering the φ, ν-dependent radial stress p_1, have the forms

$$\sigma_{im} = -\varrho_{2e}^{me} \left(\xi_{i11m} + \xi_{i12m} \, r^{c_{3m}-1} - \xi_{i13m} \ln r \right), \quad i = 1, 2, 3, \tag{6.358}$$

$$\begin{aligned} w_{im} = \frac{s_{11m} \left(\varrho_{2e}^{me}\right)^2}{2} & \left[\xi_{i11m}^2 + \xi_{i12m}^2 \, r^{2(c_{3m}-1)} + \xi_{i13m}^2 \ln^2 r + 2\,\xi_{i11m}\,\xi_{i12m}\, r^{c_{3m}-1} \right. \\ & \left. - \, 2\,\xi_{i11m}\,\xi_{i13m} \ln r - 2\,\xi_{i12m}\,\xi_{i13m}\, r^{c_{3m}-1} \ln r \right], \\ i = 1, 2, 3, & \end{aligned} \tag{6.359}$$

where the elastic modulus s_{11m} and the coefficients ξ_{i11m}, ξ_{i12m}, ξ_{i13m} $(i=1,2,3)$ are given by Eqs. (3.23) and (6.40)–(6.42), respectively. With regard to the φ, ν-independent radial stress p_1, considering Eqs. (3.46), (3.47), (6.342), (6.343), (6.346), (6.347), we get

$$\sigma_{im} = -p_1 \, \varrho_{2e}^{me} \left(\xi_{i1m} + \xi_{i2m} \, r^{c_{3m}-1} - \xi_{i3m} \ln r \right), \quad i = 1, 2, 3, \tag{6.360}$$

$$\begin{aligned} w_{im} = \frac{s_{11m} \left(p_1 \, \varrho_{2e}^{me}\right)^2}{2} & \left[\xi_{i1m}^2 + \xi_{i2m}^2 \, r^{2(c_{3m}-1)} + \xi_{i3m}^2 \ln^2 r + 2\,\xi_{i1m}\,\xi_{i2m}\, r^{c_{3m}-1} \right. \\ & \left. -2\,\xi_{i1m}\,\xi_{i3m} \ln r - 2\,\xi_{i2m}\,\xi_{i3m}\, r^{c_{3m}-1} \ln r \right], \\ i = 1, 2, 3, & \end{aligned} \tag{6.361}$$

where the coefficients ξ_{i1m}, ξ_{i2m}, ξ_{i3m} $(i=1,2,3)$ are given by Eqs. (6.370)–(6.372), respectively.

Appendix. Considering the elastic modulus s_{44m} and the coefficients c_{1m}, c_{2m}, c_{3m}, $C_m^{(1)}$, a'_{1i}, a'_{2i}, a'_{3i} $(i=1,2,3)$ (see Eqs. (3.25), (3.38)–(3.40), (6.5), (12.26)–(12.34)), the coefficients χ_{11m}, χ_{12m}, $\omega_{21m}, \ldots, \omega_{26m}$, ξ_{i1m}, ξ_{i2m}, ξ_{i3m} $(i=1,2,3)$ have the forms

$$\chi_{11m} = \frac{1}{3} - \ln R_2, \tag{6.362}$$

$$\chi_{12m} = C_m^{(1)} R_2^{c_{3m}-1}, \tag{6.363}$$

$$\omega_{21m} = \frac{2}{9}\left(3\left[\kappa_{1m}^{(1)}\right]^2 (c_{1m} + 2c_{2m}) + \frac{1}{s_{44m}}\left\{\left[\frac{\partial\kappa_{1m}^{(1)}}{\partial\varphi}\right]^2 + \left[\frac{\partial\kappa_{1m}^{(1)}}{\partial\nu}\right]^2\right\}\right), \qquad (6.364)$$

$$\omega_{22m} = \frac{4C_m^{(1)}}{3}\left\{\kappa_{1m}^{(1)}\kappa_{3m}^{(1)} (c_{1m} + 2c_{2m})(1 - c_{3m}) + \frac{1}{s_{44m}}\left[\frac{\partial\kappa_{1m}^{(1)}}{\partial\varphi}\frac{\partial\kappa_{3m}^{(1)}}{\partial\varphi} + \frac{\partial\kappa_{1m}^{(1)}}{\partial\nu}\frac{\partial\kappa_{3m}^{(1)}}{\partial\nu}\right]\right\}, \qquad (6.365)$$

$$\omega_{23m} = \left[C_m^{(1)}\right]^2\left(\left[\kappa_{3m}^{(1)}\right]^2\left[c_{3m}^2 (c_{1m} + c_{2m}) + 2(c_{1m} - 2c_{2m}c_{3m})\right] + \frac{2}{s_{44m}}\left\{\left[\frac{\partial\kappa_{3m}^{(1)}}{\partial\varphi}\right]^2 + \left[\frac{\partial\kappa_{3m}^{(1)}}{\partial\nu}\right]^2\right\}\right), \qquad (6.366)$$

$$\omega_{24m} = -\frac{4}{3s_{44m}}\left\{\left[\frac{\partial\kappa_{1m}^{(1)}}{\partial\varphi}\right]^2 + \left[\frac{\partial\kappa_{1m}^{(1)}}{\partial\nu}\right]^2\right\}, \qquad (6.367)$$

$$\omega_{25m} = 3\left[\kappa_{1m}^{(1)}\right]^2 (c_{1m} - c_{2m}) + \frac{2}{s_{44m}}\left\{\left[\frac{\partial\kappa_{1m}^{(1)}}{\partial\varphi}\right]^2 + \left[\frac{\partial\kappa_{1m}^{(1)}}{\partial\nu}\right]^2\right\}, \qquad (6.368)$$

$$\omega_{26m} = -2C_m^{(1)}\left\{\kappa_{1m}^{(1)}\kappa_{3m}^{(1)} (c_{1m} - c_{2m})(c_{3m} + 2) + \frac{2}{s_{44m}}\left[\frac{\partial\kappa_{1m}^{(1)}}{\partial\varphi}\frac{\partial\kappa_{3m}^{(1)}}{\partial\varphi} + \frac{\partial\kappa_{1m}^{(1)}}{\partial\nu}\frac{\partial\kappa_{3m}^{(1)}}{\partial\nu}\right]\right\}, \qquad (6.369)$$

$$\xi_{i1m} = \frac{1}{3}\left\{\kappa_{1m}^{(1)}\left(a'_{2i} + a'_{3i} - 2a'_{1i}\right)(c_{1m} + 2c_{2m}) + \frac{1}{s_{44m}}\left[\left(a'_{1i} + a'_{2i}\right)\frac{\partial\kappa_{1m}^{(1)}}{\partial\varphi} + \left(a'_{1i} + a'_{3i}\right)\frac{\partial\kappa_{1m}^{(1)}}{\partial\nu}\right]\right\},$$
$$i = 1, 2, 3, \qquad (6.370)$$

$$\xi_{i2m} = C_m^{(1)}\left(\kappa_{3m}^{(1)}\left\{a'_{1i}\left[c_{3m}(c_{1m} + c_{2m}) - 2c_{2m}\right] + \left(a'_{2i} + a'_{3i}\right)(c_{1m} - c_{2m}c_{3m})\right\} + \frac{1}{s_{44m}}\left[\left(a'_{1i} + a'_{2i}\right)\frac{\partial\kappa_{3m}^{(1)}}{\partial\varphi} + \left(a'_{1i} + a'_{3i}\right)\frac{\partial\kappa_{3m}^{(1)}}{\partial\nu}\right]\right),$$
$$i = 1, 2, 3, \qquad (6.371)$$

$$\xi_{i3m} = \kappa_{1m}^{(1)} \left(a'_{1i} + a'_{2i} + a'_{3i}\right) (c_{1m} - c_{2m}) + \frac{1}{s_{44m}} \left[\left(a'_{1i} + a'_{2i}\right) \frac{\partial \kappa_{1m}^{(1)}}{\partial \varphi} + \left(a'_{1i} + a'_{3i}\right) \frac{\partial \kappa_{1m}^{(1)}}{\partial \nu} \right], \quad i = 1, 2, 3. \quad (6.372)$$

B. Integration constants $C_{1m} \neq 0$, $C_{2m} = 0$, $C_{3m} \neq 0$

With regard to the boundary conditions (4.11), (4.12) and the analysis presented in Section 6.2.3, considering the terms u_{1rm} and u_{3rm} (see Eqs. (6.6), (6.8)), the former as a function of the variable $r \in \langle R_2, r_c \rangle$ and the latter as a constant function regarding the variable r, respectively, the radial displacement u_{1m} of the infinitesimal spherical cap in the cell matrix for the φ, ν-dependent or φ, ν-independent radial stress p_1, regarding Eqs. (3.31), (6.4), has the form

$$u_{1m} = -p_1 \, \varrho_{2e}^{me} \left[\kappa_{1m}^{(1)} r \left(\frac{1}{3} - \ln r \right) + \frac{3\kappa_{3m}^{(1)}}{2} \right], \quad (6.373)$$

where the radial stress p_1 acting on the particle-envelope boundary and the coefficient ϱ_{2e}^{me} are analysed at the beginning of Section 6.4.2.2. The φ, ν-dependent distance r_c, as a function of the experimentally determined parameters v, R_{1ex}, d_{1ex}, d_{2ex}, d_{3ex} (see Section 2.2.1), included in the coefficients $\kappa_{1m}^{(1)}$, $\kappa_{3m}^{(1)}$ is derived in Section 2.3.

With regard to Section 6.4.2.2.B, the coefficients $\psi_{11\eta m}^{(1)}$, $\psi_{12\eta m}^{(1)}$ $(\eta = \varphi, \nu)$, $\omega_{21m}, \ldots, \omega_{26m}$, ξ_{i1m}, ξ_{i2m}, ξ_{i3m} $(i = 1,2,3)$ and $\omega_{11m}, \ldots, \omega_{16m}$, ξ_{i11m}, ξ_{i12m}, ξ_{i13m} $(i = 1,2,3)$ mentioned below (see Eqs. (6.376), (6.377), (6.382), (6.383), (6.388), (6.392), (6.393) and (6.386), (6.390), (6.391)) depend on $\kappa_{1m}^{(1)}$, $\kappa_{3m}^{(1)}$ and $\kappa_{1m}^{(1)}$, $\kappa_{3m}^{(1)}$, $\psi_{11\eta m}^{(1)}$, $\psi_{12\eta m}^{(1)}$, respectively, where the coefficients $\kappa_{1m}^{(1)}$, $\kappa_{3m}^{(1)}$ (see Eq. (12.35)) are functions of χ_{11m}, χ_{12m}, χ_{21m}, χ_{22m} given by Eqs. (6.362), (6.394), (6.30), (6.58), respectively.

Consequently, considering the φ, ν-dependent or φ, ν-independent radial stress p_1, the radial and tangential stresses, σ_{11m} and σ_{22m}, σ_{33m}, respectively, acting in the cell matrix, regarding Eqs. (3.31), (3.32), (6.373), are derived as

$$\sigma_{11m} = -p_1 \, \varrho_{2e}^{me} \left\{ -\kappa_{1m}^{(1)} \left[\frac{2}{3} (c_{1m} + 2c_{2m}) + (c_{1m} - c_{2m}) \ln r \right] - \frac{3\kappa_{3m}^{(1)} c_{2m}}{r} \right\}, \quad (6.374)$$

$$\sigma_{22m} = \sigma_{33m} = -p_1 \, \varrho_{2e}^{me} \left\{ \kappa_{1m}^{(1)} \left[\frac{1}{3} (c_{1m} + 2c_{2m}) - (c_{1m} - c_{2m}) \ln r \right] + \frac{3\kappa_{3m}^{(1)} c_{1m}}{2r} \right\}, \quad (6.375)$$

where the coefficients c_{1m}, c_{2m} are given by Eqs. (3.38), (3.39), respectively.

With regard to Eqs. (3.33), (3.34), (6.373), the shear stresses σ_{12m}, σ_{13m} acting in the cell matrix, considering the φ,ν-dependent radial stress p_1, are derived as

$$\sigma_{12m} = -\frac{\varrho_{2e}^{me}}{s_{44m}}\left[\psi_{11\varphi m}^{(1)}\left(\frac{1}{3}-\ln r\right)+\frac{3\psi_{12\varphi m}^{(1)}}{2r}\right], \tag{6.376}$$

$$\sigma_{13m} = -\frac{\varrho_{2e}^{me}}{s_{44m}}\left[\psi_{11\nu m}^{(1)}\left(\frac{1}{3}-\ln r\right)+\frac{3\psi_{12\nu m}^{(1)}}{2r}\right], \tag{6.377}$$

where the elastic modulus s_{44m} and the coefficients $\psi_{11\eta m}^{(1)}$, $\psi_{12\eta m}^{(1)}$ $(\eta=\varphi,\nu)$ are given by Eqs. (3.25) and (12.36), respectively. Considering the φ,ν-independent radial stress p_1, we get

$$\sigma_{12m} = -\frac{p_1\,\varrho_{2e}^{me}}{s_{44m}}\left[\frac{\partial\kappa_{1m}^{(1)}}{\partial\varphi}\left(\frac{1}{3}-\ln r\right)+\frac{\partial\kappa_{3m}^{(1)}}{\partial\varphi}\,\frac{3}{2r}\right], \tag{6.378}$$

$$\sigma_{13m} = -\frac{p_1\,\varrho_{2e}^{me}}{s_{44m}}\left[\frac{\partial\kappa_{1m}^{(1)}}{\partial\nu}\left(\frac{1}{3}-\ln r\right)+\frac{\partial\kappa_{3m}^{(1)}}{\partial\nu}\,\frac{3}{2r}\right]. \tag{6.379}$$

Finally, considering the φ,ν-dependent or φ,ν-independent radial stress p_1, the radial and tangential strains, ε_{11m} and ε_{22m}, ε_{33m}, respectively, acting in the cell matrix, regarding Eqs. (3.1)–(3.3), (6.373), have the forms[12]

$$\varepsilon_{11m} = p_1\,\varrho_{2e}^{me}\kappa_{1m}^{(1)}\left(\frac{2}{3}+\ln r\right), \tag{6.380}$$

$$\varepsilon_{22m} = \varepsilon_{33m} = -p_1\,\varrho_{2e}^{me}\left[\kappa_{1m}^{(1)}\left(\frac{1}{3}-\ln r\right)+\frac{3\kappa_{3m}^{(1)}}{2r}\right]. \tag{6.381}$$

With regard to Eqs. (3.4), (3.5), (6.373), the shear strains σ_{12m}, σ_{13m} acting in the cell matrix, considering the φ,ν-dependent radial stress p_1, are derived as

$$\varepsilon_{12m} = -\varrho_{2e}^{me}\left[\psi_{11\varphi m}^{(1)}\left(\frac{1}{3}-\ln r\right)+\frac{3\psi_{12\varphi m}^{(1)}}{2r}\right], \tag{6.382}$$

$$\varepsilon_{13m} = -\varrho_{2e}^{me}\left[\psi_{11\nu m}^{(1)}\left(\frac{1}{3}-\ln r\right)+\frac{3\psi_{12\nu m}^{(1)}}{2r}\right], \tag{6.383}$$

and considering the φ,ν-independent radial stress p_1, we get

$$\varepsilon_{12m} = -p_1\,\varrho_{2e}^{me}\left[\frac{\partial\kappa_{1m}^{(1)}}{\partial\varphi}\left(\frac{1}{3}-\ln r\right)+\frac{\partial\kappa_{3m}^{(1)}}{\partial\varphi}\,\frac{3}{2r}\right], \tag{6.384}$$

[12]To avoid misunderstanding, a sign on the right side in Eq. (6.380) is really plus, in contrast to minus sign on the right sides in Eqs. (6.381)–(6.385).

$$\varepsilon_{13m} = -p_1\,\varrho_{2e}^{me}\left[\frac{\partial \kappa_{1m}^{(1)}}{\partial \nu}\left(\frac{1}{3} - \ln r\right) + \frac{\partial \kappa_{3m}^{(1)}}{\partial \nu}\,\frac{3}{2r}\right]. \tag{6.385}$$

With regard to the φ, ν-dependent radial stress p_1, considering Eqs. (3.41), (3.42), (6.374)–(6.377), (6.380)–(6.383), the thermal-stress induced elastic energy density w_m and the thermal-stress induced elastic energy W_m, accumulated in an arbitrary point of the cell matrix and in the volume of the cell matrix, respectively, are derived as

$$w_m = \frac{(\varrho_{2e}^{me})^2}{2}\left[\omega_{11m} + \frac{\omega_{12m}}{r} + \frac{\omega_{13m}}{r^2} + \omega_{14m}\ln r + \omega_{15m}\ln^2 r + \omega_{16m}\frac{\ln r}{r}\right], \tag{6.386}$$

$$W_m = 8\int_0^{\pi/2}\int_0^{\pi/2}\int_{R_2}^{r_c} w_m\, r^2 dr d\varphi d\nu = (\varrho_{2e}^{me})^2\int_0^{\pi/2}\int_0^{\pi/2}\Omega_{12m} d\varphi d\nu, \tag{6.387}$$

where the coefficient Ω_{12m} given by Eq. (6.63) is a function of $\omega_{11m}, \ldots, \omega_{16m}$ given by Eqs. (6.33), (6.60), (6.61), (6.36), (6.37), (6.62), respectively.

With regard to the φ, ν-independent radial stress p_1, considering Eqs. (3.41), (3.42), (6.374), (6.375), (6.378)–(6.381), (6.384), (6.385), we get

$$w_m = \frac{(p_1\,\varrho_{2e}^{me})^2}{2}\left[\omega_{21m} + \frac{\omega_{22m}}{r} + \frac{\omega_{23m}}{r^2} + \omega_{24m}\ln r + \omega_{25m}\ln^2 r + \omega_{26m}\frac{\ln r}{r}\right], \tag{6.388}$$

$$W_m = 8\int_0^{\pi/2}\int_0^{\pi/2}\int_{R_2}^{r_c} w_m\, r^2\, dr\, d\varphi\, d\nu = (p_1\,\varrho_{2e}^{me})^2\int_0^{\pi/2}\int_0^{\pi/2}\Omega_{22m}\, d\varphi\, d\nu, \tag{6.389}$$

where the coefficient Ω_{22m} given by Eq. (6.63) is a function of $\omega_{21m}, \ldots, \omega_{26m}$ given by Eqs. (6.364), (6.395), (6.396), (6.367), (6.368), (6.397), respectively.

With regard to the determination of the thermal-stress induced strengthening presented in Volume III, the stress σ_{im} acting in a direction of the axis x_i and the elastic energy density w_{im} induced by the stress σ_{im} ($i = 1,2,3$) (see Eqs. (3.46), (3.47), (6.374)–(6.377)), both related to an arbitrary point of the cell matrix, considering the φ, ν-dependent radial stress p_1, have the forms

$$\sigma_{im} = -\varrho_{2e}^{me}\left(\xi_{i11m} + \frac{\xi_{i12m}}{r} - \xi_{i13m}\ln r\right), \quad i = 1, 2, 3, \tag{6.390}$$

$$w_{im} = \frac{s_{11m}\,(\varrho_{2e}^{me})^2}{2}\left[\xi_{i11m}^2 + \frac{\xi_{i12m}^2}{r^2} + \xi_{i13m}^2\ln^2 r + \frac{2\,\xi_{i11m}\,\xi_{i12m}}{r} - 2\,\xi_{i11m}\,\xi_{i13m}\ln r - 2\,\xi_{i12m}\,\xi_{i13m}\frac{\ln r}{r}\right],$$

$$i = 1, 2, 3, \tag{6.391}$$

where the elastic modulus s_{11m} and the coefficients ξ_{i11m}, ξ_{i12m} ξ_{i13m} $(i=1,2,3)$ are given by Eqs. (3.23) and (6.40), (6.64), (6.42), respectively. With regard to the φ, ν-independent radial stress p_1, considering Eqs. (3.46), (3.47), (6.374), (6.375), (6.378), (6.379), we get

$$\sigma_{im} = -p_1\, \varrho_{2e}^{me} \left(\xi_{i1m} + \frac{\xi_{i2m}}{r} - \xi_{i3m} \ln r \right), \quad i = 1, 2, 3, \tag{6.392}$$

$$\begin{aligned} w_{im} = {} & \frac{s_{11m} \left(p_1\, \varrho_{2e}^{me}\right)^2}{2} \\ & \times \left[\xi_{i1m}^2 + \frac{\xi_{i2m}^2}{r^2} + \xi_{i3m}^2 \ln^2 r + \frac{2\, \xi_{i1m} \xi_{i2m}}{r} - 2\, \xi_{i1m} \xi_{i3m} \ln r - 2\, \xi_{i2m} \xi_{i3m} \frac{\ln r}{r} \right], \\ & i = 1, 2, 3, \end{aligned} \tag{6.393}$$

where the coefficients ξ_{i1m}, ξ_{i2m} ξ_{i3m} $(i=1,2,3)$ are given by Eqs. (6.370), (6.398), (6.372), respectively.

Appendix. Considering the elastic modulus s_{44m} and the coefficients c_{1m}, c_{2m} a_{1i}', a_{2i}', a_{3i}' $(i=1,2,3)$ (see Eqs. (3.25), (3.38), (3.39), (12.26)–(12.34)), the coefficients χ_{12m}, ω_{22m}, ω_{23m}, ω_{26m}, ξ_{i2m} $(i=1,2,3)$ have the forms

$$\chi_{12m} = \frac{3}{2R_2}, \tag{6.394}$$

$$\omega_{22m} = 2 \left\{ \kappa_{1m}^{(1)} \kappa_{3m}^{(1)} \left(c_{1m} + 2c_{2m}\right) + \frac{1}{s_{44m}} \left[\frac{\partial \kappa_{1m}^{(1)}}{\partial \varphi} \frac{\partial \kappa_{3m}^{(1)}}{\partial \varphi} + \frac{\partial \kappa_{1m}^{(1)}}{\partial \nu} \frac{\partial \kappa_{3m}^{(1)}}{\partial \nu} \right] \right\}, \tag{6.395}$$

$$\omega_{23m} = \frac{9}{2} \left(c_{1m} \left[\kappa_{3m}^{(1)} \right]^2 + \frac{1}{s_{44m}} \left\{ \left[\frac{\partial \kappa_{3m}^{(1)}}{\partial \varphi} \right]^2 + \left[\frac{\partial \kappa_{3m}^{(1)}}{\partial \nu} \right]^2 \right\} \right), \tag{6.396}$$

$$\omega_{26m} = -6 \left\{ \kappa_{1m}^{(1)} \kappa_{3m}^{(1)} \left(c_{1m} - c_{2m}\right) + \frac{1}{s_{44m}} \left[\frac{\partial \kappa_{1m}^{(1)}}{\partial \varphi} \frac{\partial \kappa_{3m}^{(1)}}{\partial \varphi} + \frac{\partial \kappa_{1m}^{(1)}}{\partial \nu} \frac{\partial \kappa_{3m}^{(1)}}{\partial \nu} \right] \right\}, \tag{6.397}$$

$$\begin{aligned} \xi_{i2m} = {} & \frac{3}{2} \left\{ \kappa_{3m}^{(1)} \left[c_{1m} \left(a_{2i}' + a_{3i}'\right) - 2a_{1i}'\, c_{2m} \right] \right. \\ & \left. + \frac{1}{s_{44m}} \left[\left(a_{1i}' + a_{2i}'\right) \frac{\partial \kappa_{3m}^{(1)}}{\partial \varphi} + \left(a_{1i}' + a_{3i}'\right) \frac{\partial \kappa_{3m}^{(1)}}{\partial \nu} \right] \right\}, \\ & i = 1, 2, 3. \end{aligned} \tag{6.398}$$

C. Integration constants $C_{1m} = 0$, $C_{2m} \neq 0$, $C_{3m} \neq 0$

With regard to the boundary conditions (4.11), (4.12) and the analysis presented in Section 6.2.3, considering the terms u_{2rm} and u_{3rm} (see Eqs. (6.7), (6.8)), the former as a function of the variable $r \in \langle R_2, r_c \rangle$ and the latter as a constant function regarding the variable r, respectively, the radial displacement u_{1m} of the infinitesimal spherical cap in the cell matrix for the φ, ν-dependent or φ, ν-independent radial stress p_1, regarding Eqs. (3.31), (6.4), has the form

$$u_{1m} = -p_1\, \varrho_{2e}^{me} \left[\kappa_{1m}^{(1)} C_m^{(1)}\, r^{c_{3m}} + \frac{3\kappa_{3m}^{(1)}}{2} \right], \tag{6.399}$$

where the coefficients $c_{3m} < 0$, $C_m^{(1)}$ are given by Eqs. (3.40), (6.5), respectively. The radial stress p_1 acting on the particle-envelope boundary and the coefficient ϱ_{2e}^{me} are analysed at the beginning of Section 6.4.2.2. The φ, ν-dependent distance r_c, as a function of the experimentally determined parameters v, R_{1ex}, d_{1ex}, d_{2ex}, d_{3ex} (see Section 2.2.1), included in the coefficients $\kappa_{1m}^{(1)}$, $\kappa_{3m}^{(1)}$ is derived in Section 2.3.

With regard to Section 6.4.2.2.C, the coefficients $\psi_{11\eta m}^{(1)}$, $\psi_{12\eta m}^{(1)}$ $(\eta = \varphi, \nu)$, ω_{21m}, ω_{22m}, ω_{23m}, ξ_{i1m}, ξ_{i2m} $(i = 1,2,3)$ and ω_{11m}, ω_{12m}, ω_{13m}, ξ_{i11m}, ξ_{i12m} $(i = 1,2,3)$ mentioned below (see Eqs. (6.402), (6.403), (6.406), (6.407), (6.414), (6.418), (6.419) and (6.412), (6.416), (6.417)) depend on $\kappa_{1m}^{(1)}$, $\kappa_{3m}^{(1)}$ and $\kappa_{1m}^{(1)}$, $\kappa_{3m}^{(1)}$, $\psi_{11\eta m}^{(1)}$, $\psi_{12\eta m}^{(1)}$, respectively, where the coefficients $\kappa_{1m}^{(1)}$, $\kappa_{3m}^{(1)}$ (see Eq. (12.35)) are functions of χ_{11m}, χ_{12m}, χ_{21m}, χ_{22m} given by Eqs. (6.420), (6.394), (6.80), (6.58), respectively.

Consequently, considering the φ, ν-dependent or φ, ν-independent radial stress p_1, the radial and tangential stresses, σ_{11m} and σ_{22m}, σ_{33m}, respectively, acting in the cell matrix, regarding Eqs. (3.31), (3.32), (6.399), are derived as

$$\sigma_{11m} = -p_1\, \varrho_{2e}^{me} \left\{ \kappa_{1m}^{(1)} C_m^{(1)} \left[c_{3m} \left(c_{1m} + c_{2m} \right) - 2c_{2m} \right] r^{c_{3m}-1} - \frac{3\kappa_{3m}^{(1)} c_{2m}}{r} \right\}, \tag{6.400}$$

$$\sigma_{22m} = \sigma_{33m} = -p_1\, \varrho_{2e}^{me} \left[\kappa_{1m}^{(1)} C_m^{(1)} \left(c_{1m} - c_{2m} c_{3m} \right) r^{c_{3m}-1} + \frac{3\kappa_{3m}^{(1)} c_{1m}}{2r} \right], \tag{6.401}$$

where the coefficients c_{1m}, c_{2m} are given by Eqs. (3.38), (3.39), respectively.

With regard to Eqs. (3.33), (3.34), (6.399), the shear stresses σ_{12m}, σ_{13m} acting in the cell matrix, considering the φ, ν-dependent radial stress p_1, are derived as

$$\sigma_{12m} = -\frac{\varrho_{2e}^{me}}{s_{44m}} \left[\psi_{11\varphi m}^{(1)} C_m^{(1)}\, r^{c_{3m}-1} + \frac{3\psi_{12\varphi m}^{(1)}}{2r} \right], \tag{6.402}$$

$$\sigma_{13m} = -\frac{\varrho_{2e}^{me}}{s_{44m}} \left[\psi_{11\nu m}^{(1)} C_m^{(1)}\, r^{c_{3m}-1} + \frac{3\psi_{12\nu m}^{(1)}}{2r} \right], \tag{6.403}$$

where the elastic modulus s_{44m} and the coefficients $\psi_{11\eta m}^{(1)}$, $\psi_{12\eta m}^{(1)}$ $(\eta = \varphi, \nu)$ are given by Eqs. (3.25) and (12.36), respectively. Considering the φ, ν-independent radial stress p_1, we get

$$\sigma_{12m} = -\frac{p_1\, \varrho_{2e}^{me}}{s_{44m}} \left[\frac{\partial \kappa_{1m}^{(1)}}{\partial \varphi} C_m^{(1)}\, r^{c_{3m}-1} + \frac{\partial \kappa_{3m}^{(1)}}{\partial \varphi} \frac{3}{2r} \right], \tag{6.404}$$

$$\sigma_{13m} = -\frac{p_1\, \varrho_{2e}^{me}}{s_{44m}} \left[\frac{\partial \kappa_{1m}^{(1)}}{\partial \nu} C_m^{(1)}\, r^{c_{3m}-1} + \frac{\partial \kappa_{3m}^{(1)}}{\partial \nu} \frac{3}{2r} \right]. \tag{6.405}$$

Finally, considering the φ, ν-dependent or φ, ν-independent radial stress p_1, the radial and tangential strains, ε_{11m} and ε_{22m}, ε_{33m}, respectively, acting in the cell matrix, regarding Eqs. (3.1)–(3.3), (6.399), have the forms

$$\varepsilon_{11m} = -p_1\, \varrho_{2e}^{me} \kappa_{1m}^{(1)} C_m^{(1)} c_{3m}\, r^{c_{3m}-1}, \tag{6.406}$$

$$\varepsilon_{22m} = \varepsilon_{33m} = -p_1\, \varrho_{2e}^{me} \left[\kappa_{1m}^{(1)} C_m^{(1)}\, r^{c_{3m}-1} + \frac{3\kappa_{3m}^{(1)}}{2r} \right]. \tag{6.407}$$

With regard to Eqs. (3.4), (3.5), (6.399), the shear strains σ_{12m}, σ_{13m} acting in the cell matrix, considering the φ, ν-dependent radial stress p_1, are derived as

$$\varepsilon_{12m} = -\varrho_{2e}^{me} \left[\psi_{11\varphi m}^{(1)} C_m^{(1)}\, r^{c_{3m}-1} + \frac{3\psi_{12\varphi m}^{(1)}}{2r} \right], \tag{6.408}$$

$$\varepsilon_{13m} = -\varrho_{2e}^{me} \left[\psi_{11\nu m}^{(1)} C_m^{(1)}\, r^{c_{3m}-1} + \frac{3\psi_{12\nu m}^{(1)}}{2r} \right]. \tag{6.409}$$

and considering the φ, ν-independent radial stress p_1, we get

$$\varepsilon_{12m} = -p_1\, \varrho_{2e}^{me} \left[\frac{\partial \kappa_{1m}^{(1)}}{\partial \varphi} C_m^{(1)}\, r^{c_{3m}-1} + \frac{\partial \kappa_{3m}^{(1)}}{\partial \varphi} \frac{3}{2r} \right], \tag{6.410}$$

$$\varepsilon_{13m} = -p_1\, \varrho_{2e}^{me} \left[\frac{\partial \kappa_{1m}^{(1)}}{\partial \nu} C_m^{(1)}\, r^{c_{3m}-1} + \frac{\partial \kappa_{3m}^{(1)}}{\partial \nu} \frac{3}{2r} \right]. \tag{6.411}$$

With regard to the φ, ν-dependent radial stress p_1, considering Eqs. (3.41), (3.42), (6.400)–(6.403), (6.406)–(6.409), the thermal-stress induced elastic energy density w_m and the thermal-stress induced elastic energy W_m, accumulated in an arbitrary point of the cell matrix and in the volume of the cell matrix, respectively, are derived as

$$w_m = \frac{\left(\varrho_{2e}^{me}\right)^2}{2} \left[\frac{\omega_{11m}}{r^2} + \omega_{12m}\, r^{c_{3m}-2} + \omega_{13m}\, r^{2(c_{3m}-1)} \right], \tag{6.412}$$

$$W_m = 8 \int_0^{\pi/2} \int_0^{\pi/2} \int_{R_2}^{r_c} w_m \, r^2 dr d\varphi d\nu = (\varrho_{2e}^{me})^2 \int_0^{\pi/2} \int_0^{\pi/2} \Omega_{12m} d\varphi d\nu, \tag{6.413}$$

where the coefficient Ω_{12m} given by Eq. (6.85) is a function of ω_{11m}, ω_{12m}, ω_{13m} given by Eqs. (6.82)–(6.84), respectively.

With regard to the φ, ν-independent radial stress p_1, considering Eqs. (3.41), (3.42), (6.400), (6.401), (6.404)–(6.407), (6.410), (6.411), we get

$$w_m = \frac{(p_1 \, \varrho_{2e}^{me})^2}{2} \left[\frac{\omega_{21m}}{r^2} + \omega_{22m} \, r^{c_{3m}-2} + \omega_{23m} \, r^{2(c_{3m}-1)} \right], \tag{6.414}$$

$$W_m = 8 \int_0^{\pi/2} \int_0^{\pi/2} \int_{R_2}^{r_c} w_m \, r^2 \, dr \, d\varphi \, d\nu = (p_1 \, \varrho_{2e}^{me})^2 \int_0^{\pi/2} \int_0^{\pi/2} \Omega_{22m} \, d\varphi \, d\nu, \tag{6.415}$$

where the coefficient Ω_{22m} given by Eq. (6.83) is a function of ω_{21m}, ω_{22m}, ω_{23m} given by Eqs. (6.421)–(6.423), respectively.

With regard to the determination of the thermal-stress induced strengthening presented in Volume III, the stress σ_{im} acting in a direction of the axis x_i and the elastic energy density w_{im} induced by the stress σ_{im} ($i=1,2,3$) (see Eqs. (3.46), (3.47), (6.400)–(6.403)), both related to an arbitrary point of the cell matrix, considering the φ, ν-dependent radial stress p_1, have the forms

$$\sigma_{im} = -\varrho_{2e}^{me} \left(\frac{\xi_{i\,11m}}{r} + \xi_{i\,12m} \, r^{c_{3m}-1} \right), \quad i = 1, 2, 3, \tag{6.416}$$

$$w_{im} = \frac{s_{11m} \, (\varrho_{2e}^{me})^2}{2} \left[\frac{\xi_{i\,11m}^2}{r^2} + \xi_{i\,12m}^2 \, r^{2(c_{3m}-1)} + 2 \, \xi_{i\,11m} \, \xi_{i\,12m} \, r^{c_{3m}-2} \right], \quad i = 1, 2, 3, \tag{6.417}$$

where the elastic modulus s_{11m} and the coefficients $\xi_{i\,11m}$, $\xi_{i\,12m}$ ($i=1,2,3$) are given by Eqs. (3.23) and (6.86), (6.87), respectively. With regard to the φ, ν-independent radial stress p_1, considering Eqs. (3.46), (3.47), (6.400), (6.401), (6.404), (6.405), we get

$$\sigma_{im} = -p_1 \, \varrho_{2e}^{me} \left(\frac{\xi_{i1m}}{r} + \xi_{i2m} \, r^{c_{3m}-1} \right), \quad i = 1, 2, 3, \tag{6.418}$$

$$w_{im} = \frac{s_{11m} \, (p_1 \, \varrho_{2e}^{me})^2}{2} \left[\frac{\xi_{i1m}^2}{r^2} + \xi_{i2m}^2 \, r^{2(c_{3m}-1)} + 2 \, \xi_{i1m} \, \xi_{i2m} \, r^{c_{3m}-2} \right], \quad i = 1, 2, 3, \tag{6.419}$$

where the coefficients ξ_{i1m}, ξ_{i2m} ($i=1,2,3$) are given by Eqs. (6.424), (6.425), respectively.

Appendix. Considering the elastic modulus s_{44m} and the coefficients c_{1m}, c_{2m}, c_{3m}, $C_m^{(1)}$, a'_{1i}, a'_{2i}, a'_{3i} $(i=1,2,3)$ (see Eqs. (3.25), (3.38)–(3.40), (6.5), (12.26)–(12.34)), the coefficients χ_{11m}, ω_{21m}, ω_{22m}, ω_{23m}, ξ_{i1m}, ξ_{i2m} $(i=1,2,3)$ have the forms

$$\chi_{11m} = C_m^{(1)} R_2^{c_{3m}-1}, \tag{6.420}$$

$$\omega_{21m} = \frac{9}{2}\left(c_{1m}\left[\kappa_{3m}^{(1)}\right]^2 + \frac{1}{s_{44m}}\left\{\left[\frac{\partial\kappa_{3m}^{(1)}}{\partial\varphi}\right]^2 + \left[\frac{\partial\kappa_{3m}^{(1)}}{\partial\nu}\right]^2\right\}\right), \tag{6.421}$$

$$\omega_{22m} = 6C_m^{(1)}\left\{\kappa_{1m}^{(1)}\kappa_{3m}^{(1)}\left(c_{1m} - c_{2m}c_{3m}\right) + \frac{1}{s_{44m}}\left[\frac{\partial\kappa_{1m}^{(1)}}{\partial\varphi}\frac{\partial\kappa_{3m}^{(1)}}{\partial\varphi} + \frac{\partial\kappa_{1m}^{(1)}}{\partial\nu}\frac{\partial\kappa_{3m}^{(1)}}{\partial\nu}\right]\right\}, \tag{6.422}$$

$$\omega_{23m} = \left[C_m^{(1)}\right]^2\left(\left[\kappa_{1m}^{(1)}\right]^2\left[c_{3m}^2\left(c_{1m} + c_{2m}\right) + 2\left(c_{1m} - 2c_{2m}c_{3m}\right)\right] + \frac{2}{s_{44m}}\left\{\left[\frac{\partial\kappa_{1m}^{(1)}}{\partial\varphi}\right]^2 + \left[\frac{\partial\kappa_{1m}^{(1)}}{\partial\nu}\right]^2\right\}\right), \tag{6.423}$$

$$\xi_{i1m} = \frac{3}{2}\left\{\kappa_{3m}^{(1)}\left[c_{1m}\left(a'_{2i} + a'_{3i}\right) - 2a'_{1i}\,c_{2m}\right] + \frac{1}{s_{44m}}\left[\left(a'_{1i} + a'_{2i}\right)\frac{\partial\kappa_{3m}^{(1)}}{\partial\varphi} + \left(a'_{1i} + a'_{3i}\right)\frac{\partial\kappa_{3m}^{(1)}}{\partial\nu}\right]\right\},\quad i = 1,2,3, \tag{6.424}$$

$$\xi_{i2m} = C_m^{(1)}\left(\kappa_{1m}^{(1)}\left\{a'_{1i}\left[c_{3m}\left(c_{1m} + c_{2m}\right) - 2c_{2m}\right] + \left(a'_{2i} + a'_{3i}\right)\left(c_{1m} - c_{2m}c_{3m}\right)\right\} + \frac{1}{s_{44m}}\left[\left(a'_{1i} + a'_{2i}\right)\frac{\partial\kappa_{1m}^{(1)}}{\partial\varphi} + \left(a'_{1i} + a'_{3i}\right)\frac{\partial\kappa_{1m}^{(1)}}{\partial\nu}\right]\right),\quad i = 1,2,3. \tag{6.425}$$

D. Integration constants $C_{1m} \neq 0$, $C_{2m} \neq 0$, $C_{3m} \neq 0$

With regard to the boundary conditions (4.11), (4.12), (4.14) and the analysis presented in Section 6.2.3, considering the terms u_{1rm}, u_{2rm} and u_{3rm} (see Eqs. (6.6)–(6.8)) to represent functions of the variable $r \in \langle R_2, r_c\rangle$ and a constant function regarding the variable r, respectively, the radial displacement u_{1m} of the infinitesimal spherical cap in the cell matrix for the φ, ν-dependent or φ, ν-independent radial stress p_1, regarding Eqs. (3.31), (6.4), has the form

$$u_{1m} = -p_1\, \varrho_{2e}^{me} \left[\kappa_{4m}^{(2)} r \left(\frac{1}{3} - \ln r \right) + \kappa_{5m}^{(2)} C_m^{(1)}\, r^{c_{3m}} + \frac{3\kappa_{6m}^{(2)}}{2} \right], \tag{6.426}$$

where the coefficients $c_{3m} < 0$, $C_m^{(1)}$ are given by Eqs. (3.40), (6.5), respectively. The radial stress p_1 acting on the particle-envelope boundary and the coefficient ϱ_{2e}^{me} are analysed at the beginning of Section 6.4.2.2. The φ, ν-dependent distance r_c, as a function of the experimentally determined parameters v, R_{1ex}, d_{1ex}, d_{2ex}, d_{3ex} (see Section 2.2.1), included in the coefficients $\kappa_{4m}^{(2)}$, $\kappa_{5m}^{(2)}$, $\kappa_{6m}^{(2)}$ is derived in Section 2.3.

With regard to Section 6.4.2.2.D, the coefficients $\psi_{11\eta m}^{(2)}$, $\psi_{12\eta m}^{(2)}$, $\psi_{13\eta m}^{(2)}$ $(\eta = \varphi, \nu)$, $\omega_{21m}, \ldots, \omega_{210m}$, $\xi_{i1m}, \ldots, \xi_{i4m}$ $(i = 1,2,3)$ and $\omega_{11m}, \ldots, \omega_{110m}$, $\xi_{i11m}, \ldots, \xi_{i14m}$ $(i = 1,2,3)$ mentioned below (see Eqs. (6.429), (6.430), (6.435), (6.436), (6.441), (6.445), (6.446) and (6.439), (6.443), (6.444)) depend on $\kappa_{4m}^{(2)}$, $\kappa_{5m}^{(2)}$, $\kappa_{6m}^{(2)}$ and $\kappa_{4m}^{(2)}$, $\kappa_{5m}^{(2)}$, $\kappa_{6m}^{(2)}$, $\psi_{11\eta m}^{(2)}$, $\psi_{12\eta m}^{(2)}$, $\psi_{13\eta m}^{(2)}$, respectively, where the coefficients $\kappa_{4m}^{(2)}$, $\kappa_{5m}^{(2)}$, $\kappa_{6m}^{(2)}$ (see Eq. (12.39)) are functions of χ_{11m}, χ_{12m}, χ_{13m}, χ_{21m}, χ_{22m}, χ_{23m}, χ_{31m}, χ_{32m}, χ_{33m} given by Eqs. (6.362), (6.363), (6.447), (6.30), (6.31), (6.103)–(6.106), respectively.

Consequently, considering the φ, ν-dependent or φ, ν-independent radial stress p_1, the radial and tangential stresses, σ_{11m} and σ_{22m}, σ_{33m}, respectively, acting in the cell matrix, regarding Eqs. (3.31), (3.32), (6.426), are derived as

$$\sigma_{11m} = -p_1\, \varrho_{2e}^{me} \left\{ -\kappa_{4m}^{(2)} \left[\frac{2}{3} (c_{1m} + 2c_{2m}) + (c_{1m} - c_{2m}) \ln r \right] \right.$$
$$\left. + \kappa_{5m}^{(2)} C_m^{(1)} \left[c_{3m} (c_{1m} + c_{2m}) - 2c_{2m} \right] r^{c_{3m}-1} - \frac{3\kappa_{6m}^{(2)} c_{2m}}{r} \right\}, \tag{6.427}$$

$$\sigma_{22m} = \sigma_{33m} = -p_1\, \varrho_{2e}^{me} \left\{ \kappa_{4m}^{(2)} \left[\frac{1}{3} (c_{1m} + 2c_{2m}) - (c_{1m} - c_{2m}) \ln r \right] \right.$$
$$\left. + \kappa_{5m}^{(2)} C_m^{(1)} (c_{1m} - c_{2m} c_{3m})\, r^{c_{3m}-1} + \frac{3\kappa_{6m}^{(2)} c_{1m}}{2r} \right\}, \tag{6.428}$$

where the coefficients c_{1m}, c_{2m} are given by Eqs. (3.38), (3.39), respectively.

With regard to Eqs. (3.33), (3.34), (6.426), the shear stresses σ_{12m}, σ_{13m} acting in the cell matrix, considering the φ, ν-dependent radial stress p_1, are derived as

$$\sigma_{12m} = -\frac{\varrho_{2e}^{me}}{s_{44m}} \left[\psi_{11\varphi m}^{(2)} \left(\frac{1}{3} - \ln r \right) + \psi_{12\varphi m}^{(2)} C_m^{(1)}\, r^{c_{3m}-1} + \frac{3\psi_{13\varphi m}^{(2)}}{2r} \right], \tag{6.429}$$

$$\sigma_{13m} = -\frac{\varrho_{2e}^{me}}{s_{44m}} \left[\psi_{11\nu m}^{(2)} \left(\frac{1}{3} - \ln r \right) + \psi_{12\nu m}^{(2)} C_m^{(1)} r^{c_{3m}-1} + \frac{3\psi_{13\nu m}^{(2)}}{2r} \right], \tag{6.430}$$

where the elastic modulus s_{44m} and the coefficients $\psi_{11\eta m}^{(2)}$, $\psi_{12\eta m}^{(2)}$, $\psi_{13\eta m}^{(2)}$ $(\eta = \varphi, \nu)$ are given by Eqs. (3.25) and (12.40), respectively. Considering the φ, ν-independent radial stress p_1, we get

$$\sigma_{12m} = -\frac{p_1 \varrho_{2e}^{me}}{s_{44m}} \left[\frac{\partial \kappa_{4m}^{(2)}}{\partial \varphi} \left(\frac{1}{3} - \ln r \right) + \frac{\partial \kappa_{5m}^{(2)}}{\partial \varphi} C_m^{(1)} r^{c_{3m}-1} + \frac{\partial \kappa_{6m}^{(2)}}{\partial \varphi} \frac{3}{2r} \right], \tag{6.431}$$

$$\sigma_{13m} = -\frac{p_1 \varrho_{2e}^{me}}{s_{44m}} \left[\frac{\partial \kappa_{4m}^{(2)}}{\partial \nu} \left(\frac{1}{3} - \ln r \right) + \frac{\partial \kappa_{5m}^{(2)}}{\partial \nu} C_m^{(1)} r^{c_{3m}-1} + \frac{\partial \kappa_{6m}^{(2)}}{\partial \nu} \frac{3}{2r} \right]. \tag{6.432}$$

Finally, considering the φ, ν-dependent or φ, ν-independent radial stress p_1, the radial and tangential strains, ε_{11m} and ε_{22m}, ε_{33m}, respectively, acting in the cell matrix, regarding Eqs. (3.1)–(3.3), (6.426), have the forms

$$\varepsilon_{11m} = -p_1 \varrho_{2e}^{me} \left[-\kappa_{4m}^{(2)} \left(\frac{2}{3} + \ln r \right) + \kappa_{5m}^{(2)} C_m^{(1)} c_{3m} r^{c_{3m}-1} \right], \tag{6.433}$$

$$\varepsilon_{22m} = \varepsilon_{33m} = -p_1 \varrho_{2e}^{me} \left[\kappa_{4m}^{(2)} \left(\frac{1}{3} - \ln r \right) + \kappa_{5m}^{(2)} C_m^{(1)} r^{c_{3m}-1} + \frac{3\kappa_{6m}^{(2)}}{2r} \right]. \tag{6.434}$$

With regard to Eqs. (3.4), (3.5), (6.426), the shear strains σ_{12m}, σ_{13m} acting in the cell matrix, considering the φ, ν-dependent radial stress p_1, are derived as

$$\varepsilon_{12m} = -\varrho_{2e}^{me} \left[\psi_{11\varphi m}^{(2)} \left(\frac{1}{3} - \ln r \right) + \psi_{12\varphi m}^{(2)} C_m^{(1)} r^{c_{3m}-1} + \frac{3\psi_{13\varphi m}^{(2)}}{2r} \right], \tag{6.435}$$

$$\varepsilon_{13m} = -\varrho_{2e}^{me} \left[\psi_{11\nu m}^{(2)} \left(\frac{1}{3} - \ln r \right) + \psi_{12\nu m}^{(2)} C_m^{(1)} r^{c_{3m}-1} + \frac{3\psi_{13\nu m}^{(2)}}{2r} \right], \tag{6.436}$$

and considering the φ, ν-independent radial stress p_1, we get

$$\varepsilon_{12m} = -p_1 \varrho_{2e}^{me} \left[\frac{\partial \kappa_{4m}^{(2)}}{\partial \varphi} \left(\frac{1}{3} - \ln r \right) + \frac{\partial \kappa_{5m}^{(2)}}{\partial \varphi} C_m^{(1)} r^{c_{3m}-1} + \frac{\partial \kappa_{6m}^{(2)}}{\partial \varphi} \frac{3}{2r} \right], \tag{6.437}$$

$$\varepsilon_{13m} = -p_1 \varrho_{2e}^{me} \left[\frac{\partial \kappa_{4m}^{(2)}}{\partial \nu} \left(\frac{1}{3} - \ln r \right) + \frac{\partial \kappa_{5m}^{(2)}}{\partial \nu} C_m^{(1)} r^{c_{3m}-1} + \frac{\partial \kappa_{6m}^{(2)}}{\partial \nu} \frac{3}{2r} \right]. \tag{6.438}$$

With regard to the φ, ν-dependent radial stress p_1, considering Eqs. (3.41), (3.42), (6.427)–(6.430), (6.433)–(6.436), the thermal-stress induced elastic energy density w_m and the thermal-stress induced elastic energy W_m, accumulated in an arbitrary point of the cell matrix and in the volume of the cell matrix, respectively, are derived as

$$w_m = \frac{(\varrho_{2e}^{me})^2}{2}\left[\omega_{11m} + \frac{\omega_{12m}}{r} + \frac{\omega_{13m}}{r^2} + \omega_{14m}\, r^{c_{3m}-1} + \omega_{15m}\, r^{c_{3m}-2} + \omega_{16m}\, r^{2(c_{3m}-1)} + \omega_{17m}\ln r + \omega_{18m}\ln^2 r + \frac{\omega_{19m}\ln r}{r} + \omega_{110m}\, r^{c_{3m}-1}\ln r\right], \quad (6.439)$$

$$W_m = 8\int_0^{\pi/2}\int_0^{\pi/2}\int_{R_2}^{r_c} w_m\, r^2 dr d\varphi d\nu = (\varrho_{2e}^{me})^2\int_0^{\pi/2}\int_0^{\pi/2}\Omega_{12m} d\varphi d\nu, \quad (6.440)$$

where the coefficient Ω_{12m} given by Eq. (6.118) is a function of $\omega_{11m}, \ldots, \omega_{110m}$ Eqs. (6.108)–(6.117), respectively.

With regard to the φ, ν-independent radial stress p_1, considering Eqs. (3.41), (3.42), (6.427), (6.428), (6.431)–(6.434), (6.437), (6.438), we get

$$w_m = \frac{(p_1\,\varrho_{2e}^{me})^2}{2}\left[\omega_{21m} + \frac{\omega_{22m}}{r} + \frac{\omega_{23m}}{r^2} + \omega_{24m}\, r^{c_{3m}-1} + \omega_{25m}\, r^{c_{3m}-2} + \omega_{26m}\, r^{2(c_{3m}-1)} + \omega_{27m}\ln r + \omega_{28m}\ln^2 r + \frac{\omega_{29m}\ln r}{r} + \omega_{210m}\, r^{c_{3m}-1}\ln r\right], \quad (6.441)$$

$$W_m = 8\int_0^{\pi/2}\int_0^{\pi/2}\int_{R_2}^{r_c} w_m\, r^2\, dr\, d\varphi\, d\nu = (p_1\,\varrho_{2e}^{me})^2\int_0^{\pi/2}\int_0^{\pi/2}\Omega_{22m}\, d\varphi\, d\nu, \quad (6.442)$$

where the coefficient Ω_{22m} given by Eq. (6.118) is a function of $\omega_{21m}, \ldots, \omega_{210m}$ given by Eqs. (6.448)–(6.457), respectively.

With regard to the determination of the thermal-stress induced strengthening presented in Volume III, the stress σ_{im} acting in a direction of the axis x_i and the elastic energy density w_{im} induced by the stress σ_{im} ($i = 1,2,3$) (see Eqs. (3.46), (3.47), (6.427)–(6.430)), both related to an arbitrary point of the cell matrix, considering the φ, ν-dependent radial stress p_1, have the forms

$$\sigma_{im} = -\varrho_{2e}^{me}\left(\xi_{i11m} + \frac{\xi_{i12m}}{r} + \xi_{i13m}\, r^{c_{3m}-1} - \xi_{i14m}\ln r\right), \quad i = 1, 2, 3, \quad (6.443)$$

$$w_{im} = \frac{s_{11m}\left(\varrho_{2e}^{me}\right)^2}{2}\left[\xi_{i\,11m}^2 + \frac{\xi_{i\,12m}^2}{r^2} + \xi_{i\,13m}^2\, r^{2(c_{3m}-1)} + \xi_{i\,14m}^2 \ln^2 r + \frac{2\,\xi_{i\,11m}\,\xi_{i\,12m}}{r} + 2\,\xi_{i\,11m}\,\xi_{i\,13m}\, r^{c_{3m}-1} - 2\xi_{i\,11m}\,\xi_{i\,14m}\ln r + 2\xi_{i\,12m}\,\xi_{i\,13m}\, r^{c_{3m}-2} - 2\,\xi_{i\,12m}\,\xi_{i\,14m}\frac{\ln r}{r} - 2\,\xi_{i\,13m}\,\xi_{i\,14m}\, r^{c_{3m}-1}\ln r\right],$$
$$i = 1, 2, 3, \tag{6.444}$$

where the elastic modulus s_{11m} and the coefficients $\xi_{i\,11m}, \ldots, \xi_{i\,14m}$ ($i=1,2,3$) are given by Eqs. (3.23) and (6.119)–(6.122), respectively. With regard to the φ, ν-independent radial stress p_1, considering Eqs. (3.46), (3.47), (6.427), (6.428), (6.431), (6.432), we get

$$\sigma_{im} = -p_1\,\varrho_{2e}^{me}\left(\xi_{i1m} + \frac{\xi_{i2m}}{r} + \xi_{i3m}\, r^{c_{3m}-1} - \xi_{i4m}\ln r\right), \quad i = 1, 2, 3, \tag{6.445}$$

$$w_{im} = \frac{s_{11m}\left(p_1\,\varrho_{2e}^{me}\right)^2}{2}\left[\xi_{i1m}^2 + \frac{\xi_{i2m}^2}{r^2} + \xi_{i3m}^2\, r^{2(c_{3m}-1)} + \xi_{i4m}^2 \ln^2 r + \frac{2\,\xi_{i1m}\,\xi_{i2m}}{r} + 2\,\xi_{i1m}\,\xi_{i3m}\, r^{c_{3m}-1} - 2\,\xi_{i1m}\xi_{i4m}\ln r + 2\,\xi_{i2m}\,\xi_{i3m}\, r^{c_{3m}-2} - 2\,\xi_{i2m}\xi_{i4m}\frac{\ln r}{r} - 2\,\xi_{i3m}\xi_{i4m}\, r^{c_{3m}-1}\ln r\right],$$
$$i = 1, 2, 3, \tag{6.446}$$

where the coefficients $\xi_{i1m}, \ldots, \xi_{i4m}$ ($i=1,2,3$) are given by Eqs. (6.458)–(6.461), respectively.

Appendix. Considering the elastic modulus s_{44m} and the coefficients c_{1m}, c_{2m}, a'_{1i}, a'_{2i}, a'_{3i} ($i=1,2,3$) (see Eqs. (3.25), (3.38), (3.39), (12.26)–(12.34)), the coefficients χ_{13m}, $\omega_{21m}, \ldots, \omega_{210m}$, $\xi_{i1m}, \ldots, \xi_{i4m}$ ($i=1,2,3$) have the forms

$$\chi_{13m} = \frac{3}{2R_2}, \tag{6.447}$$

$$\omega_{21m} = \frac{2}{9}\left(3\left[\kappa_{4m}^{(2)}\right]^2 (c_{1m} + 2c_{2m}) + \frac{1}{s_{44m}}\left\{\left[\frac{\partial\kappa_{4m}^{(2)}}{\partial\varphi}\right]^2 + \left[\frac{\partial\kappa_{4m}^{(2)}}{\partial\nu}\right]^2\right\}\right), \tag{6.448}$$

$$\omega_{22m} = 2\left\{\kappa_{4m}^{(2)}\kappa_{6m}^{(2)} (c_{1m} + 2c_{2m}) + \frac{1}{s_{44m}}\left[\frac{\partial\kappa_{4m}^{(2)}}{\partial\varphi}\,\frac{\partial\kappa_{6m}^{(2)}}{\partial\varphi} + \frac{\partial\kappa_{4m}^{(2)}}{\partial\nu}\,\frac{\partial\kappa_{6m}^{(2)}}{\partial\nu}\right]\right\}, \tag{6.449}$$

$$\omega_{23m} = \frac{9}{2}\left(c_{1m}\left[\kappa_{6m}^{(2)}\right]^2 + \frac{1}{s_{44m}}\left\{\left[\frac{\partial \kappa_{6m}^{(2)}}{\partial \varphi}\right]^2 + \left[\frac{\partial \kappa_{6m}^{(2)}}{\partial \nu}\right]^2\right\}\right), \tag{6.450}$$

$$\omega_{24m} = \frac{4C_m^{(1)}}{3}\left\{\kappa_{4m}^{(2)}\kappa_{5m}^{(2)}\left[c_{1m}\left(c_{3m}+1\right)+2c_{2m}\right] + \frac{1}{s_{44m}}\left[\frac{\partial \kappa_{4m}^{(2)}}{\partial \varphi}\frac{\partial \kappa_{5m}^{(2)}}{\partial \varphi} + \frac{\partial \kappa_{4m}^{(2)}}{\partial \nu}\frac{\partial \kappa_{5m}^{(2)}}{\partial \nu}\right]\right\}, \tag{6.451}$$

$$\omega_{25m} = 6C_m^{(1)}\left\{\kappa_{5m}^{(2)}\kappa_{6m}^{(2)}\left(c_{1m}-c_{2m}c_{3m}\right) + \frac{1}{s_{44m}}\left[\frac{\partial \kappa_{5m}^{(2)}}{\partial \varphi}\frac{\partial \kappa_{6m}^{(2)}}{\partial \varphi} + \frac{\partial \kappa_{5m}^{(2)}}{\partial \nu}\frac{\partial \kappa_{6m}^{(2)}}{\partial \nu}\right]\right\}, \tag{6.452}$$

$$\omega_{26m} = \left[C_m^{(1)}\right]^2\left(\left[\kappa_{5m}^{(2)}\right]^2\left[c_{3m}^2\left(c_{1m}+c_{2m}\right)+2\left(c_{1m}-2c_{2m}c_{3m}\right)\right] + \frac{2}{s_{44m}}\left\{\left[\frac{\partial \kappa_{5m}^{(2)}}{\partial \varphi}\right]^2 + \left[\frac{\partial \kappa_{5m}^{(2)}}{\partial \nu}\right]^2\right\}\right), \tag{6.453}$$

$$\omega_{27m} = -\frac{4}{3s_{44m}}\left\{\left[\frac{\partial \kappa_{4m}^{(2)}}{\partial \varphi}\right]^2 + \left[\frac{\partial \kappa_{4m}^{(2)}}{\partial \nu}\right]^2\right\}, \tag{6.454}$$

$$\omega_{28m} = 3\left[\kappa_{4m}^{(2)}\right]^2\left(c_{1m}-c_{2m}\right) + \frac{2}{s_{44m}}\left\{\left[\frac{\partial \kappa_{4m}^{(2)}}{\partial \varphi}\right]^2 + \left[\frac{\partial \kappa_{4m}^{(2)}}{\partial \nu}\right]^2\right\}, \tag{6.455}$$

$$\omega_{29m} = -6\left\{\kappa_{4m}^{(2)}\kappa_{6m}^{(2)}\left(c_{1m}-c_{2m}\right) + \frac{1}{s_{44m}}\left[\frac{\partial \kappa_{4m}^{(2)}}{\partial \varphi}\frac{\partial \kappa_{6m}^{(2)}}{\partial \varphi} + \frac{\partial \kappa_{4m}^{(2)}}{\partial \nu}\frac{\partial \kappa_{6m}^{(2)}}{\partial \nu}\right]\right\}, \tag{6.456}$$

$$\omega_{210m} = -2C_m^{(1)}\left\{\kappa_{4m}^{(2)}\kappa_{5m}^{(2)}\left[c_{1m}\left(2-c_{3m}\right)-c_{2m}\left(3c_{3m}+2\right)\right] + \frac{2}{s_{44m}}\left[\frac{\partial \kappa_{4m}^{(2)}}{\partial \varphi}\frac{\partial \kappa_{5m}^{(2)}}{\partial \varphi} + \frac{\partial \kappa_{4m}^{(2)}}{\partial \nu}\frac{\partial \kappa_{5m}^{(2)}}{\partial \nu}\right]\right\}, \tag{6.457}$$

$$\xi_{i\,1m} = \frac{1}{3}\left\{\kappa_{4m}^{(2)}\left(a'_{2i}+a'_{3i}-2a'_{1i}\right)\left(c_{1m}+2c_{2m}\right) + \frac{1}{s_{44m}}\left[\left(a'_{1i}+a'_{2i}\right)\frac{\partial \kappa_{4m}^{(2)}}{\partial \varphi} + \left(a'_{1i}+a'_{3i}\right)\frac{\partial \kappa_{4m}^{(2)}}{\partial \nu}\right]\right\},$$
$$i = 1, 2, 3, \tag{6.458}$$

$$\xi_{i\,2m} = \frac{3}{2}\left\{\kappa_{6m}^{(2)}\left[c_{1m}\left(a'_{2i}+a'_{3i}\right)-2a'_{1i}\,c_{2m}\right]\right.$$
$$\left.+\frac{1}{s_{44m}}\left[\left(a'_{1i}+a'_{2i}\right)\frac{\partial\kappa_{6m}^{(2)}}{\partial\varphi}+\left(a'_{1i}+a'_{3i}\right)\frac{\partial\kappa_{6m}^{(2)}}{\partial\nu}\right]\right\},$$
$$i=1,2,3, \tag{6.459}$$

$$\xi_{i\,3m} = C_m^{(1)}\left(\kappa_{5m}^{(2)}\left\{a'_{1i}\left[c_{3m}\left(c_{1m}+c_{2m}\right)-2c_{2m}\right]+\left(a'_{2i}+a'_{3i}\right)\left(c_{1m}-c_{2m}c_{3m}\right)\right\}\right.$$
$$\left.+\frac{1}{s_{44m}}\left[\left(a'_{1i}+a'_{2i}\right)\frac{\partial\kappa_{5m}^{(2)}}{\partial\varphi}+\left(a'_{1i}+a'_{3i}\right)\frac{\partial\kappa_{5m}^{(2)}}{\partial\nu}\right]\right),$$
$$i=1,2,3, \tag{6.460}$$

$$\xi_{i\,4m} = \kappa_{4m}^{(2)}\left(a'_{1i}+a'_{2i}+a'_{3i}\right)\left(c_{1m}-c_{2m}\right)$$
$$+\frac{1}{s_{44m}}\left[\left(a'_{1i}+a'_{2i}\right)\frac{\partial\kappa_{4m}^{(2)}}{\partial\varphi}+\left(a'_{1i}+a'_{3i}\right)\frac{\partial\kappa_{4m}^{(2)}}{\partial\nu}\right],\quad i=1,2,3. \tag{6.461}$$

6.5 One-particle-(envelope)-matrix system. Conditions $\varepsilon_{11tm}-\varepsilon_{11tp}=f\left(\varphi,\nu\right)$, $\varepsilon_{11te}-\varepsilon_{11tp}=f\left(\varphi,\nu\right)$, $\varepsilon_{11tm}-\varepsilon_{11te}=f\left(\varphi,\nu\right)$

In contrast to Section 9.5, the analytical model of the thermal stresses and the corresponding quantities presented in Section 6.5, the latter represented by w_q, W_q, σ_{iq}, w_{iq} ($i=1,2,3$) (see Eqs. (3.41), (3.46), (3.47)), consider the φ,ν-dependent differences in the phase-transformation induced radial strains, $\varepsilon_{11tm}-\varepsilon_{11tp}=f\left(\varphi,\nu\right)$, and $\varepsilon_{11te}-\varepsilon_{11tp}=f\left(\varphi,\nu\right)$, $\varepsilon_{11tm}-\varepsilon_{11te}=f\left(\varphi,\nu\right)$, (see Eqs. (12.23)), resulting in the φ,ν-dependent differences $\beta_m-\beta_p$, and $\beta_e-\beta_p$, $\beta_m-\beta_e$ (see Eqs. (3.48)–(3.50)), respectively, and consequently resulting in the φ,ν-dependent radial stresses p_1 and p_2 (see Eqs. (10.1)–(10.5)) acting on the particle-matrix, particle-envelope and matrix-envelope boundaries related to the one-particle-matrix and one-particle-envelope-matrix systems, respectively. Accordingly, the analytical model results from a system of the differential equations (3.35), (3.36) with the solution (6.4).

6.5.1 Thermal stresses in spherical particle and envelope

Formulae for the spherical particle and for the spherical envelope in the one-particle-(envelope)-matrix system are identical with those presented in Sections 5.5.1, 5.6.1.2 and 5.6.2, 6.4.1, 7.5.1, respectively, where the coefficient ϱ_{1+im} ($i=0,1$) in formulae (10.1)–(10.3), (10.5) for the radial stresses p_1, p_2 is given by Eqs. (5.178), (6.476), considering the Castigliano's theorem (see Section 3.1.9). Section 5.5.1 is related to the one-particle-matrix and one-particle-envelope-matrix systems with regard to the conditions $\beta_p \neq \beta_m$ and $\beta_p \neq \beta_e \neq \beta_m$, $\beta_p \neq \beta_e = \beta_m$ (see Eqs. (3.48)–(3.50)), respectively.

6.5.2 Thermal stresses in infinite matrix

6.5.2.1 Conditions $\beta_p \neq \beta_m$, $\beta_p \neq \beta_e \neq \beta_m$, $\beta_p = \beta_e \neq \beta_m$

Considering Eqs. (6.462)–(6.479) containing the subscript $i = 0{,}1$, the thermal stresses and the corresponding quantities for the infinite matrix presented in Section 6.5.2.1 are derived

1. with regard to the condition $\beta_p \neq \beta_m$ related to the one-particle-matrix system, and consequently the subscript $i = 0$,
2. with regard to the conditions $\beta_p \neq \beta_e \neq \beta_m$, $\beta_p = \beta_e \neq \beta_m$ related to the one-particle-envelope-matrix system, and consequently the subscript $i = 1$.

With regard to the boundary conditions (4.9), (4.16) to determine one integration constant only, considering the analysis presented in Section 6.2.4 to define u_{1m} as a decreasing function of the variable $r \in \langle R_{1+i}, \infty)$ $(i = 0{,}1)$, and thus the integration constant $C_{1m} = 0$, the radial displacement u_{1m} of the infinitesimal spherical cap in the infinite matrix, regarding Eqs. (3.31), (6.4), has the form

$$u_{1m} = -p_{1+i}\, \varrho_{1+im}\, r \left(\frac{r}{R_{1+i}} \right)^{c_{3m}-1}, \quad i = 0, 1, \tag{6.462}$$

where the coefficients ϱ_{1+im} $(i = 0{,}1)$, $c_{3m} < 0$ are given by Eqs. (6.476), (3.40), respectively. Provided that $\beta_p \neq \beta_m$ and $\beta_p \neq \beta_e \neq \beta_m$ or $\beta_p = \beta_e \neq \beta_m$, the radial stresses p_1, p_2 are given by Eqs. (10.1) and (10.2), (10.3) or (10.5), respectively.

Consequently, the radial stress σ_{11m}, the tangential and shear stresses σ_{22m}, σ_{33m} and σ_{12m}, σ_{13m} acting in the infinite matrix, respectively, regarding Eqs. (3.31)–(3.34), (6.462), are derived as

$$\sigma_{11m} = -p_{1+i} \left(\frac{r}{R_{1+i}} \right)^{c_{3m}-1}, \quad i = 0, 1, \tag{6.463}$$

$$\sigma_{22m} = \sigma_{33m} = -p_{1+i}\, \varrho_{1+im} \left(c_{1m} - c_{2m} c_{3m} \right) \left(\frac{r}{R_{1+i}} \right)^{c_{3m}-1}, \quad i = 0, 1, \tag{6.464}$$

$$\sigma_{12m} = -\frac{\varrho_{1+im}}{s_{44m}} \frac{\partial p_{1+i}}{\partial \varphi} \left(\frac{r}{R_{1+i}} \right)^{c_{3m}-1}, \quad i = 0, 1, \tag{6.465}$$

$$\sigma_{13m} = -\frac{\varrho_{1+im}}{s_{44m}} \frac{\partial p_{1+i}}{\partial \nu} \left(\frac{r}{R_{1+i}} \right)^{c_{3m}-1}, \quad i = 0, 1, \tag{6.466}$$

where the elastic modulus s_{44m} is given by Eq. (3.25).

Finally, the radial strain ε_{11m}, the tangential and shear strains ε_{22m}, ε_{33m} and ε_{12m}, ε_{13m} in the infinite matrix, respectively, regarding Eqs. (3.1)–(3.5), (6.462), have the forms

$$\varepsilon_{11m} = -p_{1+i}\, \varrho_{1+im}\, c_{3m} \left(\frac{r}{R_{1+i}} \right)^{c_{3m}-1}, \quad i = 0, 1, \tag{6.467}$$

$$\varepsilon_{22m} = \varepsilon_{33m} = -p_{1+i} \varrho_{1+im} \left(\frac{r}{R_{1+i}} \right)^{c_{3m}-1}, \quad i = 0, 1, \tag{6.468}$$

$$\varepsilon_{12m} = -\varrho_{1+im} \frac{\partial p_{1+i}}{\partial \varphi} \left(\frac{r}{R_{1+i}} \right)^{c_{3m}-1}, \quad i = 0, 1, \tag{6.469}$$

$$\varepsilon_{13m} = -\varrho_{1+im} \frac{\partial p_{1+i}}{\partial \nu} \left(\frac{r}{R_{1+i}} \right)^{c_{3m}-1}, \quad i = 0, 1. \tag{6.470}$$

With regard to the determination of the compressive or tensile radial stress $p_{1+i} > 0$ or $p_{1+i} < 0$ acting on the particle-matrix, matrix-envelope boundary, thus for $r = R_1$, $r = R_2$, respectively, considering Eqs. (3.58), (3.63), (3.64), the tangential strains on the particle-matrix and matrix-envelope boundaries, $(\varepsilon_{22m})_{r=R_1}$ and $(\varepsilon_{22m})_{r=R_2}$, respectively, are required to be derived in the form

$$(\varepsilon_{22m})_{r=R_{1+i}} = -p_{1+i}\, \varrho_{1+im}, \quad i = 0, 1, \tag{6.471}$$

where the coefficient ϱ_{1+im} ($i=0,1$) is given by Eq. (6.476).

Considering Eqs. (3.41), (3.42), (6.463)–(6.470), the thermal-stress induced elastic energy density w_m and the thermal-stress induced elastic energy $W_m^{(1)}$ [13], accumulated in an arbitrary point of the infinite matrix and in the volume of the infinite matrix, respectively, have the forms

$$w_m = \frac{\omega_{1+im}\, \varrho_{1+im}^2}{2} \left(\frac{r}{R_{1+i}} \right)^{2(c_{3m}-1)}, \quad i = 0, 1, \tag{6.472}$$

$$W_m^{(1)} = 8 \int_0^{\pi/2} \int_0^{\pi/2} \int_{R_{1+i}}^{\infty} w_m\, r^2 dr d\varphi d\nu = \int_0^{\pi/2} \int_0^{\pi/2} \Omega_{1+i\,1+im} d\varphi d\nu, \quad i = 0, 1, \tag{6.473}$$

where the coefficient $\Omega_{1+i\,1+im}$ ($i=0,1$) given by Eq. (6.478) is a function of ω_{1+im} given by Eq. (6.477).

With regard to the determination of the thermal-stress induced strengthening presented in Volume III, the stress σ_{jm} acting in a direction of the axis x_j and the

[13]The superscript (1) is related to the *one*-particle-(envelope)-matrix system.

elastic energy density w_{jm} induced by the stress σ_{jm} ($j=1,2,3$) (see Eqs. (3.46), (3.47), (6.463)–(6.466)), both related to an arbitrary point of the infinite matrix, are derived as

$$\sigma_{jm} = -\xi_{j\,1+im}\,\varrho_{1+im}\left(\frac{r}{R_{1+i}}\right)^{c_{3m}-1}, \quad i=0,1; \quad j=1,2,3, \tag{6.474}$$

$$w_{jm} = \frac{s_{11m}\left(\xi_{j\,1+im}\,\varrho_{1+im}\right)^2}{2}\left(\frac{r}{R_{1+i}}\right)^{2(c_{3m}-1)}, \quad i=0,1; \quad j=1,2,3, \tag{6.475}$$

where the elastic modulus s_{11m} and the coefficient $\xi_{j\,1+im}$ ($i=0,1$; $j=1,2,3$) are given by Eqs. (3.23) and (6.479), respectively.

Appendix. Considering the elastic modulus s_{44m}, the coefficients c_{1m}, c_{2m}, c_{3m}, a'_{1j}, a'_{2j}, a'_{3j} ($j=1,2,3$) (see Eqs. (3.25), (3.38)–(3.40), (12.26)–(12.34)), the coefficients ϱ_{1+im} [14], ω_{1+im}, $\Omega_{1+i\,1+jm}$ [15] ($i,j=0,1$), $\xi_{j\,1+im}$ ($i=0,1$; $j=1,2,3$) have the forms

$$\varrho_{1+im} = \frac{1}{c_{3m}\left(c_{1m}+c_{2m}\right)-2c_{2m}}, \quad i=0,1, \tag{6.476}$$

$$\begin{aligned}\omega_{1+im} &= p_{1+i}^2\left[c_{3m}^2\left(c_{1m}+c_{2m}\right)+2\left(c_{1m}-2c_{2m}c_{3m}\right)\right] \\ &+ \frac{2}{s_{44m}}\left[\left(\frac{\partial p_{1+i}}{\partial\varphi}\right)^2+\left(\frac{\partial p_{1+i}}{\partial\nu}\right)^2\right], \quad i=0,1,\end{aligned} \tag{6.477}$$

$$\Omega_{1+i\,1+jm} = \frac{4\,\omega_{1+im}\,\varrho_{1+im}^2 R_{1+j}^3}{\left|2c_{3m}+1\right|}, \quad i,j=0,1, \tag{6.478}$$

$$\begin{aligned}\xi_{j\,1+im} &= p_{1+i}\left[\frac{a'_{1j}}{\varrho_{1+im}}+\left(a'_{2j}+a'_{3j}\right)\left(c_{1m}-c_{2m}c_{3m}\right)\right] \\ &+ \frac{1}{s_{44m}}\left[\left(a'_{1j}+a'_{2j}\right)\frac{\partial p_{1+i}}{\partial\varphi}+\left(a'_{1j}+a'_{3j}\right)\frac{\partial p_{1+i}}{\partial\nu}\right], \\ & i=0,1; \quad j=1,2,3.\end{aligned} \tag{6.479}$$

[14] The subscript $1+i$ in Eq. (6.476) is used due to uniform notations applied in formulae for the tangential strain $\left(\varepsilon_{22q}\right)_{r=R_{1+i}}$ ($i=0,1$; $q=p,e,m$), e.g. as presented in Eq. (6.471).

[15] With regard to $c_{2m} < -2$ (see Eq. (3.40)) considering the Poisson's number $\mu_m < 0.5$ for real materials [1, p 38], the term $-\left(2c_{3m}+1\right) > 0$ resulting in the integration (see Eq. (6.473)) is replaced by $\left|2c_{3m}+1\right|$. Additionally, the coefficient Ω_{21m} is determined but not used in formulae for the elastic energy W_m of the infinite matrix.

6.5.2.2 Condition $\beta_p \neq \beta_e = \beta_m$

The formulae in Section 6.5.2.2 consider the φ, ν-dependent radial stress $p_1 = (\beta_e - \beta_p) / (\varrho_{1p} - \varrho_{1e}^{me})$ (see Eq. (10.4)) acting on the particle-envelope boundary and derived from Eq. (3.63), where the φ, ν-independent coefficients ϱ_{1p} and ϱ_{1e}^{me}, along with ϱ_{2e}^{me} included in the formulae in Section 5.7.2.2, are given Eqs. (5.36) or (9.50), and (5.106), (6.207), (6.228), (6.246), (7.230) or (9.63), respectively. The φ, ν-dependence of the radial stress p_1 results from the φ, ν-dependent difference $\varepsilon_{11te} - \varepsilon_{11tp} = f(\varphi, \nu)$ (see Eq. (12.23)), resulting in the φ, ν-dependent difference $\beta_e - \beta_p$ (see Eqs. (3.48)–(3.50)).

With regard to the boundary conditions (4.12), (4.16) to determine one integration constant only, considering the analysis presented in Section 6.2.4 to define u_{1m} as a decreasing function of the variable $r \in \langle R_2, \infty)$, and thus the integration constant $C_{1m} = 0$, the radial displacement u_{1m} of the infinitesimal spherical cap in the infinite matrix, regarding Eqs. (3.31), (6.4), has the form

$$u_{1m} = -p_1 \varrho_{2e}^{me} \varrho_{1m} r \left(\frac{r}{R_2} \right)^{c_{3m}-1}, \tag{6.480}$$

where the coefficients ϱ_{1m}, $c_{3m} < 0$ are given by Eqs. (6.476), (3.40), respectively.

Consequently, the radial stress σ_{11m}, the tangential and shear stresses σ_{22m}, σ_{33m} and σ_{12m}, σ_{13m} acting in the infinite matrix, respectively, regarding Eqs. (3.31)–(3.34), (6.480), are derived as

$$\sigma_{11m} = -p_1 \varrho_{2e}^{me} \left(\frac{r}{R_2} \right)^{c_{3m}-1}, \tag{6.481}$$

$$\sigma_{22m} = \sigma_{33m} = -p_1 \varrho_{2e}^{me} \varrho_{1m} (c_{1m} - c_{2m} c_{3m}) \left(\frac{r}{R_2} \right)^{c_{3m}-1}, \tag{6.482}$$

$$\sigma_{12m} = -\frac{\varrho_{2e}^{me} \varrho_{1m}}{s_{44m}} \frac{\partial p_1}{\partial \varphi} \left(\frac{r}{R_2} \right)^{c_{3m}-1}, \tag{6.483}$$

$$\sigma_{13m} = -\frac{\varrho_{2e}^{me} \varrho_{1m}}{s_{44m}} \frac{\partial p_1}{\partial \nu} \left(\frac{r}{R_2} \right)^{c_{3m}-1}, \tag{6.484}$$

where the elastic modulus s_{44m} is given by Eq. (3.25).

Finally, the radial strain ε_{11m}, the tangential and shear strains ε_{22m}, ε_{33m} and ε_{12m}, ε_{13m} in the infinite matrix, respectively, regarding Eqs. (3.1)–(3.5), (6.480), have the forms

$$\varepsilon_{11m} = -p_1 \varrho_{2e}^{me} \varrho_{1m} c_{3m} \left(\frac{r}{R_2} \right)^{c_{3m}-1}, \tag{6.485}$$

$$\varepsilon_{22m} = \varepsilon_{33m} = -p_1 \varrho_{2e}^{me} \varrho_{1m} \left(\frac{r}{R_2} \right)^{c_{3m}-1}, \tag{6.486}$$

$$\varepsilon_{12m} = -\varrho_{2e}^{me}\,\varrho_{1m}\,\frac{\partial p_1}{\partial \varphi}\left(\frac{r}{R_2}\right)^{c_{3m}-1}, \tag{6.487}$$

$$\varepsilon_{13m} = -\varrho_{2e}^{me}\,\varrho_{1m}\,\frac{\partial p_1}{\partial \nu}\left(\frac{r}{R_2}\right)^{c_{3m}-1}. \tag{6.488}$$

Considering Eqs. (3.41), (3.42), (6.481)–(6.488), the thermal-stress induced elastic energy density w_m and the thermal-stress induced elastic energy $W_m^{(1)}$ [16], accumulated in an arbitrary point of the infinite matrix and in the volume of the infinite matrix, respectively, have the forms

$$w_m = \frac{\omega_{1m}\left(\varrho_{2e}^{me}\,\varrho_{1m}\right)^2}{2}\left(\frac{r}{R_2}\right)^{2(c_{3m}-1)}, \tag{6.489}$$

$$W_m^{(1)} = 8\int_0^{\pi/2}\int_0^{\pi/2}\int_{R_2}^{\infty} w_m\, r^2 dr d\varphi d\nu = \left(\varrho_{2e}^{me}\right)^2\int_0^{\pi/2}\int_0^{\pi/2}\Omega_{12m} d\varphi d\nu, \tag{6.490}$$

where the coefficient Ω_{12m} given by Eq. (6.478) is a function of ω_{1m} given by Eq. (6.477).

With regard to the determination of the thermal-stress induced strengthening presented in Volume III, the stress σ_{im} acting in a direction of the axis x_i and the elastic energy density w_{im} induced by the stress σ_{im} ($i=1,2,3$) (see Eqs. (3.46), (3.47), (6.481)–(6.484)), both related to an arbitrary point of the infinite matrix, are derived as

$$\sigma_{im} = -\xi_{i1m}\,\varrho_{2e}^{me}\,\varrho_{1m}\left(\frac{r}{R_2}\right)^{c_{3m}-1}, \quad i = 1, 2, 3, \tag{6.491}$$

$$w_{im} = \frac{s_{11m}\left(\xi_{i1m}\,\varrho_{2e}^{me}\,\varrho_{1m}\right)^2}{2}\left(\frac{r}{R_2}\right)^{2(c_{3m}-1)}, \quad i = 1, 2, 3, \tag{6.492}$$

where the elastic modulus s_{11m} and the coefficient ξ_{i1m} ($i=1,2,3$) are given by Eqs. (3.23) and (6.479), respectively.

[16] The superscript (1) is related to the *one*-particle-(envelope)-matrix system.

Chapter 7

Isotropic multi- and one-particle-(envelope)-matrix systems. Solutions 5, 6

Chapter 7 presents two different mathematical techniques in Sections 7.1 and 7.2 applied to a system of the differential equations (3.35), (3.36), resulting in two different linear differential equations (7.2) and (7.11) with non-zero right sides, with two different solutions for the radial displacement u_{1q} of the infinitesimal spherical cap given by Eqs. (7.3) and (7.12), representing linear combinations of the particular solutions $C_{1q}\, a_{1q}\, u_{1rq}$, $C_{2q}\, a_{2q}\, u_{2rq}$, $C_{2q}\, a_{2q}\, u_{3rq}$ and $C_{1q}\, a_{1q}\, u_{1rq}$, $C_{2q}\, a_{2q}\, u_{2rq}$ for $a_{1q} = 1/3$, $a_{2q} = C_q^{(2)}$, $a_{3q} = 1/2$ and $a_{1q} = 1/3$, $a_{2q} = C_q^{(3)}$ (see Eqs. (4.1), (6.7), (7.5), (7.6)), respectively. Consequently, one, two or three of the terms u_{1rq}, u_{2rq}, u_{3rq} are considered for the determination of the integration constants C_{1q}, C_{2q}, C_{3q}.

With regard to the analyses presented in Sections 6.2.1–6.2.3, 7.3 to consider the boundary conditions defined in Section 4.1, the two different solutions are suitable for the determination of the thermal stresses and the corresponding quantities, the latter represented by w_q, W_q, σ_{iq}, w_{iq} $(i=1,2,3)$ (see Eqs. (3.41), (3.46), (3.47)), in the spherical envelope $(q=e)$ and the cell matrix $(q=m)$ as two of components of composite materials.

As a consequence of the reduction of the coefficients $a_q^{(2)}$, $a_q^{(3)}$ (see Eqs. (5.14), (5.19)) within the determination of the integration constants C_{1q}, C_{2q}, C_{3q} using the boundary conditions, formulae for the thermal stresses and for the corresponding quantities, presented in Sections 7.4.1.A, 7.5.1.1.A, 7.5.2.1.A, 7.5.2.2.A and related to both solutions (7.3), (7.12), exhibit the same form, except for the coefficient $C_q = C_q^{(i)}$ $(i=2,3;\ q=e,m)$ substituted by $C_q^{(2)}$, $C_q^{(3)}$ (see Eqs. (7.4), (7.13)). Finally, formulae in Sections 7.4.1.D, 7.5.2.1.D, 7.5.2.2.D consider the assumed boundary condition (4.14) regarding the cell matrix.

The radial displacement u_{1q}, the thermal stresses and the corresponding quantities are functions of the angles φ, ν included in the φ,ν-dependent distance r_c as a function of the experimentally determined parameters v, R_{1ex}, d_{1ex}, d_{2ex}, d_{3ex} (see

Section 2.2.1, 2.3). As a consequence of symmetry of the multi-particle-(envelope)-matrix system resulting from the matrix infinity and from the periodical distribution of the spherical particles shown in Fig. 2.1a,b, the angles φ, ν are both from the interval $\langle 0, \pi/2 \rangle$.

To derive the elastic energy density W induced by the thermal stresses, formulae for the integration of the r-dependent integrand wr^2 (see Eq. (3.42)) are presented in Section 12.4.4. Consequently, the integral in the formula $W = \int_0^{\pi/2} \int_0^{\pi/2} \Omega \, d\varphi \, d\nu$ (see Eqs. (7.27), (7.55), (7.83), (7.110), (7.150), (7.176), (7.201), (7.227), (7.248), (7.263), (7.284), (7.300), (7.315), (7.330), (7.348), (7.350), (7.380), (7.382), (7.410), (7.412), (7.441), (7.443)), related to the integrand represented by the φ, ν-dependent coefficient Ω and to the variables $\varphi, \nu \in \langle 0, \pi/2 \rangle$, can be derived numerically for a real composite material, considering Eq. (1.5).

With regard to the one-particle-matrix and one-particle-envelope-matrix systems provided that $\varepsilon_{11tm} - \varepsilon_{11tp} = f(\varphi, \nu)$ and $\varepsilon_{11te} - \varepsilon_{11tp} = f(\varphi, \nu)$, $\varepsilon_{11tm} - \varepsilon_{11te} = f(\varphi, \nu)$, the term $u_{2rm} = r^{c_{3m}}$ (see Eq. (6.7)), included in the solution (7.3), representing a decreasing function of the variable $r \in \langle R_1, \infty)$ and $r \in \langle R_2, \infty)$, respectively, is considered for the determination of the thermal stresses and the corresponding quantities in the infinite matrix described by formulae presented in Section 6.5.2. Finally, considering the term u_{1rm} to represent a decreasing function of the variable $r \in \langle R_1, \infty \rangle$ and $r \in \langle R_2, \infty \rangle$ for $R_{1+i} \geq e^{1/3} \approx 1.39561$ $(i = 0,1)$ (see Eqs. (7.5), (7.8)), respectively, we get $(\sigma_{iim})_{r \to \infty} \propto (\ln r)_{r \to \infty} \to \infty$, $(\sigma_{12+jm})_{r \to \infty} \propto (\ln r)_{r \to \infty} \to \infty$ $(i = 1,2,3;\ j = 0,1)$ (see Eqs. (3.31)–(3.34)) what is physically unacceptable, and additionally the boundary condition (4.17) is not fulfilled.

7.1 Mathematical techniques 5

Considering the right sides of Eqs. (3.35), (6.2), we get

$$r \, \frac{\partial U_{1q}}{\partial r} = c_{3q} \left(C_{1q} \, r + C_{2q} \, r^{c_{3q}} + C_{3q} \right), \tag{7.1}$$

where the function U_{1q} and the coefficient c_{3q} are given by Eq. (3.37) and (3.40), respectively, and C_{1q}, C_{2q}, C_{3q} are integration constants. Substituting Eq. (7.1) to Eq. (5.12), the differential equation (5.12) is transformed to the form

$$\frac{\partial^3 u_{1q}}{\partial r^3} + \frac{4}{r} \frac{\partial^2 u_{1q}}{\partial r^2} = -a_q^{(2)} \left(\frac{C_{1q}}{r^2} + C_{2q} \, r^{c_{3q} - 3} + \frac{C_{3q}}{r^3} \right), \tag{7.2}$$

representing a linear differential equation of the third rank regarding the variable r with a non-zero right side, and the coefficient $a_q^{(2)}$ is given by Eq. (5.14).

Using the Wronskian's method [5] presented in Section 12.4.5, representing a short introduction to problems of linear differential equations with non-zero right sides, the solution of the differential equation (7.2) for the radial displacement u_{1q} of the infinitesimal spherical cap (see Figs. 3.2, 3.3) is derived as

$$u_{1q} = a_q^{(2)} \left[\frac{C_{1q}\, r}{3} \left(\frac{4}{3} - \ln r \right) + C_{2q}\, C_q^{(2)}\, r^{c_{3q}} + \frac{C_{3q}}{2} \left(\frac{1}{2} + \ln r \right) \right], \tag{7.3}$$

where the coefficient $C_q^{(2)}$ has the form

$$C_q^{(2)} = \frac{1}{2c_{3q}} - \frac{1}{3\,(c_{3q} - 1)} - \frac{1}{6\,(c_{3q} + 2)}. \tag{7.4}$$

The integration constants C_{1q}, C_{2q}, C_{3q} are derived from the boundary conditions (see Sections 4.1.1–4.1.3), considering one, two or three of the terms u_{1rq}, u_{2rq}, u_{3rq}, regarding Eq. (4.1). The term $u_{2rq} = r^{c_{3q}}$ to represent a decreasing function of the variable r due to $c_{3q} < 0$ (see Eq. (3.40)) is given by Eq. (6.7), and the terms u_{1rq}, u_{3rq} are derived as

$$u_{1rq} = r \left(\frac{4}{3} - \ln r \right), \tag{7.5}$$

$$u_{3rq} = \frac{1}{2} + \ln r, \tag{7.6}$$

where u_{3rq} is an increasing function of the variable r. The term u_{1rq} represents a decreasing function of the variable r on the condition [6, p 368]

$$\frac{\partial u_{1rq}}{\partial r} = \frac{1}{3} - \ln r < 0, \tag{7.7}$$

resulting in the condition

$$r > R = e^{1/3} \approx 1.39561, \tag{7.8}$$

where $e \approx 2.71828$ is the Euler's number [6, p 247]. Similarly, on the condition

$$r < R = e^{1/3} \approx 1.39561, \tag{7.9}$$

the term u_{1rq} represents an increasing function of the variable r.

With regard to the analyses presented in Sections 6.2, 7.3 and concerning the integration constants C_{1q}, C_{2q}, C_{3q}, the subscript q in formulae in Sections 7.1, 7.2 is related to material parameters of the spherical envelope ($q = e$) and the cell matrix ($q = m$).

7.2 Mathematical techniques 6

Performing $r\,[\partial\text{Eq.}(6.2)/\partial r]$, the differential equation (6.2) is transformed to the form

$$r\,\frac{\partial U_{1q}}{\partial r} = C_{1q}\,r + C_{2q}c_{3q}\,r^{c_{3q}}, \tag{7.10}$$

where the function U_{1q} and the coefficient c_{3q} are given by Eq. (3.37) and (3.40), respectively, and C_{1q}, C_{2q} are integration constants. Substituting Eq. (7.10) to Eq. (5.12), the differential equation (5.12) is transformed to the form

$$\frac{\partial^3 u_{1q}}{\partial r^3} + \frac{4}{r}\,\frac{\partial^2 u_{1q}}{\partial r^2} = -a_q^{(3)}\left(\frac{C_{1q}}{r^2} + C_{2q}c_{3q}\,r^{c_{3q}-3}\right), \tag{7.11}$$

representing a linear differential equation of the third rank regarding the variable r with a non-zero right side, and the coefficient $a_q^{(3)}$ is given by Eq. (5.19).

Using the Wronskian's method [5] presented in Section 12.4.5, representing a short introduction to problems of linear differential equations with non-zero right sides, the solution of the differential equation (7.11) for the radial displacement u_{1q} of the infinitesimal spherical cap (see Figs. 3.2, 3.3) is derived as

$$u_{1q} = a_q^{(3)}\left[\frac{C_{1q}\,r}{3}\left(\frac{4}{3} - \ln r\right) + C_{2q}\,C_q^{(3)}\,r^{c_{3q}}\right], \tag{7.12}$$

where the coefficient $C_q^{(3)}$, as a function of $C_q^{(2)}$ given by Eq. (7.4), has the form

$$C_q^{(3)} = c_{3q}\,C_q^{(2)}. \tag{7.13}$$

7.3 Analysis of solution

Resulting from the analysis presented in Sections 6.2.1–6.2.3, the subscript q in Sections 7.1, 7.2 is related to material parameters of the spherical envelope ($q = e$) and the cell matrix ($q = m$), replacing the conditions (6.12), (6.13) by (7.14), (7.15) in the forms

$$R_1 \geq R \approx 1.39561, \tag{7.14}$$

$$R_2 \leq R \approx 1.39561. \tag{7.15}$$

Additionally, in contrast to the constant function u_{3rq} given by Eq. (6.8), the term u_{3rq} given by Eq. (7.6) to represent an increasing function of the variable r, is not considered for the determination of the thermal stresses and the corresponding quantities in the spherical envelope, provided that $\beta_p \neq \beta_e = \beta_m$ and $\varepsilon_{11te} - \varepsilon_{11tp} = f(\varphi, \nu)$ (see Eqs. (3.48)–(3.50), (12.23)), regarding the integration constants $C_{3e} \neq 0$, $C_{1e} = C_{2e} = 0$.

7.4 Multi-particle-matrix system

The variable r related to the thermal stresses and to the corresponding quantities in the cell matrix is from the interval $\langle R_1, r_c \rangle$, where the φ, ν-dependent distance r_c as a function of the experimentally determined parameters v, R_{1ex}, d_{1ex}, d_{2ex}, d_{3ex} (see Section 2.2.1) is derived in Section 2.3. With regard to Eq. (3.58), the compressive or tensile radial stress $p_1 > 0$ or $p_1 < 0$ acting on the particle-matrix boundary, thus for $r = R_1$, respectively, is given by Eq. (10.1). The determination of the radial stress p_1, as a consequence of the condition $\beta_p \neq \beta_m$ (see Eqs. (3.48)–(3.50)), considers the tangential strains $(\varepsilon_{22p})_{r=R_1}$ and $(\varepsilon_{22m})_{r=R_1}$ in the spherical particle and in the cell matrix on the particle-matrix boundary (see Eq. (3.58)), resulting in the coefficients ϱ_{1p} and ϱ_{1m} (see Chapter 10), respectively.

7.4.1 Thermal stresses in cell matrix

A. Integration constants $C_{1m} \neq 0$, $C_{2m} \neq 0$, $C_{3m} = 0$

With regard to the boundary conditions (4.9), (4.11) and the analysis presented in Section 6.2.3, considering the terms u_{1rm} and u_{2rm} (see Eqs. (7.5), (6.7)) to represent functions of the variable $r \in \langle R_1, r_c \rangle$, the radial displacement u_{1m} of the infinitesimal spherical cap in the cell matrix, regarding Eqs. (3.31), (7.3), has the form

$$u_{1m} = -p_1 \left[\frac{\kappa_{1m}^{(1)} r}{3} \left(\frac{4}{3} - \ln r \right) + \kappa_{3m}^{(1)} C_m \, r^{c_{3m}} \right], \tag{7.16}$$

where the coefficients $c_{3m} < 0$, $C_m = C_m^{(i)}$ $(i=2,3)$ and the radial stress p_1 acting on the particle-matrix boundary are given by Eqs. (3.40), (7.4), (7.13) and (10.1), respectively. The φ, ν-dependent distance r_c, as a function of the experimentally determined parameters v, R_{1ex}, d_{1ex}, d_{2ex}, d_{3ex} (see Section 2.2.1), included in the coefficients $\kappa_{1m}^{(1)}$, $\kappa_{3m}^{(1)}$ is derived in Section 2.3.

With regard to Section 7.4.1.A, the coefficients $\psi_{11\eta m}^{(1)}$, $\psi_{12\eta m}^{(1)}$ $(\eta = \varphi, \nu)$, ϱ_{1m} and $\omega_{11m}, \ldots, \omega_{16m}$, ξ_{i11m}, ξ_{i12m}, ξ_{i13m} $(i=1,2,3)$ mentioned below (see Eqs. (7.19), (7.20), (7.23)–(7.26), (7.28), (7.29)) depend on $\kappa_{1m}^{(1)}$, $\kappa_{3m}^{(1)}$ and $\kappa_{1m}^{(1)}$, $\kappa_{3m}^{(1)}$, $\psi_{11\eta m}^{(1)}$, $\psi_{12\eta m}^{(1)}$, respectively, where the coefficients $\kappa_{1m}^{(1)}$, $\kappa_{3m}^{(1)}$ (see Eq. (12.35)) are functions of χ_{11m}, χ_{12m}, χ_{21m}, χ_{22m} given by Eqs. (7.30)–(7.33), respectively.

Consequently, the radial stress σ_{11m}, the tangential and shear stresses σ_{22m}, σ_{33m} and σ_{12m}, σ_{13m} acting in the cell matrix, respectively, regarding Eqs. (3.31)–(3.34), (7.16), are derived as

$$\sigma_{11m} = -p_1 \left\{ \frac{\kappa_{1m}^{(1)}}{3} \left[\frac{1}{3} \left(c_{1m} - 7c_{2m} \right) - \left(c_{1m} - c_{2m} \right) \ln r \right] \right.$$
$$\left. + \kappa_{3m}^{(1)} C_m \left[c_{3m} \left(c_{1m} + c_{2m} \right) - 2c_{2m} \right] r^{c_{3m}-1} \right\}, \tag{7.17}$$

$$\sigma_{22m} = \sigma_{33m} = -p_1 \left\{ \frac{\kappa_{1m}^{(1)}}{3} \left[\frac{1}{3} \left(4c_{1m} - c_{2m} \right) - \left(c_{1m} - c_{2m} \right) \ln r \right] \right.$$
$$\left. + \kappa_{3m}^{(1)} C_m \left(c_{1m} - c_{2m} c_{3m} \right) r^{c_{3m}-1} \right\}, \tag{7.18}$$

$$\sigma_{12m} = -\frac{1}{s_{44m}} \left[\frac{\psi_{11\varphi m}^{(1)}}{3} \left(\frac{4}{3} - \ln r \right) + \psi_{12\varphi m}^{(1)} C_m \, r^{c_{3m}-1} \right], \tag{7.19}$$

$$\sigma_{13m} = -\frac{1}{s_{44m}} \left[\frac{\psi_{11\nu m}^{(1)}}{3} \left(\frac{4}{3} - \ln r \right) + \psi_{12\nu m}^{(1)} C_m \, r^{c_{3m}-1} \right], \tag{7.20}$$

where the elastic modulus s_{44m} and the coefficients c_{1m}, c_{2m}, $\psi_{11\eta m}^{(1)}$, $\psi_{12\eta m}^{(1)}$ $(\eta = \varphi, \nu)$ are given by Eqs. (3.25) and (3.38), (3.39), (12.36), respectively.

Finally, the radial strain ε_{11m}, the tangential and shear strains ε_{22m}, ε_{33m} and ε_{12m}, ε_{13m} in the cell matrix, respectively, regarding Eqs. (3.1)–(3.5), (7.16), have the forms

$$\varepsilon_{11m} = -p_1 \left[\frac{\kappa_{1m}^{(1)}}{3} \left(\frac{1}{3} - \ln r \right) + \kappa_{3m}^{(1)} C_m c_{3m} \, r^{c_{3m}-1} \right], \tag{7.21}$$

$$\varepsilon_{22m} = \varepsilon_{33m} = -p_1 \left[\frac{\kappa_{1m}^{(1)}}{3} \left(\frac{4}{3} - \ln r \right) + \kappa_{3m}^{(1)} C_m \, r^{c_{3m}-1} \right], \tag{7.22}$$

$$\varepsilon_{12m} = -\left[\frac{\psi_{11\varphi m}^{(1)}}{3} \left(\frac{4}{3} - \ln r \right) + \psi_{12\varphi m}^{(1)} C_m \, r^{c_{3m}-1} \right], \tag{7.23}$$

$$\varepsilon_{13m} = -\left[\frac{\psi_{11\nu m}^{(1)}}{3} \left(\frac{4}{3} - \ln r \right) + \psi_{12\nu m}^{(1)} C_m \, r^{c_{3m}-1} \right]. \tag{7.24}$$

With regard to the determination of the compressive or tensile radial stress $p_1 > 0$ or $p_1 < 0$ acting on the particle-matrix boundary, thus for $r = R_1$, respectively, considering Eq. (3.58), the tangential strain on the particle-matrix boundary, $(\varepsilon_{22m})_{r=R_1}$, is required to be derived in the form

$$(\varepsilon_{22m})_{r=R_1} = -p_1 \, \varrho_{1m}, \tag{7.25}$$

where the coefficient ϱ_{1m} is given by Eq. (7.34).

Considering Eqs. (3.41), (3.42), (7.17)–(7.24), the thermal-stress induced elastic energy density w_m and the thermal-stress induced elastic energy W_m, accumulated in an arbitrary point of the cell matrix and in the volume of the cell matrix, respectively, have the forms

$$w_m = \frac{1}{2}\left[\omega_{11m} + \omega_{12m}\, r^{c_{3m}-1} + \omega_{13m}\, r^{2(c_{3m}-1)} + \omega_{14m} \ln r \right. \\ \left. + \omega_{15m} \ln^2 r + \omega_{16m}\, r^{c_{3m}-1} \ln r \right], \tag{7.26}$$

$$W_m = 8 \int_0^{\pi/2} \int_0^{\pi/2} \int_{R_1}^{r_c} w_m\, r^2 dr d\varphi d\nu = \int_0^{\pi/2} \int_0^{\pi/2} \Omega_{11m} d\varphi d\nu, \tag{7.27}$$

where the coefficient Ω_{11m} given by Eq. (6.39) is a function of $\omega_{11m}, \ldots, \omega_{16m}$ given by Eqs. (7.35)–(7.40), respectively.

With regard to the determination of the thermal-stress induced strengthening presented in Volume III, the stress σ_{im} acting in a direction of the axis x_i and the elastic energy density w_{im} induced by the stress σ_{im} $(i=1,2,3)$ (see Eqs. (3.46), (3.47), (7.17)–(7.20)), both related to an arbitrary point of the cell matrix, are derived as

$$\sigma_{im} = -\left(\xi_{i11m} + \xi_{i12m}\, r^{c_{3m}-1} - \xi_{i13m} \ln r\right), \quad i = 1, 2, 3, \tag{7.28}$$

$$w_{im} = \frac{s_{11m}}{2}\left[\xi_{i11m}^2 + \xi_{i12m}^2\, r^{2(c_{3m}-1)} + \xi_{i13m}^2 \ln^2 r + 2\,\xi_{i11m}\,\xi_{i12m}\, r^{c_{3m}-1} \right. \\ \left. - 2\,\xi_{i11m}\,\xi_{i13m} \ln r - 2\,\xi_{i12m}\,\xi_{i13m}\, r^{c_{3m}-1} \ln r\right], \\ i = 1, 2, 3, \tag{7.29}$$

where the elastic modulus s_{11m} and the coefficients ξ_{i11m}, ξ_{i12m}, ξ_{i13m} $(i=1,2,3)$ are given by Eqs. (3.23) and (7.41)–(7.43), respectively.

Appendix. Considering the elastic modulus s_{44m} and the coefficients c_{1m}, c_{2m}, c_{3m}, $C_m = C_m^{(i)}$ $(i=2,3)$, a'_{1i}, a'_{2i}, a'_{3i} $(i=1,2,3)$, (see Eqs. (3.25), (3.38)–(3.40), (7.4), (7.13), (12.26)–(12.34)), the coefficients $\chi_{1+i\,1+jm}$, ϱ_{1+im}, $\omega_{1+i\,1m}, \ldots, \omega_{1+i\,6m}$, $\Omega_{1+i\,1+jm}$[1] $(i,j=0,1)$, $\xi_{i1+j\,1m}$, $\xi_{i1+j\,2m}$, $\xi_{i1+j\,3m}$ $(i=1,2,3;\ j=0,1)$ have the forms

$$\chi_{11m} = \frac{1}{3}\left[\frac{1}{3}(c_{1m} - 7c_{2m}) - (c_{1m} - c_{2m}) \ln R_1\right], \tag{7.30}$$

[1]The coefficient Ω_{21m} is determined but not used in formulae for the elastic energy W_m of the cell matrix.

$$\chi_{12m} = C_m \left[c_{3m}\left(c_{1m} + c_{2m}\right) - 2c_{2m}\right] R_1^{c_{3m}-1}, \tag{7.31}$$

$$\chi_{21m} = \frac{r_c}{3}\left(\frac{4}{3} - \ln r_c\right), \tag{7.32}$$

$$\chi_{22m} = C_m\, r_c^{c_{3m}}, \tag{7.33}$$

$$\varrho_{1+im} = \frac{\kappa_{1m}^{(1)}}{3}\left(\frac{4}{3} - \ln R_{1+i}\right) + \kappa_{3m}^{(1)} C_m R_{1+i}^{c_{3m}-1}, \quad i = 0, 1, \tag{7.34}$$

$$\omega_{1+i\,1m} = \frac{1}{81}\left(3p_{1+i}^2\left[\kappa_{1m}^{(1)}\right]^2 (11c_{1m} - 5c_{2m}) + \frac{32}{s_{44m}}\left\{\left[\psi_{1+i\,1\varphi m}^{(1)}\right]^2 + \left[\psi_{1+i\,1\nu m}^{(1)}\right]^2\right\}\right),$$
$$i = 0, 1, \tag{7.35}$$

$$\omega_{1+i\,2m} = \frac{2C_m}{9}\left\{p_{1+i}^2 \kappa_{1m}^{(1)}\kappa_{3m}^{(1)}\left[c_{1m}\left(c_{3m} + 8\right) - c_{2m}\left(7c_{3m} + 2\right)\right]\right.$$
$$\left. + \frac{8}{s_{44m}}\left[\psi_{1+i\,1\varphi m}^{(1)}\psi_{1+i\,2\varphi m}^{(1)} + \psi_{1+i\,1\nu m}^{(1)}\psi_{1+i\,2\nu m}^{(1)}\right]\right\}, \quad i=0,1, \tag{7.36}$$

$$\omega_{1+i\,3m} = C_m^2\left(p_{1+i}^2\left[\kappa_{3m}^{(1)}\right]^2\left[c_{3m}^2\left(c_{1m} + c_{2m}\right) + 2\left(c_{1m} - 2c_{2m}c_{3m}\right)\right]\right.$$
$$\left. + \frac{2}{s_{44m}}\left\{\left[\psi_{1+i\,2\varphi m}^{(1)}\right]^2 + \left[\psi_{1+i\,2\nu m}^{(1)}\right]^2\right\}\right), \quad i = 0, 1, \tag{7.37}$$

$$\omega_{1+i\,4m} = -\frac{2}{27}\left(9p_{1+i}^2\left[\kappa_{1m}^{(1)}\right]^2 (c_{1m} - c_{2m}) + \frac{8}{s_{44m}}\left\{\left[\psi_{1+i\,1\varphi m}^{(1)}\right]^2 + \left[\psi_{1+i\,1\nu m}^{(1)}\right]^2\right\}\right),$$
$$i = 0, 1, \tag{7.38}$$

$$\omega_{1+i\,5m} = \frac{1}{9}\left(3p_{1+i}^2\left[\kappa_{1m}^{(1)}\right]^2 (c_{1m} - c_{2m}) + \frac{2}{s_{44m}}\left\{\left[\psi_{1+i\,1\varphi m}^{(1)}\right]^2 + \left[\psi_{1+i\,1\nu m}^{(1)}\right]^2\right\}\right),$$
$$i = 0, 1, \tag{7.39}$$

$$\omega_{1+i\,6m} = -\frac{2C_m}{3}\left\{p_{1+i}^2 \kappa_{1m}^{(1)}\kappa_{3m}^{(1)}\left(c_{1m} - c_{2m}\right)\left(c_{3m} + 2\right)\right.$$
$$\left. + \frac{2}{s_{44m}}\left[\psi_{1+i\,1\varphi m}^{(1)}\psi_{1+i\,2\varphi m}^{(1)} + \psi_{1+i\,1\nu m}^{(1)}\psi_{1+i\,2\nu m}^{(1)}\right]\right\},$$
$$i = 0, 1, \tag{7.40}$$

$$\xi_{i\,1+j\;1m} = \frac{1}{9}\left\{p_{1+j}\kappa_{1m}^{(1)}\left[a'_{1i}\left(c_{1m} - 7c_{2m}\right) + \left(a'_{2i} + a'_{3i}\right)\left(4c_{1m} - c_{2m}\right)\right]\right.$$
$$\left. + \frac{4}{s_{44m}}\left[\left(a'_{1i} + a'_{2i}\right)\psi_{1+j\,1\varphi m}^{(1)} + \left(a'_{1i} + a'_{3i}\right)\psi_{1+j\,1\nu m}^{(1)}\right]\right\},$$
$$i = 1,2,3; \quad j = 0,1, \tag{7.41}$$

$$\xi_{i\,1+j\;2m} = C_m\left(p_{1+j}\kappa_{3m}^{(1)}\left\{a'_{1i}\left[c_{3m}\left(c_{1m} + c_{2m}\right) - 2c_{2m}\right] + \left(a'_{2i} + a'_{3i}\right)\left(c_{1m} - c_{2m}c_{3m}\right)\right\}\right.$$
$$\left. + \frac{1}{s_{44m}}\left[\left(a'_{1i} + a'_{2i}\right)\psi_{1+j\,2\varphi m}^{(1)} + \left(a'_{1i} + a'_{3i}\right)\psi_{1+j\,2\nu m}^{(1)}\right]\right),$$
$$i = 1,2,3; \quad j = 0,1, \tag{7.42}$$

$$\xi_{i\,1+j\;3m} = \frac{1}{3}\left\{p_{1+j}\kappa_{1m}^{(1)}\left(a'_{1i} + a'_{2i} + a'_{3i}\right)\left(c_{1m} - c_{2m}\right)\right.$$
$$\left. + \frac{1}{s_{44m}}\left[\left(a'_{1i} + a'_{2i}\right)\psi_{1+j\,1\varphi m}^{(1)} + \left(a'_{1i} + a'_{3i}\right)\psi_{1+j\,1\nu m}^{(1)}\right]\right\},$$
$$i = 1,2,3; \quad j = 0,1. \tag{7.43}$$

B. Integration constants $C_{1m} \neq 0$, $C_{2m} = 0$, $C_{3m} \neq 0$

With regard to the boundary conditions (4.9), (4.11) and the analysis presented in Section 6.2.3, considering the terms u_{1rm} and u_{3rm} (see Eqs. (7.5), (7.6)) to represent functions of the variable $r \in \langle R_1, r_c \rangle$, the radial displacement u_{1m} of the infinitesimal spherical cap in the cell matrix, regarding Eqs. (3.31), (7.3), has the form

$$u_{1m} = -p_1\left[\frac{\kappa_{1m}^{(1)} r}{3}\left(\frac{4}{3} - \ln r\right) + \frac{\kappa_{3m}^{(1)}}{2}\left(\frac{1}{2} + \ln r\right)\right], \tag{7.44}$$

where the radial stress p_1 acting on the particle-matrix boundary is given by Eq. (10.1). The φ,ν-dependent distance r_c, as a function of the experimentally determined parameters v, R_{1ex}, d_{1ex}, d_{2ex}, d_{3ex} (see Section 2.2.1), included in the coefficients $\kappa_{1m}^{(1)}$, $\kappa_{3m}^{(1)}$ is derived in Section 2.3.

With regard to Section 7.4.1.B, the coefficients $\psi_{11\eta m}^{(1)}$, $\psi_{12\eta m}^{(1)}$ $(\eta = \varphi, \nu)$, ϱ_{1m} and $\omega_{11m}, \ldots, \omega_{19m}$, $\xi_{i11m}, \ldots, \xi_{i14m}$ $(i = 1,2,3)$ mentioned below (see Eqs. (7.47), (7.48), (7.51)–(7.54), (7.56), (7.57)) depend on $\kappa_{1m}^{(1)}$, $\kappa_{3m}^{(1)}$ and $\kappa_{1m}^{(1)}$, $\kappa_{3m}^{(1)}$, $\psi_{11\eta m}^{(1)}$, $\psi_{12\eta m}^{(1)}$, respectively, where the coefficients $\kappa_{1m}^{(1)}$, $\kappa_{3m}^{(1)}$ (see Eq. (12.35)) are functions of χ_{11m}, χ_{12m}, χ_{21m}, χ_{22m} given by Eqs. (7.30), (7.58), (7.32), (7.59), respectively.

Consequently, the radial stress σ_{11m}, the tangential and shear stresses σ_{22m}, σ_{33m} and σ_{12m}, σ_{13m} acting in the cell matrix, respectively, regarding Eqs. (3.31)–(3.34), (7.44), are derived as

$$\sigma_{11m} = -p_1 \left\{ \frac{\kappa_{1m}^{(1)}}{3} \left[\frac{1}{3} (c_{1m} - 7c_{2m}) - (c_{1m} - c_{2m}) \ln r \right] \right.$$
$$\left. + \frac{\kappa_{3m}^{(1)}}{2r} (c_{1m} - 2c_{2m} \ln r) \right\}, \tag{7.45}$$

$$\sigma_{22m} = \sigma_{33m} = -p_1 \left\{ \frac{\kappa_{1m}^{(1)}}{3} \left[\frac{1}{3} (4c_{1m} - c_{2m}) - (c_{1m} - c_{2m}) \ln r \right] \right.$$
$$\left. + \frac{\kappa_{3m}^{(1)}}{2r} \left[c_{1m} \left(\frac{1}{2} + \ln r \right) - c_{2m} \right] \right\}, \tag{7.46}$$

$$\sigma_{12m} = -\frac{1}{s_{44m}} \left[\frac{\psi_{11\varphi m}^{(1)}}{3} \left(\frac{4}{3} - \ln r \right) + \frac{\psi_{12\varphi m}^{(1)}}{2r} \left(\frac{1}{2} + \ln r \right) \right], \tag{7.47}$$

$$\sigma_{13m} = -\frac{1}{s_{44m}} \left[\frac{\psi_{11\nu m}^{(1)}}{3} \left(\frac{4}{3} - \ln r \right) + \frac{\psi_{12\nu m}^{(1)}}{2r} \left(\frac{1}{2} + \ln r \right) \right], \tag{7.48}$$

where the elastic modulus s_{44m} and the coefficients c_{1m}, c_{2m}, $\psi_{11\eta m}^{(1)}$, $\psi_{12\eta m}^{(1)}$ $(\eta = \varphi, \nu)$ are given by Eqs. (3.25) and (3.38), (3.39), (12.36), respectively.

Finally, the radial strain ε_{11m}, the tangential and shear strains ε_{22m}, ε_{33m} and ε_{12m}, ε_{13m} in the cell matrix, respectively, regarding Eqs. (3.1)–(3.5), (7.44), have the forms

$$\varepsilon_{11m} = -p_1 \left[\frac{\kappa_{1m}^{(1)}}{3} \left(\frac{1}{3} - \ln r \right) + \frac{\kappa_{3m}^{(1)}}{2r} \right], \tag{7.49}$$

$$\varepsilon_{22m} = \varepsilon_{33m} = -p_1 \left[\frac{\kappa_{1m}^{(1)}}{3} \left(\frac{4}{3} - \ln r \right) + \frac{\kappa_{3m}^{(1)}}{2r} \left(\frac{1}{2} + \ln r \right) \right], \tag{7.50}$$

$$\varepsilon_{12m} = -\left[\frac{\psi_{11\varphi m}^{(1)}}{3} \left(\frac{4}{3} - \ln r \right) + \frac{\psi_{12\varphi m}^{(1)}}{2r} \left(\frac{1}{2} + \ln r \right) \right], \tag{7.51}$$

$$\varepsilon_{13m} = -\left[\frac{\psi_{11\nu m}^{(1)}}{3} \left(\frac{4}{3} - \ln r \right) + \frac{\psi_{12\nu m}^{(1)}}{2r} \left(\frac{1}{2} + \ln r \right) \right]. \tag{7.52}$$

With regard to the determination of the compressive or tensile radial stress $p_1 > 0$ or $p_1 < 0$ acting on the particle-matrix boundary, thus for $r = R_1$, respectively, considering Eq. (3.58), the tangential strain on the particle-matrix boundary, $(\varepsilon_{22m})_{r=R_1}$, is required to be derived in the form

$$(\varepsilon_{22m})_{r=R_1} = -p_1 \, \varrho_{1m}, \tag{7.53}$$

where the coefficient ϱ_{1m} is given by Eq. (7.60).

Considering Eqs. (3.41), (3.42), (7.45)–(7.52), the thermal-stress induced elastic energy density w_m and the thermal-stress induced elastic energy W_m, accumulated in an arbitrary point of the cell matrix and in the volume of the cell matrix, respectively, have the forms

$$w_m = \frac{1}{2}\left(\omega_{11m} + \frac{\omega_{12m}}{r} + \frac{\omega_{13m}}{r^2} + \omega_{14m}\ln r + \omega_{15m}\ln^2 r + \omega_{16m}\frac{\ln r}{r} + \omega_{17m}\frac{\ln r}{r^2} + \omega_{18m}\frac{\ln^2 r}{r} + \omega_{19m}\frac{\ln^2 r}{r^2}\right), \quad (7.54)$$

$$W_m = 8\int_0^{\pi/2}\int_0^{\pi/2}\int_{R_1}^{r_c} w_m r^2 dr d\varphi d\nu = \int_0^{\pi/2}\int_0^{\pi/2} \Omega_{11m} d\varphi d\nu, \quad (7.55)$$

where the coefficient Ω_{11m} given by Eq. (7.67) is a function of $\omega_{11m}, \ldots, \omega_{19m}$ given by Eqs. (7.35), (7.61), (7.62), (7.38), (7.39), (7.63)–(7.66), respectively.

With regard to the determination of the thermal-stress induced strengthening presented in Volume III, the stress σ_{im} acting in a direction of the axis x_i and the elastic energy density w_{im} induced by the stress σ_{im} $(i=1,2,3)$ (see Eqs. (3.46), (3.47), (7.45)–(7.48)), both related to an arbitrary point of the cell matrix, are derived as

$$\sigma_{im} = -\left(\xi_{i11m} + \frac{\xi_{i12m}}{r} - \xi_{i13m}\ln r + \xi_{i14m}\frac{\ln r}{r}\right), \quad i = 1, 2, 3, \quad (7.56)$$

$$w_{im} = \frac{s_{11m}}{2}\left(\xi_{i11m}^2 + \frac{\xi_{i12m}^2}{r^2} + \xi_{i13m}^2\ln^2 r + \xi_{i14m}^2\frac{\ln^2 r}{r^2} + \frac{2\xi_{i11m}\xi_{i12m}}{r} - 2\xi_{i11m}\xi_{i13m}\ln r + 2\xi_{i11m}\xi_{i14m}\frac{\ln r}{r} - 2\xi_{i12m}\xi_{i13m}\frac{\ln r}{r} + 2\xi_{i12m}\xi_{i14m}\frac{\ln r}{r^2} - 2\xi_{i13m}\xi_{i14m}\frac{\ln^2 r}{r}\right), \quad i = 1, 2, 3, \quad (7.57)$$

where the elastic modulus s_{11m} and the coefficients $\xi_{i11m}, \ldots, \xi_{i14m}$ $(i=1,2,3)$ are given by Eqs. (3.23) and (7.68)–(7.71), respectively.

Appendix. Considering the elastic modulus s_{44m} and the coefficients c_{1m}, c_{2m}, c_{3m}, a'_{1i}, a'_{2i}, a'_{3i} $(i=1,2,3)$ (see Eqs. (3.25), (3.38)–(3.40), (12.26)–(12.34)), the coefficients $\chi_{1+i\,2m}$, ϱ_{1+im}, $\omega_{1+i\,2m}$, $\omega_{1+i\,3m}$, $\omega_{1+i\,6m}, \ldots, \omega_{1+i\,9m}$, $\Omega_{1+i\;1+jm}$[2] $(i, j=0,1)$, $\xi_{i\,1+j\;1m}, \ldots, \xi_{i\,1+j\;4m}$ $(i=1,2,3;\ j=0,1)$ have the forms

[2]The coefficient Ω_{21m} is determined but not used in formulae for the elastic energy W_m of the cell matrix.

$$\chi_{12m} = \frac{c_{1m} - 2c_{2m} \ln R_1}{2R_1}, \tag{7.58}$$

$$\chi_{22m} = \frac{1}{2}\left(\frac{1}{2} + \ln r_c\right), \tag{7.59}$$

$$\varrho_{1+im} = \frac{\kappa_{1m}^{(1)}}{3}\left(\frac{4}{3} - \ln R_{1+i}\right) + \frac{\kappa_{3m}^{(1)}}{2R_{1+i}}\left(\frac{1}{2} + \ln R_{1+i}\right), \quad i = 0, 1, \tag{7.60}$$

$$\omega_{1+i\,2m} = \frac{1}{9}\left\{p_{1+i}^2 \kappa_{1m}^{(1)} \kappa_{3m}^{(1)} \left(5c_{1m} - 8c_{2m}\right) + \frac{4}{s_{44m}}\left[\psi_{1+i\,1\varphi m}^{(1)} \psi_{1+i\,2\varphi m}^{(1)} + \psi_{1+i\,1\nu m}^{(1)} \psi_{1+i\,2\nu m}^{(1)}\right]\right\}, \quad i = 0, 1, \tag{7.61}$$

$$\omega_{1+i\,3m} = \frac{1}{8}\left(p_{1+i}^2 \left[\kappa_{3m}^{(1)}\right]^2 \left(3c_{1m} - 2c_{2m}\right) + \frac{1}{s_{44m}}\left\{\left[\psi_{1+i\,2\varphi m}^{(1)}\right]^2 + \left[\psi_{1+i\,2\nu m}^{(1)}\right]^2\right\}\right), \quad i = 0, 1, \tag{7.62}$$

$$\omega_{1+i\,6m} = \frac{1}{9}\left\{2p_{1+i}^2 \kappa_{1m}^{(1)} \kappa_{3m}^{(1)} \left(c_{1m} + 2c_{2m}\right) + \frac{5}{s_{44m}}\left[\psi_{1+i\,1\varphi m}^{(1)} \psi_{1+i\,2\varphi m}^{(1)} + \psi_{1+i\,1\nu m}^{(1)} \psi_{1+i\,2\nu m}^{(1)}\right]\right\}, \quad i = 0, 1, \tag{7.63}$$

$$\omega_{1+i\,7m} = \frac{1}{2}\left(p_{1+i}^2 \left[\kappa_{3m}^{(1)}\right]^2 \left(c_{1m} - 2c_{2m}\right) + \frac{1}{s_{44m}}\left\{\left[\psi_{1+i\,2\varphi m}^{(1)}\right]^2 + \left[\psi_{1+i\,2\nu m}^{(1)}\right]^2\right\}\right), \quad i = 0, 1, \tag{7.64}$$

$$\omega_{1+i\,8m} = -\frac{2}{3}\left\{p_{1+i}^2 \kappa_{1m}^{(1)} \kappa_{3m}^{(1)} \left(c_{1m} - c_{2m}\right) + \frac{1}{s_{44m}}\left[\psi_{1+i\,1\varphi m}^{(1)} \psi_{1+i\,2\varphi m}^{(1)} + \psi_{1+i\,1\nu m}^{(1)} \psi_{1+i\,2\nu m}^{(1)}\right]\right\}, \quad i = 0, 1, \tag{7.65}$$

$$\omega_{1+i\,9m} = \frac{1}{2}\left(p_{1+i}^2 \left[\kappa_{3m}^{(1)}\right]^2 c_{1m} + \frac{1}{s_{44m}}\left\{\left[\psi_{1+i\,2\varphi m}^{(1)}\right]^2 + \left[\psi_{1+i\,2\nu m}^{(1)}\right]^2\right\}\right), \quad i = 0, 1, \tag{7.66}$$

$$\Omega_{1+i\,1+jm} = 4\left(\frac{\omega_{1+i\,1m}}{3}\left(r_c^3 - R_{1+j}^3\right) + \frac{\omega_{1+i\,2m}}{2}\left(r_c^2 - R_{1+j}^2\right)\right.$$
$$+ \omega_{1+i\,3m}\left(r_c - R_{1+j}\right) + \frac{\omega_{1+i\,4m}}{3}\left[r_c^3\left(\ln r_c - \frac{1}{3}\right) - R_{1+j}^3\left(\ln R_{1+j} - \frac{1}{3}\right)\right]$$
$$+ \frac{\omega_{1+i\,5m}}{3}\left\{r_c^3\left[\left(\ln r_c - \frac{1}{3}\right)^2 + \frac{1}{9}\right] - R_{1+j}^3\left[\left(\ln R_{1+j} - \frac{1}{3}\right)^2 + \frac{1}{9}\right]\right\}$$
$$+ \frac{\omega_{1+i\,6m}}{2}\left[r_c^2\left(\ln r_c - \frac{1}{2}\right) - R_{1+j}^2\left(\ln R_{1+j} - \frac{1}{2}\right)\right]$$
$$+ \omega_{1+i\,7m}\left[r_c\left(\ln r_c - 1\right) - R_{1+j}\left(\ln R_{1+j} - 1\right)\right]$$
$$+ \frac{\omega_{1+i\,8m}}{2}\left\{r_c^2\left[\left(\ln r_c - \frac{1}{2}\right)^2 + \frac{1}{4}\right] - R_{1+j}^2\left[\left(\ln R_{1+j} - \frac{1}{2}\right)^2 + \frac{1}{4}\right]\right\}$$
$$\left. + \omega_{1+i\,9m}\left\{r_c\left[\left(\ln r_c - 1\right)^2 + 1\right] - R_{1+j}\left[\left(\ln R_{1+j} - 1\right)^2 + 1\right]\right\}\right),$$
$$i, j = 0, 1, \tag{7.67}$$

$$\xi_{i\,1+j\;1m} = \frac{1}{9}\left\{p_{1+j}\kappa_{1m}^{(1)}\left[a'_{1i}\left(c_{1m} - 7c_{2m}\right) + \left(a'_{2i} + a'_{3i}\right)\left(4c_{1m} - c_{2m}\right)\right]\right.$$
$$\left. + \frac{4}{s_{44m}}\left[\left(a'_{1i} + a'_{2i}\right)\psi_{1+j\,1\varphi m}^{(1)} + \left(a'_{1i} + a'_{3i}\right)\psi_{1+j\,1\nu m}^{(1)}\right]\right\},$$
$$i = 1, 2, 3; \quad j = 0, 1, \tag{7.68}$$

$$\xi_{i\,1+j\;2m} = \frac{1}{4}\left\{p_{1+j}\kappa_{3m}^{(1)}\left[2a'_{1i}\,c_{1m} + \left(a'_{2i} + a'_{3i}\right)\left(c_{1m} - 2c_{2m}\right)\right]\right.$$
$$\left. + \frac{1}{s_{44m}}\left[\left(a'_{1i} + a'_{2i}\right)\psi_{1+j\,2\varphi m}^{(1)} + \left(a'_{1i} + a'_{3i}\right)\psi_{1+j\,2\nu m}^{(1)}\right]\right\},$$
$$i = 1, 2, 3; \quad j = 0, 1, \tag{7.69}$$

$$\xi_{i\,1+j\;3m} = \frac{1}{3}\left\{p_{1+j}\kappa_{1m}^{(1)}\left(a'_{1i} + a'_{2i} + a'_{3i}\right)\left(c_{1m} - c_{2m}\right)\right.$$
$$\left. + \frac{1}{s_{44m}}\left[\left(a'_{1i} + a'_{2i}\right)\psi_{1+j\,1\varphi m}^{(1)} + \left(a'_{1i} + a'_{3i}\right)\psi_{1+j\,1\nu m}^{(1)}\right]\right\},$$
$$i = 1, 2, 3; \quad j = 0, 1. \tag{7.70}$$

$$\xi_{i\,1+j\;4m} = \frac{1}{2}\left\{p_{1+j}\kappa_{3m}^{(1)}\left[c_{1m}\left(a'_{2i} + a'_{3i}\right) - 2a'_{1i}\,c_{2m}\right]\right.$$
$$\left. + \frac{1}{s_{44m}}\left[\left(a'_{1i} + a'_{2i}\right)\psi_{1+j\,2\varphi m}^{(1)} + \left(a'_{1i} + a'_{3i}\right)\psi_{1+j\,2\nu m}^{(1)}\right]\right\},$$
$$i = 1, 2, 3; \quad j = 0, 1. \tag{7.71}$$

C. Integration constants $C_{1m} = 0$, $C_{2m} \neq 0$, $C_{3m} \neq 0$

With regard to the boundary conditions (4.9), (4.11) and the analysis presented in Section 6.2.3, considering the terms u_{2rm} and u_{3rm} (see Eqs. (6.7), (7.6)) to represent functions of the variable $r \in \langle R_1, r_c \rangle$, the radial displacement u_{1m} of the infinitesimal spherical cap in the cell matrix, regarding Eqs. (3.31), (7.3), has the form

$$u_{1m} = -p_1 \left[\kappa_{1m}^{(1)} C_m^{(2)} r^{c_{3m}} + \frac{\kappa_{3m}^{(1)}}{2} \left(\frac{1}{2} + \ln r \right) \right], \tag{7.72}$$

where the coefficients $c_{3m} < 0$, $C_m^{(2)}$ and the radial stress p_1 acting on the particle-matrix boundary are given by Eqs. (3.40), (7.4) and (10.1), respectively. The φ, ν-dependent distance r_c, as a function of the experimentally determined parameters v, R_{1ex}, d_{1ex}, d_{2ex}, d_{3ex} (see Section 2.2.1), included in the coefficients $\kappa_{1m}^{(1)}$, $\kappa_{3m}^{(1)}$ is derived in Section 2.3.

With regard to Section 7.4.1.C, the coefficients $\psi_{11\eta m}^{(1)}$, $\psi_{12\eta m}^{(1)}$ $(\eta = \varphi, \nu)$, ϱ_{1m} and $\omega_{11m}, \ldots, \omega_{16m}$, ξ_{i11m}, ξ_{i12m}, ξ_{i13m} $(i = 1,2,3)$ mentioned below (see Eqs. (7.75), (7.76), (7.79)–(7.82), (7.84), (7.85)) depend on $\kappa_{1m}^{(1)}$, $\kappa_{3m}^{(1)}$ and $\kappa_{1m}^{(1)}$, $\kappa_{3m}^{(1)}$, $\psi_{11\eta m}^{(1)}$, $\psi_{12\eta m}^{(1)}$, respectively, where the coefficients $\kappa_{1m}^{(1)}$, $\kappa_{3m}^{(1)}$ (see Eq. (12.35)) are functions of χ_{11m}, χ_{12m}, χ_{21m}, χ_{22m} given by Eqs. (7.86), (7.58), (7.87), (7.59), respectively.

Consequently, the radial stress σ_{11m}, the tangential and shear stresses σ_{22m}, σ_{33m} and σ_{12m}, σ_{13m} acting in the cell matrix, respectively, regarding Eqs. (3.31)–(3.34), (7.72), are derived as

$$\sigma_{11m} = -p_1 \left\{ \kappa_{1m}^{(1)} C_m^{(2)} \left[c_{3m} \left(c_{1m} + c_{2m} \right) - 2 c_{2m} \right] r^{c_{3m} - 1} + \frac{\kappa_{3m}^{(1)}}{2r} \left(c_{1m} - 2 c_{2m} \ln r \right) \right\}, \tag{7.73}$$

$$\sigma_{22m} = \sigma_{33m} = -p_1 \left\{ \kappa_{1m}^{(1)} C_m^{(2)} \left(c_{1m} - c_{2m} c_{3m} \right) r^{c_{3m} - 1} + \frac{\kappa_{3m}^{(1)}}{2r} \left[c_{1m} \left(\frac{1}{2} + \ln r \right) - c_{2m} \right] \right\}, \tag{7.74}$$

$$\sigma_{12m} = -\frac{1}{s_{44m}} \left[\psi_{11\varphi m}^{(1)} C_m^{(2)} r^{c_{3m} - 1} + \frac{\psi_{12\varphi m}^{(1)}}{2r} \left(\frac{1}{2} + \ln r \right) \right], \tag{7.75}$$

$$\sigma_{13m} = -\frac{1}{s_{44m}} \left[\psi_{11\nu m}^{(1)} C_m^{(2)} r^{c_{3m} - 1} + \frac{\psi_{12\nu m}^{(1)}}{2r} \left(\frac{1}{2} + \ln r \right) \right], \tag{7.76}$$

where the elastic modulus s_{44m} and the coefficients c_{1m}, c_{2m}, $\psi^{(1)}_{11\eta m}$, $\psi^{(1)}_{12\eta m}$ $(\eta=\varphi,\nu)$ are given by Eqs. (3.25) and (3.38), (3.39), (12.36), respectively.

Finally, the radial strain ε_{11m}, the tangential and shear strains ε_{22m}, ε_{33m} and ε_{12m}, ε_{13m} in the cell matrix, respectively, regarding Eqs. (3.1)–(3.5), (7.72), have the forms

$$\varepsilon_{11m} = -p_1 \left[\kappa^{(1)}_{1m} C^{(2)}_m c_{3m}\, r^{c_{3m}-1} + \frac{\kappa^{(1)}_{3m}}{2r} \right], \tag{7.77}$$

$$\varepsilon_{22m} = \varepsilon_{33m} = -p_1 \left[\kappa^{(1)}_{1m} C^{(2)}_m\, r^{c_{3m}-1} + \frac{\kappa^{(1)}_{3m}}{2r} \left(\frac{1}{2} + \ln r \right) \right], \tag{7.78}$$

$$\varepsilon_{12m} = -\left[\psi^{(1)}_{11\varphi m} C^{(2)}_m\, r^{c_{3m}-1} + \frac{\psi^{(1)}_{12\varphi m}}{2r} \left(\frac{1}{2} + \ln r \right) \right], \tag{7.79}$$

$$\varepsilon_{13m} = -\left[\psi^{(1)}_{11\nu m} C^{(2)}_m\, r^{c_{3m}-1} + \frac{\psi^{(1)}_{12\nu m}}{2r} \left(\frac{1}{2} + \ln r \right) \right]. \tag{7.80}$$

With regard to the determination of the compressive or tensile radial stress $p_1 > 0$ or $p_1 < 0$ acting on the particle-matrix boundary, thus for $r = R_1$, respectively, considering Eq. (3.58), the tangential strain on the particle-matrix boundary, $(\varepsilon_{22m})_{r=R_1}$, is required to be derived in the form

$$(\varepsilon_{22m})_{r=R_1} = -p_1\, \varrho_{1m}, \tag{7.81}$$

where the coefficient ϱ_{1m} is given by Eq. (7.88).

Considering Eqs. (3.41), (3.42), (7.73)–(7.80), the thermal-stress induced elastic energy density w_m and the thermal-stress induced elastic energy W_m, accumulated in an arbitrary point of the cell matrix and in the volume of the cell matrix, respectively, have the forms

$$w_m = \frac{1}{2} \left[\frac{\omega_{11m}}{r^2} + \omega_{12m}\, r^{c_{3m}-2} + \omega_{13m}\, r^{2(c_{3m}-1)} + \omega_{14m} \frac{\ln r}{r^2} + \omega_{15m}\, r^{c_{3m}-2} \ln r + \omega_{16m} \frac{\ln^2 r}{r^2} \right], \tag{7.82}$$

$$W_m = 8 \int_0^{\pi/2} \int_0^{\pi/2} \int_{R_1}^{r_c} w_m\, r^2 dr d\varphi d\nu = \int_0^{\pi/2} \int_0^{\pi/2} \Omega_{11m} d\varphi d\nu, \tag{7.83}$$

where the coefficient Ω_{11m} given by Eq. (7.95) is a function of $\omega_{11m}, \ldots, \omega_{16m}$ given by Eqs. (7.89)–(7.94), respectively.

With regard to the determination of the thermal-stress induced strengthening presented in Volume III, the stress σ_{im} acting in a direction of the axis x_i and the elastic energy density w_{im} induced by the stress σ_{im} ($i=1,2,3$) (see Eqs. (3.46), (3.47), (7.73)–(7.76)), both related to an arbitrary point of the cell matrix, are derived as

$$\sigma_{im} = -\left(\frac{\xi_{i\,11m}}{r} + \xi_{i\,12m}\, r^{c_{3m}-1} + \xi_{i\,13m}\frac{\ln r}{r}\right), \quad i = 1, 2, 3, \tag{7.84}$$

$$w_{im} = \frac{s_{11m}}{2}\left[\frac{\xi^2_{i\,11m}}{r^2} + \xi^2_{i\,12m}\, r^{2(c_{3m}-1)} + \xi^2_{i\,13m}\frac{\ln^2 r}{r^2} + 2\,\xi_{i\,11m}\,\xi_{i\,12m}\, r^{c_{3m}-2}\right.$$
$$\left. + \, 2\,\xi_{i\,11m}\,\xi_{i\,13m}\frac{\ln r}{r^2} + 2\,\xi_{i\,12m}\,\xi_{i\,13m}\, r^{c_{3m}-2}\ln r\right],$$
$$i = 1, 2, 3, \tag{7.85}$$

where the elastic modulus s_{11m} and the coefficients $\xi_{i\,11m}$, $\xi_{i\,12m}$, $\xi_{i\,13m}$ ($i=1,2,3$) are given by Eqs. (3.23) and (7.96)–(7.98), respectively.

Appendix. Considering the elastic modulus s_{44m} and the coefficients c_{1m}, c_{2m}, c_{3m}, $C_m^{(2)}$, a'_{1i}, a'_{2i}, a'_{3i} ($i=1,2,3$), (see Eqs. (3.25), (3.38)–(3.40), (7.4), (12.26)–(12.34)), the coefficients $\chi_{1+i\,1m}$, ϱ_{1+im}, $\omega_{1+i\,1m}, \ldots, \omega_{1+i\,6m}$, $\Omega_{1+i\;1+jm}$[3] ($i, j=0,1$), $\xi_{i\,1+j\;1m}$, $\xi_{i\,1+j\;2m}$, $\xi_{i\,1+j\;3m}$ ($i=1,2,3$; $j=0,1$) have the forms

$$\chi_{11m} = C_m^{(2)}\left[c_{3m}\left(c_{1m} + c_{2m}\right) - 2c_{2m}\right] R_1^{c_{3m}-1}, \tag{7.86}$$

$$\chi_{21m} = C_m^{(2)} r_c^{c_{3m}}, \tag{7.87}$$

$$\varrho_{1+im} = \kappa_{1m}^{(1)} C_m^{(2)} R_{1+i}^{c_{3m}-1} + \frac{\kappa_{3m}^{(1)}}{2R_{1+i}}\left(\frac{1}{2} + \ln R_{1+i}\right), \quad i = 0, 1, \tag{7.88}$$

$$\omega_{1+i\,1m} = \frac{1}{8}\left(p^2_{1+i}\left[\kappa_{3m}^{(1)}\right]^2 (3c_{1m} - 2c_{2m}) + \frac{1}{s_{44m}}\left\{\left[\psi^{(1)}_{1+i\,2\varphi m}\right]^2 + \left[\psi^{(1)}_{1+i\,2\nu m}\right]^2\right\}\right),$$
$$i = 0, 1, \tag{7.89}$$

$$\omega_{1+i\,2m} = C_m^{(2)}\left\{p^2_{1+i}\kappa_{1m}^{(1)}\kappa_{3m}^{(1)}\left[c_{1m}\left(c_{3m} + 1\right) - 2c_{2m}\right]\right.$$
$$\left. + \frac{1}{s_{44m}}\left[\psi^{(1)}_{1+i\,1\varphi m}\psi^{(1)}_{1+i\,2\varphi m} + \psi^{(1)}_{1+i\,1\nu m}\psi^{(1)}_{1+i\,2\nu m}\right]\right\},$$
$$i = 0, 1, \tag{7.90}$$

[3]The coefficient Ω_{21m} is determined but not used in formulae for the elastic energy W_m of the cell matrix.

$$\omega_{1+i\,3m} = \left[C_m^{(2)}\right]^2 \left(p_{1+i}^2 \left[\kappa_{1m}^{(1)}\right]^2 \left[c_{3m}^2 \left(c_{1m} + c_{2m}\right) + 2\left(c_{1m} - 2c_{2m}c_{3m}\right)\right] + \frac{2}{s_{44m}} \left\{\left[\psi_{1+i\,1\varphi m}^{(1)}\right]^2 + \left[\psi_{1+i\,1\nu m}^{(1)}\right]^2\right\}\right), \quad i=0,1, \tag{7.91}$$

$$\omega_{1+i\,4m} = \frac{1}{2}\left(p_{1+i}^2 \left[\kappa_{3m}^{(1)}\right]^2 \left(c_{1m} - 2c_{2m}\right) + \frac{1}{s_{44m}} \left\{\left[\psi_{1+i\,2\varphi m}^{(1)}\right]^2 + \left[\psi_{1+i\,2\nu m}^{(1)}\right]^2\right\}\right), \quad i = 0, 1, \tag{7.92}$$

$$\omega_{1+i\,5m} = 2C_m^{(2)} \left\{p_{1+i}^2 \kappa_{1m}^{(1)} \kappa_{3m}^{(1)} \left(c_{1m} - c_{2m}c_{3m}\right) + \frac{1}{s_{44m}} \left[\psi_{1+i\,1\varphi m}^{(1)} \psi_{1+i\,2\varphi m}^{(1)} + \psi_{1+i\,1\nu m}^{(1)} \psi_{1+i\,2\nu m}^{(1)}\right]\right\}, \quad i = 0, 1, \tag{7.93}$$

$$\omega_{1+i\,6m} = \frac{1}{2}\left(p_{1+i}^2 \left[\kappa_{3m}^{(1)}\right]^2 c_{1m} + \frac{1}{s_{44m}} \left\{\left[\psi_{1+i\,2\varphi m}^{(1)}\right]^2 + \left[\psi_{1+i\,2\nu m}^{(1)}\right]^2\right\}\right), \quad i = 0, 1, \tag{7.94}$$

$$\begin{aligned}\Omega_{1+i\;1+jm} = 4 \Bigg(& \omega_{1+i\,1m} \left(r_c - R_{1+j}\right) + \frac{\omega_{1+i\,2m}}{c_{3m}+1} \left(r_c^{c_{3m}+1} - R_{1+j}^{c_{3m}+1}\right) \\ &+ \frac{\omega_{1+i\,3m}}{2c_{3m}+1} \left(r_c^{2c_{3m}+1} - R_{1+j}^{2c_{3m}+1}\right) \\ &+ \omega_{1+i\,4m} \left[r_c \left(\ln r_c - 1\right) - R_{1+j} \left(\ln R_{1+j} - 1\right)\right] \\ &+ \frac{\omega_{1+i\,5m}}{c_{3m}+1} \left[r_c^{c_{3m}+1} \left(\ln r_c - \frac{1}{c_{3m}+1}\right) - R_{1+j}^{c_{3m}+1} \left(\ln R_{1+j} - \frac{1}{c_{3m}+1}\right)\right] \\ &+ \omega_{1+i\,6m} \left\{r_c \left[\left(\ln r_c - 1\right)^2 + 1\right] - R_{1+j} \left[\left(\ln R_{1+j} - 1\right)^2 + 1\right]\right\}\Bigg), \\ & i, j = 0, 1,\end{aligned} \tag{7.95}$$

$$\xi_{i\,1+j\;1m} = \frac{1}{4}\left\{p_{1+j}\kappa_{3m}^{(1)} \left[2a'_{1i} c_{1m} + \left(a'_{2i} + a'_{3i}\right) \left(c_{1m} - 2c_{2m}\right)\right] + \frac{1}{s_{44m}} \left[\left(a'_{1i} + a'_{2i}\right) \psi_{1+j\,2\varphi m}^{(1)} + \left(a'_{1i} + a'_{3i}\right) \psi_{1+j\,2\nu m}^{(1)}\right]\right\}, \quad i = 1, 2, 3; \quad j = 0, 1, \tag{7.96}$$

$$\begin{aligned}\xi_{i\,1+j\;2m} = C_m^{(2)} \Bigg(& p_{1+j}\kappa_{1m}^{(1)} \left\{a'_{1i} \left[c_{3m} \left(c_{1m} + c_{2m}\right) - 2c_{2m}\right] + \left(a'_{2i} + a'_{3i}\right) \left(c_{1m} - c_{2m}c_{3m}\right)\right\} \\ &+ \frac{1}{s_{44m}} \left[\left(a'_{1i} + a'_{2i}\right) \psi_{1+j\,1\varphi m}^{(1)} + \left(a'_{1i} + a'_{3i}\right) \psi_{1+j\,1\nu m}^{(1)}\right]\Bigg), \\ & i = 1, 2, 3; \quad j = 0, 1.\end{aligned} \tag{7.97}$$

$$\xi_{i\,1+j\,3m} = \frac{1}{2}\left\{p_{1+j}\kappa_{3m}^{(1)}\left[c_{1m}\left(a'_{2i}+a'_{3i}\right)-2a'_{1i}\,c_{2m}\right]\right.$$
$$\left.+\frac{1}{s_{44m}}\left[\left(a'_{1i}+a'_{2i}\right)\psi_{1+j\,2\varphi m}^{(1)}+\left(a'_{1i}+a'_{3i}\right)\psi_{1+j\,2\nu m}^{(1)}\right]\right\},$$
$$i=1,2,3;\quad j=0,1. \tag{7.98}$$

D. Integration constants $C_{1m}\neq 0$, $C_{2m}\neq 0$, $C_{3m}\neq 0$

With regard to the boundary conditions (4.9), (4.11), (4.14) and the analysis presented in Section 6.2.3, considering the terms u_{1rm}, u_{2rm} and u_{3rm} (see Eqs. (7.5), (6.7) (7.6)) to represent functions of the variable $r\in\langle R_1, r_c\rangle$, the radial displacement u_{1m} of the infinitesimal spherical cap in the cell matrix, regarding Eqs. (3.31), (7.3), has the form

$$u_{1m} = -p_1\left[\frac{\kappa_{4m}^{(2)}r}{3}\left(\frac{4}{3}-\ln r\right)+\kappa_{5m}^{(2)}C_m^{(2)}\,r^{c_{3m}}+\frac{\kappa_{6m}^{(2)}}{2}\left(\frac{1}{2}+\ln r\right)\right], \tag{7.99}$$

where the coefficients $c_{3m}<0$, $C_m^{(2)}$ and the radial stress p_1 acting on the particle-matrix boundary are given by Eqs. (3.40), (7.4) and (10.1), respectively. The φ,ν-dependent distance r_c, as a function of the experimentally determined parameters v, R_{1ex}, d_{1ex}, d_{2ex}, d_{3ex} (see Section 2.2.1), included in the coefficients $\kappa_{4m}^{(2)}$, $\kappa_{5m}^{(2)}$, $\kappa_{6m}^{(2)}$ is derived in Section 2.3.

With regard to Section 7.4.1.D, the coefficients $\psi_{11\eta m}^{(2)}$, $\psi_{12\eta m}^{(2)}$, $\psi_{13\eta m}^{(2)}$ $(\eta=\varphi,\nu)$, ϱ_{1m} and $\omega_{11m},\ldots,\omega_{114m}$, $\xi_{i11m},\ldots,\xi_{i15m}$ $(i=1,2,3)$ mentioned below (see Eqs. (7.102), (7.103), (7.106)–(7.109), (7.111), (7.112)) depend on $\kappa_{4m}^{(2)}$, $\kappa_{5m}^{(2)}$, $\kappa_{6m}^{(2)}$ and $\kappa_{4m}^{(2)}$, $\kappa_{5m}^{(2)}$, $\kappa_{6m}^{(2)}$, $\psi_{11\eta m}^{(2)}$, $\psi_{12\eta m}^{(2)}$, $\psi_{13\eta m}^{(2)}$, respectively, where the coefficients $\kappa_{4m}^{(2)}$, $\kappa_{5m}^{(2)}$, $\kappa_{6m}^{(2)}$ (see Eq. (12.39)) are functions of χ_{11m}, χ_{12m}, χ_{13m}, χ_{21m}, χ_{22m}, χ_{23m}, χ_{31m}, χ_{32m}, χ_{33m} given by Eqs. (7.30), (7.31), (7.113), (7.32), (7.33), (7.114)–(7.117), respectively.

Consequently, the radial stress σ_{11m}, the tangential and shear stresses σ_{22m}, σ_{33m} and σ_{12m}, σ_{13m} acting in the cell matrix, respectively, regarding Eqs. (3.31)–(3.34), (7.99), are derived as

$$\sigma_{11m} = -p_1\left\{\frac{\kappa_{4m}^{(2)}}{3}\left[\frac{1}{3}\left(c_{1m}-7c_{2m}\right)-\left(c_{1m}-c_{2m}\right)\ln r\right]\right.$$
$$+\kappa_{5m}^{(2)}C_m^{(2)}\left[c_{3m}\left(c_{1m}+c_{2m}\right)-2c_{2m}\right]r^{c_{3m}-1}$$
$$\left.+\frac{\kappa_{6m}^{(2)}}{2r}\left(c_{1m}-2c_{2m}\ln r\right)\right\}, \tag{7.100}$$

$$\sigma_{22m} = \sigma_{33m} = -p_1 \left\{ \frac{\kappa_{4m}^{(2)}}{3} \left[\frac{1}{3} \left(4c_{1m} - c_{2m} \right) - \left(c_{1m} - c_{2m} \right) \ln r \right] + \kappa_{5m}^{(2)} C_m^{(2)} \left(c_{1m} - c_{2m} c_{3m} \right) r^{c_{3m}-1} + \frac{\kappa_{6m}^{(2)}}{2r} \left[c_{1m} \left(\frac{1}{2} + \ln r \right) - c_{2m} \right] \right\}, \tag{7.101}$$

$$\sigma_{12m} = -\frac{1}{s_{44m}} \left[\frac{\Psi_{11\varphi m}^{(2)}}{3} \left(\frac{4}{3} - \ln r \right) + \Psi_{12\varphi m}^{(2)} C_m^{(2)} r^{c_{3m}-1} + \frac{\Psi_{13\varphi m}^{(2)}}{2r} \left(\frac{1}{2} + \ln r \right) \right], \tag{7.102}$$

$$\sigma_{13m} = -\frac{1}{s_{44m}} \left[\frac{\Psi_{11\nu m}^{(2)}}{3} \left(\frac{4}{3} - \ln r \right) + \Psi_{12\nu m}^{(2)} C_m^{(2)} r^{c_{3m}-1} + \frac{\Psi_{13\nu m}^{(2)}}{2r} \left(\frac{1}{2} + \ln r \right) \right], \tag{7.103}$$

where the elastic modulus s_{44m} and the coefficients c_{1m}, c_{2m}, $\psi_{11\eta m}^{(2)}$, $\psi_{12\eta m}^{(2)}$, $\psi_{13\eta m}^{(2)}$ $(\eta = \varphi, \nu)$ are given by Eqs. (3.25) and (3.38), (3.39), (12.40), respectively.

Finally, the radial strain ε_{11m}, the tangential and shear strains ε_{22m}, ε_{33m} and ε_{12m}, ε_{13m} in the cell matrix, respectively, regarding Eqs. (3.1)–(3.5), (7.99), have the forms

$$\varepsilon_{11m} = -p_1 \left[\frac{\kappa_{4m}^{(2)}}{3} \left(\frac{1}{3} - \ln r \right) + \kappa_{5m}^{(2)} C_m^{(2)} c_{3m} r^{c_{3m}-1} + \frac{\kappa_{6m}^{(2)}}{2r} \right], \tag{7.104}$$

$$\varepsilon_{22m} = \varepsilon_{33m} = -p_1 \left[\frac{\kappa_{4m}^{(2)}}{3} \left(\frac{4}{3} - \ln r \right) + \kappa_{5m}^{(2)} C_m^{(2)} r^{c_{3m}-1} + \frac{\kappa_{6m}^{(2)}}{2r} \left(\frac{1}{2} + \ln r \right) \right], \tag{7.105}$$

$$\varepsilon_{12m} = -\left[\frac{\Psi_{11\varphi m}^{(2)}}{3} \left(\frac{4}{3} - \ln r \right) + \Psi_{12\varphi m}^{(2)} C_m^{(2)} r^{c_{3m}-1} + \frac{\Psi_{13\varphi m}^{(2)}}{2r} \left(\frac{1}{2} + \ln r \right) \right], \tag{7.106}$$

$$\varepsilon_{13m} = -\left[\frac{\Psi_{11\nu m}^{(2)}}{3} \left(\frac{4}{3} - \ln r \right) + \Psi_{12\nu m}^{(2)} C_m^{(2)} r^{c_{3m}-1} + \frac{\Psi_{13\nu m}^{(2)}}{2r} \left(\frac{1}{2} + \ln r \right) \right]. \tag{7.107}$$

With regard to the determination of the compressive or tensile radial stress $p_1 > 0$ or $p_1 < 0$ acting on the particle-matrix boundary, thus for $r = R_1$, respectively, considering Eq. (3.58), the tangential strain on the particle-matrix boundary, $(\varepsilon_{22m})_{r=R_1}$, is required to be derived in the form

$$(\varepsilon_{22m})_{r=R_1} = -p_1\,\varrho_{1m}, \tag{7.108}$$

where the coefficient ϱ_{1m} is given by Eq. (7.118).

Considering Eqs. (3.41), (3.42), (7.100)–(7.107), the thermal-stress induced elastic energy density w_m and the thermal-stress induced elastic energy W_m, accumulated in an arbitrary point of the cell matrix and in the volume of the cell matrix, respectively, have the forms

$$\begin{aligned} w_m = \frac{1}{2}\Big[& \omega_{11m} + \frac{\omega_{12m}}{r} + \frac{\omega_{13m}}{r^2} + \omega_{14m}\,r^{c_{3m}-1} + \omega_{15m}\,r^{c_{3m}-2} + \omega_{16m}\,r^{2(c_{3m}-1)} \\ & + \omega_{17m}\ln r + \omega_{18m}\ln^2 r + \frac{\omega_{19m}\ln r}{r} + \frac{\omega_{110m}\ln r}{r^2} + \frac{\omega_{111m}\ln^2 r}{r} \\ & + \omega_{112m}\frac{\ln^2 r}{r^2} + \omega_{113m}\,r^{c_{3m}-1}\ln r + \omega_{114m}\,r^{c_{3m}-2}\ln r\Big], \end{aligned} \tag{7.109}$$

$$W_m = 8\int_0^{\pi/2}\int_0^{\pi/2}\int_{R_1}^{r_c} w_m\,r^2 dr d\varphi d\nu = \int_0^{\pi/2}\int_0^{\pi/2}\Omega_{11m} d\varphi d\nu, \tag{7.110}$$

where the coefficient Ω_{11m} given by Eq. (7.133) is a function of $\omega_{11m},\ldots,\omega_{114m}$ Eqs. (7.119)–(7.132), respectively.

With regard to the determination of the thermal-stress induced strengthening presented in Volume III, the stress σ_{im} acting in a direction of the axis x_i and the elastic energy density w_{im} induced by the stress σ_{im} $(i=1,2,3)$ (see Eqs. (3.46), (3.47), (7.100)–(7.103)), both related to an arbitrary point of the cell matrix, are derived as

$$\begin{aligned} \sigma_{im} = & -\left(\xi_{i11m} + \frac{\xi_{i12m}}{r} + \xi_{i13m}\,r^{c_{3m}-1} - \xi_{i14m}\ln r + \xi_{i15m}\frac{\ln r}{r}\right), \\ & i = 1,2,3, \end{aligned} \tag{7.111}$$

$$\begin{aligned} w_{im} = \frac{s_{11m}}{2}\Big[& \xi_{i11m}^2 + \frac{\xi_{i12m}^2}{r^2} + \xi_{i13m}^2\,r^{2(c_{3m}-1)} + \xi_{i14m}^2\ln^2 r + \xi_{i15m}^2\frac{\ln^2 r}{r^2} \\ & + \frac{2\,\xi_{i11m}\,\xi_{i12m}}{r} + 2\,\xi_{i11m}\,\xi_{i13m}\,r^{c_{3m}-1} - 2\,\xi_{i11m}\,\xi_{i14m}\ln r \\ & + 2\left(\xi_{i11m}\,\xi_{i15m} - \xi_{i12m}\,\xi_{i14m}\right)\frac{\ln r}{r} + 2\,\xi_{i12m}\,\xi_{i13m}\,r^{c_{3m}-2} \end{aligned}$$

$$+ 2\,\xi_{i\,12m}\,\xi_{i\,15m}\frac{\ln r}{r^2} - 2\,\xi_{i\,13m}\,\xi_{i\,14m}\,r^{c_{3m}-1}\ln r$$
$$\left. + 2\,\xi_{i\,13m}\,\xi_{i\,15m}\,r^{c_{3m}-2}\ln r - 2\,\xi_{i\,14m}\,\xi_{i\,15m}\frac{\ln^2 r}{r}\right],$$
$$i = 1, 2, 3, \tag{7.112}$$

where the elastic modulus s_{11m} and the coefficients $\xi_{i\,11m}, \ldots, \xi_{i\,15m}$ $(i=1,2,3)$ are given by Eqs. (3.23) and (7.134)–(7.138), respectively.

Appendix. Considering the elastic modulus s_{44m} and the coefficients c_{1m}, c_{2m}, c_{3m}, $C_m^{(2)}$, a'_{1i}, a'_{2i}, a'_{3i} $(i=1,2,3)$ (see Eqs. (3.25), (3.38)–(3.40), (7.4), (12.26)–(12.34)), the coefficients χ_{ijm} $(i,j=1,2,3)$, $\varrho_{\,1+im}$, $\omega_{1+i\,1m}, \ldots, \omega_{1+i\,14m}$, $\Omega_{1+i\;1+jm}$[4] $(i,j=0,1)$, $\xi_{i\,1+j\;1m}, \ldots, \xi_{i\,1+j\;5m}$ $(i=1,2,3;\ j=0,1)$ have the forms

$$\chi_{13m} = \frac{c_{1m} - 2c_{2m}\ln R_1}{2R_1}, \tag{7.113}$$

$$\chi_{23m} = \frac{1}{2}\left(\frac{1}{2} + \ln r_c\right), \tag{7.114}$$

$$\chi_{31m} = \frac{1}{3}\left(\frac{1}{3} - \ln r_c\right), \tag{7.115}$$

$$\chi_{32m} = C_m^{(2)} c_{3m}\, r_c^{c_{3m}-1}, \tag{7.116}$$

$$\chi_{33m} = \frac{1}{2r_c}, \tag{7.117}$$

$$\varrho_{\,1+im} = \frac{\kappa_{4m}^{(2)}}{3}\left(\frac{4}{3} - \ln R_{1+i}\right) + \kappa_{5m}^{(2)} C_m^{(2)} R_{1+i}^{c_{3m}-1} + \frac{\kappa_{6m}^{(2)}}{2R_{1+i}}\left(\frac{1}{2} + \ln R_{1+i}\right),$$
$$i = 0, 1, \tag{7.118}$$

$$\omega_{1+i\,1m} = \frac{1}{81}\left(3p_{1+i}^2\left[\kappa_{4m}^{(2)}\right]^2 (11c_{1m} - 5c_{2m}) + \frac{32}{s_{44m}}\left\{\left[\Psi_{1+i1\varphi m}^{(2)}\right]^2 + \left[\Psi_{1+i1\nu m}^{(2)}\right]^2\right\}\right),$$
$$i = 0, 1, \tag{7.119}$$

$$\omega_{1+i\,2m} = \frac{1}{9}\left\{p_{1+i}^2\kappa_{4m}^{(2)}\kappa_{6m}^{(2)}(5c_{1m} - 8c_{2m})\right.$$
$$\left. + \frac{4}{s_{44m}}\left[\Psi_{1+i1\varphi m}^{(2)}\Psi_{1+i3\varphi m}^{(2)} + \Psi_{1+i1\nu m}^{(2)}\Psi_{1+i3\nu m}^{(2)}\right]\right\}, \quad i = 0, 1, \tag{7.120}$$

[4]The coefficient Ω_{21m} is determined but not used in formulae for the elastic energy W_m of the cell matrix.

$$\omega_{1+i\,3m} = \frac{1}{8}\left(p_{1+i}^2\left[\kappa_{6m}^{(2)}\right]^2(3c_{1m}-2c_{2m}) + \frac{1}{s_{44m}}\left\{\left[\Psi_{1+i3\varphi m}^{(2)}\right]^2 + \left[\Psi_{1+i3\nu m}^{(2)}\right]^2\right\}\right),$$
$$i = 0, 1, \tag{7.121}$$

$$\omega_{1+i\,4m} = \frac{2C_m^{(2)}}{9}\left\{9p_{1+i}^2\kappa_{4m}^{(2)}\kappa_{5m}^{(2)}(c_{1m}-c_{2m}) + \frac{8}{s_{44m}}\left[\Psi_{1+i1\varphi m}^{(2)}\Psi_{1+i2\varphi m}^{(2)} + \Psi_{1+i1\nu m}^{(2)}\Psi_{1+i2\nu m}^{(2)}\right]\right\},$$
$$i = 0, 1, \tag{7.122}$$

$$\omega_{1+i\,5m} = C_m^{(2)}\left\{2p_{1+i}^2\kappa_{5m}^{(2)}\kappa_{6m}^{(2)}(c_{1m}-c_{2m}) + \frac{1}{s_{44m}}\left[\Psi_{1+i2\varphi m}^{(2)}\Psi_{1+i3\varphi m}^{(2)} + \Psi_{1+i2\nu m}^{(2)}\Psi_{1+i3\nu m}^{(2)}\right]\right\},$$
$$i = 0, 1, \tag{7.123}$$

$$\omega_{1+i\,6m} = \left[C_m^{(2)}\right]^2\left(3p_{1+i}^2\left[\kappa_{5m}^{(2)}\right]^2(c_{1m}-c_{2m}) + \frac{2}{s_{44m}}\left\{\left[\Psi_{1+i2\varphi m}^{(2)}\right]^2 + \left[\Psi_{1+i2\nu m}^{(2)}\right]^2\right\}\right),$$
$$i = 0, 1, \tag{7.124}$$

$$\omega_{1+i\,7m} = -\frac{2}{3}\left(p_{1+i}^2\left[\kappa_{4m}^{(2)}\right]^2(c_{1m}-c_{2m}) + \frac{8}{9s_{44m}}\left\{\left[\Psi_{1+i1\varphi m}^{(2)}\right]^2 + \left[\Psi_{1+i1\nu m}^{(2)}\right]^2\right\}\right),$$
$$i = 0, 1, \tag{7.125}$$

$$\omega_{1+i\,8m} = \frac{1}{3}\left(p_{1+i}^2\left[\kappa_{4m}^{(2)}\right]^2(c_{1m}-c_{2m}) + \frac{2}{3s_{44m}}\left\{\left[\Psi_{1+i1\varphi m}^{(2)}\right]^2 + \left[\Psi_{1+i1\nu m}^{(2)}\right]^2\right\}\right),$$
$$i = 0, 1, \tag{7.126}$$

$$\omega_{1+i\,9m} = \frac{1}{9}\left\{2p_{1+i}^2\kappa_{4m}^{(2)}\kappa_{6m}^{(2)}(c_{1m}+2c_{2m}) + \frac{5}{s_{44m}}\left[\Psi_{1+i1\varphi m}^{(2)}\Psi_{1+i3\varphi m}^{(2)} + \Psi_{1+i1\nu m}^{(2)}\Psi_{1+i3\nu m}^{(2)}\right]\right\}, \quad i = 0, 1, \tag{7.127}$$

$$\omega_{1+i\,10m} = \frac{1}{2}\left(p_{1+i}^2\left[\kappa_{6m}^{(2)}\right]^2(c_{1m}-2c_{2m}) + \frac{1}{s_{44m}}\left\{\left[\Psi_{1+i3\varphi m}^{(2)}\right]^2 + \left[\Psi_{1+i3\nu m}^{(2)}\right]^2\right\}\right),$$
$$i = 0, 1, \tag{7.128}$$

$$\omega_{1+i\,11m} = -\frac{2}{3}\left\{p_{1+i}^2\kappa_{4m}^{(2)}\kappa_{6m}^{(2)}\left(c_{1m}-c_{2m}\right) + \frac{1}{s_{44m}}\left[\Psi_{1+i1\varphi m}^{(2)}\Psi_{1+i3\varphi m}^{(2)} + \Psi_{1+i1\nu m}^{(2)}\Psi_{1+i3\nu m}^{(2)}\right]\right\},$$
$$i = 0, 1, \qquad (7.129)$$

$$\omega_{1+i\,12m} = \frac{1}{2}\left(p_{1+i}^2\left[\kappa_{6m}^{(2)}\right]^2 c_{1m} + \frac{1}{s_{44m}}\left\{\left[\Psi_{1+i3\varphi m}^{(2)}\right]^2 + \left[\Psi_{1+i3\nu m}^{(2)}\right]^2\right\}\right),$$
$$i = 0, 1, \qquad (7.130)$$

$$\omega_{1+i\,13m} = -\frac{2C_m^{(2)}}{3}\left\{3p_{1+i}^2\kappa_{4m}^{(2)}\kappa_{5m}^{(2)}\left(c_{1m}-c_{2m}\right) + \frac{2}{s_{44m}}\left[\Psi_{1+i1\varphi m}^{(2)}\Psi_{1+i2\varphi m}^{(2)} + \Psi_{1+i1\nu m}^{(2)}\Psi_{1+i2\nu m}^{(2)}\right]\right\},$$
$$i = 0, 1, \qquad (7.131)$$

$$\omega_{1+i\,14m} = 2C_m^{(2)}\left\{p_{1+i}^2\kappa_{5m}^{(2)}\kappa_{6m}^{(2)}\left(c_{1m}-c_{2m}\right) + \frac{1}{s_{44m}}\left[\Psi_{1+i2\varphi m}^{(2)}\Psi_{1+i3\varphi m}^{(2)} + \Psi_{1+i2\nu m}^{(2)}\Psi_{1+i3\nu m}^{(2)}\right]\right\},$$
$$i = 0, 1, \qquad (7.132)$$

$$\begin{aligned}
\Omega_{1+i1+jm} &= 4\left(\frac{\omega_{1+i1m}}{3}\left(r_c^3 - R_{1+j}^3\right) + \frac{\omega_{1+i2m}}{2}\left(r_c^2 - R_{1+j}^2\right)\right.\\
&+ \omega_{1+i3m}\left(r_c - R_{1+j}\right) + \frac{\omega_{1+i4m}}{c_{3m}+2}\left(r_c^{c_{3m}+2} - R_{1+j}^{c_{3m}+2}\right)\\
&+ \frac{\omega_{1+i5m}}{c_{3m}+1}\left(r_c^{c_{3m}+1} - R_{1+j}^{c_{3m}+1}\right) + \frac{\omega_{1+i6m}}{2c_{3m}+1}\left(r_c^{2c_{3m}+1} - R_{1+j}^{2c_{3m}+1}\right)\\
&+ \frac{\omega_{1+i7m}}{3}\left[r_c^3\left(\ln r_c - \frac{1}{3}\right) - R_{1+j}^3\left(\ln R_{1+j} - \frac{1}{3}\right)\right]\\
&+ \frac{\omega_{1+i8m}}{3}\left\{r_c^3\left[\ln^2 r_c - \frac{2}{3}\left(\ln r_c - \frac{1}{3}\right)\right] - R_{1+j}^3\left[\ln^2 R_{1+j} - \frac{2}{3}\left(\ln R_{1+j} - \frac{1}{3}\right)\right]\right\}\\
&+ \frac{\omega_{1+i9m}}{2}\left[r_c^2\left(\ln r_c - \frac{1}{2}\right) - R_{1+j}^2\left(\ln R_{1+j} - \frac{1}{2}\right)\right]\\
&+ \omega_{1+i10m}\left[r_c\left(\ln r_c - 1\right) - R_{1+j}\left(\ln R_{1+j} - 1\right)\right]\\
&+ \frac{\omega_{1+i11m}}{2}\left[r_c^2\left(\ln^2 r_c - \ln r_c + \frac{1}{2}\right) - R_{1+j}^2\left(\ln^2 R_{1+j} - \ln R_{1+j} + \frac{1}{2}\right)\right]\\
&+ \omega_{1+i12m}\left[r_c\left(\ln^2 r_c - 2\ln r_c + 2\right) - R_{1+j}\left(\ln^2 R_{1+j} - 2\ln R_{1+j} + 2\right)\right]\\
&+ \frac{\omega_{1+i13m}}{c_{3m}+2}\left[r_c^{c_{3m}+2}\left(\ln r_c - \frac{1}{c_{3m}+2}\right) - R_{1+j}^{c_{3m}+2}\left(\ln R_{1+j} - \frac{1}{c_{3m}+2}\right)\right]
\end{aligned}$$

$$+ \frac{\omega_{1+i14m}}{c_{3m}+1}\left[r_c^{c_{3m}+1}\left(\ln r_c - \frac{1}{c_{3m}+1}\right) - R_{1+j}^{c_{3m}+1}\left(\ln R_{1+j} - \frac{1}{c_{3m}+1}\right)\right]\Bigg),$$
$$i, j = 0, 1, \tag{7.133}$$

$$\xi_{i\,1+j\;1m} = \frac{1}{9}\left\{p_{1+j}\kappa_{4m}^{(2)}\left[a'_{1i}\left(c_{1m} - 7c_{2m}\right) + \left(a'_{2i} + a'_{3i}\right)\left(4c_{1m} - c_{2m}\right)\right]\right.$$
$$\left. + \frac{4}{s_{44m}}\left[\left(a'_{1i} + a'_{2i}\right)\psi_{1+j\,1\varphi m}^{(2)} + \left(a'_{1i} + a'_{3i}\right)\psi_{1+j\,1\nu m}^{(2)}\right]\right\},$$
$$i = 1, 2, 3; \quad j = 0, 1, \tag{7.134}$$

$$\xi_{i\,1+j\;2m} = \frac{1}{2}\left\{p_{1+j}\kappa_{6m}^{(2)}\left[a'_{1i}\,c_{1m} + \left(a'_{2i} + a'_{3i}\right)\left(\frac{c_{1m}}{2} - c_{2m}\right)\right]\right.$$
$$\left. + \frac{1}{2s_{44m}}\left[\left(a'_{1i} + a'_{2i}\right)\psi_{1+j\,3\varphi m}^{(2)} + \left(a'_{1i} + a'_{3i}\right)\psi_{1+j\,3\nu m}^{(2)}\right]\right\},$$
$$i = 1, 2, 3; \quad j = 0, 1, \tag{7.135}$$

$$\xi_{i\,1+j\;3m} = C_m^{(2)}\Bigg(p_{1+j}\kappa_{5m}^{(2)}\left\{a'_{1i}\left[c_{3m}\left(c_{1m}+c_{2m}\right)-2c_{2m}\right]+\left(a'_{2i} + a'_{3i}\right)\left(c_{1m} - c_{2m}c_{3m}\right)\right\}$$
$$+ \frac{1}{s_{44m}}\left[\left(a'_{1i} + a'_{2i}\right)\psi_{1+j\,2\varphi m}^{(2)} + \left(a'_{1i} + a'_{3i}\right)\psi_{1+j\,2\nu m}^{(2)}\right]\Bigg),$$
$$i = 1, 2, 3; \quad j = 0, 1, \tag{7.136}$$

$$\xi_{i\,1+j\;4m} = \frac{1}{3}\left\{p_{1+j}\kappa_{4m}^{(2)}\left(a'_{1i} + a'_{2i} + a'_{3i}\right)\left(c_{1m} - c_{2m}\right)\right.$$
$$\left. + \frac{1}{s_{44m}}\left[\left(a'_{1i} + a'_{2i}\right)\psi_{1+j\,1\varphi m}^{(2)} + \left(a'_{1i} + a'_{3i}\right)\psi_{1+j\,1\nu m}^{(2)}\right]\right\},$$
$$i = 1, 2, 3; \quad j = 0, 1, \tag{7.137}$$

$$\xi_{i\,1+j\;5m} = p_{1+j}\kappa_{6m}^{(2)}\left[\frac{c_{1m}}{2}\left(a'_{2i} + a'_{3i}\right) - a'_{1i}\,c_{2m}\right]$$
$$+ \frac{1}{2s_{44m}}\left[\left(a'_{1i} + a'_{2i}\right)\psi_{1+j\,3\varphi m}^{(2)} + \left(a'_{1i} + a'_{3i}\right)\psi_{1+j\,3\nu m}^{(2)}\right],$$
$$i = 1, 2, 3; \quad j = 0, 1. \tag{7.138}$$

7.5 Multi-particle-envelope-matrix system

The variable r related to the thermal stresses and to the corresponding quantities in the spherical envelope with the radii $R_1 < R_2$ and in the cell matrix is from the intervals $\langle R_1, R_2\rangle$ and $\langle R_2, r_c\rangle$, respectively, where the φ, ν-dependent distance r_c

as a function of the experimentally determined parameters v, R_{1ex}, d_{1ex}, d_{2ex}, d_{3ex} (see Section 2.2.1) is derived in Section 2.3.

The compressive or tensile radial stresses $p_1 > 0$ or $p_1 < 0$ and $p_2 > 0$ or $p_2 < 0$, as consequences of the conditions $\beta_p \neq \beta_m$ and $\beta_m \neq \beta_e$, acting on the particle-envelope and matrix-envelope boundaries, thus for $r = R_1$ and $r = R_2$, respectively, are given by Eqs. (10.2)–(10.5) as dependent on the following conditions concerning the coefficients β_p, β_e, β_m (see Eqs. (3.48)–(3.50)).

With regard to the condition $\beta_p \neq \beta_e \neq \beta_m$, the radial stresses p_1, p_2 are determined by the tangential strains $(\varepsilon_{22p})_{r=R_1}$, $(\varepsilon_{22e})_{r=R_{1+i}}$ $(i=0,1)$ and $(\varepsilon_{22m})_{r=R_2}$ in the spherical particle, in the spherical envelope and in the cell matrix, resulting in the coefficients ϱ_{1p}, ϱ_{1+ie} and ϱ_{2m}, respectively (see Eqs. (3.59), (3.60), (10.2), (10.3)).

Similarly, provided that $\beta_p \neq \beta_e = \beta_m$ and regarding the φ, ν-dependent difference $\varepsilon_{11te} - \varepsilon_{11tp} = f(\varphi, \nu)$ (see Eq. (12.23)), the radial stress p_1 is determined by the tangential strains $(\varepsilon_{22p})_{r=R_1}$ and $(\varepsilon_{22e})_{r=R_1}$, resulting in the coefficients ϱ_{1p} and ϱ_{1e}^{me}, respectively (see Eqs. (3.59), (10.4)).

Finally, provided that $\beta_p = \beta_e \neq \beta_m$, the radial stress p_2 is determined by the tangential strains $(\varepsilon_{22m})_{r=R_2}$ and $(\varepsilon_{22e})_{r=R_2}$, resulting in the coefficients ϱ_{2m} and ϱ_{2e}^{pe}, respectively (see Eqs. (3.60), (10.5)).

7.5.1 Thermal stresses in spherical envelope

7.5.1.1 Condition $\beta_p \neq \beta_e \neq \beta_m$

A. Integration constants $C_{1e} \neq 0$, $C_{2e} \neq 0$, $C_{3e} = 0$

With regard to the boundary conditions (4.7), (4.8) and the analysis presented in Section 6.2.2, considering the terms u_{1re} and u_{2re} (see Eqs. (7.5), (6.7)) to represent functions of the variable $r \in \langle R_1, R_2 \rangle$, the radial displacement u_{1e} of the infinitesimal spherical cap in the spherical envelope, regarding Eqs. (3.31), (7.3), has the form

$$u_{1e} = -\left\{ \left[p_1 \kappa_{1e}^{(1)} - p_2 \kappa_{2e}^{(1)} \right] \frac{r}{3} \left(\frac{4}{3} - \ln r \right) + \left[p_1 \kappa_{3e}^{(1)} - p_2 \kappa_{4e}^{(1)} \right] C_e \, r^{c_{3e}} \right\}, \quad (7.139)$$

where the coefficients $c_{3e} < 0$, $C_e = C_e^{(i)}$ $(i=2,3)$ and the radial stresses p_1, p_2 acting on the particle-envelope, matrix-envelope boundaries are given by Eqs. (3.40), (7.4), (7.13) and (10.2), (10.3), respectively.

With regard to Section 7.5.1.1.A, the coefficients $\varrho_{1+i\,1+je}$ $(i,j=0,1)$, $\omega_{1e}, \ldots, \omega_{6e}$, ξ_{i1e}, ξ_{i2e}, ξ_{i3e} $(i=1,2,3)$ mentioned below (see Eqs. (7.148), (7.149), (7.151), (7.152)) depend on $\kappa_{1e}^{(1)}, \ldots, \kappa_{4e}^{(1)}$ (see Eq. (12.35)) representing functions of χ_{11e}, χ_{12e}, χ_{21e}, χ_{22e} given by Eqs. (7.153), (7.154).

Consequently, the radial stress σ_{11e}, the tangential and shear stresses σ_{22e}, σ_{33e} and σ_{12e}, σ_{13e} acting in the spherical envelope, respectively, regarding Eqs. (3.31)–

(3.34), (7.139), are derived as

$$\sigma_{11e} = -\left\{\frac{1}{3}\left[p_1\kappa_{1e}^{(1)} - p_2\kappa_{2e}^{(1)}\right]\left[\frac{1}{3}\left(c_{1e} - 7c_{2e}\right) - \left(c_{1e} - c_{2e}\right)\ln r\right]\right.$$
$$\left. + \left[p_1\kappa_{3e}^{(1)} - p_2\kappa_{4e}^{(1)}\right] C_e\left[c_{3e}\left(c_{1e} + c_{2e}\right) - 2c_{2e}\right] r^{c_{3e}-1}\right\}, \quad (7.140)$$

$$\sigma_{22e} = \sigma_{33e} = -\left\{\frac{1}{3}\left[p_1\kappa_{1e}^{(1)} - p_2\kappa_{2e}^{(1)}\right]\left[\frac{1}{3}\left(4c_{1e} - c_{2e}\right) - \left(c_{1e} - c_{2e}\right)\ln r\right]\right.$$
$$\left. + \left[p_1\kappa_{3e}^{(1)} - p_2\kappa_{4e}^{(1)}\right] C_e\left(c_{1e} - c_{2e}c_{3e}\right) r^{c_{3e}-1}\right\}, \quad (7.141)$$

$$\sigma_{12e} = -\frac{1}{s_{44e}}\left\{\frac{1}{3}\left[\kappa_{1e}^{(1)}\frac{\partial p_1}{\partial\varphi} - \kappa_{2e}^{(1)}\frac{\partial p_2}{\partial\varphi}\right]\left(\frac{4}{3} - \ln r\right)\right.$$
$$\left. + \left[\kappa_{3e}^{(1)}\frac{\partial p_1}{\partial\varphi} - \kappa_{4e}^{(1)}\frac{\partial p_2}{\partial\varphi}\right] C_e\, r^{c_{3e}-1}\right\}, \quad (7.142)$$

$$\sigma_{13e} = -\frac{1}{s_{44e}}\left\{\frac{1}{3}\left[\kappa_{1e}^{(1)}\frac{\partial p_1}{\partial\nu} - \kappa_{2e}^{(1)}\frac{\partial p_2}{\partial\nu}\right]\left(\frac{4}{3} - \ln r\right)\right.$$
$$\left. + \left[\kappa_{3e}^{(1)}\frac{\partial p_1}{\partial\nu} - \kappa_{4e}^{(1)}\frac{\partial p_2}{\partial\nu}\right] C_e\, r^{c_{3e}-1}\right\}, \quad (7.143)$$

where the elastic modulus s_{44e} and the coefficients c_{1e}, c_{2e} are given by Eqs. (3.25) and (3.38), (3.39), respectively.

Finally, the radial strain ε_{11e}, the tangential and shear strains ε_{22e}, ε_{33e} and ε_{12e}, ε_{13e} in the spherical envelope, respectively, regarding Eqs. (3.26)–(3.30), (7.139), have the forms

$$\varepsilon_{11e} = -\left\{\frac{1}{3}\left[p_1\kappa_{1e}^{(1)} - p_2\kappa_{2e}^{(1)}\right]\left(\frac{1}{3} - \ln r\right) + \left[p_1\kappa_{3e}^{(1)} - p_2\kappa_{4e}^{(1)}\right] C_e c_{3e}\, r^{c_{3e}-1}\right\}, \quad (7.144)$$

$$\varepsilon_{22e} = \varepsilon_{33e} = -\left\{\frac{1}{3}\left[p_1\kappa_{1e}^{(1)} - p_2\kappa_{2e}^{(1)}\right]\left(\frac{4}{3} - \ln r\right)\right.$$
$$\left. + \left[p_1\kappa_{3e}^{(1)} - p_2\kappa_{4e}^{(1)}\right] C_e\, r^{c_{3e}-1}\right\}, \quad (7.145)$$

$$\varepsilon_{12e} = -\left\{\frac{1}{3}\left[\kappa_{1e}^{(1)}\frac{\partial p_1}{\partial\varphi} - \kappa_{2e}^{(1)}\frac{\partial p_2}{\partial\varphi}\right]\left(\frac{4}{3} - \ln r\right)\right.$$
$$\left. + \left[\kappa_{3e}^{(1)}\frac{\partial p_1}{\partial\varphi} - \kappa_{4e}^{(1)}\frac{\partial p_2}{\partial\varphi}\right] C_e\, r^{c_{3e}-1}\right\}, \quad (7.146)$$

$$\varepsilon_{13e} = -\left\{\frac{1}{3}\left[\kappa_{1e}^{(1)}\frac{\partial p_1}{\partial \nu} - \kappa_{2e}^{(1)}\frac{\partial p_2}{\partial \nu}\right]\left(\frac{4}{3} - \ln r\right) + \left[\kappa_{3e}^{(1)}\frac{\partial p_1}{\partial \nu} - \kappa_{4e}^{(1)}\frac{\partial p_2}{\partial \nu}\right] C_e\, r^{c_{3e}-1}\right\}. \tag{7.147}$$

With regard to the determination of the compressive or tensile radial stresses $p_1 > 0$ or $p_1 < 0$ and $p_2 > 0$ or $p_2 < 0$ acting on the particle-envelope and matrix-envelope boundaries, thus for $r = R_1$ and $r = R_2$, respectively, considering Eqs. (3.59), (3.60), the tangential strains on the particle-envelope and matrix-envelope boundaries, $(\varepsilon_{22e})_{r=R_1}$ and $(\varepsilon_{22e})_{r=R_2}$, respectively, are required to be derived in the form

$$(\varepsilon_{22e})_{r=R_{1+i}} = -\sum_{j=0}^{1} p_{1+j}\,\varrho_{1+i\;1+je}, \quad i = 0, 1, \tag{7.148}$$

where the coefficient $\varrho_{1+i\;1+je}$ $(i, j=0,1)$ is given by Eq. (7.155).

Considering Eqs. (3.41), (3.42), (7.140)–(7.147), the thermal-stress induced elastic energy density w_e and the thermal-stress induced elastic energy W_e, accumulated in an arbitrary point of the spherical envelope and in the volume of the spherical envelope, respectively, have the forms

$$w_e = \frac{1}{2}\Big[\omega_{1e} + \omega_{2e}\, r^{c_{3e}-1} + \omega_{3e}\, r^{2(c_{3e}-1)} + \omega_{4e}\ln r + \omega_{5e}\ln^2 r + \omega_{6e}\, r^{c_{3e}-1}\ln r\Big], \tag{7.149}$$

$$W_e = 8\int_0^{\pi/2}\int_0^{\pi/2}\int_{R_1}^{R_2} w_e\, r^2 dr\, d\varphi\, d\nu = \int_0^{\pi/2}\int_0^{\pi/2} \Omega_e\, d\varphi\, d\nu, \tag{7.150}$$

where the coefficient Ω_e given by Eq. (6.146) is a function of $\omega_{1e}, \ldots, \omega_{6e}$ given by Eqs. (7.156)–(7.161), respectively.

With regard to the determination of the thermal-stress induced strengthening presented in Volume III, the stress σ_{ie} acting in a direction of the axis x_i and the elastic energy density w_{ie} induced by the stress σ_{ie} $(i=1,2,3)$ (see Eqs. (3.46), (3.47), (7.140)–(7.143)), both related to an arbitrary point of the spherical envelope, are derived as

$$\sigma_{ie} = -\left(\xi_{i1e} + \xi_{i2e}\, r^{c_{3e}-1} - \xi_{i3e}\ln r\right), \quad i = 1, 2, 3, \tag{7.151}$$

$$w_{ie} = \frac{s_{11e}}{2}\Big[\xi_{i1e}^2 + \xi_{i2e}^2\, r^{2(c_{3e}-1)} + \xi_{i3e}^2\ln^2 r + 2\,\xi_{i1e}\,\xi_{i2e}\, r^{c_{3e}-1} - 2\,\xi_{i1e}\,\xi_{i3e}\ln r - 2\,\xi_{i2e}\,\xi_{i3e}\, r^{c_{3e}-1}\ln r\Big], \quad i = 1, 2, 3, \tag{7.152}$$

where the elastic modulus s_{11e} and the coefficients ξ_{i1e}, ξ_{i2e}, ξ_{i3e} $(i=1,2,3)$ are given by Eqs. (3.23) and (7.162)–(7.164), respectively.

Appendix. Considering the elastic modulus s_{44e} and the coefficients c_{1e}, c_{2e}, c_{3e}, $C_e = C_e^{(i)}$ $(i=2,3)$, a'_{1i}, a'_{2i}, a'_{3i} $(i=1,2,3)$ (see Eqs. (3.25), (3.38)–(3.40), (7.4), (7.13), (12.26)–(12.34)), the coefficients $\chi_{1+i\,1e}$, $\chi_{1+i\,2e}$, $\varrho_{1+i\;1+je}$ $(i,j=0,1)$, $\omega_{1e},\ldots,\omega_{6e}$, Ω_e, $\xi_{i\,1e}$, $\xi_{i\,2e}$, $\xi_{i\,3e}$ $(i=1,2,3)$ have the forms

$$\chi_{1+i\,1e} = \frac{1}{3}\left[\frac{1}{3}\left(c_{1e} - 7c_{2e}\right) - \left(c_{1e} - c_{2e}\right)\ln R_{1+i}\right], \quad i = 0, 1, \tag{7.153}$$

$$\chi_{1+i\,2e} = C_e\left[c_{3e}\left(c_{1e} + c_{2e}\right) - 2c_{2e}\right] R_{1+i}^{c_{3e}-1}, \quad i = 0, 1, \tag{7.154}$$

$$\varrho_{1+i\;1+je} = \left(\delta_{0j} - \delta_{1j}\right)\left[\frac{\kappa_{1+je}^{(1)}}{3}\left(\frac{4}{3} - \ln R_{1+i}\right) + \kappa_{3+je}^{(1)} C_e R_{1+i}^{c_{3e}-1}\right],$$
$$i,j = 0,1, \tag{7.155}$$

$$\omega_{1e} = \frac{1}{81}\left(3\left[p_1\kappa_{1e}^{(1)} - p_2\kappa_{2e}^{(1)}\right]^2\left(11c_{1e} - 5c_{2e}\right)\right.$$
$$\left. + \frac{32}{s_{44e}}\left\{\left[\kappa_{1e}^{(1)}\frac{\partial p_1}{\partial\varphi} - \kappa_{2e}^{(1)}\frac{\partial p_2}{\partial\varphi}\right]^2 + \left[\kappa_{1e}^{(1)}\frac{\partial p_1}{\partial\nu} - \kappa_{2e}^{(1)}\frac{\partial p_2}{\partial\nu}\right]^2\right\}\right), \tag{7.156}$$

$$\omega_{2e} = \frac{2C_e}{9}\left(\left[p_1\kappa_{1e}^{(1)} - p_2\kappa_{2e}^{(1)}\right]\left[p_1\kappa_{3e}^{(1)} - p_2\kappa_{4e}^{(1)}\right]\left[c_{1e}\left(c_{3e} + 8\right) - c_{2e}\left(7c_{3e} + 2\right)\right]\right.$$
$$+ \frac{8}{s_{44e}}\left\{\left[\kappa_{1e}^{(1)}\frac{\partial p_1}{\partial\varphi} - \kappa_{2e}^{(1)}\frac{\partial p_2}{\partial\varphi}\right]\left[\kappa_{3e}^{(1)}\frac{\partial p_1}{\partial\varphi} - \kappa_{4e}^{(1)}\frac{\partial p_2}{\partial\varphi}\right]\right.$$
$$\left.\left. + \left[\kappa_{1e}^{(1)}\frac{\partial p_1}{\partial\nu} - \kappa_{2e}^{(1)}\frac{\partial p_2}{\partial\nu}\right]\left[\kappa_{3e}^{(1)}\frac{\partial p_1}{\partial\nu} - \kappa_{4e}^{(1)}\frac{\partial p_2}{\partial\nu}\right]\right\}\right), \tag{7.157}$$

$$\omega_{3e} = C_e^2\left(\left[p_1\kappa_{3e}^{(1)} - p_2\kappa_{4e}^{(1)}\right]^2\left[c_{3e}^2\left(c_{1e} + c_{2e}\right) + 2\left(c_{1e} - 2c_{2e}c_{3e}\right)\right]\right.$$
$$\left. + \frac{2}{s_{44e}}\left\{\left[\kappa_{3e}^{(1)}\frac{\partial p_1}{\partial\varphi} - \kappa_{4e}^{(1)}\frac{\partial p_2}{\partial\varphi}\right]^2 + \left[\kappa_{3e}^{(1)}\frac{\partial p_1}{\partial\nu} - \kappa_{4e}^{(1)}\frac{\partial p_2}{\partial\nu}\right]^2\right\}\right), \tag{7.158}$$

$$\omega_{4e} = -\frac{2}{27}\left(9\left[p_1\kappa_{1e}^{(1)} - p_2\kappa_{2e}^{(1)}\right]^2\left(c_{1e} - c_{2e}\right)\right.$$
$$\left. + \frac{8}{s_{44e}}\left\{\left[\kappa_{1e}^{(1)}\frac{\partial p_1}{\partial\varphi} - \kappa_{2e}^{(1)}\frac{\partial p_2}{\partial\varphi}\right]^2 + \left[\kappa_{1e}^{(1)}\frac{\partial p_1}{\partial\nu} - \kappa_{2e}^{(1)}\frac{\partial p_2}{\partial\nu}\right]^2\right\}\right), \tag{7.159}$$

$$\omega_{5e} = \frac{1}{9}\left(3\left[p_1\kappa_{1e}^{(1)} - p_2\kappa_{2e}^{(1)}\right]^2 (c_{1e} - c_{2e}) + \frac{2}{s_{44e}}\left\{\left[\kappa_{1e}^{(1)}\frac{\partial p_1}{\partial \varphi} - \kappa_{2e}^{(1)}\frac{\partial p_2}{\partial \varphi}\right]^2 + \left[\kappa_{1e}^{(1)}\frac{\partial p_1}{\partial \nu} - \kappa_{2e}^{(1)}\frac{\partial p_2}{\partial \nu}\right]^2\right\}\right), \quad (7.160)$$

$$\omega_{6e} = -\frac{2C_e}{3}\left(\left[p_1\kappa_{1e}^{(1)} - p_2\kappa_{2e}^{(1)}\right]\left[p_1\kappa_{3e}^{(1)} - p_2\kappa_{4e}^{(1)}\right](c_{1e} - c_{2e})(c_{3e} + 2) + \frac{2}{s_{44e}}\left\{\left[\kappa_{1e}^{(1)}\frac{\partial p_1}{\partial \varphi} - \kappa_{2e}^{(1)}\frac{\partial p_2}{\partial \varphi}\right]\left[\kappa_{3e}^{(1)}\frac{\partial p_1}{\partial \varphi} - \kappa_{4e}^{(1)}\frac{\partial p_2}{\partial \varphi}\right] + \left[\kappa_{1e}^{(1)}\frac{\partial p_1}{\partial \nu} - \kappa_{2e}^{(1)}\frac{\partial p_2}{\partial \nu}\right]\left[\kappa_{3e}^{(1)}\frac{\partial p_1}{\partial \nu} - \kappa_{4e}^{(1)}\frac{\partial p_2}{\partial \nu}\right]\right\}\right), \quad (7.161)$$

$$\xi_{i1e} = \frac{1}{9}\left(\left[p_1\kappa_{1e}^{(1)} - p_2\kappa_{2e}^{(1)}\right]\left[a'_{1i}(c_{1e} - 7c_{2e}) + (a'_{2i} + a'_{3i})(4c_{1e} - c_{2e})\right] + \frac{4}{s_{44e}}\left\{(a'_{1i} + a'_{2i})\left[\kappa_{1e}^{(1)}\frac{\partial p_1}{\partial \varphi} - \kappa_{2e}^{(1)}\frac{\partial p_2}{\partial \varphi}\right] + (a'_{1i} + a'_{3i})\left[\kappa_{1e}^{(1)}\frac{\partial p_1}{\partial \nu} - \kappa_{2e}^{(1)}\frac{\partial p_2}{\partial \nu}\right]\right\}, \quad i = 1, 2, 3, \quad (7.162)$$

$$\xi_{i2e} = C_e\left(\left[p_1\kappa_{3e}^{(1)} - p_2\kappa_{4e}^{(1)}\right]\left\{a'_{1i}\left[c_{3e}(c_{1e} + c_{2e}) - 2c_{2e}\right] + (a'_{2i} + a'_{3i})(c_{1e} - c_{2e}c_{3e})\right\} + \frac{1}{s_{44e}}\left\{(a'_{1i} + a'_{2i})\left[\kappa_{3e}^{(1)}\frac{\partial p_1}{\partial \varphi} - \kappa_{4e}^{(1)}\frac{\partial p_2}{\partial \varphi}\right] + (a'_{1i} + a'_{3i})\left[\kappa_{3e}^{(1)}\frac{\partial p_1}{\partial \nu} - \kappa_{4e}^{(1)}\frac{\partial p_2}{\partial \nu}\right]\right\}\right), \quad i = 1, 2, 3, \quad (7.163)$$

$$\xi_{i3e} = \frac{1}{3}\left(\left[p_1\kappa_{1e}^{(1)} - p_2\kappa_{2e}^{(1)}\right](a'_{1i} + a'_{2i} + a'_{3i})(c_{1e} - c_{2e}) + \frac{1}{s_{44e}}\left\{(a'_{1i} + a'_{2i})\left[\kappa_{1e}^{(1)}\frac{\partial p_1}{\partial \varphi} - \kappa_{2e}^{(1)}\frac{\partial p_2}{\partial \varphi}\right] + (a'_{1i} + a'_{3i})\left[\kappa_{1e}^{(1)}\frac{\partial p_1}{\partial \nu} - \kappa_{2e}^{(1)}\frac{\partial p_2}{\partial \nu}\right]\right\}, \quad i = 1, 2, 3. \quad (7.164)$$

B. Integration constants $C_{1e} \neq 0$, $C_{2e} = 0$, $C_{3e} \neq 0$

With regard to the boundary conditions (4.7), (4.8) and the analysis presented in Section 6.2.2, considering the terms u_{1re} and u_{3re} (see Eqs. (7.5), (7.6)) to represent functions of the variable $r \in \langle R_1, R_2 \rangle$, the radial displacement u_{1e} of the infinitesimal spherical cap in the spherical envelope, regarding Eqs. (3.31), (7.3), has the form

$$u_{1e}=-\left\{\left[p_1\kappa_{1e}^{(1)}-p_2\kappa_{2e}^{(1)}\right]\frac{r}{3}\left(\frac{4}{3}-\ln r\right)+\frac{1}{2}\left[p_1\kappa_{3e}^{(1)}-p_2\kappa_{4e}^{(1)}\right]\left(\frac{1}{2}+\ln r\right)\right\}, \quad (7.165)$$

where the radial stresses p_1, p_2 acting on the particle-envelope, matrix-envelope boundaries are given by Eqs. (10.2), (10.3), respectively.

With regard to Section 7.5.1.1.B, the coefficients $\varrho_{1+i\,1+je}$ $(i,j=0,1)$, $\omega_{1e},\ldots,\omega_{9e}$, $\xi_{i1e},\ldots,\xi_{i4e}$ $(i=1,2,3)$ mentioned below (see Eqs. (7.174), (7.175), (7.177), (7.178)) depend on $\kappa_{1e}^{(1)},\ldots,\kappa_{4e}^{(1)}$ (see Eq. (12.35)) representing functions of χ_{11e}, χ_{12e}, χ_{21e}, χ_{22e} given by Eqs. (7.153), (7.179).

Consequently, the radial stress σ_{11e}, the tangential and shear stresses σ_{22e}, σ_{33e} and σ_{12e}, σ_{13e} acting in the spherical envelope, respectively, regarding Eqs. (3.31)–(3.34), (7.165), are derived as

$$\begin{aligned}\sigma_{11e} = -&\left\{\frac{1}{3}\left[p_1\kappa_{1e}^{(1)} - p_2\kappa_{2e}^{(1)}\right]\left[\frac{1}{3}\left(c_{1e} - 7c_{2e}\right) - \left(c_{1e} - c_{2e}\right)\ln r\right]\right.\\ &\left.+\left[p_1\kappa_{3e}^{(1)} - p_2\kappa_{4e}^{(1)}\right]\frac{1}{2r}\left(c_{1e} - 2c_{2e}\ln r\right)\right\},\end{aligned} \quad (7.166)$$

$$\begin{aligned}\sigma_{22e} = \sigma_{33e} = -&\left\{\frac{1}{3}\left[p_1\kappa_{1e}^{(1)} - p_2\kappa_{2e}^{(1)}\right]\left[\frac{1}{3}\left(4c_{1e} - c_{2e}\right) - \left(c_{1e} - c_{2e}\right)\ln r\right]\right.\\ &\left.+\left[p_1\kappa_{3e}^{(1)} - p_2\kappa_{4e}^{(1)}\right]\frac{1}{2r}\left[c_{1e}\left(\frac{1}{2} + \ln r\right) - c_{2e}\right]\right\},\end{aligned} \quad (7.167)$$

$$\begin{aligned}\sigma_{12e} = -\frac{1}{s_{44e}}&\left\{\frac{1}{3}\left[\kappa_{1e}^{(1)}\frac{\partial p_1}{\partial\varphi} - \kappa_{2e}^{(1)}\frac{\partial p_2}{\partial\varphi}\right]\left(\frac{4}{3} - \ln r\right)\right.\\ &\left.+\left[\kappa_{3e}^{(1)}\frac{\partial p_1}{\partial\varphi} - \kappa_{4e}^{(1)}\frac{\partial p_2}{\partial\varphi}\right]\frac{1}{2r}\left(\frac{1}{2} + \ln r\right)\right\},\end{aligned} \quad (7.168)$$

$$\begin{aligned}\sigma_{13e} = -\frac{1}{s_{44e}}&\left\{\frac{1}{3}\left[\kappa_{1e}^{(1)}\frac{\partial p_1}{\partial\nu} - \kappa_{2e}^{(1)}\frac{\partial p_2}{\partial\nu}\right]\left(\frac{4}{3} - \ln r\right)\right.\\ &\left.+\left[\kappa_{3e}^{(1)}\frac{\partial p_1}{\partial\nu} - \kappa_{4e}^{(1)}\frac{\partial p_2}{\partial\nu}\right]\frac{1}{2r}\left(\frac{1}{2} + \ln r\right)\right\},\end{aligned} \quad (7.169)$$

where the elastic modulus s_{44e} and the coefficients c_{1e}, c_{2e} are given by Eqs. (3.25) and (3.38), (3.39), respectively.

Finally, the radial strain ε_{11e}, the tangential and shear strains ε_{22e}, ε_{33e} and ε_{12e}, ε_{13e} in the spherical envelope, respectively, regarding Eqs. (3.26)–(3.30), (7.165), have the forms

$$\varepsilon_{11e} = -\left\{\frac{1}{3}\left[p_1\kappa_{1e}^{(1)} - p_2\kappa_{2e}^{(1)}\right]\left(\frac{1}{3} - \ln r\right) + \frac{1}{2r}\left[p_1\kappa_{3e}^{(1)} - p_2\kappa_{4e}^{(1)}\right]\right\}, \quad (7.170)$$

$$\varepsilon_{22e} = \varepsilon_{33e} = -\left\{\frac{1}{3}\left[p_1\kappa_{1e}^{(1)} - p_2\kappa_{2e}^{(1)}\right]\left(\frac{4}{3} - \ln r\right) + \left[p_1\kappa_{3e}^{(1)} - p_2\kappa_{4e}^{(1)}\right]\frac{1}{2r}\left(\frac{1}{2} + \ln r\right)\right\}, \tag{7.171}$$

$$\varepsilon_{12e} = -\left\{\frac{1}{3}\left[\kappa_{1e}^{(1)}\frac{\partial p_1}{\partial \varphi} - \kappa_{2e}^{(1)}\frac{\partial p_2}{\partial \varphi}\right]\left(\frac{4}{3} - \ln r\right) + \left[\kappa_{3e}^{(1)}\frac{\partial p_1}{\partial \varphi} - \kappa_{4e}^{(1)}\frac{\partial p_2}{\partial \varphi}\right]\frac{1}{2r}\left(\frac{1}{2} + \ln r\right)\right\}, \tag{7.172}$$

$$\varepsilon_{13e} = -\left\{\frac{1}{3}\left[\kappa_{1e}^{(1)}\frac{\partial p_1}{\partial \nu} - \kappa_{2e}^{(1)}\frac{\partial p_2}{\partial \nu}\right]\left(\frac{4}{3} - \ln r\right) + \left[\kappa_{3e}^{(1)}\frac{\partial p_1}{\partial \nu} - \kappa_{4e}^{(1)}\frac{\partial p_2}{\partial \nu}\right]\frac{1}{2r}\left(\frac{1}{2} + \ln r\right)\right\}. \tag{7.173}$$

With regard to the determination of the compressive or tensile radial stresses $p_1 > 0$ or $p_1 < 0$ and $p_2 > 0$ or $p_2 < 0$ acting on the particle-envelope and matrix-envelope boundaries, thus for $r = R_1$ and $r = R_2$, respectively, considering Eqs. (3.59), (3.60), the tangential strains on the particle-envelope and matrix-envelope boundaries, $(\varepsilon_{22e})_{r=R_1}$ and $(\varepsilon_{22e})_{r=R_2}$, respectively, are required to be derived in the form

$$(\varepsilon_{22e})_{r=R_{1+i}} = -\sum_{j=0}^{1} p_{1+j}\varrho_{1+i\,1+je}, \quad i = 0, 1, \tag{7.174}$$

where the coefficient $\varrho_{1+i\,1+je}$ $(i, j = 0{,}1)$ is given by Eq. (7.180).

Considering Eqs. (3.41), (3.42), (7.166)–(7.173), the thermal-stress induced elastic energy density w_e and the thermal-stress induced elastic energy W_e, accumulated in an arbitrary point of the spherical envelope and in the volume of the spherical envelope, respectively, have the forms

$$w_e = \frac{1}{2}\left(\omega_{1e} + \frac{\omega_{2e}}{r} + \frac{\omega_{3e}}{r^2} + \omega_{4e}\ln r + \omega_{5e}\ln^2 r + \omega_{6e}\frac{\ln r}{r} + \omega_{7e}\frac{\ln r}{r^2} + \omega_{8e}\frac{\ln^2 r}{r} + \omega_{9e}\frac{\ln^2 r}{r^2}\right), \tag{7.175}$$

$$W_e = 8\int_0^{\pi/2}\int_0^{\pi/2}\int_{R_1}^{R_2} w_e r^2 dr d\varphi d\nu = \int_0^{\pi/2}\int_0^{\pi/2} \Omega_e d\varphi d\nu, \tag{7.176}$$

where the coefficient Ω_e given by Eq. (7.187) is a function of $\omega_{1e}, \ldots, \omega_{9e}$ given by Eqs. (7.156), (7.181), (7.182), (7.159), (7.160), (7.183)–(7.186), respectively.

With regard to the determination of the thermal-stress induced strengthening presented in Volume III, the stress σ_{ie} acting in a direction of the axis x_i and the elastic energy density w_{ie} induced by the stress σ_{ie} ($i = 1,2,3$) (see Eqs. (3.46), (3.47), (7.166)–(7.169)), both related to an arbitrary point of the spherical envelope, are derived as

$$\sigma_{ie} = -\left(\xi_{i1e} + \frac{\xi_{i2e}}{r} - \xi_{i3e}\ln r + \xi_{i4e}\frac{\ln r}{r}\right), \quad i = 1, 2, 3, \tag{7.177}$$

$$w_{ie} = \frac{s_{11e}}{2}\left[\xi_{i1e}^2 + \frac{\xi_{i2e}^2}{r^2} + \xi_{i3e}^2\ln^2 r + \xi_{i4e}^2\frac{\ln^2 r}{r^2} + \frac{2\,\xi_{i1e}\,\xi_{i2e}}{r} - 2\,\xi_{i1e}\,\xi_{i3e}\ln r \right.$$
$$\left. + 2\left(\xi_{i1e}\,\xi_{i4e} - \xi_{i2e}\,\xi_{i3e}\right)\frac{\ln r}{r} + 2\,\xi_{i2e}\,\xi_{i4e}\frac{\ln r}{r^2} - 2\,\xi_{i3e}\,\xi_{i4e}\frac{\ln^2 r}{r}\right],$$
$$i = 1, 2, 3, \tag{7.178}$$

where the elastic modulus s_{11e} and the coefficients $\xi_{i1e}, \ldots, \xi_{i4e}$ ($i = 1,2,3$) are given by Eqs. (3.23) and (7.162), (7.188), (7.164), (7.189), respectively.

Appendix. Considering the elastic modulus s_{44e} and the coefficients c_{1e}, c_{2e}, c_{3e}, a'_{1i}, a'_{2i}, a'_{3i} ($i = 1,2,3$) (see Eqs. (3.25), (3.38)–(3.40), (12.26)–(12.34)), the coefficients $\chi_{1+i\,2e}$, $\varrho_{1+i\,1+je}$ ($i, j = 0,1$), ω_{2e}, ω_{3e}, $\omega_{6e}, \ldots, \omega_{9e}$, Ω_e, ξ_{i2e}, ξ_{i4e} ($i = 1,2,3$) have the forms

$$\chi_{1+i\,2e} = \frac{c_{1e} - 2c_{2e}\ln R_{1+i}}{2R_{1+i}}, \quad i = 0, 1, \tag{7.179}$$

$$\varrho_{1+i\,1+je} = \left(\delta_{0j} - \delta_{1j}\right)\left[\frac{\kappa_{1+je}^{(1)}}{3}\left(\frac{4}{3} - \ln R_{1+i}\right) + \frac{\kappa_{3+je}^{(1)}}{2R_{1+i}}\left(\frac{1}{2} + \ln R_{1+i}\right)\right],$$
$$i, j = 0, 1, \tag{7.180}$$

$$\omega_{2e} = \frac{1}{9}\left(\left[p_1\kappa_{1e}^{(1)} - p_2\kappa_{2e}^{(1)}\right]\left[p_1\kappa_{3e}^{(1)} - p_2\kappa_{4e}^{(1)}\right](5c_{1e} - 8c_{2e})\right.$$
$$+ \frac{4}{s_{44e}}\left\{\left[\kappa_{1e}^{(1)}\frac{\partial p_1}{\partial\varphi} - \kappa_{2e}^{(1)}\frac{\partial p_2}{\partial\varphi}\right]\left[\kappa_{3e}^{(1)}\frac{\partial p_1}{\partial\varphi} - \kappa_{4e}^{(1)}\frac{\partial p_2}{\partial\varphi}\right]\right.$$
$$\left.\left. + \left[\kappa_{1e}^{(1)}\frac{\partial p_1}{\partial\nu} - \kappa_{2e}^{(1)}\frac{\partial p_2}{\partial\nu}\right]\left[\kappa_{3e}^{(1)}\frac{\partial p_1}{\partial\nu} - \kappa_{4e}^{(1)}\frac{\partial p_2}{\partial\nu}\right]\right\}\right), \tag{7.181}$$

$$\omega_{3e} = \frac{1}{8}\left(\left[p_1\kappa_{3e}^{(1)} - p_2\kappa_{4e}^{(1)}\right]^2(3c_{1e} - 2c_{2e})\right.$$
$$\left. + \frac{1}{s_{44e}}\left\{\left[\kappa_{3e}^{(1)}\frac{\partial p_1}{\partial\varphi} - \kappa_{4e}^{(1)}\frac{\partial p_2}{\partial\varphi}\right]^2 + \left[\kappa_{3e}^{(1)}\frac{\partial p_1}{\partial\nu} - \kappa_{4e}^{(1)}\frac{\partial p_2}{\partial\nu}\right]^2\right\}\right), \tag{7.182}$$

$$\omega_{6e} = \frac{1}{9}\left(2\left[p_1\kappa_{1e}^{(1)} - p_2\kappa_{2e}^{(1)}\right]\left[p_1\kappa_{3e}^{(1)} - p_2\kappa_{4e}^{(1)}\right](c_{1e} + 2c_{2e}) + \frac{5}{s_{44e}}\left\{\left[\kappa_{1e}^{(1)}\frac{\partial p_1}{\partial\varphi} - \kappa_{2e}^{(1)}\frac{\partial p_2}{\partial\varphi}\right]\left[\kappa_{3e}^{(1)}\frac{\partial p_1}{\partial\varphi} - \kappa_{4e}^{(1)}\frac{\partial p_2}{\partial\varphi}\right] + \left[\kappa_{1e}^{(1)}\frac{\partial p_1}{\partial\nu} - \kappa_{2e}^{(1)}\frac{\partial p_2}{\partial\nu}\right]\left[\kappa_{3e}^{(1)}\frac{\partial p_1}{\partial\nu} - \kappa_{4e}^{(1)}\frac{\partial p_2}{\partial\nu}\right]\right\}\right), \quad (7.183)$$

$$\omega_{7e} = \frac{1}{2}\left(\left[p_1\kappa_{3e}^{(1)} - p_2\kappa_{4e}^{(1)}\right]^2(c_{1e} - 2c_{2e}) + \frac{1}{s_{44e}}\left\{\left[\kappa_{3e}^{(1)}\frac{\partial p_1}{\partial\varphi} - \kappa_{4e}^{(1)}\frac{\partial p_2}{\partial\varphi}\right]^2 + \left[\kappa_{3e}^{(1)}\frac{\partial p_1}{\partial\nu} - \kappa_{4e}^{(1)}\frac{\partial p_2}{\partial\nu}\right]^2\right\}\right), \quad (7.184)$$

$$\omega_{8e} = -\frac{2}{3}\left(\left[p_1\kappa_{1e}^{(1)} - p_2\kappa_{2e}^{(1)}\right]\left[p_1\kappa_{3e}^{(1)} - p_2\kappa_{4e}^{(1)}\right](c_{1e} - c_{2e}) + \frac{1}{s_{44e}}\left\{\left[\kappa_{1e}^{(1)}\frac{\partial p_1}{\partial\varphi} - \kappa_{2e}^{(1)}\frac{\partial p_2}{\partial\varphi}\right]\left[\kappa_{3e}^{(1)}\frac{\partial p_1}{\partial\varphi} - \kappa_{4e}^{(1)}\frac{\partial p_2}{\partial\varphi}\right] + \left[\kappa_{1e}^{(1)}\frac{\partial p_1}{\partial\nu} - \kappa_{2e}^{(1)}\frac{\partial p_2}{\partial\nu}\right]\left[\kappa_{3e}^{(1)}\frac{\partial p_1}{\partial\nu} - \kappa_{4e}^{(1)}\frac{\partial p_2}{\partial\nu}\right]\right\}\right), \quad (7.185)$$

$$\omega_{9e} = \frac{1}{2}\left(\left[p_1\kappa_{3e}^{(1)} - p_2\kappa_{4e}^{(1)}\right]^2 c_{1e} + \frac{1}{s_{44e}}\left\{\left[\kappa_{3e}^{(1)}\frac{\partial p_1}{\partial\varphi} - \kappa_{4e}^{(1)}\frac{\partial p_2}{\partial\varphi}\right]^2 + \left[\kappa_{3e}^{(1)}\frac{\partial p_1}{\partial\nu} - \kappa_{4e}^{(1)}\frac{\partial p_2}{\partial\nu}\right]^2\right\}\right), \quad (7.186)$$

$$\begin{aligned}\Omega_e = 4\Bigg(&\frac{\omega_{1e}}{3}\left(R_2^3 - R_1^3\right) + \frac{\omega_{2e}}{2}\left(r_c^2 - R_1^2\right)\\
&+ \omega_{3e}\left(R_2 - R_1\right) + \frac{\omega_{4e}}{3}\left[R_2^3\left(\ln R_2 - \frac{1}{3}\right) - R_1^3\left(\ln R_1 - \frac{1}{3}\right)\right]\\
&+ \frac{\omega_{5e}}{3}\left\{R_2^3\left[\left(\ln R_2 - \frac{1}{3}\right)^2 + \frac{1}{9}\right] - R_1^3\left[\left(\ln R_1 - \frac{1}{3}\right)^2 + \frac{1}{9}\right]\right\}\\
&+ \frac{\omega_{6e}}{2}\left[R_2^2\left(\ln R_2 - \frac{1}{2}\right) - R_1^2\left(\ln R_1 - \frac{1}{2}\right)\right]\\
&+ \omega_{7e}\left[R_2\left(\ln R_2 - 1\right) - R_1\left(\ln R_1 - 1\right)\right]\\
&+ \frac{\omega_{8e}}{2}\left\{R_2^2\left[\left(\ln R_2 - \frac{1}{2}\right)^2 + \frac{1}{4}\right] - R_1^2\left[\left(\ln R_1 - \frac{1}{2}\right)^2 + \frac{1}{4}\right]\right\}\\
&+ \omega_{9e}\left\{R_2\left[\left(\ln R_2 - 1\right)^2 + 1\right] - R_1\left[\left(\ln R_1 - 1\right)^2 + 1\right]\right\}\Bigg),\\
&i, j = 0, 1,\end{aligned} \quad (7.187)$$

$$\xi_{i2e} = \frac{1}{4}\left(\left[p_1\kappa_{3e}^{(1)} - p_2\kappa_{4e}^{(1)}\right]\left[2a'_{1i}\,c_{1e} + \left(a'_{2i} + a'_{3i}\right)\left(c_{1e} - 2c_{2e}\right)\right]\right.$$
$$+ \frac{1}{s_{44e}}\left\{\left(a'_{1i} + a'_{2i}\right)\left[\kappa_{3e}^{(1)}\frac{\partial p_1}{\partial\varphi} - \kappa_{4e}^{(1)}\frac{\partial p_2}{\partial\varphi}\right]\right.$$
$$\left.\left. + \left(a'_{1i} + a'_{3i}\right)\left[\kappa_{3e}^{(1)}\frac{\partial p_1}{\partial\nu} - \kappa_{4e}^{(1)}\frac{\partial p_2}{\partial\nu}\right]\right\}\right), \quad i = 1,2,3, \tag{7.188}$$

$$\xi_{i4e} = \frac{1}{2}\left(\left[p_1\kappa_{3e}^{(1)} - p_2\kappa_{4e}^{(1)}\right]\left[c_{1e}\left(a'_{2i} + a'_{3i}\right) - 2a'_{1i}\,c_{2e}\right]\right.$$
$$+ \frac{1}{s_{44e}}\left\{\left(a'_{1i} + a'_{2i}\right)\left[\kappa_{3e}^{(1)}\frac{\partial p_1}{\partial\varphi} - \kappa_{4e}^{(1)}\frac{\partial p_2}{\partial\varphi}\right]\right.$$
$$\left.\left. + \left(a'_{1i} + a'_{3i}\right)\left[\kappa_{3e}^{(1)}\frac{\partial p_1}{\partial\nu} - \kappa_{4e}^{(1)}\frac{\partial p_2}{\partial\nu}\right]\right\}\right), \quad i = 1,2,3. \tag{7.189}$$

C. Integration constants $C_{1e} = 0$, $C_{2e} \neq 0$, $C_{3e} \neq 0$

With regard to the boundary conditions (4.7), (4.8) and the analysis presented in Section 6.2.2, considering the terms u_{2re} and u_{3re} (see Eqs. (6.7), (7.6)) to represent functions of the variable $r \in \langle R_1, R_2\rangle$, the radial displacement u_{1e} of the infinitesimal spherical cap in the spherical envelope, regarding Eqs. (3.31), (7.3), has the form

$$u_{1e} = -\left\{\left[p_1\kappa_{1e}^{(1)} - p_2\kappa_{2e}^{(1)}\right]C_e^{(2)}\,r^{c_{3e}} + \frac{1}{2}\left[p_1\kappa_{3e}^{(1)} - p_2\kappa_{4e}^{(1)}\right]\left(\frac{1}{2} + \ln r\right)\right\}, \tag{7.190}$$

where the coefficients $c_{3e} < 0$, $C_e^{(2)}$ and the radial stresses p_1, p_2 acting on the particle-envelope, matrix-envelope boundaries are given by Eqs. (3.40), (7.4) and (10.2), (10.3), respectively.

With regard to Section 7.5.1.1.C, the coefficients $\varrho_{1+i\;1+je}$ $(i,j=0,1)$, $\omega_{1e},\ldots,\omega_{6e}$, ξ_{i1e}, ξ_{i2e}, ξ_{i3e} $(i=1,2,3)$ mentioned below (see Eqs. (7.199), (7.200), (7.202), (7.203)) depend on $\kappa_{1e}^{(1)},\ldots,\kappa_{4e}^{(1)}$ (see Eq. (12.35)) representing functions of χ_{11e}, χ_{12e}, χ_{21e}, χ_{22e} given by Eqs. (7.179), (7.204).

Consequently, the radial stress σ_{11e}, the tangential and shear stresses σ_{22e}, σ_{33e} and σ_{12e}, σ_{13e} acting in the spherical envelope, respectively, regarding Eqs. (3.31)–(3.34), (7.190), are derived as

$$\sigma_{11e} = -\left\{\left[p_1\kappa_{1e}^{(1)} - p_2\kappa_{2e}^{(1)}\right]C_e^{(2)}\left[c_{3e}\left(c_{1e} + c_{2e}\right) - 2c_{2e}\right]r^{c_{3e}-1}\right.$$
$$\left. + \left[p_1\kappa_{3e}^{(1)} - p_2\kappa_{4e}^{(1)}\right]\frac{1}{2r}\left(c_{1e} - 2c_{2e}\ln r\right)\right\}, \tag{7.191}$$

$$\sigma_{22e} = \sigma_{33e} = -\left\{\left[p_1\kappa_{1e}^{(1)} - p_2\kappa_{2e}^{(1)}\right]C_e^{(2)}\left(c_{1e} - c_{2e}c_{3e}\right)r^{c_{3e}-1}\right.$$
$$\left. + \left[p_1\kappa_{3e}^{(1)} - p_2\kappa_{4e}^{(1)}\right]\frac{1}{2r}\left[c_{1e}\left(\frac{1}{2} + \ln r\right) - c_{2e}\right]\right\}, \tag{7.192}$$

$$\sigma_{12e} = -\frac{1}{s_{44e}}\left\{\left[\kappa_{1e}^{(1)}\frac{\partial p_1}{\partial \varphi} - \kappa_{2e}^{(1)}\frac{\partial p_2}{\partial \varphi}\right] C_e^{(2)}\, r^{c_{3e}-1} + \left[\kappa_{3e}^{(1)}\frac{\partial p_1}{\partial \varphi} - \kappa_{4e}^{(1)}\frac{\partial p_2}{\partial \varphi}\right]\frac{1}{2r}\left(\frac{1}{2} + \ln r\right)\right\}, \tag{7.193}$$

$$\sigma_{13e} = -\frac{1}{s_{44e}}\left\{\left[\kappa_{1e}^{(1)}\frac{\partial p_1}{\partial \nu} - \kappa_{2e}^{(1)}\frac{\partial p_2}{\partial \nu}\right] C_e^{(2)}\, r^{c_{3e}-1} + \left[\kappa_{3e}^{(1)}\frac{\partial p_1}{\partial \nu} - \kappa_{4e}^{(1)}\frac{\partial p_2}{\partial \nu}\right]\frac{1}{2r}\left(\frac{1}{2} + \ln r\right)\right\}, \tag{7.194}$$

where the elastic modulus s_{44e} and the coefficients c_{1e}, c_{2e} are given by Eqs. (3.25) and (3.38), (3.39), respectively.

Finally, the radial strain ε_{11e}, the tangential and shear strains ε_{22e}, ε_{33e} and ε_{12e}, ε_{13e} in the spherical envelope, respectively, regarding Eqs. (3.26)–(3.30), (7.190), have the forms

$$\varepsilon_{11e} = -\left\{\left[p_1\kappa_{1e}^{(1)} - p_2\kappa_{2e}^{(1)}\right] C_e^{(2)} c_{3e}\, r^{c_{3e}-1} + \frac{1}{2r}\left[p_1\kappa_{3e}^{(1)} - p_2\kappa_{4e}^{(1)}\right]\right\}, \tag{7.195}$$

$$\varepsilon_{22e} = \varepsilon_{33e} = -\left\{\left[p_1\kappa_{1e}^{(1)} - p_2\kappa_{2e}^{(1)}\right] C_e^{(2)}\, r^{c_{3e}-1} + \left[p_1\kappa_{3e}^{(1)} - p_2\kappa_{4e}^{(1)}\right]\frac{1}{2r}\left(\frac{1}{2} + \ln r\right)\right\}, \tag{7.196}$$

$$\varepsilon_{12e} = -\left\{\left[\kappa_{1e}^{(1)}\frac{\partial p_1}{\partial \varphi} - \kappa_{2e}^{(1)}\frac{\partial p_2}{\partial \varphi}\right] C_e^{(2)}\, r^{c_{3e}-1} + \left[\kappa_{3e}^{(1)}\frac{\partial p_1}{\partial \varphi} - \kappa_{4e}^{(1)}\frac{\partial p_2}{\partial \varphi}\right]\frac{1}{2r}\left(\frac{1}{2} + \ln r\right)\right\}, \tag{7.197}$$

$$\varepsilon_{13e} = -\left\{\left[\kappa_{1e}^{(1)}\frac{\partial p_1}{\partial \nu} - \kappa_{2e}^{(1)}\frac{\partial p_2}{\partial \nu}\right] C_e^{(2)}\, r^{c_{3e}-1} + \left[\kappa_{3e}^{(1)}\frac{\partial p_1}{\partial \nu} - \kappa_{4e}^{(1)}\frac{\partial p_2}{\partial \nu}\right]\frac{1}{2r}\left(\frac{1}{2} + \ln r\right)\right\}. \tag{7.198}$$

With regard to the determination of the compressive or tensile radial stresses $p_1 > 0$ or $p_1 < 0$ and $p_2 > 0$ or $p_2 < 0$ acting on the particle-envelope and matrix-envelope boundaries, thus for $r = R_1$ and $r = R_2$, respectively, considering Eqs. (3.59), (3.60), the tangential strains on the particle-envelope and matrix-envelope boundaries, $(\varepsilon_{22e})_{r=R_1}$ and $(\varepsilon_{22e})_{r=R_2}$, respectively, are required to be derived in the form

$$(\varepsilon_{22e})_{r=R_{1+i}} = -\sum_{j=0}^{1} p_{1+j}\varrho_{1+i\,1+je}, \quad i=0,1, \tag{7.199}$$

where the coefficient $\varrho_{1+i\,1+je}$ $(i,j=0,1)$ is given by Eq. (7.205).

Considering Eqs. (3.41), (3.42), (7.191)–(7.198), the thermal-stress induced elastic energy density w_e and the thermal-stress induced elastic energy W_e, accumulated in an arbitrary point of the spherical envelope and in the volume of the spherical envelope, respectively, have the forms

$$w_e = \frac{1}{2}\left[\frac{\omega_{1e}}{r^2} + \omega_{2e}\, r^{c_{3e}-2} + \omega_{3e}\, r^{2(c_{3e}-1)} + \omega_{4e}\frac{\ln r}{r^2} + \omega_{5e}\, r^{c_{3e}-2}\ln r + \omega_{6e}\frac{\ln^2 r}{r^2}\right], \tag{7.200}$$

$$W_e = 8\int_0^{\pi/2}\int_0^{\pi/2}\int_{R_1}^{R_2} w_e\, r^2 dr d\varphi d\nu = \int_0^{\pi/2}\int_0^{\pi/2} \Omega_e d\varphi d\nu, \tag{7.201}$$

where the coefficient Ω_e given by Eq. (7.212) is a function of $\omega_{1e},\ldots,\omega_{6e}$ given by Eqs. (7.206)–(7.211), respectively.

With regard to the determination of the thermal-stress induced strengthening presented in Volume III, the stress σ_{ie} acting in a direction of the axis x_i and the elastic energy density w_{ie} induced by the stress σ_{ie} $(i=1,2,3)$ (see Eqs. (3.46), (3.47), (7.191)–(7.194)), both related to an arbitrary point of the spherical envelope, are derived as

$$\sigma_{ie} = -\left(\frac{\xi_{i1e}}{r} + \xi_{i2e}\, r^{c_{3e}-1} + \xi_{i3e}\frac{\ln r}{r}\right), \quad i=1,2,3, \tag{7.202}$$

$$w_{ie} = \frac{s_{11e}}{2}\left[\frac{\xi_{i1e}^2}{r^2} + \xi_{i2e}^2\, r^{2(c_{3e}-1)} + \xi_{i3e}^2\frac{\ln^2 r}{r^2} + 2\,\xi_{i1e}\,\xi_{i2e}\, r^{c_{3e}-2} + 2\,\xi_{i1e}\,\xi_{i3e}\frac{\ln r}{r^2} + 2\,\xi_{i2e}\,\xi_{i3e}\, r^{c_{3e}-2}\ln r\right], \quad i=1,2,3, \tag{7.203}$$

where the elastic modulus s_{11e} and the coefficients ξ_{i1e}, ξ_{i2e}, ξ_{i3e} $(i=1,2,3)$ are given by Eqs. (3.23) and (7.213)–(7.215), respectively.

Appendix. Considering the elastic modulus s_{44e} and the coefficients c_{1e}, c_{2e}, c_{3e}, $C_e^{(2)}$, a'_{1i}, a'_{2i}, a'_{3i} $(i=1,2,3)$ (see Eqs. (3.25), (3.38)–(3.40), (7.4), (12.26)–(12.34)), the coefficients $\chi_{1+i\,1e}$, $\varrho_{1+i\,1+je}$ $(i,j=0,1)$, $\omega_{1e},\ldots,\omega_{6e}$, Ω_e, $\xi_{i\,1e}$, $\xi_{i\,2e}$, $\xi_{i\,3e}$ $(i=1,2,3)$ have the forms

$$\chi_{1+i\,1e} = C_e^{(2)} \left[c_{3e}\left(c_{1e}+c_{2e}\right)-2c_{2e}\right] R_{1+i}^{c_{3e}-1}, \quad i=0,1, \tag{7.204}$$

$$\varrho_{\,1+i\;1+je} = \left(\delta_{0j}-\delta_{1j}\right)\left[\kappa_{1+je}^{(1)} C_e^{(2)} R_{1+i}^{c_{3e}-1} + \frac{\kappa_{3+je}^{(1)}}{2R_{1+i}}\left(\frac{1}{2}+\ln R_{1+i}\right)\right], \quad i,j=0,1, \tag{7.205}$$

$$\omega_{1e} = \frac{1}{8}\left(\left[p_1\kappa_{3e}^{(1)} - p_2\kappa_{4e}^{(1)}\right]^2 \left(3c_{1e}-2c_{2e}\right) + \frac{1}{s_{44e}}\left\{\left[\kappa_{3e}^{(1)}\frac{\partial p_1}{\partial\varphi} - \kappa_{4e}^{(1)}\frac{\partial p_2}{\partial\varphi}\right]^2 + \left[\kappa_{3e}^{(1)}\frac{\partial p_1}{\partial\nu} - \kappa_{4e}^{(1)}\frac{\partial p_2}{\partial\nu}\right]^2\right\}\right), \tag{7.206}$$

$$\begin{aligned}\omega_{2e} = C_e^{(2)}\Bigg(&\left[p_1\kappa_{1e}^{(1)} - p_2\kappa_{2e}^{(1)}\right]\left[p_1\kappa_{3e}^{(1)} - p_2\kappa_{4e}^{(1)}\right]\left[c_{1e}\left(c_{3e}+1\right)-2c_{2e}\right] \\ &+ \frac{1}{s_{44e}}\left\{\left[\kappa_{1e}^{(1)}\frac{\partial p_1}{\partial\varphi} - \kappa_{2e}^{(1)}\frac{\partial p_2}{\partial\varphi}\right]\left[\kappa_{3e}^{(1)}\frac{\partial p_1}{\partial\varphi} - \kappa_{4e}^{(1)}\frac{\partial p_2}{\partial\varphi}\right]\right. \\ &\left.+ \left[\kappa_{1e}^{(1)}\frac{\partial p_1}{\partial\nu} - \kappa_{2e}^{(1)}\frac{\partial p_2}{\partial\nu}\right]\left[\kappa_{3e}^{(1)}\frac{\partial p_1}{\partial\nu} - \kappa_{4e}^{(1)}\frac{\partial p_2}{\partial\nu}\right]\right\}\Bigg),\end{aligned} \tag{7.207}$$

$$\omega_{3e} = \left[C_e^{(2)}\right]^2\left(\left[p_1\kappa_{1e}^{(1)} - p_2\kappa_{2e}^{(1)}\right]^2\left[c_{3e}^2\left(c_{1e}+c_{2e}\right)+2\left(c_{1e}-2c_{2e}c_{3e}\right)\right] + \frac{2}{s_{44e}}\left\{\left[\kappa_{1e}^{(1)}\frac{\partial p_1}{\partial\varphi} - \kappa_{2e}^{(1)}\frac{\partial p_2}{\partial\varphi}\right]^2 + \left[\kappa_{1e}^{(1)}\frac{\partial p_1}{\partial\nu} - \kappa_{2e}^{(1)}\frac{\partial p_2}{\partial\nu}\right]^2\right\}\right), \tag{7.208}$$

$$\omega_{4e} = \frac{1}{2}\left(\left[p_1\kappa_{3e}^{(1)} - p_2\kappa_{4e}^{(1)}\right]^2\left(c_{1e}-2c_{2e}\right) + \frac{1}{s_{44e}}\left\{\left[\kappa_{3e}^{(1)}\frac{\partial p_1}{\partial\varphi} - \kappa_{4e}^{(1)}\frac{\partial p_2}{\partial\varphi}\right]^2 + \left[\kappa_{3e}^{(1)}\frac{\partial p_1}{\partial\nu} - \kappa_{4e}^{(1)}\frac{\partial p_2}{\partial\nu}\right]^2\right\}\right), \tag{7.209}$$

$$\begin{aligned}\omega_{5e} = 2C_e^{(2)}\Bigg(&\left[p_1\kappa_{1e}^{(1)} - p_2\kappa_{2e}^{(1)}\right]\left[p_1\kappa_{3e}^{(1)} - p_2\kappa_{4e}^{(1)}\right]\left(c_{1e}-c_{2e}c_{3e}\right) \\ &+ \frac{1}{s_{44e}}\left\{\left[\kappa_{1e}^{(1)}\frac{\partial p_1}{\partial\varphi} - \kappa_{2e}^{(1)}\frac{\partial p_2}{\partial\varphi}\right]\left[\kappa_{3e}^{(1)}\frac{\partial p_1}{\partial\varphi} - \kappa_{4e}^{(1)}\frac{\partial p_2}{\partial\varphi}\right]\right. \\ &\left.+ \left[\kappa_{1e}^{(1)}\frac{\partial p_1}{\partial\nu} - \kappa_{2e}^{(1)}\frac{\partial p_2}{\partial\nu}\right]\left[\kappa_{3e}^{(1)}\frac{\partial p_1}{\partial\nu} - \kappa_{4e}^{(1)}\frac{\partial p_2}{\partial\nu}\right]\right\}\Bigg),\end{aligned} \tag{7.210}$$

$$\omega_{6e} = \frac{1}{2}\left(\left[p_1\kappa_{3e}^{(1)} - p_2\kappa_{4e}^{(1)}\right]^2 c_{1e} + \frac{1}{s_{44e}}\left\{\left[\kappa_{3e}^{(1)}\frac{\partial p_1}{\partial \varphi} - \kappa_{4e}^{(1)}\frac{\partial p_2}{\partial \varphi}\right]^2 + \left[\kappa_{3e}^{(1)}\frac{\partial p_1}{\partial \nu} - \kappa_{4e}^{(1)}\frac{\partial p_2}{\partial \nu}\right]^2\right\}\right), \quad (7.211)$$

$$\begin{aligned}\Omega_e = 4\Bigg(&\omega_{1e}\left(R_2 - R_1\right) + \frac{\omega_{2e}}{c_{3e}+1}\left(R_2^{c_{3e}+1} - R_1^{c_{3e}+1}\right) \\ &+ \frac{\omega_{3e}}{2c_{3e}+1}\left(R_2^{2c_{3e}+1} - R_1^{2c_{3e}+1}\right) + \omega_{4e}\left[R_2\left(\ln R_2 - 1\right) - R_1\left(\ln R_1 - 1\right)\right] \\ &+ \frac{\omega_{5e}}{c_{3e}+1}\left[R_2^{c_{3e}+1}\left(\ln R_2 - \frac{1}{c_{3e}+1}\right) - R_1^{c_{3e}+1}\left(\ln R_1 - \frac{1}{c_{3e}+1}\right)\right] \\ &+ \omega_{6e}\left\{R_2\left[\left(\ln R_2 - 1\right)^2 + 1\right] - R_1\left[\left(\ln R_1 - 1\right)^2 + 1\right]\right\}\Bigg), \end{aligned} \quad (7.212)$$

$$\begin{aligned}\xi_{i1e} = \frac{1}{4}\Bigg(&\left[p_1\kappa_{3e}^{(1)} - p_2\kappa_{4e}^{(1)}\right]\left[2a'_{1i}c_{1e} + \left(a'_{2i} + a'_{3i}\right)\left(c_{1e} - 2c_{2e}\right)\right] \\ &+ \frac{1}{s_{44e}}\left\{\left(a'_{1i} + a'_{2i}\right)\left[\kappa_{3e}^{(1)}\frac{\partial p_1}{\partial \varphi} - \kappa_{4e}^{(1)}\frac{\partial p_2}{\partial \varphi}\right]\right. \\ &\left.+ \left(a'_{1i} + a'_{3i}\right)\left[\kappa_{3e}^{(1)}\frac{\partial p_1}{\partial \nu} - \kappa_{4e}^{(1)}\frac{\partial p_2}{\partial \nu}\right]\right\}\Bigg), \quad i = 1, 2, 3, \end{aligned} \quad (7.213)$$

$$\begin{aligned}\xi_{i2e} = C_e^{(2)}\Bigg(&\left[p_1\kappa_{1e}^{(1)} - p_2\kappa_{2e}^{(1)}\right]\left\{a'_{1i}\left[c_{3e}\left(c_{1e} + c_{2e}\right) - 2c_{2e}\right] + \left(a'_{2i} + a'_{3i}\right)\left(c_{1e} - c_{2e}c_{3e}\right)\right\} \\ &+ \frac{1}{s_{44e}}\left\{\left(a'_{1i} + a'_{2i}\right)\left[\kappa_{1e}^{(1)}\frac{\partial p_1}{\partial \varphi} - \kappa_{2e}^{(1)}\frac{\partial p_2}{\partial \varphi}\right]\right. \\ &\left.+ \left(a'_{1i} + a'_{3i}\right)\left[\kappa_{1e}^{(1)}\frac{\partial p_1}{\partial \nu} - \kappa_{2e}^{(1)}\frac{\partial p_2}{\partial \nu}\right]\right\}\Bigg), \quad i = 1, 2, 3, \end{aligned} \quad (7.214)$$

$$\begin{aligned}\xi_{i3e} = \frac{1}{2}\Bigg(&\left[p_1\kappa_{3e}^{(1)} - p_2\kappa_{4e}^{(1)}\right]\left[c_{1e}\left(a'_{2i} + a'_{3i}\right) - 2a'_{1i}c_{2e}\right] \\ &+ \frac{1}{s_{44e}}\left\{\left(a'_{1i} + a'_{2i}\right)\left[\kappa_{3e}^{(1)}\frac{\partial p_1}{\partial \varphi} - \kappa_{4e}^{(1)}\frac{\partial p_2}{\partial \varphi}\right]\right. \\ &\left.+ \left(a'_{1i} + a'_{3i}\right)\left[\kappa_{3e}^{(1)}\frac{\partial p_1}{\partial \nu} - \kappa_{4e}^{(1)}\frac{\partial p_2}{\partial \nu}\right]\right\}\Bigg), \quad i = 1, 2, 3. \end{aligned} \quad (7.215)$$

7.5.1.2 Conditions $\beta_p \neq \beta_e = \beta_m$ and $\varepsilon_{11te} - \varepsilon_{11tp} = f(\varphi, \nu)$

The thermal stresses and the corresponding quantities presented in Section 7.5.1.2 and related to Eq. (6.4) are derived on the condition $\beta_p \neq \beta_e = \beta_m$, considering the φ, ν-dependent difference $\varepsilon_{11te} - \varepsilon_{11tp} = f(\varphi, \nu)$ (see Eq. (12.23)), resulting in the

φ, ν-dependent difference $\beta_e - \beta_p$ (see Eqs. (3.48)–(3.50)) and consequently in the φ, ν-dependent radial stress $p_1 = (\beta_e - \beta_p) / (\varrho_{1p} - \varrho_{1e}^{me})$ (see Eq. (10.4)) acting on the particle-envelope boundary, derived from Eq. (3.63) and included in the formulae in Section 7.5.1.2, where the φ, ν-independent coefficients ϱ_{1p} and ϱ_{1e}^{me} are given by Eqs. (5.36) and (5.106), (6.207), (6.228), (6.246), (7.230), respectively, considering the Castigliano's theorem (see Section 3.1.9).

With regard to the boundary condition (4.7) and the analysis presented in Section 6.2.2, considering the term u_{2re} (see Eq. (6.7)) to represent a decreasing function of the variable $r \in \langle R_1, R_2 \rangle$, the thermal stresses and the corresponding quantities for the spherical envelope in the multi-particle-envelope-matrix system, provided that $\beta_p \neq \beta_e = \beta_m$ and $\varepsilon_{11te} - \varepsilon_{11tp} = f(\varphi, \nu)$, and regarding the integration constants $C_{2e} \neq 0$, $C_{1e} = C_{3e} = 0$, are presented in the part B of Section 6.4.1.2 (see Eqs. (6.214)–(6.231)).

A. Integration constants $C_{1e} \neq 0$, $C_{2e} = C_{3e} = 0$

With regard to the boundary condition (4.7) and the analyses presented in Sections 6.2.2, 7.3, considering the term u_{1re} (see Eq. (7.5)) to represent a decreasing function of the variable $r \in \langle R_1, R_2 \rangle$ provided that $R_1 > R \approx 1.39561$ (see Eqs. (7.8), (7.14)), the radial displacement u_{1e} of the infinitesimal spherical cap in the spherical envelope, regarding Eqs. (3.31), (7.3), has the form

$$u_{1e} = -\frac{p_1 r (4 - 3 \ln r)}{c_{1e} - 7c_{2e} - 3(c_{1e} - c_{2e}) \ln R_1}, \tag{7.216}$$

where the coefficients c_{1e}, c_{2e} and the radial stress p_1 acting on the particle-envelope boundary are given by Eqs. (3.38), (3.39) and (10.4), respectively.

Against the extreme value of the radius $R_1 \geq e^{1/3} \approx 1.39561$ (see Eq. (7.14)) in comparison with radii of particles in real composite materials [1], regarding completeness of the presented topic, the thermal stresses and the corresponding quantities in the spherical envelope related to the term u_{1re} are presented as a curiosity only.

Consequently, the radial stress σ_{11e}, the tangential and shear stresses σ_{22e}, σ_{33e} and σ_{12e}, σ_{13e} acting in the spherical envelope, respectively, regarding Eqs. (3.31)–(3.34), (7.216), are derived as

$$\sigma_{11e} = -\frac{p_1 [c_{1e} - 7c_{2e} - 3(c_{1e} - c_{2e}) \ln r]}{c_{1e} - 7c_{2e} - 3(c_{1e} - c_{2e}) \ln R_1}, \tag{7.217}$$

$$\sigma_{22e} = \sigma_{33e} = -\frac{p_1 [4c_{1e} - c_{2e} - 3(c_{1e} - c_{2e}) \ln r]}{c_{1e} - 7c_{2e} - 3(c_{1e} - c_{2e}) \ln R_1}, \tag{7.218}$$

$$\sigma_{12e} = -\frac{4 - 3 \ln r}{s_{44e} [c_{1e} - 7c_{2e} - 3(c_{1e} - c_{2e}) \ln R_1]} \frac{\partial p_1}{\partial \varphi}, \tag{7.219}$$

$$\sigma_{13e} = -\frac{4 - 3\ln r}{s_{44e}\left[c_{1e} - 7c_{2e} - 3\left(c_{1e} - c_{2e}\right)\ln R_1\right]}\frac{\partial p_1}{\partial \nu}, \tag{7.220}$$

where the elastic modulus s_{44e} is given by Eq. (3.25).

Finally, the radial strain ε_{11e}, the tangential and shear strains ε_{22e}, ε_{33e} and ε_{12e}, ε_{13e} in the spherical envelope, respectively, regarding Eqs. (3.26)–(3.30), (7.216), have the forms

$$\varepsilon_{11e} = -\frac{p_1\left(1 - 3\ln r\right)}{c_{1e} - 7c_{2e} - 3\left(c_{1e} - c_{2e}\right)\ln R_1}, \tag{7.221}$$

$$\varepsilon_{22e} = \varepsilon_{33e} = -\frac{p_1\left(4 - 3\ln r\right)}{c_{1e} - 7c_{2e} - 3\left(c_{1e} - c_{2e}\right)\ln R_1}, \tag{7.222}$$

$$\varepsilon_{12e} = -\frac{4 - 3\ln r}{c_{1e} - 7c_{2e} - 3\left(c_{1e} - c_{2e}\right)\ln R_1}\frac{\partial p_1}{\partial \varphi}, \tag{7.223}$$

$$\varepsilon_{13e} = -\frac{4 - 3\ln r}{c_{1e} - 7c_{2e} - 3\left(c_{1e} - c_{2e}\right)\ln R_1}\frac{\partial p_1}{\partial \nu}. \tag{7.224}$$

With regard to the determination of the compressive or tensile radial stress $p_1 > 0$ or $p_1 < 0$ acting on the particle-matrix boundary, thus for $r = R_1$, respectively, considering Eq. (3.59), and to the boundary condition for the cell matrix provided that $\beta_p \neq \beta_e = \beta_m$, the tangential strains on the particle-envelope and matrix-envelope boundaries, $\left(\varepsilon_{22e}\right)_{r=R_1}$ and $\left(\varepsilon_{22e}\right)_{r=R_2}$, respectively, are required to be derived in the form

$$\left(\varepsilon_{22e}\right)_{r=R_{1+i}} = -p_1\,\varrho_{1+ie}^{me}, \quad i = 0, 1, \tag{7.225}$$

where the coefficient ϱ_{1+ie}^{me} $(i = 0,1)$ is given by Eq. (7.230).

Considering Eqs. (3.41), (3.42), (7.217)–(7.224), the thermal-stress induced elastic energy density w_e and the thermal-stress induced elastic energy W_e, accumulated in an arbitrary point of the spherical envelope and in the volume of the spherical envelope, respectively, have the forms

$$w_e = \frac{\omega_{1e} + \omega_{2e}\ln r + \omega_{3e}\ln^2 r}{2\left[c_{1e} - 7c_{2e} - 3\left(c_{1e} - c_{2e}\right)\ln R_1\right]^2}, \tag{7.226}$$

$$W_e = 8\int_0^{\pi/2}\int_0^{\pi/2}\int_{R_1}^{R_2} w_e\, r^2 dr d\varphi d\nu = \int_0^{\pi/2}\int_0^{\pi/2}\Omega_{1e} d\varphi d\nu, \tag{7.227}$$

where the coefficient Ω_{1e} given by Eq. (7.234) is a function of ω_{1e}, ω_{2e}, ω_{3e} given by Eqs. (7.231)–(7.233), respectively.

With regard to the determination of the thermal-stress induced strengthening presented in Volume III, the stress σ_{ie} acting in a direction of the axis x_i and the elastic energy density w_{ie} induced by the stress σ_{ie} $(i=1,2,3)$ (see Eqs. (3.46), (3.47), (7.217)–(7.220)), both related to an arbitrary point of the spherical envelope, are derived as

$$\sigma_{ie} = -\frac{\xi_{i1e} - \xi_{i2e}\ln r}{c_{1e} - 7c_{2e} - 3\left(c_{1e} - c_{2e}\right)\ln R_1}, \quad i = 1, 2, 3, \tag{7.228}$$

$$w_{ie} = \frac{s_{11e}\left(\xi_{i1e}^2 + \xi_{i2e}^2\ln^2 r - 2\,\xi_{i1e}\xi_{i2e}\ln r\right)}{2\left[c_{1e} - 7c_{2e} - 3\left(c_{1e} - c_{2e}\right)\ln R_1\right]^2}, \quad i = 1, 2, 3, \tag{7.229}$$

where the elastic modulus s_{11e} and the coefficients ξ_{i1e}, ξ_{i2e} $(i=1,2,3)$ are given by Eqs. (3.23) and (7.235), (7.236), respectively.

Appendix. Considering the elastic modulus s_{44e} and the coefficients c_{1e}, c_{2e}, a'_{1i}, a'_{2i}, a'_{3i} $(i=1,2,3)$ (see Eqs. (3.25), (3.38), (3.39), (12.26)–(12.34)), the coefficients ϱ^{me}_{1+ie}, ω_{1+3ie}, ω_{2+3ie}, ω_{3+3ie}, Ω_{1+ie} $(i=0,1)$, ξ_{i1+2je}, ξ_{i2+2je} $(i=1,2,3;$ $j=0,1)$ have the forms

$$\varrho^{me}_{1+ie} = \frac{4 - 3\ln R_{1+i}}{c_{1e} - 7c_{2e} - 3\left(c_{1e} - c_{2e}\right)\ln R_1}, \quad i = 0, 1, \tag{7.230}$$

$$\omega_{1+3ie} = 3p_{1+i}^2\left(11c_{1e} - 5c_{2e}\right) + \frac{16}{s_{44e}}\left[\left(\frac{\partial p_{1+i}}{\partial \varphi}\right)^2 + \left(\frac{\partial p_{1+i}}{\partial \nu}\right)^2\right], \quad i = 0, 1, \tag{7.231}$$

$$\omega_{2+3ie} = -6\left\{9p_{1+i}^2\left(c_{1e} - c_{2e}\right) + \frac{4}{s_{44e}}\left[\left(\frac{\partial p_{1+i}}{\partial \varphi}\right)^2 + \left(\frac{\partial p_{1+i}}{\partial \nu}\right)^2\right]\right\},$$
$$i = 0, 1, \tag{7.232}$$

$$\omega_{3+3ie} = 9\left\{3p_{1+i}^2\left(c_{1e} - c_{2e}\right) + \frac{1}{s_{44e}}\left[\left(\frac{\partial p_{1+i}}{\partial \varphi}\right)^2 + \left(\frac{\partial p_{1+i}}{\partial \nu}\right)^2\right]\right\},$$
$$i = 0, 1, \tag{7.233}$$

$$\Omega_{1+ie} = \left[\frac{2}{c_{1e} - 7c_{2e} - 3\left(c_{1e} - c_{2e}\right)\ln R_{1+i}}\right]^2$$
$$\times\left(\frac{\omega_{1+3ie}}{3}\left(R_2^3 - R_1^3\right) + \frac{\omega_{2+3ie}}{3}\left[R_2^3\left(\ln R_2 - \frac{1}{3}\right) - R_1^3\left(\ln R_1 - \frac{1}{3}\right)\right]\right.$$
$$\left. + \frac{\omega_{3+3ie}}{3}\left\{R_2^3\left[\left(\ln R_2 - \frac{1}{3}\right)^2 + \frac{1}{9}\right] - R_1^3\left[\left(\ln R_1 - \frac{1}{3}\right)^2 + \frac{1}{9}\right]\right\}\right),$$
$$i = 0, 1, \tag{7.234}$$

$$\xi_{i\,1+2je} = p_{1+j}\left[a'_{1i}\left(c_{1e} - 7c_{2e}\right) + \left(a'_{2i} + a'_{3i}\right)\left(4c_{1e} - c_{2e}\right)\right]$$
$$+ \frac{4}{s_{44e}}\left[\left(a'_{1i} + a'_{2i}\right)\frac{\partial p_{1+j}}{\partial \varphi} + \left(a'_{1i} + a'_{3i}\right)\frac{\partial p_{1+j}}{\partial \nu}\right],$$
$$i = 1, 2, 3; \quad j = 0, 1, \tag{7.235}$$

$$\xi_{i\,2+2je} = 3\left\{p_{1+j}\left(a'_{1i} + a'_{2i} + a'_{3i}\right)\left(c_{1e} - c_{2e}\right)\right.$$
$$\left. + \frac{1}{s_{44e}}\left[\left(a'_{1i} + a'_{2i}\right)\frac{\partial p_{1+j}}{\partial \varphi} + \left(a'_{1i} + a'_{3i}\right)\frac{\partial p_{1+j}}{\partial \nu}\right]\right\},$$
$$i = 1, 2, 3; \quad j = 0, 1. \tag{7.236}$$

7.5.1.3 Condition $\beta_p = \beta_e \neq \beta_m$

A. Integration constants $C_{1e} \neq 0$, $C_{2e} = C_{3e} = 0$

With regard to the boundary condition (4.8) and the analyses presented in Sections 6.2.2, 7.3, considering the term u_{1re} (see Eq. (7.5)) to represent an increasing function of the variable $r \in \langle R_1, R_2 \rangle$ provided that $R_2 < R = e^{1/3} \approx 1.39561$ (see Eq. (7.15)), as fulfilled for the radii $R_1 < R_2$ of envelopes on the surface of particles in real composite materials [1], the radial displacement u_{1e} of the infinitesimal spherical cap in the spherical envelope, regarding Eqs. (3.31), (7.3), has the form

$$u_{1e} = -\frac{p_2\, r\left(4 - 3\ln r\right)}{c_{1e} - 7c_{2e} - 3\left(c_{1e} - c_{2e}\right)\ln R_2}, \tag{7.237}$$

where the coefficients c_{1e}, c_{2e} and the radial stress p_2 acting on the matrix-envelope boundary are given by Eqs. (3.38), (3.39) and (10.5), respectively.

Consequently, the radial stress σ_{11e}, the tangential and shear stresses σ_{22e}, σ_{33e} and σ_{12e}, σ_{13e} acting in the spherical envelope, respectively, regarding Eqs. (3.31)–(3.34), (7.237), are derived as

$$\sigma_{11e} = -\frac{p_2\left[c_{1e} - 7c_{2e} - 3\left(c_{1e} - c_{2e}\right)\ln r\right]}{c_{1e} - 7c_{2e} - 3\left(c_{1e} - c_{2e}\right)\ln R_2}, \tag{7.238}$$

$$\sigma_{22e} = \sigma_{33e} = -\frac{p_2\left[4c_{1e} - c_{2e} - 3\left(c_{1e} - c_{2e}\right)\ln r\right]}{c_{1e} - 7c_{2e} - 3\left(c_{1e} - c_{2e}\right)\ln R_2}, \tag{7.239}$$

$$\sigma_{12e} = -\frac{4 - 3\ln r}{s_{44e}\left[c_{1e} - 7c_{2e} - 3\left(c_{1e} - c_{2e}\right)\ln R_2\right]}\frac{\partial p_2}{\partial \varphi}, \tag{7.240}$$

$$\sigma_{13e} = -\frac{4 - 3\ln r}{s_{44e}\left[c_{1e} - 7c_{2e} - 3\left(c_{1e} - c_{2e}\right)\ln R_2\right]}\frac{\partial p_2}{\partial \nu}, \tag{7.241}$$

where the elastic modulus s_{44e} is given by Eq. (3.25).

Finally, the radial strain ε_{11e}, the tangential and shear strains ε_{22e}, ε_{33e} and ε_{12e}, ε_{13e} in the spherical envelope, respectively, regarding Eqs. (3.26)–(3.30), (7.237), have the forms

$$\varepsilon_{11e} = -\frac{p_2\,(1 - 3\ln r)}{c_{1e} - 7c_{2e} - 3\,(c_{1e} - c_{2e})\ln R_2}, \tag{7.242}$$

$$\varepsilon_{22e} = \varepsilon_{33e} = -\frac{p_2\,(4 - 3\ln r)}{c_{1e} - 7c_{2e} - 3\,(c_{1e} - c_{2e})\ln R_2}, \tag{7.243}$$

$$\varepsilon_{12e} = -\frac{4 - 3\ln r}{c_{1e} - 7c_{2e} - 3\,(c_{1e} - c_{2e})\ln R_2}\,\frac{\partial p_2}{\partial \varphi}, \tag{7.244}$$

$$\varepsilon_{13e} = -\frac{4 - 3\ln r}{c_{1e} - 7c_{2e} - 3\,(c_{1e} - c_{2e})\ln R_2}\,\frac{\partial p_2}{\partial \nu}. \tag{7.245}$$

With regard to the determination of the compressive or tensile radial stress $p_2 > 0$ or $p_2 < 0$ acting on the matrix-envelope boundary, thus for $r = R_2$, respectively, considering Eq. (3.60), and to the boundary condition for the spherical particle provided that $\beta_p = \beta_e \neq \beta_m$, the tangential strains on the particle-envelope and matrix-envelope boundaries, $(\varepsilon_{22e})_{r=R_1}$ and $(\varepsilon_{22e})_{r=R_2}$, respectively, are required to be derived in the form

$$(\varepsilon_{22e})_{r=R_{1+i}} = -p_2\,\varrho^{pe}_{1+ie}, \quad i = 0, 1, \tag{7.246}$$

where the coefficient ϱ^{pe}_{1+ie} ($i = 0,1$) is given by Eq. (7.251).

Considering Eqs. (3.41), (3.42), (7.238)–(7.245), the thermal-stress induced elastic energy density w_e and the thermal-stress induced elastic energy W_e, accumulated in an arbitrary point of the spherical envelope and in the volume of the spherical envelope, respectively, have the forms

$$w_e = \frac{\omega_{4e} + \omega_{5e}\ln r + \omega_{6e}\ln^2 r}{2\,[c_{1e} - 7c_{2e} - 3\,(c_{1e} - c_{2e})\ln R_2]^2}, \tag{7.247}$$

$$W_e = 8\int_0^{\pi/2}\int_0^{\pi/2}\int_{R_1}^{R_2} w_e\,r^2 dr d\varphi d\nu = \int_0^{\pi/2}\int_0^{\pi/2} \Omega_{2e} d\varphi d\nu, \tag{7.248}$$

where the coefficient Ω_{2e} given by Eq. (7.234) is a function of ω_{4e}, ω_{5e}, ω_{6e} given by Eqs. (7.231)–(7.233), respectively.

With regard to the determination of the thermal-stress induced strengthening presented in Volume III, the stress σ_{ie} acting in a direction of the axis x_i and the elastic energy density w_{ie} induced by the stress σ_{ie} ($i = 1,2,3$) (see Eqs. (3.46), (3.47),

(7.238)–(7.241)), both related to an arbitrary point of the spherical envelope, are derived as

$$\sigma_{ie} = - \frac{\xi_{i3e} - \xi_{i4e} \ln r}{c_{1e} - 7c_{2e} - 3\left(c_{1e} - c_{2e}\right) \ln R_2}, \quad i = 1, 2, 3, \tag{7.249}$$

$$w_{ie} = \frac{s_{11e}\left(\xi_{i3e}^2 + \xi_{i4e}^2 \ln^2 r - 2\,\xi_{i3e}\,\xi_{i4e} \ln r\right)}{2\left[c_{1e} - 7c_{2e} - 3\left(c_{1e} - c_{2e}\right) \ln R_2\right]^2}, \quad i = 1, 2, 3, \tag{7.250}$$

where the elastic modulus s_{11e} and the coefficients ξ_{i3e}, ξ_{i4e} $(i=1,2,3)$ are given by Eqs. (3.23) and (7.235), (7.236), respectively.

Appendix. Considering the coefficients c_{1e}, c_{2e} (see Eq. (3.38), (3.39)), the coefficient ϱ_{1+ie}^{pe} $(i=0,1)$ has the form

$$\varrho_{1+ie}^{pe} = \frac{4 - 3\ln R_{1+i}}{c_{1e} - 7c_{2e} - 3\left(c_{1e} - c_{2e}\right) \ln R_2}, \quad i = 0, 1. \tag{7.251}$$

B. Integration constants $C_{3e} \neq 0$, $C_{1e} = C_{2e} = 0$

With regard to the boundary condition (4.8) and the analyses presented in Sections 6.2.2, 7.3, considering the term u_{3re} (see Eq. (7.6)) to represent an increasing function of the variable $r \in \langle R_1, R_2 \rangle$, the radial displacement u_{1e} of the infinitesimal spherical cap in the spherical envelope, regarding Eqs. (3.31), (7.3), has the form

$$u_{1e} = - \frac{p_2 R_2 \left(1 + 2\ln r\right)}{2\left(c_{1e} - 2c_{2e} \ln R_2\right)}, \tag{7.252}$$

where the coefficients c_{1e}, c_{2e} and the radial stress p_2 acting on the matrix-envelope boundary are given by Eqs. (3.38), (3.39) and (10.5), respectively.

Consequently, the radial stress σ_{11e}, the tangential and shear stresses σ_{22e}, σ_{33e} and σ_{12e}, σ_{13e} acting in the spherical envelope, respectively, regarding Eqs. (3.31)–(3.34), (7.252), are derived as

$$\sigma_{11e} = - \frac{p_2 R_2 \left(c_{1e} - 2c_{2e} \ln r\right)}{r\left(c_{1e} - 2c_{2e} \ln R_2\right)}, \tag{7.253}$$

$$\sigma_{22e} = \sigma_{33e} = - \frac{p_2 R_2 \left[c_{1e}\left(1 + 2\ln r\right) - 2c_{2e}\right]}{2r\left(c_{1e} - 2c_{2e} \ln R_2\right)}, \tag{7.254}$$

$$\sigma_{12e} = - \frac{R_2\left(1 + 2\ln r\right)}{2s_{44e} r\left(c_{1e} - 2c_{2e} \ln R_2\right)} \frac{\partial p_2}{\partial \varphi}, \tag{7.255}$$

$$\sigma_{13e} = - \frac{R_2\left(1 + 2\ln r\right)}{2s_{44e} r\left(c_{1e} - 2c_{2e} \ln R_2\right)} \frac{\partial p_2}{\partial \nu}, \tag{7.256}$$

where the elastic modulus s_{44e} is given by Eq. (3.25).

Finally, the radial strain ε_{11e}, the tangential and shear strains ε_{22e}, ε_{33e} and ε_{12e}, ε_{13e} in the spherical envelope, respectively, regarding Eqs. (3.26)–(3.30), (7.252), have the forms

$$\varepsilon_{11e} = -\frac{p_2 R_2}{r\left(c_{1e} - 2c_{2e}\ln R_2\right)}, \tag{7.257}$$

$$\varepsilon_{22e} = \varepsilon_{33e} = -\frac{p_2 R_2 \left(1 + 2\ln r\right)}{2r\left(c_{1e} - 2c_{2e}\ln R_2\right)}, \tag{7.258}$$

$$\varepsilon_{12e} = -\frac{R_2 \left(1 + 2\ln r\right)}{2r\left(c_{1e} - 2c_{2e}\ln R_2\right)}\frac{\partial p_2}{\partial \varphi}, \tag{7.259}$$

$$\varepsilon_{13e} = -\frac{R_2 \left(1 + 2\ln r\right)}{2r\left(c_{1e} - 2c_{2e}\ln R_2\right)}\frac{\partial p_2}{\partial \nu}. \tag{7.260}$$

With regard to the determination of the compressive or tensile radial stress $p_2 > 0$ or $p_2 < 0$ acting on the matrix-envelope boundary, thus for $r = R_2$, respectively, considering Eq. (3.60), and to the boundary condition for the spherical particle provided that $\beta_p = \beta_e \neq \beta_m$, the tangential strains on the particle-envelope and matrix-envelope boundaries, $(\varepsilon_{22e})_{r=R_1}$ and $(\varepsilon_{22e})_{r=R_2}$, respectively, are required to be derived in the form

$$(\varepsilon_{22e})_{r=R_{1+i}} = -p_2\, \varrho_{1+ie}^{pe}, \quad i = 0, 1, \tag{7.261}$$

where the coefficient ϱ_{1+ie}^{pe} $(i=0,1)$ is given by Eq. (7.266).

Considering Eqs. (3.41), (3.42), (7.253)–(7.260), the thermal-stress induced elastic energy density w_e and the thermal-stress induced elastic energy W_e, accumulated in an arbitrary point of the spherical envelope and in the volume of the spherical envelope, respectively, have the forms

$$w_e = \left(\frac{R_2}{c_{1e} - 2c_{2e}\ln R_2}\right)^2 \left(\frac{\omega_{1e}}{r^2} + \omega_{2e}\frac{\ln r}{r^2} + \omega_{3e}\frac{\ln^2 r}{r^2}\right), \tag{7.262}$$

$$W_e = 8\int_0^{\pi/2}\int_0^{\pi/2}\int_{R_1}^{R_2} w_e\, r^2 dr d\varphi d\nu = \int_0^{\pi/2}\int_0^{\pi/2} \Omega_e d\varphi d\nu, \tag{7.263}$$

where the coefficient Ω_e given by Eq. (7.270) is a function of ω_{1e}, ω_{2e}, ω_{3e} given by Eqs. (7.267)–(7.269), respectively.

With regard to the determination of the thermal-stress induced strengthening presented in Volume III, the stress σ_{ie} acting in a direction of the axis x_i and the elastic energy density w_{ie} induced by the stress σ_{ie} $(i=1,2,3)$ (see Eqs. (3.46), (3.47), (7.253)–(7.256)), both related to an arbitrary point of the spherical envelope, are derived as

$$\sigma_{ie} = - \frac{R_2}{2\left(c_{1e} - 2c_{2e}\ln R_2\right)}\left(\frac{\xi_{i1e}}{r} + \xi_{i2e}\frac{\ln r}{r}\right), \quad i = 1,2,3, \tag{7.264}$$

$$w_{ie} = \frac{s_{11e}}{8}\left(\frac{R_2}{c_{1e} - 2c_{2e}\ln R_2}\right)^2\left(\frac{\xi_{i1e}^2}{r^2} + \xi_{i2e}^2\frac{\ln^2 r}{r^2} + 2\,\xi_{i1e}\,\xi_{i2e}\frac{\ln r}{r^2}\right), \quad i = 1,2,3, \tag{7.265}$$

where the elastic modulus s_{11e} and the coefficients ξ_{i1e}, ξ_{i2e} ($i=1,2,3$) is given by Eqs. (3.23) and (7.271), (7.272), respectively.

Appendix. Considering the coefficients c_{1e}, c_{2e}, a'_{1i}, a'_{2i}, a'_{3i} ($i=1,2,3$) (see Eqs. (3.38), (3.39), (12.26)–(12.34)), the coefficients ϱ^{pe}_{1+ie} ($i=0,1$), ω_{1e}, ω_{2e}, ω_{3e}, Ω_e, ξ_{i1e}, ξ_{i2e} ($i=1,2,3$) have the forms

$$\varrho^{pe}_{1+ie} = \frac{R_2\left(1 + 2\ln R_{1+i}\right)}{2R_{1+i}\left(c_{1e} - 2c_{2e}\ln R_2\right)}, \quad i = 0,1, \tag{7.266}$$

$$\omega_{1e} = \frac{1}{2}\left\{p_2^2\left(3c_{1e} - 2c_{2e}\right) + \frac{1}{s_{44e}}\left[\left(\frac{\partial p_2}{\partial \varphi}\right)^2 + \left(\frac{\partial p_2}{\partial \nu}\right)^2\right]\right\}, \tag{7.267}$$

$$\omega_{2e} = 2\left\{p_2^2\left(c_{1e} - 2c_{2e}\right) + \frac{1}{s_{44e}}\left[\left(\frac{\partial p_2}{\partial \varphi}\right)^2 + \left(\frac{\partial p_2}{\partial \nu}\right)^2\right]\right\}, \tag{7.268}$$

$$\omega_{3e} = 2\left\{p_2^2 c_{1e} + \frac{1}{s_{44e}}\left[\left(\frac{\partial p_2}{\partial \varphi}\right)^2 + \left(\frac{\partial p_2}{\partial \nu}\right)^2\right]\right\}, \tag{7.269}$$

$$\Omega_e = 8\left(\frac{R_2}{c_{1e} - 2c_{2e}\ln R_2}\right)^2\left(\omega_{1e}\left(R_2 - R_1\right) + \omega_{2e}\left[R_2\left(\ln R_2 - 1\right) - R_1\left(\ln R_1 - 1\right)\right] + \omega_{3e}\left\{R_2\left[\left(\ln R_2 - 1\right)^2 + 1\right] - R_1\left[\left(\ln R_1 - 1\right)^2 + 1\right]\right\}\right), \tag{7.270}$$

$$\xi_{i1e} = p_2\left[2a'_{1i}c_{1e} + \left(a'_{2i} + a'_{3i}\right)\left(c_{1e} - 2c_{2e}\right)\right] + \frac{1}{s_{44e}}\left[\left(a'_{1i} + a'_{2i}\right)\frac{\partial p_2}{\partial \varphi} + \left(a'_{1i} + a'_{3i}\right)\frac{\partial p_2}{\partial \nu}\right], \quad i = 1,2,3, \tag{7.271}$$

$$\xi_{i2e} = 2\left\{p_2\left[c_{1e}\left(a'_{2i} + a'_{3i}\right) - 2a'_{1i}c_{2e}\right] + \frac{1}{s_{44e}}\left[\left(a'_{1i} + a'_{2i}\right)\frac{\partial p_2}{\partial \varphi} + \left(a'_{1i} + a'_{3i}\right)\frac{\partial p_2}{\partial \nu}\right]\right\}, \quad i = 1,2,3. \tag{7.272}$$

7.5.2 Thermal stresses in cell matrix

7.5.2.1 Conditions $\beta_p \neq \beta_e \neq \beta_m$ or $\beta_p = \beta_e \neq \beta_m$

The thermal stresses and the corresponding quantities in the cell matrix presented in Section 7.5.2.1 are derived regarding the conditions $\beta_p \neq \beta_e \neq \beta_m$ or $\beta_p = \beta_e \neq \beta_m$ (see Eqs. (3.48)–(3.50)).

A. Integration constants $C_{1m} \neq 0$, $C_{2m} \neq 0$, $C_{3m} = 0$

With regard to the boundary conditions (4.10), (4.11) and the analysis presented in Section 6.2.3, considering the terms u_{1rm} and u_{2rm} (see Eqs. (7.5), (6.7)) to represent functions of the variable $r \in \langle R_2, r_c \rangle$, the radial displacement u_{1m} of the infinitesimal spherical cap in the cell matrix, regarding Eqs. (3.31), (7.3), has the form

$$u_{1m} = -p_2 \left[\frac{\kappa_{1m}^{(1)} r}{3} \left(\frac{4}{3} - \ln r \right) + \kappa_{3m}^{(1)} C_m \, r^{c_{3m}} \right], \tag{7.273}$$

where the coefficients $c_{3m} < 0$ and $C_m = C_m^{(i)}$ $(i = 2{,}3)$ are given by Eqs. (3.40) and (7.4), (7.13), respectively. Provided that $\beta_p \neq \beta_e \neq \beta_m$ or $\beta_p = \beta_e \neq \beta_m$, the radial stress p_2 acting on the matrix-envelope boundary is given by Eqs. (10.3) or (10.5), respectively. The φ, ν-dependent distance r_c, as a function of the experimentally determined parameters v, R_{1ex}, d_{1ex}, d_{2ex}, d_{3ex} (see Section 2.2.1), included in the coefficients $\kappa_{1m}^{(1)}$, $\kappa_{3m}^{(1)}$ is derived in Section 2.3.

With regard to Section 7.5.2.1.A, the coefficients $\psi_{21\eta m}^{(1)}$, $\psi_{22\eta m}^{(1)}$ $(\eta = \varphi, \nu)$, ϱ_{2m} and $\omega_{21m}, \ldots, \omega_{26m}$, ξ_{i21m}, ξ_{i22m}, ξ_{i23m} $(i = 1{,}2{,}3)$ mentioned below (see Eqs. (7.276), (7.277), (7.280)–(7.283), (7.285), (7.286)) depend on $\kappa_{1m}^{(1)}$, $\kappa_{3m}^{(1)}$ and $\kappa_{1m}^{(1)}$, $\kappa_{3m}^{(1)}$, $\psi_{21\eta m}^{(1)}$, $\psi_{22\eta m}^{(1)}$, respectively, where the coefficients $\kappa_{1m}^{(1)}$, $\kappa_{3m}^{(1)}$ (see Eq. (12.35)) are functions of χ_{11m}, χ_{12m}, χ_{21m}, χ_{22m} given by Eqs. (7.287), (7.288), (7.32), (7.33), respectively.

Consequently, the radial stress σ_{11m}, the tangential and shear stresses σ_{22m}, σ_{33m} and σ_{12m}, σ_{13m} acting in the cell matrix, respectively, regarding Eqs. (3.31)–(3.34), (7.273), are derived as

$$\sigma_{11m} = -p_2 \left\{ \frac{\kappa_{1m}^{(1)}}{3} \left[\frac{1}{3} (c_{1m} - 7c_{2m}) - (c_{1m} - c_{2m}) \ln r \right] \right. \\ \left. + \kappa_{3m}^{(1)} C_m \left[c_{3m} (c_{1m} + c_{2m}) - 2c_{2m} \right] r^{c_{3m}-1} \right\}, \tag{7.274}$$

$$\sigma_{22m} = \sigma_{33m} = -p_2 \left\{ \frac{\kappa_{1m}^{(1)}}{3} \left[\frac{1}{3} (4c_{1m} - c_{2m}) - (c_{1m} - c_{2m}) \ln r \right] \right. \\ \left. + \kappa_{3m}^{(1)} C_m \left(c_{1m} - c_{2m} c_{3m} \right) r^{c_{3m}-1} \right\}, \tag{7.275}$$

$$\sigma_{12m} = -\frac{1}{s_{44m}}\left[\frac{\psi^{(1)}_{21\varphi m}}{3}\left(\frac{4}{3} - \ln r\right) + \psi^{(1)}_{22\varphi m} C_m\, r^{c_{3m}-1}\right], \tag{7.276}$$

$$\sigma_{13m} = -\frac{1}{s_{44m}}\left[\frac{\psi^{(1)}_{21\nu m}}{3}\left(\frac{4}{3} - \ln r\right) + \psi^{(1)}_{22\nu m} C_m\, r^{c_{3m}-1}\right], \tag{7.277}$$

where the elastic modulus s_{44m} and the coefficients c_{1m}, c_{2m}, $\psi^{(1)}_{21\eta m}$, $\psi^{(1)}_{22\eta m}$ $(\eta = \varphi, \nu)$ are given by Eqs. (3.25) and (3.38), (3.39), (12.36), respectively.

Finally, the radial strain ε_{11m}, the tangential and shear strains ε_{22m}, ε_{33m} and ε_{12m}, ε_{13m} in the cell matrix, respectively, regarding Eqs. (3.1)–(3.5), (7.273), have the forms

$$\varepsilon_{11m} = -p_2\left[\frac{\kappa^{(1)}_{1m}}{3}\left(\frac{1}{3} - \ln r\right) + \kappa^{(1)}_{3m} C_m c_{3m}\, r^{c_{3m}-1}\right], \tag{7.278}$$

$$\varepsilon_{22m} = \varepsilon_{33m} = -p_2\left[\frac{\kappa^{(1)}_{1m}}{3}\left(\frac{4}{3} - \ln r\right) + \kappa^{(1)}_{3m} C_m\, r^{c_{3m}-1}\right], \tag{7.279}$$

$$\varepsilon_{12m} = -\left[\frac{\psi^{(1)}_{21\varphi m}}{3}\left(\frac{4}{3} - \ln r\right) + \psi^{(1)}_{22\varphi m} C_m\, r^{c_{3m}-1}\right], \tag{7.280}$$

$$\varepsilon_{13m} = -\left[\frac{\psi^{(1)}_{21\nu m}}{3}\left(\frac{4}{3} - \ln r\right) + \psi^{(1)}_{22\nu m} C_m\, r^{c_{3m}-1}\right]. \tag{7.281}$$

With regard to the determination of the compressive or tensile radial stress $p_2 > 0$ or $p_2 < 0$ acting on the matrix-envelope boundary, thus for $r = R_2$, respectively, considering Eq. (3.58), the tangential strain on the matrix-envelope boundary, $(\varepsilon_{22m})_{r=R_2}$, is required to be derived in the form

$$(\varepsilon_{22m})_{r=R_2} = -p_2\, \varrho_{2m}, \tag{7.282}$$

where the coefficient ϱ_{2m} is given by Eq. (7.34).

Considering Eqs. (3.41), (3.42), (7.274)–(7.281), the thermal-stress induced elastic energy density w_m and the thermal-stress induced elastic energy W_m, accumulated in an arbitrary point of the cell matrix and in the volume of the cell matrix, respectively, have the forms

$$w_m = \frac{1}{2}\Big[\omega_{21m} + \omega_{22m}\, r^{c_{3m}-1} + \omega_{23m}\, r^{2(c_{3m}-1)} + \omega_{24m} \ln r + \omega_{25m} \ln^2 r + \omega_{26m}\, r^{c_{3m}-1} \ln r\Big], \tag{7.283}$$

$$W_m = 8\int_0^{\pi/2}\int_0^{\pi/2}\int_{R_2}^{r_c} w_m\, r^2 dr d\varphi d\nu = \int_0^{\pi/2}\int_0^{\pi/2} \Omega_{22m} d\varphi d\nu, \tag{7.284}$$

where the coefficient Ω_{22m} given by Eq. (6.39) is a function of $\omega_{21m}, \ldots, \omega_{26m}$ given by Eqs. (7.35)–(7.40), respectively.

With regard to the determination of the thermal-stress induced strengthening presented in Volume III, the stress σ_{im} acting in a direction of the axis x_i and the elastic energy density w_{im} induced by the stress σ_{im} ($i = 1,2,3$) (see Eqs. (3.46), (3.47), (7.274)–(7.277)), both related to an arbitrary point of the cell matrix, are derived as

$$\sigma_{im} = -\left(\xi_{i21m} + \xi_{i22m}\, r^{c_{3m}-1} - \xi_{i23m} \ln r\right), \quad i = 1, 2, 3, \tag{7.285}$$

$$\begin{aligned} w_{im} = \frac{s_{11m}}{2} \Big[& \xi_{i21m}^2 + \xi_{i22m}^2\, r^{2(c_{3m}-1)} + \xi_{i23m}^2 \ln^2 r + 2\,\xi_{i21m}\,\xi_{i22m}\, r^{c_{3m}-1} \\ & - 2\,\xi_{i21m}\,\xi_{i23m} \ln r - 2\,\xi_{i22m}\,\xi_{i23m}\, r^{c_{3m}-1} \ln r \Big], \\ i = 1, 2, 3, \end{aligned} \tag{7.286}$$

where the elastic modulus s_{11m} and the coefficients ξ_{i21m}, ξ_{i22m}, ξ_{i23m} ($i = 1,2,3$) are given by Eqs. (3.23) and (7.41)–(7.43), respectively.

Appendix. Considering the coefficients c_{1m}, c_{2m}, c_{3m}, $C_m = C_m^{(i)}$ ($i = 2,3$) (see Eqs. (3.38)–(3.40), (7.4), (7.13), the coefficients χ_{11m}, χ_{12m} have the forms

$$\chi_{11m} = \frac{1}{3}\left[\frac{1}{3}\left(c_{1m} - 7c_{2m}\right) - \left(c_{1m} - c_{2m}\right) \ln R_2\right], \tag{7.287}$$

$$\chi_{12m} = C_m \left[c_{3m}\left(c_{1m} + c_{2m}\right) - 2c_{2m}\right] R_2^{c_{3m}-1}. \tag{7.288}$$

B. Integration constants $C_{1m} \neq 0$, $C_{2m} = 0$, $C_{3m} \neq 0$

With regard to the boundary conditions (4.10), (4.11) and the analysis presented in Section 6.2.3, considering the terms u_{1rm} and u_{3rm} (see Eqs. (7.5), (7.6)) to represent functions functions of the variable $r \in \langle R_2, r_c \rangle$, the radial displacement u_{1m} of the infinitesimal spherical cap in the cell matrix, regarding Eqs. (3.31), (7.3), has the form

$$u_{1m} = -p_2 \left[\frac{\kappa_{1m}^{(1)}\, r}{3}\left(\frac{4}{3} - \ln r\right) + \frac{\kappa_{3m}^{(1)}}{2}\left(\frac{1}{2} + \ln r\right)\right]. \tag{7.289}$$

Provided that $\beta_p \neq \beta_e \neq \beta_m$ or $\beta_p = \beta_e \neq \beta_m$, the radial stress p_2 acting on the matrix-envelope boundary is given by Eqs. (10.3) or (10.5), respectively. The φ, ν-dependent distance r_c, as a function of the experimentally determined parameters v, R_{1ex}, d_{1ex}, d_{2ex}, d_{3ex} (see Section 2.2.1), included in the coefficients $\kappa_{1m}^{(1)}$, $\kappa_{3m}^{(1)}$ is derived in Section 2.3.

With regard to Section 7.5.2.1.B, the coefficients $\psi_{21\eta m}^{(1)}$, $\psi_{22\eta m}^{(1)}$ $(\eta = \varphi, \nu)$, ϱ_{2m} and $\omega_{21m}, \ldots, \omega_{29m}$, $\xi_{i21m}, \ldots, \xi_{i24m}$ $(i = 1,2,3)$ mentioned below (see Eqs. (7.292), (7.293), (7.296)–(7.299), (7.301), (7.302)) depend on $\kappa_{1m}^{(1)}$, $\kappa_{3m}^{(1)}$ and $\kappa_{1m}^{(1)}$, $\kappa_{3m}^{(1)}$, $\psi_{21\eta m}^{(1)}$, $\psi_{22\eta m}^{(1)}$, respectively, where the coefficients $\kappa_{1m}^{(1)}$, $\kappa_{3m}^{(1)}$ (see Eq. (12.35)) are functions of χ_{11m}, χ_{12m}, χ_{21m}, χ_{22m} given by Eqs. (7.287), (7.303), (7.32), (7.59), respectively.

Consequently, the radial stress σ_{11m}, the tangential and shear stresses σ_{22m}, σ_{33m} and σ_{12m}, σ_{13m} acting in the cell matrix, respectively, regarding Eqs. (3.31)–(3.34), (7.289), are derived as

$$\sigma_{11m} = -p_2 \left\{ \frac{\kappa_{1m}^{(1)}}{3} \left[\frac{1}{3} (c_{1m} - 7c_{2m}) - (c_{1m} - c_{2m}) \ln r \right] + \frac{\kappa_{3m}^{(1)}}{2r} (c_{1m} - 2c_{2m} \ln r) \right\}, \tag{7.290}$$

$$\sigma_{22m} = \sigma_{33m} = -p_2 \left\{ \frac{\kappa_{1m}^{(1)}}{3} \left[\frac{1}{3} (4c_{1m} - c_{2m}) - (c_{1m} - c_{2m}) \ln r \right] + \frac{\kappa_{3m}^{(1)}}{2r} \left[c_{1m} \left(\frac{1}{2} + \ln r \right) - c_{2m} \right] \right\}, \tag{7.291}$$

$$\sigma_{12m} = -\frac{1}{s_{44m}} \left[\frac{\psi_{21\varphi m}^{(1)}}{3} \left(\frac{4}{3} - \ln r \right) + \frac{\psi_{22\varphi m}^{(1)}}{2r} \left(\frac{1}{2} + \ln r \right) \right], \tag{7.292}$$

$$\sigma_{13m} = -\frac{1}{s_{44m}} \left[\frac{\psi_{21\nu m}^{(1)}}{3} \left(\frac{4}{3} - \ln r \right) + \frac{\psi_{22\nu m}^{(1)}}{2r} \left(\frac{1}{2} + \ln r \right) \right], \tag{7.293}$$

where the elastic modulus s_{44m} and the coefficients c_{1m}, c_{2m}, $\psi_{21\eta m}^{(1)}$, $\psi_{22\eta m}^{(1)}$ $(\eta = \varphi, \nu)$ are given by Eqs. (3.25) and (3.38), (3.39), (12.36), respectively.

Finally, the radial strain ε_{11m}, the tangential and shear strains ε_{22m}, ε_{33m} and ε_{12m}, ε_{13m} in the cell matrix, respectively, regarding Eqs. (3.1)–(3.5), (7.289), have the forms

$$\varepsilon_{11m} = -p_2 \left[\frac{\kappa_{1m}^{(1)}}{3} \left(\frac{1}{3} - \ln r \right) + \frac{\kappa_{3m}^{(1)}}{2r} \right], \tag{7.294}$$

$$\varepsilon_{22m} = \varepsilon_{33m} = -p_2 \left[\frac{\kappa_{1m}^{(1)}}{3} \left(\frac{4}{3} - \ln r \right) + \frac{\kappa_{3m}^{(1)}}{2r} \left(\frac{1}{2} + \ln r \right) \right], \tag{7.295}$$

$$\varepsilon_{12m} = -\left[\frac{\psi_{21\varphi m}^{(1)}}{3} \left(\frac{4}{3} - \ln r \right) + \frac{\psi_{22\varphi m}^{(1)}}{2r} \left(\frac{1}{2} + \ln r \right) \right], \tag{7.296}$$

$$\varepsilon_{13m} = -\left[\frac{\psi_{21\nu m}^{(1)}}{3}\left(\frac{4}{3} - \ln r\right) + \frac{\psi_{22\nu m}^{(1)}}{2r}\left(\frac{1}{2} + \ln r\right)\right]. \tag{7.297}$$

With regard to the determination of the compressive or tensile radial stress $p_2 > 0$ or $p_2 < 0$ acting on the matrix-envelope boundary, thus for $r = R_2$, respectively, considering Eq. (3.58), the tangential strain on the matrix-envelope boundary, $(\varepsilon_{22m})_{r=R_2}$, is required to be derived in the form

$$(\varepsilon_{22m})_{r=R_2} = -p_2\, \varrho_{2m}, \tag{7.298}$$

where the coefficient ϱ_{2m} is given by Eq. (7.60).

Considering Eqs. (3.41), (3.42), (7.290)–(7.297), the thermal-stress induced elastic energy density w_m and the thermal-stress induced elastic energy W_m, accumulated in an arbitrary point of the cell matrix and in the volume of the cell matrix, respectively, have the forms

$$w_m = \frac{1}{2}\left(\omega_{21m} + \frac{\omega_{22m}}{r} + \frac{\omega_{23m}}{r^2} + \omega_{24m}\ln r + \omega_{25m}\ln^2 r + \omega_{26m}\frac{\ln r}{r} + \omega_{27m}\frac{\ln r}{r^2} + \omega_{28m}\frac{\ln^2 r}{r} + \omega_{29m}\frac{\ln^2 r}{r^2}\right), \tag{7.299}$$

$$W_m = 8\int_0^{\pi/2}\int_0^{\pi/2}\int_{R_2}^{r_c} w_m r^2 dr d\varphi d\nu = \int_0^{\pi/2}\int_0^{\pi/2} \Omega_{22m} d\varphi d\nu, \tag{7.300}$$

where the coefficient Ω_{22m} given by Eq. (7.67) is a function of $\omega_{21m}, \ldots, \omega_{29m}$ given by Eqs. (7.35), (7.61), (7.62), (7.38), (7.39), (7.63)–(7.66), respectively.

With regard to the determination of the thermal-stress induced strengthening presented in Volume III, the stress σ_{im} acting in a direction of the axis x_i and the elastic energy density w_{im} induced by the stress σ_{im} $(i = 1,2,3)$ (see Eqs. (3.46), (3.47), (7.290)–(7.293)), both related to an arbitrary point of the cell matrix, are derived as

$$\sigma_{im} = -\left(\xi_{i21m} + \frac{\xi_{i22m}}{r} - \xi_{i23m}\ln r + \xi_{i24m}\frac{\ln r}{r}\right), \quad i = 1, 2, 3, \tag{7.301}$$

$$w_{im} = \frac{s_{11m}}{2}\left(\xi_{i21m}^2 + \frac{\xi_{i22m}^2}{r^2} + \xi_{i23m}^2\ln^2 r + \xi_{i24m}^2\frac{\ln^2 r}{r^2} + \frac{2\,\xi_{i21m}\,\xi_{i22m}}{r} - 2\,\xi_{i21m}\,\xi_{i23m}\ln r + 2\,\xi_{i21m}\,\xi_{i24m}\frac{\ln r}{r} - 2\,\xi_{i22m}\,\xi_{i23m}\frac{\ln r}{r} + 2\,\xi_{i22m}\,\xi_{i24m}\frac{\ln r}{r^2} - 2\,\xi_{i23m}\,\xi_{i24m}\frac{\ln^2 r}{r}\right), \quad i = 1, 2, 3, \tag{7.302}$$

where the elastic modulus s_{11m} and the coefficients $\xi_{i21m}, \ldots, \xi_{i24m}$ $(i = 1,2,3)$ are given by Eqs. (3.23) and (7.68)–(7.71), respectively.

Appendix. Considering the coefficients c_{1m}, c_{2m} (see Eqs. (3.38), (3.39)), the coefficient χ_{12m} has the form

$$\chi_{12m} = \frac{c_{1m} - 2c_{2m} \ln R_2}{2R_2}. \tag{7.303}$$

C. Integration constants $C_{1m} = 0$, $C_{2m} \neq 0$, $C_{3m} \neq 0$

With regard to the boundary conditions (4.10), (4.11) and the analysis presented in Section 6.2.3, considering the terms u_{2rm} and u_{3rm} (see Eqs. (6.7), (7.6)) to represent functions of the variable $r \in \langle R_2, r_c \rangle$, the radial displacement u_{1m} of the infinitesimal spherical cap in the cell matrix, regarding Eqs. (3.31), (7.3), has the form

$$u_{1m} = -p_2 \left[\kappa_{1m}^{(1)} C_m^{(2)} r^{c_{3m}} + \frac{\kappa_{3m}^{(1)}}{2} \left(\frac{1}{2} + \ln r \right) \right], \tag{7.304}$$

where the coefficients $c_{3m} < 0$, $C_m^{(2)}$ are given by Eqs. (3.40), (7.4), respectively. Provided that $\beta_p \neq \beta_e \neq \beta_m$ or $\beta_p = \beta_e \neq \beta_m$, the radial stress p_2 acting on the matrix-envelope boundary is given by Eqs. (10.3) or (10.5), respectively. The φ, ν-dependent distance r_c, as a function of the experimentally determined parameters v, R_{1ex}, d_{1ex}, d_{2ex}, d_{3ex} (see Section 2.2.1), included in the coefficients $\kappa_{1m}^{(1)}$, $\kappa_{3m}^{(1)}$ is derived in Section 2.3.

With regard to Section 7.5.2.1.C, the coefficients $\psi_{21\eta m}^{(1)}$, $\psi_{22\eta m}^{(1)}$ $(\eta = \varphi, \nu)$, ϱ_{2m} and $\omega_{21m}, \ldots, \omega_{26m}$, ξ_{i21m}, ξ_{i22m}, ξ_{i23m} $(i = 1,2,3)$ mentioned below (see Eqs. (7.307), (7.308), (7.311)–(7.314), (7.316), (7.317)) depend on $\kappa_{1m}^{(1)}$, $\kappa_{3m}^{(1)}$ and $\kappa_{1m}^{(1)}$, $\kappa_{3m}^{(1)}$, $\psi_{21\eta m}^{(1)}$, $\psi_{22\eta m}^{(1)}$, respectively, where the coefficients $\kappa_{1m}^{(1)}$, $\kappa_{3m}^{(1)}$ (see Eq. (12.35)) are functions of χ_{11m}, χ_{12m}, χ_{21m}, χ_{22m} given by Eqs. (7.318), (7.303), (7.87), (7.59), respectively.

Consequently, the radial stress σ_{11m}, the tangential and shear stresses σ_{22m}, σ_{33m} and σ_{12m}, σ_{13m} acting in the cell matrix, respectively, regarding Eqs. (3.31)–(3.34), (7.304), are derived as

$$\sigma_{11m} = -p_2 \left\{ \kappa_{1m}^{(1)} C_m^{(2)} \left[c_{3m} \left(c_{1m} + c_{2m} \right) - 2c_{2m} \right] r^{c_{3m}-1} + \frac{\kappa_{3m}^{(1)}}{2r} \left(c_{1m} - 2c_{2m} \ln r \right) \right\}, \tag{7.305}$$

$$\sigma_{22m} = \sigma_{33m} = -p_2 \left\{ \kappa_{1m}^{(1)} C_m^{(2)} \left(c_{1m} - c_{2m} c_{3m} \right) r^{c_{3m}-1} + \frac{\kappa_{3m}^{(1)}}{2r} \left[c_{1m} \left(\frac{1}{2} + \ln r \right) - c_{2m} \right] \right\}, \tag{7.306}$$

$$\sigma_{12m} = -\left[\psi_{21\varphi m}^{(1)} C_m^{(2)} r^{c_{3m}-1} + \frac{\psi_{22\varphi m}^{(1)}}{2r}\left(\frac{1}{2} + \ln r\right)\right], \tag{7.307}$$

$$\sigma_{13m} = -\left[\psi_{21\nu m}^{(1)} C_m^{(2)} r^{c_{3m}-1} + \frac{\psi_{22\nu m}^{(1)}}{2r}\left(\frac{1}{2} + \ln r\right)\right], \tag{7.308}$$

where the elastic modulus s_{44m} and the coefficients c_{1m}, c_{2m}, $\psi_{21\eta m}^{(1)}$, $\psi_{22\eta m}^{(1)}$ $(\eta = \varphi, \nu)$ are given by Eqs. (3.25) and (3.38), (3.39), (12.36), respectively.

Finally, the radial strain ε_{11m}, the tangential and shear strains ε_{22m}, ε_{33m} and ε_{12m}, ε_{13m} in the cell matrix, respectively, regarding Eqs. (3.1)–(3.5), (7.304), have the forms

$$\varepsilon_{11m} = -p_2\left[\kappa_{1m}^{(1)} C_m^{(2)} c_{3m}\, r^{c_{3m}-1} + \frac{\kappa_{3m}^{(1)}}{2r}\right], \tag{7.309}$$

$$\varepsilon_{22m} = \varepsilon_{33m} = -p_2\left[\kappa_{1m}^{(1)} C_m^{(2)} r^{c_{3m}-1} + \frac{\kappa_{3m}^{(1)}}{2r}\left(\frac{1}{2} + \ln r\right)\right], \tag{7.310}$$

$$\varepsilon_{12m} = -\left[\psi_{21\varphi m}^{(1)} C_m^{(2)} r^{c_{3m}-1} + \frac{\psi_{22\varphi m}^{(1)}}{2r}\left(\frac{1}{2} + \ln r\right)\right], \tag{7.311}$$

$$\varepsilon_{13m} = -\left[\psi_{21\nu m}^{(1)} C_m^{(2)} r^{c_{3m}-1} + \frac{\psi_{22\nu m}^{(1)}}{2r}\left(\frac{1}{2} + \ln r\right)\right]. \tag{7.312}$$

With regard to the determination of the compressive or tensile radial stress $p_2 > 0$ or $p_2 < 0$ acting on the matrix-envelope boundary, thus for $r = R_2$, respectively, considering Eq. (3.58), the tangential strain on the matrix-envelope boundary, $(\varepsilon_{22m})_{r=R_2}$, is required to be derived in the form

$$(\varepsilon_{22m})_{r=R_2} = -p_2\, \varrho_{2m}, \tag{7.313}$$

where the coefficient ϱ_{2m} is given by Eq. (7.88).

Considering Eqs. (3.41), (3.42), (7.305)–(7.312), the thermal-stress induced elastic energy density w_m and the thermal-stress induced elastic energy W_m, accumulated in an arbitrary point of the cell matrix and in the volume of the cell matrix, respectively, have the forms

$$w_m = \frac{1}{2}\left[\frac{\omega_{21m}}{r^2} + \omega_{22m}\, r^{c_{3m}-2} + \omega_{23m}\, r^{2(c_{3m}-1)} + \omega_{24m}\frac{\ln r}{r^2} \right.$$
$$\left. + \, \omega_{25m}\, r^{c_{3m}-2}\ln r + \omega_{26m}\frac{\ln^2 r}{r^2}\right], \tag{7.314}$$

$$W_m = 8 \int_0^{\pi/2} \int_0^{\pi/2} \int_{R_2}^{r_c} w_m \, r^2 dr d\varphi d\nu = \int_0^{\pi/2} \int_0^{\pi/2} \Omega_{22m} d\varphi d\nu, \tag{7.315}$$

where the coefficient Ω_{22m} given by Eq. (7.95) is a function of $\omega_{21m}, \ldots, \omega_{26m}$ given by Eqs. (7.89)–(7.94), respectively.

With regard to the determination of the thermal-stress induced strengthening presented in Volume III, the stress σ_{im} acting in a direction of the axis x_i and the elastic energy density w_{im} induced by the stress σ_{im} ($i=1,2,3$) (see Eqs. (3.46), (3.47), (7.305)–(7.308)), both related to an arbitrary point of the cell matrix, are derived as

$$\sigma_{im} = -\left(\frac{\xi_{i21m}}{r} + \xi_{i22m} \, r^{c_{3m}-1} + \xi_{i23m} \frac{\ln r}{r} \right), \quad i = 1, 2, 3, \tag{7.316}$$

$$\begin{aligned} w_{im} = \frac{s_{11m}}{2} \Bigg[& \frac{\xi_{i21m}^2}{r^2} + \xi_{i22m}^2 \, r^{2(c_{3m}-1)} + \xi_{i23m}^2 \frac{\ln^2 r}{r^2} + 2 \, \xi_{i21m} \, \xi_{i22m} \, r^{c_{3m}-2} \\ & + 2 \, \xi_{i21m} \, \xi_{i23m} \frac{\ln r}{r^2} + 2 \, \xi_{i22m} \, \xi_{i23m} \, r^{c_{3m}-2} \ln r \Bigg], \\ i = 1, 2, 3, & \end{aligned} \tag{7.317}$$

where the elastic modulus s_{11m} and the coefficients ξ_{i21m}, ξ_{i22m}, ξ_{i23m} ($i=1,2,3$) are given by Eqs. (3.23) and (7.96)–(7.98), respectively.

Appendix. Considering the coefficients c_{1m}, c_{2m}, c_{3m}, $C_m^{(2)}$ (see Eqs. (3.38)–(3.40), (7.4)), the coefficient χ_{11m} has the form

$$\chi_{11m} = C_m^{(2)} \left[c_{3m} \left(c_{1m} + c_{2m} \right) - 2 c_{2m} \right] R_2^{c_{3m}-1}. \tag{7.318}$$

D. Integration constants $C_{1m} \neq 0$, $C_{2m} \neq 0$, $C_{3m} \neq 0$

With regard to the boundary conditions (4.10), (4.11), (4.14) and the analysis presented in Section 6.2.3, considering the terms u_{1rm}, u_{2rm}, u_{3rm} (see Eqs. (7.5), (6.7), (7.6)) to represent functions of the variable $r \in \langle R_2, r_c \rangle$, the radial displacement u_{1m} of the infinitesimal spherical cap in the cell matrix, regarding Eqs. (3.31), (7.3), has the form

$$u_{1m} = -p_2 \left[\frac{\kappa_{4m}^{(2)} r}{3} \left(\frac{4}{3} - \ln r \right) + \kappa_{5m}^{(2)} C_m^{(2)} \, r^{c_{3m}} + \frac{\kappa_{6m}^{(2)}}{2} \left(\frac{1}{2} + \ln r \right) \right], \tag{7.319}$$

where the coefficients $c_{3m} < 0$, $C_m^{(2)}$ are given by Eqs. (3.40), (7.4), respectively. Provided that $\beta_p \neq \beta_e \neq \beta_m$ or $\beta_p = \beta_e \neq \beta_m$, the radial stress p_2 acting on the matrix-envelope boundary is given by Eqs. (10.3) or (10.5), respectively. The φ, ν-dependent distance r_c, as a function of the experimentally determined parameters v, R_{1ex}, d_{1ex}, d_{2ex}, d_{3ex} (see Section 2.2.1), included in the coefficients $\kappa_{4m}^{(2)}$, $\kappa_{5m}^{(2)}$, $\kappa_{6m}^{(2)}$ is derived in Section 2.3.

With regard to Section 7.5.2.1.D, the coefficients $\psi_{21\eta m}^{(2)}$, $\psi_{22\eta m}^{(2)}$, $\psi_{23\eta m}^{(2)}$ $(\eta = \varphi, \nu)$, ϱ_{2m} and $\omega_{21m}, \ldots, \omega_{214m}$, $\xi_{i21m}, \ldots, \xi_{i25m}$ $(i = 1,2,3)$ mentioned below (see Eqs. (7.322), (7.323), (7.326)–(7.329), (7.331), (7.332)) depend on $\kappa_{4m}^{(2)}$, $\kappa_{5m}^{(2)}$, $\kappa_{6m}^{(2)}$ and $\kappa_{4m}^{(2)}$, $\kappa_{5m}^{(2)}$, $\kappa_{6m}^{(2)}$, $\psi_{21\eta m}^{(2)}$, $\psi_{22\eta m}^{(2)}$, $\psi_{23\eta m}^{(2)}$, respectively, where the coefficients $\kappa_{4m}^{(2)}$, $\kappa_{5m}^{(2)}$, $\kappa_{6m}^{(2)}$ (see Eq. (12.39)) are functions of χ_{11m}, χ_{12m}, χ_{13m}, χ_{21m}, χ_{22m}, χ_{23m}, χ_{31m}, χ_{32m}, χ_{33m} given by Eqs. (7.287), (7.288), (7.333), (7.32), (7.33), (7.114)–(7.117), respectively.

Consequently, the radial stress σ_{11m}, the tangential and shear stresses σ_{22m}, σ_{33m} and σ_{12m}, σ_{13m} acting in the cell matrix, respectively, regarding Eqs. (3.31)–(3.34), (7.319), are derived as

$$\sigma_{11m} = -p_2 \left\{ \frac{\kappa_{4m}^{(2)}}{3} \left[\frac{1}{3} (c_{1m} - 7c_{2m}) - (c_{1m} - c_{2m}) \ln r \right] + \kappa_{5m}^{(2)} C_m^{(2)} \left[c_{3m} (c_{1m} + c_{2m}) - 2c_{2m} \right] r^{c_{3m}-1} + \frac{\kappa_{6m}^{(2)}}{2r} (c_{1m} - 2c_{2m} \ln r) \right\}, \quad (7.320)$$

$$\sigma_{22m} = \sigma_{33m} = -p_2 \left\{ \frac{\kappa_{4m}^{(2)}}{3} \left[\frac{1}{3} (4c_{1m} - c_{2m}) - (c_{1m} - c_{2m}) \ln r \right] + \kappa_{5m}^{(2)} C_m^{(2)} (c_{1m} - c_{2m} c_{3m}) r^{c_{3m}-1} + \frac{\kappa_{6m}^{(2)}}{2r} \left[c_{1m} \left(\frac{1}{2} + \ln r \right) - c_{2m} \right] \right\}, \quad (7.321)$$

$$\sigma_{12m} = -\frac{1}{s_{44m}} \left[\frac{\Psi_{21\varphi m}^{(2)}}{3} \left(\frac{4}{3} - \ln r \right) + \Psi_{22\varphi m}^{(2)} C_m^{(2)} r^{c_{3m}-1} + \frac{\Psi_{23\varphi m}^{(2)}}{2r} \left(\frac{1}{2} + \ln r \right) \right], \quad (7.322)$$

$$\sigma_{13m} = -\frac{1}{s_{44m}} \left[\frac{\Psi_{21\nu m}^{(2)}}{3} \left(\frac{4}{3} - \ln r \right) + \Psi_{22\nu m}^{(2)} C_m^{(2)} r^{c_{3m}-1} + \frac{\Psi_{23\nu m}^{(2)}}{2r} \left(\frac{1}{2} + \ln r \right) \right], \quad (7.323)$$

where the elastic modulus s_{44m} and the coefficients c_{1m}, c_{2m}, $\psi^{(2)}_{21\eta m}$, $\psi^{(2)}_{22\eta m}$, $\psi^{(2)}_{23\eta m}$ $(\eta=\varphi,\nu)$ are given by Eqs. (3.25) and (3.38), (3.39), (12.40), respectively.

Finally, the radial strain ε_{11m}, the tangential and shear strains ε_{22m}, ε_{33m} and ε_{12m}, ε_{13m} in the cell matrix, respectively, regarding Eqs. (3.1)–(3.5), (7.319), have the forms

$$\varepsilon_{11m} = -p_2 \left[\frac{\kappa^{(2)}_{4m}}{3} \left(\frac{1}{3} - \ln r \right) + \kappa^{(2)}_{5m} C^{(2)}_m c_{3m} \, r^{c_{3m}-1} + \frac{\kappa^{(2)}_{6m}}{2r} \right], \tag{7.324}$$

$$\varepsilon_{22m} = \varepsilon_{33m} = -p_2 \left[\frac{\kappa^{(2)}_{4m}}{3} \left(\frac{4}{3} - \ln r \right) + \kappa^{(2)}_{5m} C^{(2)}_m \, r^{c_{3m}-1} + \frac{\kappa^{(2)}_{6m}}{2r} \left(\frac{1}{2} + \ln r \right) \right], \tag{7.325}$$

$$\varepsilon_{12m} = -\left[\frac{\Psi^{(2)}_{21\varphi m}}{3} \left(\frac{4}{3} - \ln r \right) + \Psi^{(2)}_{22\varphi m} C^{(2)}_m \, r^{c_{3m}-1} + \frac{\Psi^{(2)}_{23\varphi m}}{2r} \left(\frac{1}{2} + \ln r \right) \right], \tag{7.326}$$

$$\varepsilon_{13m} = -\left[\frac{\Psi^{(2)}_{21\nu m}}{3} \left(\frac{4}{3} - \ln r \right) + \Psi^{(2)}_{22\nu m} C^{(2)}_m \, r^{c_{3m}-1} + \frac{\Psi^{(2)}_{23\nu m}}{2r} \left(\frac{1}{2} + \ln r \right) \right]. \tag{7.327}$$

With regard to the determination of the compressive or tensile radial stress $p_2 > 0$ or $p_2 < 0$ acting on the matrix-envelope boundary, thus for $r = R_2$, respectively, considering Eq. (3.58), the tangential strain on the matrix-envelope boundary, $(\varepsilon_{22m})_{r=R_2}$, is required to be derived in the form

$$(\varepsilon_{22m})_{r=R_2} = -p_2 \, \varrho_{2m}, \tag{7.328}$$

where the coefficient ϱ_{2m} is given by Eq. (7.118).

Considering Eqs. (3.41), (3.42), (7.320)–(7.327), the thermal-stress induced elastic energy density w_m and the thermal-stress induced elastic energy W_m, accumulated in an arbitrary point of the cell matrix and in the volume of the cell matrix, respectively, have the forms

$$\begin{aligned} w_m = \frac{1}{2} \Bigg[& \omega_{21m} + \frac{\omega_{22m}}{r} + \frac{\omega_{23m}}{r^2} + \omega_{24m} \, r^{c_{3m}-1} + \omega_{25m} \, r^{c_{3m}-2} + \omega_{26m} \, r^{2(c_{3m}-1)} \\ & + \omega_{27m} \ln r + \omega_{28m} \ln^2 r + \frac{\omega_{29m} \ln r}{r} + \frac{\omega_{210m} \ln r}{r^2} + \frac{\omega_{211m} \ln^2 r}{r} \\ & + \omega_{212m} \frac{\ln^2 r}{r^2} + \omega_{213m} \, r^{c_{3m}-1} \ln r + \omega_{214m} \, r^{c_{3m}-2} \ln r \Bigg], \end{aligned} \tag{7.329}$$

$$W_m = 8 \int_0^{\pi/2} \int_0^{\pi/2} \int_{R_2}^{r_c} w_m \, r^2 dr d\varphi d\nu = \int_0^{\pi/2} \int_0^{\pi/2} \Omega_{22m} d\varphi d\nu, \tag{7.330}$$

where the coefficient Ω_{22m} given by Eq. (7.133) is a function of $\omega_{21m}, \ldots, \omega_{214m}$ Eqs. (7.119)–(7.132), respectively.

With regard to the determination of the thermal-stress induced strengthening presented in Volume III, the stress σ_{im} acting in a direction of the axis x_i and the elastic energy density w_{im} induced by the stress σ_{im} $(i=1,2,3)$ (see Eqs. (3.46), (3.47), (7.320)–(7.323)), both related to an arbitrary point of the cell matrix, are derived as

$$\sigma_{im} = -\left(\xi_{i\,21m} + \frac{\xi_{i\,22m}}{r} + \xi_{i\,23m}\, r^{c_{3m}-1} - \xi_{i\,24m} \ln r + \xi_{i\,25m} \frac{\ln r}{r}\right), \quad i = 1,2,3, \tag{7.331}$$

$$\begin{aligned} w_{im} = \frac{s_{11m}}{2} \Bigg[& \xi_{i\,21m}^2 + \frac{\xi_{i\,22m}^2}{r^2} + \xi_{i\,23m}^2\, r^{2(c_{3m}-1)} + \xi_{i\,24m}^2 \ln^2 r + \xi_{i\,25m}^2 \frac{\ln^2 r}{r^2} \\ & + \frac{2\,\xi_{i\,21m}\,\xi_{i\,22m}}{r} + 2\,\xi_{i\,21m}\,\xi_{i\,23m}\, r^{c_{3m}-1} - 2\,\xi_{i\,21m}\,\xi_{i\,24m} \ln r \\ & + 2\left(\xi_{i\,21m}\,\xi_{i\,25m} - \xi_{i\,22}\,\xi_{i\,24m}\right) \frac{\ln r}{r} + 2\,\xi_{i\,22m}\,\xi_{i\,23m}\, r^{c_{3m}-2} \\ & + 2\,\xi_{i\,22m}\,\xi_{i\,25m} \frac{\ln r}{r^2} - 2\,\xi_{i\,23m}\,\xi_{i\,24m}\, r^{c_{3m}-1} \ln r \\ & + 2\,\xi_{i\,23m}\,\xi_{i\,25m}\, r^{c_{3m}-2} \ln r - 2\,\xi_{i\,24m}\,\xi_{i\,25m} \frac{\ln^2 r}{r} \Bigg], \\ & i = 1,2,3, \end{aligned} \tag{7.332}$$

where the elastic modulus s_{11m} and the coefficients $\xi_{i\,21m}, \ldots, \xi_{i\,25m}$ $(i=1,2,3)$ are given by Eqs. (3.23) and (7.134)–(7.138), respectively.

Appendix. Considering the coefficients c_{1m}, c_{2m} (see Eqs. (3.38), (3.39)), the coefficient χ_{13m} has the form

$$\chi_{13m} = \frac{c_{1m} - 2c_{2m} \ln R_2}{2R_2}. \tag{7.333}$$

7.5.2.2 Conditions $\beta_p \neq \beta_e = \beta_m$, and $\varepsilon_{11te} - \varepsilon_{11tp} = f(\varphi, \nu)$ or $\varepsilon_{11te} - \varepsilon_{11tp} \neq f(\varphi, \nu)$

The formulae in Section 7.5.2.2 consider the φ, ν-dependent or φ, ν-independent radial stress $p_1 = (\beta_e - \beta_p) / (\varrho_{1p} - \varrho_{1e}^{me})$ (see Eq. (10.4)) acting on the particle-envelope boundary and derived from Eq. (3.63), where the φ, ν-independent coefficients ϱ_{1p} and ϱ_{1e}^{me}, along with ϱ_{2e}^{me} included in the formulae in Section 7.5.2.2, are given by Eqs. (5.36) or (9.50), and (5.106), (6.207), (6.228), (6.246), (7.230) or (9.63), respectively.

The φ, ν-dependence of the radial stress p_1 results from the condition $\varepsilon_{11te} - \varepsilon_{11tp} = f(\varphi, \nu)$, resulting in the φ, ν-dependent difference $\beta_e - \beta_p$ (see Eqs. (3.48)–(3.50)), where at least one of the phase-transformation induced radial strains ε_{11tp}, ε_{11te} is φ, ν-dependent (see Eq. (12.23)). The φ, ν-independence of the radial stress p_1 results from the condition $\varepsilon_{11te} - \varepsilon_{11tp} \neq f(\varphi, \nu)$ (see Eqs. (12.23), (12.24)), regarding simplification, equivalent to those defined in Items 1, 2 in Section 9.1 and resulting in the φ, ν-independent difference $\beta_e - \beta_p$.

A. Integration constants $C_{1m} \neq 0$, $C_{2m} \neq 0$, $C_{3m} = 0$

With regard to the boundary conditions (4.11), (4.12) and the analysis presented in Section 6.2.3, considering the terms u_{1rm} and u_{2rm} (see Eqs. (7.5), (6.7)) to represent functions of the variable $r \in \langle R_2, r_c \rangle$, the radial displacement u_{1m} of the infinitesimal spherical cap in the cell matrix for the φ, ν-dependent or φ, ν-independent radial stress p_1, regarding Eqs. (3.31), (7.3), has the form

$$u_{1m} = -p_1\, \varrho_{2e}^{me} \left[\frac{\kappa_{1m}^{(1)} r}{3} \left(\frac{4}{3} - \ln r \right) + \kappa_{3m}^{(1)} C_m\, r^{c_{3m}} \right], \tag{7.334}$$

where the coefficients $c_{3m} < 0$ and $C_m = C_m^{(i)}$ $(i = 2{,}3)$ are given by Eqs. (3.40) and (7.4), (7.13), respectively. The radial stress p_1 acting on the particle-envelope boundary and the coefficient ϱ_{2e}^{me} are analysed at the beginning of Section 6.4.2.2. The φ, ν-dependent distance r_c, as a function of the experimentally determined parameters v, R_{1ex}, d_{1ex}, d_{2ex}, d_{3ex} (see Section 2.2.1), included in the coefficients $\kappa_{1m}^{(1)}$, $\kappa_{3m}^{(1)}$ is derived in Section 2.3.

With regard to Section 7.5.2.2.A, the coefficients $\psi_{11\eta m}^{(1)}$, $\psi_{12\eta m}^{(1)}$ $(\eta = \varphi, \nu)$, $\omega_{21m}, \ldots, \omega_{26m}$, ξ_{i1m}, ξ_{i2m}, ξ_{i3m} $(i = 1{,}2{,}3)$ and $\omega_{11m}, \ldots, \omega_{16m}$, ξ_{i11m}, ξ_{i12m}, ξ_{i13m} $(i = 1{,}2{,}3)$ mentioned below (see Eqs. (7.337), (7.338), (7.341), (7.342), (7.349), (7.353), (7.354) and (7.347), (7.351), (7.352)) depend on $\kappa_{1m}^{(1)}$, $\kappa_{3m}^{(1)}$ and $\kappa_{1m}^{(1)}$, $\kappa_{3m}^{(1)}$, $\psi_{11\eta m}^{(1)}$, $\psi_{12\eta m}^{(1)}$, respectively, where the coefficients $\kappa_{1m}^{(1)}$, $\kappa_{3m}^{(1)}$ (see Eq. (12.35)) are functions of χ_{11m}, χ_{12m}, χ_{21m}, χ_{22m} given by Eqs. (7.355), (7.356), (7.32), (7.33), respectively.

Consequently, considering the φ, ν-dependent or φ, ν-independent radial stress p_1, the radial and tangential stresses, σ_{11m} and σ_{22m}, σ_{33m}, respectively, acting in the cell matrix, regarding Eqs. (3.31), (3.32), (7.334), are derived as

$$\sigma_{11m} = -p_1\, \varrho_{2e}^{me} \left\{ \frac{\kappa_{1m}^{(1)}}{3} \left[\frac{1}{3} (c_{1m} - 7c_{2m}) - (c_{1m} - c_{2m}) \ln r \right] \right.$$
$$\left. + \kappa_{3m}^{(1)} C_m \left[c_{3m} (c_{1m} + c_{2m}) - 2c_{2m} \right] r^{c_{3m} - 1} \right\}, \tag{7.335}$$

$$\sigma_{22m} = \sigma_{33m} = -p_1\, \varrho_{2e}^{me} \left\{ \frac{\kappa_{1m}^{(1)}}{3} \left[\frac{1}{3} (4c_{1m} - c_{2m}) - (c_{1m} - c_{2m}) \ln r \right] + \kappa_{3m}^{(1)} C_m \left(c_{1m} - c_{2m} c_{3m} \right) r^{c_{3m}-1} \right\}, \tag{7.336}$$

where the coefficients c_{1m}, c_{2m} are given by Eqs. (3.38), (3.39), respectively.

With regard to Eqs. (3.33), (3.34), (7.334), the shear stresses σ_{12m}, σ_{13m} acting in the cell matrix, considering the φ, ν-dependent radial stress p_1, are derived as

$$\sigma_{12m} = -\varrho_{2e}^{me} \left[\frac{\psi_{11\varphi m}^{(1)}}{3} \left(\frac{4}{3} - \ln r \right) + \psi_{12\varphi m}^{(1)} C_m\, r^{c_{3m}-1} \right], \tag{7.337}$$

$$\sigma_{13m} = -\varrho_{2e}^{me} \left[\frac{\psi_{11\nu m}^{(1)}}{3} \left(\frac{4}{3} - \ln r \right) + \psi_{12\nu m}^{(1)} C_m\, r^{c_{3m}-1} \right], \tag{7.338}$$

where the elastic modulus s_{44m} and the coefficients $\psi_{11\eta m}^{(1)}$, $\psi_{12\eta m}^{(1)}$ $(\eta = \varphi, \nu)$ are given by Eqs. (3.25) and (12.36), respectively. Considering the φ, ν-independent radial stress p_1, we get

$$\sigma_{12m} = -\frac{p_1\, \varrho_{2e}^{me}}{s_{44m}} \left[\frac{1}{3} \frac{\partial \kappa_{1m}^{(1)}}{\partial \varphi} \left(\frac{4}{3} - \ln r \right) + \frac{\partial \kappa_{3m}^{(1)}}{\partial \varphi} C_m\, r^{c_{3m}-1} \right], \tag{7.339}$$

$$\sigma_{13m} = -\frac{p_1\, \varrho_{2e}^{me}}{s_{44m}} \left[\frac{1}{3} \frac{\partial \kappa_{1m}^{(1)}}{\partial \nu} \left(\frac{4}{3} - \ln r \right) + \frac{\partial \kappa_{3m}^{(1)}}{\partial \nu} C_m\, r^{c_{3m}-1} \right], \tag{7.340}$$

Finally, considering the φ, ν-dependent or φ, ν-independent radial stress p_1, the radial and tangential strains, ε_{11m} and ε_{22m}, ε_{33m}, respectively, acting in the cell matrix, regarding Eqs. (3.1)–(3.3), (7.334), have the forms

$$\varepsilon_{11m} = -p_1\, \varrho_{2e}^{me} \left[\frac{\kappa_{1m}^{(1)}}{3} \left(\frac{1}{3} - \ln r \right) + \kappa_{3m}^{(1)} C_m c_{3m}\, r^{c_{3m}-1} \right], \tag{7.341}$$

$$\varepsilon_{22m} = \varepsilon_{33m} = -p_1\, \varrho_{2e}^{me} \left[\frac{\kappa_{1m}^{(1)}}{3} \left(\frac{4}{3} - \ln r \right) + \kappa_{3m}^{(1)} C_m\, r^{c_{3m}-1} \right]. \tag{7.342}$$

With regard to Eqs. (3.4), (3.5), (7.334), the shear strains σ_{12m}, σ_{13m} acting in the cell matrix, considering the φ, ν-dependent radial stress p_1, are derived as

$$\varepsilon_{12m} = -\varrho_{2e}^{me} \left[\frac{\psi_{11\varphi m}^{(1)}}{3} \left(\frac{4}{3} - \ln r \right) + \psi_{12\varphi m}^{(1)} C_m\, r^{c_{3m}-1} \right], \tag{7.343}$$

$$\varepsilon_{13m} = -\varrho_{2e}^{me}\left[\frac{\psi_{11\nu m}^{(1)}}{3}\left(\frac{4}{3} - \ln r\right) + \psi_{12\nu m}^{(1)} C_m\, r^{c_{3m}-1}\right], \tag{7.344}$$

and considering the φ, ν-independent radial stress p_1, we get

$$\varepsilon_{12m} = -p_1\, \varrho_{2e}^{me}\left[\frac{1}{3}\frac{\partial \kappa_{1m}^{(1)}}{\partial \varphi}\left(\frac{4}{3} - \ln r\right) + \frac{\partial \kappa_{3m}^{(1)}}{\partial \varphi} C_m\, r^{c_{3m}-1}\right], \tag{7.345}$$

$$\varepsilon_{13m} = -p_1\, \varrho_{2e}^{me}\left[\frac{1}{3}\frac{\partial \kappa_{1m}^{(1)}}{\partial \nu}\left(\frac{4}{3} - \ln r\right) + \frac{\partial \kappa_{3m}^{(1)}}{\partial \nu} C_m\, r^{c_{3m}-1}\right]. \tag{7.346}$$

With regard to the φ, ν-dependent radial stress p_1, considering Eqs. (3.41), (3.42), (7.335)–(7.338), (7.341)–(7.344), the thermal-stress induced elastic energy density w_m and the thermal-stress induced elastic energy W_m, accumulated in an arbitrary point of the cell matrix and in the volume of the cell matrix, respectively, are derived as

$$w_m = \frac{1}{2}\left[\omega_{11m} + \omega_{12m}\, r^{c_{3m}-1} + \omega_{13m}\, r^{2(c_{3m}-1)} + \omega_{14m} \ln r + \omega_{15m} \ln^2 r + \omega_{16m}\, r^{c_{3m}-1} \ln r\right], \tag{7.347}$$

$$W_m = 8\int_0^{\pi/2}\int_0^{\pi/2}\int_{R_2}^{r_c} w_m\, r^2 dr d\varphi d\nu = \left(\varrho_{2e}^{me}\right)^2 \int_0^{\pi/2}\int_0^{\pi/2} \Omega_{12m} d\varphi d\nu, \tag{7.348}$$

where the coefficient Ω_{12m} given by Eq. (6.39) is a function of $\omega_{11m}, \ldots, \omega_{16m}$ given by Eqs. (7.35)–(7.40), respectively.

With regard to the φ, ν-independent radial stress p_1, considering Eqs. (3.41), (3.42), (7.335), (7.336), (7.339)–(7.342), (7.345), (7.346), we get

$$w_m = \frac{1}{2}\left(p_1\, \varrho_{2e}^{me}\right)^2\left[\omega_{21m} + \omega_{22m}\, r^{c_{3m}-1} + \omega_{23m}\, r^{2(c_{3m}-1)} + \omega_{24m} \ln r + \omega_{25m} \ln^2 r + \omega_{26m}\, r^{c_{3m}-1} \ln r\right], \tag{7.349}$$

$$W_m = 8\int_0^{\pi/2}\int_0^{\pi/2}\int_{R_2}^{r_c} w_m\, r^2\, dr\, d\varphi\, d\nu = \left(p_1\, \varrho_{2e}^{me}\right)^2 \int_0^{\pi/2}\int_0^{\pi/2} \Omega_{22m}\, d\varphi\, d\nu, \tag{7.350}$$

where the coefficient Ω_{22m} given by Eq. (7.40) is a function of $\omega_{21m}, \ldots, \omega_{26m}$ given by Eqs. (7.357)–(7.362), respectively.

With regard to the determination of the thermal-stress induced strengthening presented in Volume III, the stress σ_{im} acting in a direction of the axis x_i and the elastic energy density w_{im} induced by the stress σ_{im} $(i=1,2,3)$ (see Eqs. (3.46),

(3.47), (7.335)–(7.338)), both related to an arbitrary point of the cell matrix, considering the φ, ν-dependent radial stress p_1, have the forms

$$\sigma_{im} = -\varrho_{2e}^{me} \left(\xi_{i11m} + \xi_{i12m} r^{c_{3m}-1} - \xi_{i13m} \ln r \right), \quad i = 1,2,3, \tag{7.351}$$

$$\begin{aligned} w_{im} = \frac{s_{11m}}{2} \Big[& \xi_{i11m}^2 + \xi_{i12m}^2 r^{2(c_{3m}-1)} + \xi_{i13m}^2 \ln^2 r + 2\,\xi_{i11m}\,\xi_{i12m}\, r^{c_{3m}-1} \\ & - 2\,\xi_{i11m}\,\xi_{i13m} \ln r - 2\,\xi_{i12m}\,\xi_{i13m}\, r^{c_{3m}-1} \ln r \Big], \\ & i = 1,2,3, \end{aligned} \tag{7.352}$$

where the elastic modulus s_{11m} and the coefficients ξ_{i11m}, ξ_{i12m}, ξ_{i13m} $(i=1,2,3)$ are given by Eqs. (3.23) and (7.41)–(7.43), respectively. With regard to the φ, ν-independent radial stress p_1, considering Eqs. (3.46), (3.47), (7.335), (7.336), (7.339), (7.340), we get

$$\sigma_{im} = -p_1\, \varrho_{2e}^{me} \left(\xi_{i1m} + \xi_{i2m} r^{c_{3m}-1} - \xi_{i3m} \ln r \right), \quad i = 1,2,3, \tag{7.353}$$

$$\begin{aligned} w_{im} = \frac{s_{11m}}{2} \Big[& \xi_{i1m}^2 + \xi_{i2m}^2 r^{2(c_{3m}-1)} + \xi_{i3m}^2 \ln^2 r + 2\,\xi_{i1m}\,\xi_{i2m}\, r^{c_{3m}-1} \\ & - 2\,\xi_{i1m}\,\xi_{i3m} \ln r - 2\,\xi_{i2m}\,\xi_{i3m}\, r^{c_{3m}-1} \ln r \Big], \\ & i = 1,2,3, \end{aligned} \tag{7.354}$$

where the coefficients ξ_{i1m}, ξ_{i2m}, ξ_{i3m} $(i=1,2,3)$ are given by Eqs. (7.363)–(7.365), respectively.

Appendix. Considering the elastic modulus s_{44m} and the coefficients c_{1m}, c_{2m}, c_{3m}, $C_m = C_m^{(i)}$ $(i=2,3)$, a'_{1i}, a'_{2i}, a'_{3i} $(i=1,2,3)$ (see Eqs. (3.25), (3.38)–(3.40), (7.4), (7.13), (12.26)–(12.34)), the coefficients χ_{11m}, χ_{12m}, $\omega_{21m}, \ldots, \omega_{26m}$, ξ_{i1m}, ξ_{i2m}, ξ_{i3m} $(i=1,2,3)$ have the forms

$$\chi_{11m} = \frac{1}{3} \left(\frac{4}{3} - \ln R_2 \right), \tag{7.355}$$

$$\chi_{12m} = C_m R_2^{c_{3m}-1}, \tag{7.356}$$

$$\omega_{21m} = \frac{1}{81} \left(3 \left[\kappa_{1m}^{(1)} \right]^2 (11 c_{1m} - 5 c_{2m}) + \frac{32}{s_{44m}} \left\{ \left[\frac{\partial \kappa_{1m}^{(1)}}{\partial \varphi} \right]^2 + \left[\frac{\partial \kappa_{1m}^{(1)}}{\partial \nu} \right]^2 \right\} \right), \tag{7.357}$$

$$\omega_{22m} = \frac{2C_m}{9}\left\{\kappa_{1m}^{(1)}\kappa_{3m}^{(1)}\left[c_{1m}\left(c_{3m}+8\right) - c_{2m}\left(7c_{3m}+2\right)\right] + \frac{8}{s_{44m}}\left[\frac{\partial\kappa_{1m}^{(1)}}{\partial\varphi}\frac{\partial\kappa_{3m}^{(1)}}{\partial\varphi} + \frac{\partial\kappa_{1m}^{(1)}}{\partial\nu}\frac{\partial\kappa_{3m}^{(1)}}{\partial\nu}\right]\right\}, \tag{7.358}$$

$$\omega_{23m} = C_m^2\left(\left[\kappa_{3m}^{(1)}\right]^2\left[c_{3m}^2\left(c_{1m}+c_{2m}\right) + 2\left(c_{1m} - 2c_{2m}c_{3m}\right)\right] + \frac{2}{s_{44m}}\left\{\left[\frac{\partial\kappa_{3m}^{(1)}}{\partial\varphi}\right]^2 + \left[\frac{\partial\kappa_{3m}^{(1)}}{\partial\nu}\right]^2\right\}\right), \tag{7.359}$$

$$\omega_{24m} = -\frac{2}{27}\left(9\left[\kappa_{1m}^{(1)}\right]^2\left(c_{1m}-c_{2m}\right) + \frac{8}{s_{44m}}\left\{\left[\frac{\partial\kappa_{1m}^{(1)}}{\partial\varphi}\right]^2 + \left[\frac{\partial\kappa_{1m}^{(1)}}{\partial\nu}\right]^2\right\}\right), \tag{7.360}$$

$$\omega_{25m} = \frac{1}{9}\left(3\left[\kappa_{1m}^{(1)}\right]^2\left(c_{1m}-c_{2m}\right) + \frac{2}{s_{44m}}\left\{\left[\frac{\partial\kappa_{1m}^{(1)}}{\partial\varphi}\right]^2 + \left[\frac{\partial\kappa_{1m}^{(1)}}{\partial\nu}\right]^2\right\}\right), \tag{7.361}$$

$$\omega_{26m} = -\frac{2C_m}{3}\left\{\kappa_{1m}^{(1)}\kappa_{3m}^{(1)}\left(c_{1m}-c_{2m}\right)\left(c_{3m}+2\right) + \frac{2}{s_{44m}}\left[\frac{\partial\kappa_{1m}^{(1)}}{\partial\varphi}\frac{\partial\kappa_{3m}^{(1)}}{\partial\varphi} + \frac{\partial\kappa_{1m}^{(1)}}{\partial\nu}\frac{\partial\kappa_{3m}^{(1)}}{\partial\nu}\right]\right\}, \tag{7.362}$$

$$\xi_{i1m} = \frac{1}{9}\left\{\kappa_{1m}^{(1)}\left[a'_{1i}\left(c_{1m} - 7c_{2m}\right) + \left(a'_{2i}+a'_{3i}\right)\left(4c_{1m}-c_{2m}\right)\right] + \frac{4}{s_{44m}}\left[\left(a'_{1i}+a'_{2i}\right)\frac{\partial\kappa_{1m}^{(1)}}{\partial\varphi} + \left(a'_{1i}+a'_{3i}\right)\frac{\partial\kappa_{1m}^{(1)}}{\partial\nu}\right]\right\},$$
$$i = 1, 2, 3, \tag{7.363}$$

$$\xi_{i2m} = C_m\left(\kappa_{3m}^{(1)}\left\{a'_{1i}\left[c_{3m}\left(c_{1m}+c_{2m}\right) - 2c_{2m}\right] + \left(a'_{2i}+a'_{3i}\right)\left(c_{1m} - c_{2m}c_{3m}\right)\right\} + \frac{1}{s_{44m}}\left[\left(a'_{1i}+a'_{2i}\right)\frac{\partial\kappa_{3m}^{(1)}}{\partial\varphi} + \left(a'_{1i}+a'_{3i}\right)\frac{\partial\kappa_{3m}^{(1)}}{\partial\nu}\right]\right),$$
$$i = 1, 2, 3, \tag{7.364}$$

$$\xi_{i3m} = \frac{1}{3}\left\{\kappa_{1m}^{(1)}\left(a'_{1i}+a'_{2i}+a'_{3i}\right)\left(c_{1m}-c_{2m}\right) + \frac{1}{s_{44m}}\left[\left(a'_{1i}+a'_{2i}\right)\frac{\partial\kappa_{1m}^{(1)}}{\partial\varphi} + \left(a'_{1i}+a'_{3i}\right)\frac{\partial\kappa_{1m}^{(1)}}{\partial\nu}\right]\right\},$$
$$i = 1, 2, 3. \tag{7.365}$$

B. Integration constants $C_{1m} \neq 0$, $C_{2m} = 0$, $C_{3m} \neq 0$

With regard to the boundary conditions (4.11), (4.12) and the analysis presented in Section 6.2.3, considering the terms u_{1rm} and u_{3rm} (see Eqs. (7.5), (7.6)) to represent functions of the variable $r \in \langle R_2, r_c \rangle$, the radial displacement u_{1m} of the infinitesimal spherical cap in the cell matrix for the φ, ν-dependent or φ, ν-independent radial stress p_1, regarding Eqs. (3.31), (7.3), has the form

$$u_{1m} = -p_1 \, \varrho_{2e}^{me} \left[\frac{\kappa_{1m}^{(1)} r}{3} \left(\frac{4}{3} - \ln r \right) + \frac{\kappa_{3m}^{(1)}}{2} \left(\frac{1}{2} + \ln r \right) \right], \tag{7.366}$$

where the radial stress p_1 acting on the particle-envelope boundary and the coefficient ϱ_{2e}^{me} are analysed at the beginning of Section 6.4.2.2. The φ, ν-dependent distance r_c, as a function of the experimentally determined parameters v, R_{1ex}, d_{1ex}, d_{2ex}, d_{3ex} (see Section 2.2.1), included in the coefficients $\kappa_{1m}^{(1)}$, $\kappa_{3m}^{(1)}$ is derived in Section 2.3.

With regard to Section 7.5.2.2.B, the coefficients $\psi_{11\eta m}^{(1)}$, $\psi_{12\eta m}^{(1)}$ $(\eta = \varphi, \nu)$, $\omega_{21m}, \ldots, \omega_{29m}$, $\xi_{i1m}, \ldots, \xi_{i4m}$ $(i = 1,2,3)$ and $\omega_{11m}, \ldots, \omega_{19m}$, $\xi_{i11m}, \ldots, \xi_{i14m}$ $(i = 1,2,3)$ mentioned below (see Eqs. (7.369), (7.370), (7.375), (7.376), (7.381), (7.385), (7.386) and (7.379), (7.383), (7.384)) depend on $\kappa_{1m}^{(1)}$, $\kappa_{3m}^{(1)}$ and $\kappa_{1m}^{(1)}$, $\kappa_{3m}^{(1)}$, $\psi_{11\eta m}^{(1)}$, $\psi_{12\eta m}^{(1)}$, respectively, where the coefficients $\kappa_{1m}^{(1)}$, $\kappa_{3m}^{(1)}$ (see Eq. (12.35)) are functions of χ_{11m}, χ_{12m}, χ_{21m}, χ_{22m} given by Eqs. (7.347), (7.387), (7.32), (7.59), respectively.

Consequently, considering the φ, ν-dependent or φ, ν-independent radial stress p_1, the radial and tangential stresses, σ_{11m} and σ_{22m}, σ_{33m}, respectively, acting in the cell matrix, regarding Eqs. (3.31), (3.32), (7.366), are derived as

$$\sigma_{11m} = -p_1 \, \varrho_{2e}^{me} \left\{ \frac{\kappa_{1m}^{(1)}}{3} \left[\frac{1}{3} (c_{1m} - 7 c_{2m}) - (c_{1m} - c_{2m}) \ln r \right] + \frac{\kappa_{3m}^{(1)}}{2r} (c_{1m} - 2 c_{2m} \ln r) \right\}, \tag{7.367}$$

$$\sigma_{22m} = \sigma_{33m} = -p_1 \, \varrho_{2e}^{me} \left\{ \frac{\kappa_{1m}^{(1)}}{3} \left[\frac{1}{3} (4 c_{1m} - c_{2m}) - (c_{1m} - c_{2m}) \ln r \right] + \frac{\kappa_{3m}^{(1)}}{2r} \left[c_{1m} \left(\frac{1}{2} + \ln r \right) - c_{2m} \right] \right\}, \tag{7.368}$$

where the coefficients c_{1m}, c_{2m} are given by Eqs. (3.38), (3.39), respectively.

With regard to Eqs. (3.33), (3.34), (7.366), the shear stresses σ_{12m}, σ_{13m} acting in the cell matrix, considering the φ, ν-dependent radial stress p_1, are derived as

$$\sigma_{12m} = -\frac{\varrho_{2e}^{me}}{s_{44m}} \left[\frac{\psi_{11\varphi m}^{(1)}}{3} \left(\frac{4}{3} - \ln r \right) + \frac{\psi_{12\varphi m}^{(1)}}{2r} \left(\frac{1}{2} + \ln r \right) \right], \tag{7.369}$$

$$\sigma_{13m} = -\frac{\varrho_{2e}^{me}}{s_{44m}} \left[\frac{\psi_{11\nu m}^{(1)}}{3} \left(\frac{4}{3} - \ln r \right) + \frac{\psi_{12\nu m}^{(1)}}{2r} \left(\frac{1}{2} + \ln r \right) \right], \tag{7.370}$$

where the elastic modulus s_{44m} and the coefficients $\psi_{11\eta m}^{(1)}$, $\psi_{12\eta m}^{(1)}$ $(\eta = \varphi, \nu)$ are given by Eqs. (3.25) and (12.36), respectively. Considering the φ, ν-independent radial stress p_1, we get

$$\sigma_{12m} = -\frac{p_1\, \varrho_{2e}^{me}}{s_{44m}} \left[\frac{1}{3} \frac{\partial \kappa_{1m}^{(1)}}{\partial \varphi} \left(\frac{4}{3} - \ln r \right) + \frac{1}{2r} \frac{\partial \kappa_{3m}^{(1)}}{\partial \varphi} \left(\frac{1}{2} + \ln r \right) \right], \tag{7.371}$$

$$\sigma_{13m} = -\frac{p_1\, \varrho_{2e}^{me}}{s_{44m}} \left[\frac{1}{3} \frac{\partial \kappa_{1m}^{(1)}}{\partial \nu} \left(\frac{4}{3} - \ln r \right) + \frac{1}{2r} \frac{\partial \kappa_{3m}^{(1)}}{\partial \nu} \left(\frac{1}{2} + \ln r \right) \right]. \tag{7.372}$$

Finally, considering the φ, ν-dependent or φ, ν-independent radial stress p_1, the radial and tangential strains, ε_{11m} and ε_{22m}, ε_{33m}, respectively, acting in the cell matrix, regarding Eqs. (3.1)–(3.3), (7.366), have the forms

$$\varepsilon_{11m} = -p_1\, \varrho_{2e}^{me} \left[\frac{\kappa_{1m}^{(1)}}{3} \left(\frac{1}{3} - \ln r \right) + \frac{\kappa_{3m}^{(1)}}{2r} \right], \tag{7.373}$$

$$\varepsilon_{22m} = \varepsilon_{33m} = -p_1\, \varrho_{2e}^{me} \left[\frac{\kappa_{1m}^{(1)}}{3} \left(\frac{4}{3} - \ln r \right) + \frac{\kappa_{3m}^{(1)}}{2r} \left(\frac{1}{2} + \ln r \right) \right]. \tag{7.374}$$

With regard to Eqs. (3.4), (3.5), (7.366), the shear strains σ_{12m}, σ_{13m} acting in the cell matrix, considering the φ, ν-dependent radial stress p_1, are derived as

$$\varepsilon_{12m} = -\varrho_{2e}^{me} \left[\frac{\psi_{11\varphi m}^{(1)}}{3} \left(\frac{4}{3} - \ln r \right) + \frac{\psi_{12\varphi m}^{(1)}}{2r} \left(\frac{1}{2} + \ln r \right) \right], \tag{7.375}$$

$$\varepsilon_{13m} = -\varrho_{2e}^{me} \left[\frac{\psi_{11\nu m}^{(1)}}{3} \left(\frac{4}{3} - \ln r \right) + \frac{\psi_{12\nu m}^{(1)}}{2r} \left(\frac{1}{2} + \ln r \right) \right], \tag{7.376}$$

and considering the φ, ν-independent radial stress p_1, we get

$$\varepsilon_{12m} = -p_1\, \varrho_{2e}^{me} \left[\frac{1}{3} \frac{\partial \kappa_{1m}^{(1)}}{\partial \varphi} \left(\frac{4}{3} - \ln r \right) + \frac{1}{2r} \frac{\partial \kappa_{3m}^{(1)}}{\partial \varphi} \left(\frac{1}{2} + \ln r \right) \right], \tag{7.377}$$

$$\varepsilon_{13m} = -p_1 \varrho_{2e}^{me} \left[\frac{1}{3} \frac{\partial \kappa_{1m}^{(1)}}{\partial \nu} \left(\frac{4}{3} - \ln r \right) + \frac{1}{2r} \frac{\partial \kappa_{3m}^{(1)}}{\partial \nu} \left(\frac{1}{2} + \ln r \right) \right]. \tag{7.378}$$

With regard to the φ, ν-dependent radial stress p_1, considering Eqs. (3.41), (3.42), (7.367)–(7.370), (7.373)–(7.376), the thermal-stress induced elastic energy density w_m and the thermal-stress induced elastic energy W_m, accumulated in an arbitrary point of the cell matrix and in the volume of the cell matrix, respectively, are derived as

$$w_m = \frac{\left(\varrho_{2e}^{me}\right)^2}{2} \left(\omega_{11m} + \frac{\omega_{12m}}{r} + \frac{\omega_{13m}}{r^2} + \omega_{14m} \ln r + \omega_{15m} \ln^2 r + \omega_{16m} \frac{\ln r}{r} + \omega_{17m} \frac{\ln r}{r^2} + \omega_{18m} \frac{\ln^2 r}{r} + \omega_{19m} \frac{\ln^2 r}{r^2} \right), \tag{7.379}$$

$$W_m = 8 \int_0^{\pi/2} \int_0^{\pi/2} \int_{R_2}^{r_c} w_m r^2 dr d\varphi d\nu = \left(\varrho_{2e}^{me}\right)^2 \int_0^{\pi/2} \int_0^{\pi/2} \Omega_{12m} d\varphi d\nu, \tag{7.380}$$

where the coefficient Ω_{12m} given by Eq. (7.67) is a function of $\omega_{11m}, \ldots, \omega_{19m}$ given by Eqs. (7.35), (7.61), (7.62), (7.38), (7.39), (7.63)–(7.66), respectively.

With regard to the φ, ν-independent radial stress p_1, considering Eqs. (3.41), (3.42), (7.367), (7.368), (7.371)–(7.374), (7.377), (7.378), we get

$$w_m = \frac{\left(p_1 \varrho_{2e}^{me}\right)^2}{2} \left(\omega_{21m} + \frac{\omega_{22m}}{r} + \frac{\omega_{23m}}{r^2} + \omega_{24m} \ln r + \omega_{25m} \ln^2 r + \omega_{26m} \frac{\ln r}{r} + \omega_{27m} \frac{\ln r}{r^2} + \omega_{28m} \frac{\ln^2 r}{r} + \omega_{29m} \frac{\ln^2 r}{r^2} \right), \tag{7.381}$$

$$W_m = 8 \int_0^{\pi/2} \int_0^{\pi/2} \int_{R_2}^{r_c} w_m r^2 \, dr \, d\varphi \, d\nu = \left(p_1 \varrho_{2e}^{me}\right)^2 \int_0^{\pi/2} \int_0^{\pi/2} \Omega_{22m} \, d\varphi \, d\nu, \tag{7.382}$$

where the coefficient Ω_{22m} given by Eq. (7.66) is a function of $\omega_{21m}, \ldots, \omega_{29m}$ given by Eqs. (7.357), (7.388), (7.389), (7.360), (7.361), (7.390)–(7.393), respectively.

With regard to the determination of the thermal-stress induced strengthening presented in Volume III, the stress σ_{im} acting in a direction of the axis x_i and the elastic energy density w_{im} induced by the stress σ_{im} $(i = 1,2,3)$ (see Eqs. (3.46), (3.47), (7.367)–(7.370)), both related to an arbitrary point of the cell matrix, considering the φ, ν-dependent radial stress p_1, have the forms

$$\sigma_{im} = -\varrho_{2e}^{me} \left(\xi_{i\,11m} + \frac{\xi_{i\,12m}}{r} - \xi_{i\,13m} \ln r + \xi_{i\,14m} \frac{\ln r}{r} \right), \quad i = 1, 2, 3, \tag{7.383}$$

$$w_{im} = \frac{s_{11m}\left(\varrho_{2e}^{me}\right)^2}{2}\left(\xi_{i\,11m}^2 + \frac{\xi_{i\,12m}^2}{r^2} + \xi_{i\,13m}^2 \ln^2 r + \xi_{i\,14m}^2 \frac{\ln^2 r}{r^2} + \frac{2\,\xi_{i\,11m}\,\xi_{i\,12m}}{r} - 2\,\xi_{i\,11m}\,\xi_{i\,13m}\ln r + 2\,\xi_{i\,11m}\,\xi_{i\,14m}\frac{\ln r}{r} - 2\,\xi_{i\,12m}\,\xi_{i\,13m}\frac{\ln r}{r} + 2\,\xi_{i\,12m}\,\xi_{i\,14m}\frac{\ln r}{r^2} - 2\,\xi_{i\,13m}\,\xi_{i\,14m}\frac{\ln^2 r}{r}\right),$$

$$i = 1, 2, 3, \tag{7.384}$$

where the elastic modulus s_{11m} and the coefficients $\xi_{i\,11m}, \ldots, \xi_{i\,14m}$ $(i=1,2,3)$ are given by Eqs. (3.23) and (7.68)–(7.71), respectively. With regard to the φ, ν-independent radial stress p_1, considering Eqs. (3.46), (3.47), (7.367), (7.368), (7.371), (7.372), we get

$$\sigma_{im} = -p_1\,\varrho_{2e}^{me}\left(\xi_{i1m} + \frac{\xi_{i2m}}{r} - \xi_{i3m}\ln r + \xi_{i4m}\frac{\ln r}{r}\right), \quad i = 1, 2, 3, \tag{7.385}$$

$$w_{im} = \frac{s_{11m}\left(p_1\,\varrho_{2e}^{me}\right)^2}{2}\left(\xi_{i1m}^2 + \frac{\xi_{i2m}^2}{r^2} + \xi_{i3m}^2 \ln^2 r + \xi_{i4m}^2 \frac{\ln^2 r}{r^2} + \frac{2\,\xi_{i1m}\,\xi_{i2m}}{r} - 2\,\xi_{i1m}\,\xi_{i3m}\ln r + 2\,\xi_{i1m}\,\xi_{i4m}\frac{\ln r}{r} - 2\,\xi_{i2m}\,\xi_{i3m}\frac{\ln r}{r} + 2\,\xi_{i2m}\,\xi_{i4m}\frac{\ln r}{r^2} - 2\,\xi_{i3m}\,\xi_{i4m}\frac{\ln^2 r}{r}\right),$$

$$i = 1, 2, 3, \tag{7.386}$$

where the coefficients $\xi_{i1m}, \ldots, \xi_{i4m}$ $(i=1,2,3)$ are given by Eqs. (7.363), (7.394), (7.365), (7.395), respectively.

Appendix. Considering the elastic modulus s_{44m} and the coefficients c_{1m}, c_{2m}, a'_{1i}, a'_{2i}, a'_{3i} $(i=1,2,3)$ (see Eqs. (3.25), (3.38), (3.39), (12.26)–(12.34)), the coefficients χ_{12m}, $\omega_{21m}, \ldots, \omega_{29m}$, $\xi_{i1m}, \ldots, \xi_{i4m}$ $(i=1,2,3)$ have the forms

$$\chi_{12m} = \frac{1}{2R_2}\left(\frac{1}{2} + \ln R_2\right), \tag{7.387}$$

$$\omega_{22m} = \frac{1}{9}\left\{\kappa_{1m}^{(1)}\kappa_{3m}^{(1)}\left(5c_{1m} - 8c_{2m}\right) + \frac{4}{s_{44m}}\left[\frac{\partial\kappa_{1m}^{(1)}}{\partial\varphi}\frac{\partial\kappa_{3m}^{(1)}}{\partial\varphi} + \frac{\partial\kappa_{1m}^{(1)}}{\partial\nu}\frac{\partial\kappa_{3m}^{(1)}}{\partial\nu}\right]\right\}, \tag{7.388}$$

$$\omega_{23m} = \frac{1}{8}\left(\left[\kappa_{3m}^{(1)}\right]^2\left(3c_{1m} - 2c_{2m}\right) + \frac{1}{s_{44m}}\left\{\left[\frac{\partial\kappa_{3m}^{(1)}}{\partial\varphi}\right]^2 + \left[\frac{\partial\kappa_{3m}^{(1)}}{\partial\nu}\right]^2\right\}\right), \tag{7.389}$$

$$\omega_{26m} = \frac{1}{9}\left\{2\kappa_{1m}^{(1)}\kappa_{3m}^{(1)}\left(c_{1m}+2c_{2m}\right)+\frac{5}{s_{44m}}\left[\frac{\partial\kappa_{1m}^{(1)}}{\partial\varphi}\frac{\partial\kappa_{3m}^{(1)}}{\partial\varphi}+\frac{\partial\kappa_{1m}^{(1)}}{\partial\nu}\frac{\partial\kappa_{3m}^{(1)}}{\partial\nu}\right]\right\}, \quad (7.390)$$

$$\omega_{27m} = \frac{1}{2}\left(\left[\kappa_{3m}^{(1)}\right]^2\left(c_{1m}-2c_{2m}\right)+\frac{1}{s_{44m}}\left\{\left[\frac{\partial\kappa_{3m}^{(1)}}{\partial\varphi}\right]^2+\left[\frac{\partial\kappa_{3m}^{(1)}}{\partial\nu}\right]^2\right\}\right), \quad (7.391)$$

$$\omega_{28m} = -\frac{2}{3}\left\{\kappa_{1m}^{(1)}\kappa_{3m}^{(1)}\left(c_{1m}-c_{2m}\right)+\frac{1}{s_{44m}}\left[\frac{\partial\kappa_{1m}^{(1)}}{\partial\varphi}\frac{\partial\kappa_{3m}^{(1)}}{\partial\varphi}+\frac{\partial\kappa_{1m}^{(1)}}{\partial\nu}\frac{\partial\kappa_{3m}^{(1)}}{\partial\nu}\right]\right\}, \quad (7.392)$$

$$\omega_{29m} = \frac{1}{2}\left(c_{1m}\left[\kappa_{3m}^{(1)}\right]^2+\frac{1}{s_{44m}}\left\{\left[\frac{\partial\kappa_{3m}^{(1)}}{\partial\varphi}\right]^2+\left[\frac{\partial\kappa_{3m}^{(1)}}{\partial\nu}\right]^2\right\}\right), \quad (7.393)$$

$$\begin{aligned}\xi_{i2m} &= \frac{1}{4}\left\{\kappa_{3m}^{(1)}\left[2a'_{1i}c_{1m}+\left(a'_{2i}+a'_{3i}\right)\left(c_{1m}-2c_{2m}\right)\right]\right.\\ &\left.+\frac{1}{s_{44m}}\left[\left(a'_{1i}+a'_{2i}\right)\frac{\partial\kappa_{3m}^{(1)}}{\partial\varphi}+\left(a'_{1i}+a'_{3i}\right)\frac{\partial\kappa_{3m}^{(1)}}{\partial\nu}\right]\right\},\\ i &= 1,2,3,\end{aligned} \quad (7.394)$$

$$\begin{aligned}\xi_{i4m} &= \frac{1}{2}\left\{\kappa_{3m}^{(1)}\left[c_{1m}\left(a'_{2i}+a'_{3i}\right)-2a'_{1i}c_{2m}\right]\right.\\ &\left.+\frac{1}{s_{44m}}\left[\left(a'_{1i}+a'_{2i}\right)\frac{\partial\kappa_{3m}^{(1)}}{\partial\varphi}+\left(a'_{1i}+a'_{3i}\right)\frac{\partial\kappa_{3m}^{(1)}}{\partial\nu}\right]\right\},\\ i &= 1,2,3.\end{aligned} \quad (7.395)$$

C. Integration constants $C_{1m} = 0$, $C_{2m} \neq 0$, $C_{3m} \neq 0$

With regard to the boundary conditions (4.11), (4.12) and the analysis presented in Section 6.2.3, considering the terms u_{2rm} and u_{3rm} (see Eqs. (6.7), (7.6)) to represent functions of the variable $r \in \langle R_2, r_c\rangle$, the radial displacement u_{1m} of the infinitesimal spherical cap in the cell matrix for the φ,ν-dependent or φ,ν-independent radial stress p_1, regarding Eqs. (3.31), (7.3), has the form

$$u_{1m} = -p_1\,\varrho_{2e}^{me}\left[\kappa_{1m}^{(1)}C_m^{(2)}\,r^{c_{3m}}+\frac{3\kappa_{3m}^{(1)}}{2}\right], \quad (7.396)$$

where the coefficients $c_{3m} < 0$, $C_m^{(2)}$ are given by Eqs. (3.40), (7.4), respectively. The radial stress p_1 acting on the particle-envelope boundary and the coefficient ϱ_{2e}^{me} are analysed at the beginning of Section 6.4.2.2. The φ,ν-dependent distance r_c, as a function of the experimentally determined parameters v, R_{1ex}, d_{1ex}, d_{2ex}, d_{3ex} (see Section 2.2.1), included in the coefficients $\kappa_{1m}^{(1)}$, $\kappa_{3m}^{(1)}$ is derived in Section 2.3.

With regard to Section 7.5.2.2.C, the coefficients $\psi_{11\eta m}^{(1)}$, $\psi_{12\eta m}^{(1)}$ $(\eta = \varphi, \nu)$, $\omega_{21m}, \ldots, \omega_{26m}$, ξ_{i1m}, ξ_{i2m}, ξ_{i3m} $(i = 1,2,3)$ and $\omega_{11m}, \ldots, \omega_{16m}$, ξ_{i11m}, ξ_{i12m}, ξ_{i13m} $(i = 1,2,3)$ mentioned below (see Eqs. (7.399), (7.400), (7.405), (7.406), (7.411), (7.415), (7.416) and (7.409), (7.413), (7.414)) depend on $\kappa_{1m}^{(1)}$, $\kappa_{3m}^{(1)}$ and $\kappa_{1m}^{(1)}$, $\kappa_{3m}^{(1)}$, $\psi_{11\eta m}^{(1)}$, $\psi_{12\eta m}^{(1)}$, respectively, where the coefficients $\kappa_{1m}^{(1)}$, $\kappa_{3m}^{(1)}$ (see Eq. (12.35)) are functions of χ_{11m}, χ_{12m}, χ_{21m}, χ_{22m} given by Eqs. (7.417), (7.362), (7.87), (7.59), respectively.

Consequently, considering the φ, ν-dependent or φ, ν-independent radial stress p_1, the radial and tangential stresses, σ_{11m} and σ_{22m}, σ_{33m}, respectively, acting in the cell matrix, regarding Eqs. (3.31), (3.32), (7.396), are derived as

$$\sigma_{11m} = -p_1\, \varrho_{2e}^{me} \left\{ \kappa_{1m}^{(1)} C_m^{(2)} \left[c_{3m} \left(c_{1m} + c_{2m} \right) - 2c_{2m} \right] r^{c_{3m}-1} + \frac{\kappa_{3m}^{(1)}}{2r} \left(c_{1m} - 2c_{2m} \ln r \right) \right\}, \tag{7.397}$$

$$\sigma_{22m} = \sigma_{33m} = -p_1\, \varrho_{2e}^{me} \left\{ \kappa_{1m}^{(1)} C_m^{(2)} \left(c_{1m} - c_{2m} c_{3m} \right) r^{c_{3m}-1} + \frac{\kappa_{3m}^{(1)}}{2r} \left[c_{1m} \left(\frac{1}{2} + \ln r \right) - c_{2m} \right] \right\}, \tag{7.398}$$

where the coefficients c_{1m}, c_{2m} are given by Eqs. (3.38), (3.39), respectively.

With regard to Eqs. (3.33), (3.34), (7.396), the shear stresses σ_{12m}, σ_{13m} acting in the cell matrix, considering the φ, ν-dependent radial stress p_1, are derived as

$$\sigma_{12m} = -\frac{\varrho_{2e}^{me}}{s_{44m}} \left[\psi_{11\varphi m}^{(1)} C_m^{(2)} r^{c_{3m}-1} + \frac{\psi_{12\varphi m}^{(1)}}{2r} \left(\frac{1}{2} + \ln r \right) \right], \tag{7.399}$$

$$\sigma_{13m} = -\frac{\varrho_{2e}^{me}}{s_{44m}} \left[\psi_{11\nu m}^{(1)} C_m^{(2)} r^{c_{3m}-1} + \frac{\psi_{12\nu m}^{(1)}}{2r} \left(\frac{1}{2} + \ln r \right) \right], \tag{7.400}$$

where the elastic modulus s_{44m} and the coefficients $\psi_{11\eta m}^{(1)}$, $\psi_{12\eta m}^{(1)}$ $(\eta = \varphi, \nu)$ are given by Eqs. (3.25) and (12.36), respectively. Considering the φ, ν-independent radial stress p_1, we get

$$\sigma_{12m} = -\frac{p_1\, \varrho_{2e}^{me}}{s_{44m}} \left[\frac{\partial \kappa_{1m}^{(1)}}{\partial \varphi} C_m^{(2)} r^{c_{3m}-1} + \frac{1}{2r} \frac{\partial \kappa_{3m}^{(1)}}{\partial \varphi} \left(\frac{1}{2} + \ln r \right) \right], \tag{7.401}$$

$$\sigma_{13m} = -\frac{p_1\, \varrho_{2e}^{me}}{s_{44m}} \left[\frac{\partial \kappa_{1m}^{(1)}}{\partial \nu} C_m^{(2)}\, r^{c_{3m}-1} + \frac{1}{2r} \frac{\partial \kappa_{3m}^{(1)}}{\partial \nu} \left(\frac{1}{2} + \ln r \right) \right]. \tag{7.402}$$

Finally, considering the φ, ν-dependent or φ, ν-independent radial stress p_1, the radial and tangential strains, ε_{11m} and ε_{22m}, ε_{33m}, respectively, acting in the cell matrix, regarding Eqs. (3.1)–(3.3), (7.396), have the forms

$$\varepsilon_{11m} = -p_1\, \varrho_{2e}^{me} \left[\kappa_{1m}^{(1)} C_m^{(2)} c_{3m}\, r^{c_{3m}-1} + \frac{\kappa_{3m}^{(1)}}{2r} \right], \tag{7.403}$$

$$\varepsilon_{22m} = \varepsilon_{33m} = -p_1\, \varrho_{2e}^{me} \left[\kappa_{1m}^{(1)} C_m^{(2)}\, r^{c_{3m}-1} + \frac{\kappa_{3m}^{(1)}}{2r} \left(\frac{1}{2} + \ln r \right) \right]. \tag{7.404}$$

With regard to Eqs. (3.4), (3.5), (7.396), the shear strains σ_{12m}, σ_{13m} acting in the cell matrix, considering the φ, ν-dependent radial stress p_1, are derived as

$$\varepsilon_{12m} = -\varrho_{2e}^{me} \left[\psi_{11\varphi m}^{(1)} C_m^{(2)}\, r^{c_{3m}-1} + \frac{\psi_{12\varphi m}^{(1)}}{2r} \left(\frac{1}{2} + \ln r \right) \right], \tag{7.405}$$

$$\varepsilon_{13m} = -\varrho_{2e}^{me} \left[\psi_{11\nu m}^{(1)} C_m^{(2)}\, r^{c_{3m}-1} + \frac{\psi_{12\nu m}^{(1)}}{2r} \left(\frac{1}{2} + \ln r \right) \right], \tag{7.406}$$

and considering the φ, ν-independent radial stress p_1, we get

$$\varepsilon_{12m} = -p_1\, \varrho_{2e}^{me} \left[\frac{\partial \kappa_{1m}^{(1)}}{\partial \varphi} C_m^{(2)}\, r^{c_{3m}-1} + \frac{1}{2r} \frac{\partial \kappa_{3m}^{(1)}}{\partial \varphi} \left(\frac{1}{2} + \ln r \right) \right], \tag{7.407}$$

$$\varepsilon_{13m} = -p_1\, \varrho_{2e}^{me} \left[\frac{\partial \kappa_{1m}^{(1)}}{\partial \nu} C_m^{(2)}\, r^{c_{3m}-1} + \frac{1}{2r} \frac{\partial \kappa_{3m}^{(1)}}{\partial \nu} \left(\frac{1}{2} + \ln r \right) \right]. \tag{7.408}$$

With regard to the φ, ν-dependent radial stress p_1, considering Eqs. (3.41), (3.42), (7.397)–(7.400), (7.403)–(7.406), the thermal-stress induced elastic energy density w_m and the thermal-stress induced elastic energy W_m, accumulated in an arbitrary point of the cell matrix and in the volume of the cell matrix, respectively, are derived as

$$w_m = \frac{\left(\varrho_{2e}^{me}\right)^2}{2} \left[\frac{\omega_{11m}}{r^2} + \omega_{12m}\, r^{c_{3m}-2} + \omega_{13m}\, r^{2(c_{3m}-1)} + \omega_{14m} \frac{\ln r}{r^2} + \omega_{15m}\, r^{c_{3m}-2} \ln r + \omega_{16m} \frac{\ln^2 r}{r^2} \right], \tag{7.409}$$

$$W_m = 8 \int_0^{\pi/2} \int_0^{\pi/2} \int_{R_2}^{r_c} w_m r^2 dr d\varphi d\nu = (\varrho_{2e}^{me})^2 \int_0^{\pi/2} \int_0^{\pi/2} \Omega_{12m} d\varphi d\nu, \tag{7.410}$$

where the coefficient Ω_{12m} given by Eq. (7.95) is a function of $\omega_{11m}, \ldots, \omega_{16m}$ given by Eqs. (7.89)–(7.94), respectively.

With regard to the φ, ν-independent radial stress p_1, considering Eqs. (3.41), (3.42), (7.397), (7.398), (7.401)–(7.404), (7.407), (7.408), we get

$$w_m = \frac{(p_1 \varrho_{2e}^{me})^2}{2} \left[\frac{\omega_{21m}}{r^2} + \omega_{22m} r^{c_{3m}-2} + \omega_{23m} r^{2(c_{3m}-1)} + \omega_{24m} \frac{\ln r}{r^2} \right.$$
$$\left. + \omega_{25m} r^{c_{3m}-2} \ln r + \omega_{26m} \frac{\ln^2 r}{r^2} \right], \tag{7.411}$$

$$W_m = 8 \int_0^{\pi/2} \int_0^{\pi/2} \int_{R_2}^{r_c} w_m r^2 \, dr \, d\varphi \, d\nu = (p_1 \varrho_{2e}^{me})^2 \int_0^{\pi/2} \int_0^{\pi/2} \Omega_{22m} \, d\varphi \, d\nu, \tag{7.412}$$

where the coefficient Ω_{22m} given by Eq. (7.94) is a function of $\omega_{21m}, \ldots, \omega_{26m}$ given by Eqs. (7.418)–(7.423), respectively.

With regard to the determination of the thermal-stress induced strengthening presented in Volume III, the stress σ_{im} acting in a direction of the axis x_i and the elastic energy density w_{im} induced by the stress σ_{im} $(i=1,2,3)$ (see Eqs. (3.46), (3.47), (7.397)–(7.400)), both related to an arbitrary point of the cell matrix, considering the φ, ν-dependent radial stress p_1, have the forms

$$\sigma_{im} = -\varrho_{2e}^{me} \left(\frac{\xi_{i11m}}{r} + \xi_{i12m} r^{c_{3m}-1} + \xi_{i13m} \frac{\ln r}{r} \right), \quad i = 1, 2, 3, \tag{7.413}$$

$$w_{im} = \frac{s_{11m} (\varrho_{2e}^{me})^2}{2} \left[\frac{\xi_{i11m}^2}{r^2} + \xi_{i12m}^2 r^{2(c_{3m}-1)} + \xi_{i13m}^2 \frac{\ln^2 r}{r^2} + 2 \xi_{i11m} \xi_{i12m} r^{c_{3m}-2} \right.$$
$$\left. + 2 \xi_{i11m} \xi_{i13m} \frac{\ln r}{r^2} + 2 \xi_{i12m} \xi_{i13m} r^{c_{3m}-2} \ln r \right],$$
$$i = 1, 2, 3, \tag{7.414}$$

where the elastic modulus s_{11m} and the coefficients ξ_{i11m}, ξ_{i12m}, ξ_{i13m} $(i=1,2,3)$ are given by Eqs. (3.23) and (7.96)–(7.98), respectively. With regard to the φ, ν-independent radial stress p_1, considering Eqs. (3.46), (3.47), (7.397), (7.398), (7.401), (7.402), we get

$$\sigma_{im} = -p_1 \varrho_{2e}^{me} \left(\frac{\xi_{i1m}}{r} + \xi_{i2m} r^{c_{3m}-1} + \xi_{i3m} \frac{\ln r}{r} \right), \quad i = 1, 2, 3, \tag{7.415}$$

$$w_{im} = \frac{s_{11m}\left(p_1\, \varrho_{2e}^{me}\right)^2}{2}\left[\frac{\xi_{i1m}^2}{r^2} + \xi_{i2m}^2\, r^{2(c_{3m}-1)} + \xi_{i3m}^2\frac{\ln^2 r}{r^2} + 2\,\xi_{i1m}\,\xi_{i2m}\, r^{c_{3m}-2} \right.$$
$$\left. + 2\,\xi_{i1m}\,\xi_{i3m}\frac{\ln r}{r^2} + 2\,\xi_{i2m}\,\xi_{i3m}\, r^{c_{3m}-2}\ln r\right],$$
$$i = 1, 2, 3, \tag{7.416}$$

where the coefficients ξ_{i1m}, ξ_{i2m}, ξ_{i3m} $(i=1,2,3)$ are given by Eqs. (7.424)–(7.426), respectively.

Appendix. Considering the elastic modulus s_{44m} and the coefficients c_{1m}, c_{2m}, c_{3m}, $C_m^{(2)}$, a'_{1i}, a'_{2i}, a'_{3i} $(i=1,2,3)$ (see Eqs. (3.25), (3.38)–(3.40), (7.4), (12.26)–(12.34)), the coefficients χ_{11m}, $\omega_{21m}, \ldots, \omega_{26m}$, ξ_{i1m}, ξ_{i2m}, ξ_{i3m} $(i=1,2,3)$ have the forms

$$\chi_{11m} = C_m^{(2)} R_2^{c_{3m}-1}, \tag{7.417}$$

$$\omega_{21m} = \frac{1}{8}\left(\left[\kappa_{3m}^{(1)}\right]^2 (3c_{1m} - 2c_{2m}) + \frac{1}{s_{44m}}\left\{\left[\frac{\partial\kappa_{3m}^{(1)}}{\partial\varphi}\right]^2 + \left[\frac{\partial\kappa_{3m}^{(1)}}{\partial\nu}\right]^2\right\}\right), \tag{7.418}$$

$$\omega_{22m} = C_m^{(2)}\left\{\kappa_{1m}^{(1)}\kappa_{3m}^{(1)}\left[c_{1m}\left(c_{3m}+1\right) - 2c_{2m}\right]\right.$$
$$\left. + \frac{1}{s_{44m}}\left[\frac{\partial\kappa_{1m}^{(1)}}{\partial\varphi}\frac{\partial\kappa_{3m}^{(1)}}{\partial\varphi} + \frac{\partial\kappa_{1m}^{(1)}}{\partial\nu}\frac{\partial\kappa_{3m}^{(1)}}{\partial\nu}\right]\right\}, \tag{7.419}$$

$$\omega_{23m} = \left[C_m^{(2)}\right]^2\left(\left[\kappa_{1m}^{(1)}\right]^2\left[c_{3m}^2\left(c_{1m}+c_{2m}\right) + 2\left(c_{1m} - 2c_{2m}c_{3m}\right)\right]\right.$$
$$\left. + \frac{2}{s_{44m}}\left\{\left[\frac{\partial\kappa_{1m}^{(1)}}{\partial\varphi}\right]^2 + \left[\frac{\partial\kappa_{1m}^{(1)}}{\partial\nu}\right]^2\right\}\right), \tag{7.420}$$

$$\omega_{24m} = \frac{1}{2}\left(\left[\kappa_{3m}^{(1)}\right]^2 (c_{1m} - 2c_{2m}) + \frac{1}{s_{44m}}\left\{\left[\frac{\partial\kappa_{3m}^{(1)}}{\partial\varphi}\right]^2 + \left[\frac{\partial\kappa_{3m}^{(1)}}{\partial\nu}\right]^2\right\}\right), \tag{7.421}$$

$$\omega_{25m} = 2C_m^{(2)}\left\{\kappa_{1m}^{(1)}\kappa_{3m}^{(1)}\left(c_{1m} - c_{2m}c_{3m}\right)\right.$$
$$\left. + \frac{1}{s_{44m}}\left[\frac{\partial\kappa_{1m}^{(1)}}{\partial\varphi}\frac{\partial\kappa_{3m}^{(1)}}{\partial\varphi} + \frac{\partial\kappa_{1m}^{(1)}}{\partial\nu}\frac{\partial\kappa_{3m}^{(1)}}{\partial\nu}\right]\right\}, \tag{7.422}$$

$$\omega_{26m} = \frac{1}{2}\left(c_{1m}\left[\kappa_{3m}^{(1)}\right]^2 + \frac{1}{s_{44m}}\left\{\left[\frac{\partial\kappa_{3m}^{(1)}}{\partial\varphi}\right]^2 + \left[\frac{\partial\kappa_{3m}^{(1)}}{\partial\nu}\right]^2\right\}\right), \tag{7.423}$$

$$\begin{aligned}\xi_{i1m} = \frac{1}{4}\Bigg\{&\kappa_{3m}^{(1)}\left[2a'_{1i}c_{1m} + \left(a'_{2i} + a'_{3i}\right)\left(c_{1m} - 2c_{2m}\right)\right] \\ &+ \frac{1}{s_{44m}}\left[\left(a'_{1i} + a'_{2i}\right)\frac{\partial\kappa_{3m}^{(1)}}{\partial\varphi} + \left(a'_{1i} + a'_{3i}\right)\frac{\partial\kappa_{3m}^{(1)}}{\partial\nu}\right]\Bigg\}, \\ i = 1, 2, 3,&\end{aligned} \tag{7.424}$$

$$\begin{aligned}\xi_{i2m} = C_m^{(2)}\Bigg(&\kappa_{1m}^{(1)}\left\{a'_{1i}\left[c_{3m}\left(c_{1m} + c_{2m}\right) - 2c_{2m}\right] + \left(a'_{2i} + a'_{3i}\right)\left(c_{1m} - c_{2m}c_{3m}\right)\right\} \\ &+ \frac{1}{s_{44m}}\left[\left(a'_{1i} + a'_{2i}\right)\frac{\partial\kappa_{1m}^{(1)}}{\partial\varphi} + \left(a'_{1i} + a'_{3i}\right)\frac{\partial\kappa_{1m}^{(1)}}{\partial\nu}\right]\Bigg), \\ i = 1, 2, 3,&\end{aligned} \tag{7.425}$$

$$\begin{aligned}\xi_{i3m} = \frac{1}{2}\Bigg\{&\kappa_{3m}^{(1)}\left[c_{1m}\left(a'_{2i} + a'_{3i}\right) - 2a'_{1i}c_{2m}\right] \\ &+ \frac{1}{s_{44m}}\left[\left(a'_{1i} + a'_{2i}\right)\frac{\partial\kappa_{3m}^{(1)}}{\partial\varphi} + \left(a'_{1i} + a'_{3i}\right)\frac{\partial\kappa_{3m}^{(1)}}{\partial\nu}\right]\Bigg\}, \\ i = 1, 2, 3.&\end{aligned} \tag{7.426}$$

D. Integration constants $C_{1m} \neq 0$, $C_{2m} \neq 0$, $C_{3m} \neq 0$

With regard to the boundary conditions (4.11), (4.12), (4.14) and the analysis presented in Section 6.2.3, considering the terms u_{1rm}, u_{2rm} and u_{3rm} (see Eqs. (7.5), (6.7), (7.6)) to represent functions of the variable $r \in \langle R_2, r_c\rangle$, the radial displacement u_{1m} of the infinitesimal spherical cap in the cell matrix for the φ,ν-dependent or φ,ν-independent radial stress p_1, regarding Eqs. (3.31), (7.3), has the form

$$u_{1m} = -p_1\,\varrho_{2e}^{me}\left[\frac{\kappa_{4m}^{(2)}r}{3}\left(\frac{4}{3} - \ln r\right) + \kappa_{5m}^{(2)}C_m^{(2)}\,r^{c_{3m}} + \frac{\kappa_{6m}^{(2)}}{2}\left(\frac{1}{2} + \ln r\right)\right], \tag{7.427}$$

where the coefficients $c_{3m} < 0$, $C_m^{(2)}$ are given by Eqs. (3.40), (7.4), respectively. The radial stress p_1 acting on the particle-envelope boundary and the coefficient ϱ_{2e}^{me} are analysed at the beginning of Section 6.4.2.2. The φ,ν-dependent distance r_c, as a function of the experimentally determined parameters v, R_{1ex}, d_{1ex}, d_{2ex}, d_{3ex} (see Section 2.2.1), included in the coefficients $\kappa_{4m}^{(2)}$, $\kappa_{5m}^{(2)}$, $\kappa_{6m}^{(2)}$ is derived in Section 2.3.

With regard to Section 7.5.2.2.D, the coefficients $\psi_{11\eta m}^{(2)}$, $\psi_{12\eta m}^{(2)}$, $\psi_{13\eta m}^{(2)}$ $(\eta = \varphi, \nu)$, $\omega_{21m}, \ldots, \omega_{214m}$, $\xi_{i1m}, \ldots, \xi_{i5m}$ $(i = 1,2,3)$ and $\omega_{11m}, \ldots, \omega_{114m}$, $\xi_{i11m}, \ldots, \xi_{i15m}$

($i=1,2,3$) mentioned below (see Eqs. (7.430), (7.431), (7.436), (7.437), (7.442), (7.446), (7.447) and (7.440), (7.444), (7.445)) depend on $\kappa_{4m}^{(2)}$, $\kappa_{5m}^{(2)}$, $\kappa_{6m}^{(2)}$ and $\kappa_{4m}^{(2)}$, $\kappa_{5m}^{(2)}$, $\kappa_{6m}^{(2)}$, $\psi_{11\eta m}^{(2)}$, $\psi_{12\eta m}^{(2)}$, $\psi_{13\eta m}^{(2)}$, respectively, where the coefficients $\kappa_{4m}^{(2)}$, $\kappa_{5m}^{(2)}$, $\kappa_{6m}^{(2)}$ (see Eq. (12.39)) are functions of χ_{11m}, χ_{12m}, χ_{13m}, χ_{21m}, χ_{22m}, χ_{23m}, χ_{31m}, χ_{32m}, χ_{33m} given by Eqs. (7.347), (7.348), (7.448), (7.32), (7.33), (7.114)–(7.117), respectively.

Consequently, considering the φ,ν-dependent or φ,ν-independent radial stress p_1, the radial and tangential stresses, σ_{11m} and σ_{22m}, σ_{33m}, respectively, acting in the cell matrix, regarding Eqs. (3.31), (3.32), (7.427), are derived as

$$\sigma_{11m} = -p_1\,\varrho_{2e}^{me}\left\{\frac{\kappa_{4m}^{(2)}}{3}\left[\frac{1}{3}\left(c_{1m}-7c_{2m}\right)-\left(c_{1m}-c_{2m}\right)\ln r\right]\right.$$
$$\left.+\,\kappa_{5m}^{(2)}C_m^{(2)}\left[c_{3m}\left(c_{1m}+c_{2m}\right)-2c_{2m}\right]r^{c_{3m}-1}+\frac{\kappa_{6m}^{(2)}}{2r}\left(c_{1m}-2c_{2m}\ln r\right)\right\}, \quad (7.428)$$

$$\sigma_{22m} = \sigma_{33m} = -p_1\,\varrho_{2e}^{me}\left\{\frac{\kappa_{4m}^{(2)}}{3}\left[\frac{1}{3}\left(4c_{1m}-c_{2m}\right)-\left(c_{1m}-c_{2m}\right)\ln r\right]\right.$$
$$\left.+\,\kappa_{5m}^{(2)}C_m^{(2)}\left(c_{1m}-c_{2m}c_{3m}\right)r^{c_{3m}-1}+\frac{\kappa_{6m}^{(2)}}{2r}\left[c_{1m}\left(\frac{1}{2}+\ln r\right)-c_{2m}\right]\right\}, \quad (7.429)$$

where the coefficients c_{1m}, c_{2m} are given by Eqs. (3.38), (3.39), respectively.

With regard to Eqs. (3.33), (3.34), (7.427), the shear stresses σ_{12m}, σ_{13m} acting in the cell matrix, considering the φ,ν-dependent radial stress p_1, are derived as

$$\sigma_{12m} = -\frac{\varrho_{2e}^{me}}{s_{44m}}\left[\frac{\Psi_{11\varphi m}^{(2)}}{3}\left(\frac{4}{3}-\ln r\right)\right.$$
$$\left.+\,\Psi_{12\varphi m}^{(2)}C_m^{(2)}\,r^{c_{3m}-1}+\frac{\Psi_{13\varphi m}^{(2)}}{2r}\left(\frac{1}{2}+\ln r\right)\right], \quad (7.430)$$

$$\sigma_{13m} = -\frac{\varrho_{2e}^{me}}{s_{44m}}\left[\frac{\Psi_{11\nu m}^{(2)}}{3}\left(\frac{4}{3}-\ln r\right)\right.$$
$$\left.+\,\Psi_{12\nu m}^{(2)}C_m^{(2)}\,r^{c_{3m}-1}+\frac{\Psi_{13\nu m}^{(2)}}{2r}\left(\frac{1}{2}+\ln r\right)\right], \quad (7.431)$$

where the elastic modulus s_{44m} and the coefficients $\psi_{11\eta m}^{(2)}$, $\psi_{12\eta m}^{(2)}$, $\psi_{13\eta m}^{(2)}$ $(\eta = \varphi, \nu)$ are given by Eqs. (3.25) and (12.40), respectively. Considering the φ, ν-independent radial stress p_1, we get

$$\sigma_{12m} = -\frac{p_1\,\varrho_{2e}^{me}}{s_{44m}}\left[\frac{1}{3}\frac{\partial\kappa_{4m}^{(2)}}{\partial\varphi}\left(\frac{4}{3} - \ln r\right) + \frac{\partial\kappa_{5m}^{(2)}}{\partial\varphi}\,C_m^{(2)}\,r^{c_{3m}-1} + \frac{1}{2r}\frac{\partial\kappa_{6m}^{(2)}}{\partial\varphi}\left(\frac{1}{2} + \ln r\right)\right], \tag{7.432}$$

$$\sigma_{13m} = -\frac{p_1\,\varrho_{2e}^{me}}{s_{44m}}\left[\frac{1}{3}\frac{\partial\kappa_{4m}^{(2)}}{\partial\nu}\left(\frac{4}{3} - \ln r\right) + \frac{\partial\kappa_{5m}^{(2)}}{\partial\nu}\,C_m^{(2)}\,r^{c_{3m}-1} + \frac{1}{2r}\frac{\partial\kappa_{6m}^{(2)}}{\partial\nu}\left(\frac{1}{2} + \ln r\right)\right]. \tag{7.433}$$

Finally, considering the φ, ν-dependent or φ, ν-independent radial stress p_1, the radial and tangential strains, ε_{11m} and ε_{22m}, ε_{33m}, respectively, acting in the cell matrix, regarding Eqs. (3.1)–(3.3), (7.427), have the forms

$$\varepsilon_{11m} = -p_1\,\varrho_{2e}^{me}\left[\frac{\kappa_{4m}^{(2)}}{3}\left(\frac{1}{3} - \ln r\right) + \kappa_{5m}^{(2)}C_m^{(2)}c_{3m}\,r^{c_{3m}-1} + \frac{\kappa_{6m}^{(2)}}{2r}\right], \tag{7.434}$$

$$\varepsilon_{22m} = \varepsilon_{33m} = -p_1\,\varrho_{2e}^{me}\left[\frac{\kappa_{4m}^{(2)}}{3}\left(\frac{4}{3} - \ln r\right) + \kappa_{5m}^{(2)}C_m^{(2)}\,r^{c_{3m}-1} + \frac{\kappa_{6m}^{(2)}}{2r}\left(\frac{1}{2} + \ln r\right)\right]. \tag{7.435}$$

With regard to Eqs. (3.4), (3.5), (7.427), the shear strains σ_{12m}, σ_{13m} acting in the cell matrix, considering the φ, ν-dependent radial stress p_1, are derived as

$$\varepsilon_{12m} = -\varrho_{2e}^{me}\left[\frac{\Psi_{11\varphi m}^{(2)}}{3}\left(\frac{4}{3} - \ln r\right) + \Psi_{12\varphi m}^{(2)}C_m^{(2)}\,r^{c_{3m}-1} + \frac{\Psi_{13\varphi m}^{(2)}}{2r}\left(\frac{1}{2} + \ln r\right)\right], \tag{7.436}$$

$$\varepsilon_{13m} = -\varrho_{2e}^{me} \left[\frac{\Psi_{11\nu m}^{(2)}}{3} \left(\frac{4}{3} - \ln r \right) + \Psi_{12\nu m}^{(2)} C_m^{(2)} r^{c_{3m}-1} + \frac{\Psi_{13\nu m}^{(2)}}{2r} \left(\frac{1}{2} + \ln r \right) \right], \quad (7.437)$$

and considering the φ, ν-independent radial stress p_1, we get

$$\varepsilon_{12m} = -p_1 \, \varrho_{2e}^{me} \left[\frac{1}{3} \frac{\partial \kappa_{4m}^{(2)}}{\partial \varphi} \left(\frac{4}{3} - \ln r \right) + \frac{\partial \kappa_{5m}^{(2)}}{\partial \varphi} C_m^{(2)} r^{c_{3m}-1} + \frac{1}{2r} \frac{\partial \kappa_{6m}^{(2)}}{\partial \varphi} \left(\frac{1}{2} + \ln r \right) \right], \quad (7.438)$$

$$\varepsilon_{13m} = -p_1 \, \varrho_{2e}^{me} \left[\frac{1}{3} \frac{\partial \kappa_{4m}^{(2)}}{\partial \nu} \left(\frac{4}{3} - \ln r \right) + \frac{\partial \kappa_{5m}^{(2)}}{\partial \nu} C_m^{(2)} r^{c_{3m}-1} + \frac{1}{2r} \frac{\partial \kappa_{6m}^{(2)}}{\partial \nu} \left(\frac{1}{2} + \ln r \right) \right]. \quad (7.439)$$

With regard to the φ, ν-dependent radial stress p_1, considering Eqs. (3.41), (3.42), (7.428)–(7.431), (7.434)–(7.437), the thermal-stress induced elastic energy density w_m and the thermal-stress induced elastic energy W_m, accumulated in an arbitrary point of the cell matrix and in the volume of the cell matrix, respectively, are derived as

$$w_m = \frac{\left(\varrho_{2e}^{me}\right)^2}{2} \left[\omega_{11m} + \frac{\omega_{12m}}{r} + \frac{\omega_{13m}}{r^2} + \omega_{14m} \, r^{c_{3m}-1} + \omega_{15m} \, r^{c_{3m}-2} + \omega_{16m} \, r^{2(c_{3m}-1)} + \omega_{17m} \ln r + \omega_{18m} \ln^2 r + \frac{\omega_{19m} \ln r}{r} + \frac{\omega_{110m} \ln r}{r^2} + \frac{\omega_{111m} \ln^2 r}{r} + \omega_{112m} \frac{\ln^2 r}{r^2} + \omega_{113m} \, r^{c_{3m}-1} \ln r + \omega_{114m} \, r^{c_{3m}-2} \ln r \right], \quad (7.440)$$

$$W_m = 8 \int_0^{\pi/2} \int_0^{\pi/2} \int_{R_2}^{r_c} w_m \, r^2 dr d\varphi d\nu = \left(\varrho_{2e}^{me}\right)^2 \int_0^{\pi/2} \int_0^{\pi/2} \Omega_{12m} d\varphi d\nu, \quad (7.441)$$

where the coefficient Ω_{12m} given by Eq. (7.133) is a function of $\omega_{11m}, \ldots, \omega_{114m}$ Eqs. (7.119)–(7.132), respectively.

With regard to the φ, ν-independent radial stress p_1, considering Eqs. (3.41), (3.42), (7.428), (7.429), (7.432)–(7.435), (7.438), (7.439), we get

$$w_m = \frac{(p_1\, \varrho_{2e}^{me})^2}{2} \Big[\omega_{21m} + \frac{\omega_{22m}}{r} + \frac{\omega_{23m}}{r^2} + \omega_{24m}\, r^{c_{3m}-1} + \omega_{25m}\, r^{c_{3m}-2} + \omega_{26m}\, r^{2(c_{3m}-1)} + \omega_{27m} \ln r + \omega_{28m} \ln^2 r + \omega_{29m} \frac{\ln r}{r} + \omega_{210m} \frac{\ln r}{r^2} + \omega_{211m} \frac{\ln^2 r}{r} + \omega_{212m} \frac{\ln^2 r}{r^2} + \omega_{213m}\, r^{c_{3m}-1} \ln r + \omega_{214m}\, r^{c_{3m}-2} \ln r \Big], \tag{7.442}$$

$$W_m = 8 \int_0^{\pi/2} \int_0^{\pi/2} \int_{R_2}^{r_c} w_m\, r^2\, dr\, d\varphi\, d\nu = (p_1\, \varrho_{2e}^{me})^2 \int_0^{\pi/2} \int_0^{\pi/2} \Omega_{22m}\, d\varphi\, d\nu, \tag{7.443}$$

where the coefficient Ω_{22m} given by Eq. (7.132) is a function of $\omega_{21m}, \ldots, \omega_{214m}$ given by Eqs. (7.449)–(7.462), respectively.

With regard to the determination of the thermal-stress induced strengthening presented in Volume III, the stress σ_{im} acting in a direction of the axis x_i and the elastic energy density w_{im} induced by the stress σ_{im} ($i = 1,2,3$) (see Eqs. (3.46), (3.47), (7.428)–(7.431)), both related to an arbitrary point of the cell matrix, considering the φ, ν-dependent radial stress p_1, have the forms

$$\sigma_{im} = -\varrho_{2e}^{me} \left(\xi_{i\,11m} + \frac{\xi_{i\,12m}}{r} + \xi_{i\,13m}\, r^{c_{3m}-1} - \xi_{i\,14m} \ln r + \xi_{i\,15m} \frac{\ln r}{r} \right), \quad i = 1, 2, 3, \tag{7.444}$$

$$w_{im} = \frac{s_{11m}\, (\varrho_{2e}^{me})^2}{2} \Big[\xi_{i\,11m}^2 + \frac{\xi_{i\,12m}^2}{r^2} + \xi_{i\,13m}^2\, r^{2(c_{3m}-1)} + \xi_{i\,14m}^2 \ln^2 r + \xi_{i\,15m}^2 \frac{\ln^2 r}{r^2} + \frac{2\,\xi_{i\,11m}\,\xi_{i\,12m}}{r} + 2\,\xi_{i\,11m}\,\xi_{i\,13m}\, r^{c_{3m}-1} - 2\,\xi_{i\,11m}\,\xi_{i\,14m} \ln r + 2\,(\xi_{i\,11m}\,\xi_{i\,15m} - \xi_{i\,12m}\,\xi_{i\,14m}) \frac{\ln r}{r} + 2\,\xi_{i\,12m}\,\xi_{i\,13m}\, r^{c_{3m}-2} + 2\,\xi_{i\,12m}\,\xi_{i\,15m} \frac{\ln r}{r^2} - 2\,\xi_{i\,13m}\,\xi_{i\,14m}\, r^{c_{3m}-1} \ln r + 2\,\xi_{i\,13m}\,\xi_{i\,15m}\, r^{c_{3m}-2} \ln r - 2\,\xi_{i\,14m}\,\xi_{i\,15m} \frac{\ln^2 r}{r} \Big], \quad i = 1, 2, 3, \tag{7.445}$$

where the elastic modulus s_{11m} and the coefficients $\xi_{i\,11m}, \ldots, \xi_{i\,15m}$ ($i = 1,2,3$) are given by Eqs. (3.23) and (7.134)–(7.138), respectively. With regard to the φ, ν-independent radial stress p_1, considering Eqs. (3.46), (3.47), (7.428), (7.429), (7.432), (7.433), we get

$$\sigma_{im} = -p_1\, \varrho_{2e}^{me} \left(\xi_{i1m} + \frac{\xi_{i2m}}{r} + \xi_{i3m}\, r^{c_{3m}-1} - \xi_{i4m} \ln r + \xi_{i5m} \frac{\ln r}{r} \right),$$
$$i = 1, 2, 3, \tag{7.446}$$

$$\begin{aligned} w_{im} = \frac{s_{11m}\,(p_1\, \varrho_{2e}^{me})^2}{2} \bigg[& \xi_{i1m}^2 + \frac{\xi_{i2m}^2}{r^2} + \xi_{i3m}^2\, r^{2(c_{3m}-1)} + \xi_{i4m}^2 \ln^2 r + \xi_{i5m}^2 \frac{\ln^2 r}{r^2} \\ & + \frac{2\,\xi_{i1m}\,\xi_{i2m}}{r} + 2\,\xi_{i1m}\,\xi_{i3m}\, r^{c_{3m}-1} - 2\,\xi_{i1m}\,\xi_{i4m} \ln r \\ & + 2\,(\xi_{i1m}\,\xi_{i5m} - \xi_{i2m}\,\xi_{i4m}) \frac{\ln r}{r} + 2\,\xi_{i2m}\,\xi_{i3m}\, r^{c_{3m}-2} \\ & + 2\,\xi_{i2m}\,\xi_{i5m} \frac{\ln r}{r^2} - 2\,\xi_{i3m}\,\xi_{i4m}\, r^{c_{3m}-1} \ln r \\ & + 2\,\xi_{i3m}\,\xi_{i5m}\, r^{c_{3m}-2} \ln r - 2\,\xi_{i4m}\,\xi_{i5m} \frac{\ln^2 r}{r} \bigg], \end{aligned}$$
$$i = 1, 2, 3, \tag{7.447}$$

where the coefficients $\xi_{i1m}, \ldots, \xi_{i5m}$ $(i = 1,2,3)$ are given by Eqs. (7.463)–(7.467), respectively.

Appendix. Considering the elastic modulus s_{44m} and the coefficients c_{1m}, c_{2m}, $C_m^{(2)}$, a'_{1i}, a'_{2i}, a'_{3i} $(i = 1,2,3)$ (see Eqs. (3.25), (3.38), (3.39), (7.4), (12.26)–(12.34)), the coefficients χ_{13m}, $\omega_{21m}, \ldots, \omega_{214m}$, $\xi_{i1m}, \ldots, \xi_{i5m}$ $(i = 1,2,3)$ have the forms

$$\chi_{13m} = \frac{1}{2R_2} \left(\frac{1}{2} + \ln R_2 \right), \tag{7.448}$$

$$\omega_{21m} = \frac{1}{81} \left(3 \left[\kappa_{4m}^{(2)} \right]^2 (11c_{1m} - 5c_{2m}) + \frac{32}{s_{44m}} \left\{ \left[\frac{\partial \kappa_{4m}^{(2)}}{\partial \varphi} \right]^2 + \left[\frac{\partial \kappa_{4m}^{(2)}}{\partial \nu} \right]^2 \right\} \right), \tag{7.449}$$

$$\omega_{22m} = \frac{1}{9} \left\{ \kappa_{4m}^{(2)} \kappa_{6m}^{(2)} (5c_{1m} - 8c_{2m}) + \frac{4}{s_{44m}} \left[\frac{\partial \kappa_{4m}^{(2)}}{\partial \varphi} \frac{\partial \kappa_{6m}^{(2)}}{\partial \varphi} + \frac{\partial \kappa_{4m}^{(2)}}{\partial \nu} \frac{\partial \kappa_{6m}^{(2)}}{\partial \nu} \right] \right\}, \tag{7.450}$$

$$\omega_{23m} = \frac{1}{8} \left(\left[\kappa_{6m}^{(2)} \right]^2 (3c_{1m} - 2c_{2m}) + \frac{1}{s_{44m}} \left\{ \left[\frac{\partial \kappa_{6m}^{(2)}}{\partial \varphi} \right]^2 + \left[\frac{\partial \kappa_{6m}^{(2)}}{\partial \nu} \right]^2 \right\} \right), \tag{7.451}$$

$$\begin{aligned} \omega_{24m} = \frac{2C_m^{(2)}}{9} \bigg\{ & 9\kappa_{4m}^{(2)} \kappa_{5m}^{(2)} (c_{1m} - c_{2m}) \\ & + \frac{8}{s_{44m}} \left[\frac{\partial \kappa_{4m}^{(2)}}{\partial \varphi} \frac{\partial \kappa_{5m}^{(2)}}{\partial \varphi} + \frac{\partial \kappa_{4m}^{(2)}}{\partial \nu} \frac{\partial \kappa_{5m}^{(2)}}{\partial \nu} \right] \bigg\}, \end{aligned} \tag{7.452}$$

$$\omega_{25m} = C_m^{(2)} \left\{ 2\kappa_{5m}^{(2)}\kappa_{6m}^{(2)} (c_{1m} - c_{2m}) + \frac{1}{s_{44m}} \left[\frac{\partial \kappa_{5m}^{(2)}}{\partial \varphi} \frac{\partial \kappa_{6m}^{(2)}}{\partial \varphi} + \frac{\partial \kappa_{5m}^{(2)}}{\partial \nu} \frac{\partial \kappa_{6m}^{(2)}}{\partial \nu} \right] \right\}, \quad (7.453)$$

$$\omega_{26m} = \left[C_m^{(2)} \right]^2 \left(3 \left[\kappa_{5m}^{(2)} \right]^2 (c_{1m} - c_{2m}) + \frac{2}{s_{44m}} \left\{ \left[\frac{\partial \kappa_{5m}^{(2)}}{\partial \varphi} \right]^2 + \left[\frac{\partial \kappa_{5m}^{(2)}}{\partial \nu} \right]^2 \right\} \right), \quad (7.454)$$

$$\omega_{27m} = -\frac{2}{3} \left(\left[\kappa_{4m}^{(2)} \right]^2 (c_{1m} - c_{2m}) + \frac{8}{9 s_{44m}} \left\{ \left[\frac{\partial \kappa_{4m}^{(2)}}{\partial \varphi} \right]^2 + \left[\frac{\partial \kappa_{4m}^{(2)}}{\partial \nu} \right]^2 \right\} \right), \quad (7.455)$$

$$\omega_{28m} = \frac{1}{3} \left(\left[\kappa_{4m}^{(2)} \right]^2 (c_{1m} - c_{2m}) + \frac{2}{3 s_{44m}} \left\{ \left[\frac{\partial \kappa_{4m}^{(2)}}{\partial \varphi} \right]^2 + \left[\frac{\partial \kappa_{4m}^{(2)}}{\partial \nu} \right]^2 \right\} \right), \quad (7.456)$$

$$\omega_{29m} = \frac{1}{9} \left\{ 2\kappa_{4m}^{(2)}\kappa_{6m}^{(2)} (c_{1m} + 2c_{2m}) + \frac{5}{s_{44m}} \left[\frac{\partial \kappa_{4m}^{(2)}}{\partial \varphi} \frac{\partial \kappa_{6m}^{(2)}}{\partial \varphi} + \frac{\partial \kappa_{4m}^{(2)}}{\partial \nu} \frac{\partial \kappa_{6m}^{(2)}}{\partial \nu} \right] \right\}, \quad (7.457)$$

$$\omega_{210m} = \frac{1}{2} \left(\left[\kappa_{6m}^{(2)} \right]^2 (c_{1m} - 2c_{2m}) + \frac{1}{s_{44m}} \left\{ \left[\frac{\partial \kappa_{6m}^{(2)}}{\partial \varphi} \right]^2 + \left[\frac{\partial \kappa_{6m}^{(2)}}{\partial \nu} \right]^2 \right\} \right), \quad (7.458)$$

$$\omega_{211m} = -\frac{2}{3} \left\{ \kappa_{4m}^{(2)}\kappa_{6m}^{(2)} (c_{1m} - c_{2m}) + \frac{1}{s_{44m}} \left[\frac{\partial \kappa_{4m}^{(2)}}{\partial \varphi} \frac{\partial \kappa_{6m}^{(2)}}{\partial \varphi} + \frac{\partial \kappa_{4m}^{(2)}}{\partial \nu} \frac{\partial \kappa_{6m}^{(2)}}{\partial \nu} \right] \right\}, \quad (7.459)$$

$$\omega_{212m} = \frac{1}{2} \left(c_{1m} \left[\kappa_{6m}^{(2)} \right]^2 + \frac{1}{s_{44m}} \left\{ \left[\frac{\partial \kappa_{6m}^{(2)}}{\partial \varphi} \right]^2 + \left[\frac{\partial \kappa_{6m}^{(2)}}{\partial \nu} \right]^2 \right\} \right), \quad (7.460)$$

$$\omega_{213m} = -\frac{2C_m^{(2)}}{3} \left\{ 3\kappa_{4m}^{(2)}\kappa_{5m}^{(2)} (c_{1m} - c_{2m}) \right.$$
$$\left. + \frac{2}{s_{44m}} \left[\frac{\partial \kappa_{4m}^{(2)}}{\partial \varphi} \frac{\partial \kappa_{5m}^{(2)}}{\partial \varphi} + \frac{\partial \kappa_{4m}^{(2)}}{\partial \nu} \frac{\partial \kappa_{5m}^{(2)}}{\partial \nu} \right] \right\}, \quad (7.461)$$

$$\omega_{214m} = 2C_m^{(2)} \left\{ \kappa_{5m}^{(2)}\kappa_{6m}^{(2)} (c_{1m} - c_{2m}) + \frac{1}{s_{44m}} \left[\frac{\partial \kappa_{5m}^{(2)}}{\partial \varphi} \frac{\partial \kappa_{6m}^{(2)}}{\partial \varphi} + \frac{\partial \kappa_{5m}^{(2)}}{\partial \nu} \frac{\partial \kappa_{6m}^{(2)}}{\partial \nu} \right] \right\}, \quad (7.462)$$

$$\xi_{i1m} = \frac{1}{9} \left\{ \kappa_{4m}^{(2)} \left[a'_{1i} (c_{1m} - 7c_{2m}) + \left(a'_{2i} + a'_{3i} \right) (4c_{1m} - c_{2m}) \right] \right.$$
$$\left. + \frac{4}{s_{44m}} \left[\left(a'_{1i} + a'_{2i} \right) \frac{\partial \kappa_{4m}^{(2)}}{\partial \varphi} + \left(a'_{1i} + a'_{3i} \right) \frac{\partial \kappa_{4m}^{(2)}}{\partial \nu} \right] \right\},$$
$$i = 1, 2, 3, \quad (7.463)$$

$$\xi_{i2m} = \frac{1}{2}\left\{\kappa_{6m}^{(2)}\left[a'_{1i}\,c_{1m} + \left(a'_{2i} + a'_{3i}\right)\left(\frac{c_{1m}}{2} - c_{2m}\right)\right] + \frac{1}{2s_{44m}}\left[\left(a'_{1i} + a'_{2i}\right)\frac{\partial \kappa_{6m}^{(2)}}{\partial \varphi} + \left(a'_{1i} + a'_{3i}\right)\frac{\partial \kappa_{6m}^{(2)}}{\partial \nu}\right]\right\},$$
$$i = 1, 2, 3, \tag{7.464}$$

$$\xi_{i3m} = C_m^{(2)}\left(\kappa_{5m}^{(2)}\left\{a'_{1i}\left[c_{3m}\left(c_{1m} + c_{2m}\right) - 2c_{2m}\right] + \left(a'_{2i} + a'_{3i}\right)\left(c_{1m} - c_{2m}c_{3m}\right)\right\} + \frac{1}{s_{44m}}\left[\left(a'_{1i} + a'_{2i}\right)\frac{\partial \kappa_{5m}^{(2)}}{\partial \varphi} + \left(a'_{1i} + a'_{3i}\right)\frac{\partial \kappa_{5m}^{(2)}}{\partial \nu}\right]\right),$$
$$i = 1, 2, 3, \tag{7.465}$$

$$\xi_{i4m} = \frac{1}{3}\left\{\kappa_{4m}^{(2)}\left(a'_{1i} + a'_{2i} + a'_{3i}\right)\left(c_{1m} - c_{2m}\right) + \frac{1}{s_{44m}}\left[\left(a'_{1i} + a'_{2i}\right)\frac{\partial \kappa_{4m}^{(2)}}{\partial \varphi} + \left(a'_{1i} + a'_{3i}\right)\frac{\partial \kappa_{4m}^{(2)}}{\partial \nu}\right]\right\},$$
$$i = 1, 2, 3, \tag{7.466}$$

$$\xi_{i5m} = \kappa_{6m}^{(2)}\left[\frac{c_{1m}}{2}\left(a'_{2i} + a'_{3i}\right) - a'_{1i}\,c_{2m}\right] + \frac{1}{2s_{44m}}\left[\left(a'_{1i} + a'_{2i}\right)\frac{\partial \kappa_{6m}^{(2)}}{\partial \varphi} + \left(a'_{1i} + a'_{3i}\right)\frac{\partial \kappa_{6m}^{(2)}}{\partial \nu}\right], \quad i = 1, 2, 3. \tag{7.467}$$

Chapter 8

Isotropic multi- and one-particle-(envelope)-matrix systems. Solutions 7, 8

Chapter 8 presents two different mathematical techniques in Section 8.1 applied to a system of the differential equations (3.35), (3.36), resulting in two different linear differential equations (8.2) and (8.3) with non-zero right sides, with two different solutions for the radial displacement u_{1q} of the infinitesimal spherical cap given by Eqs. (8.4) and (8.5), respectively. With regard to one integration constant, C_q, the solutions (8.4), (8.5) are accordingly considered for the determination of the thermal stresses and the corresponding quantities in the spherical envelope ($q = e$) provided that $\beta_p \neq \beta_e = \beta_m$ and $\varepsilon_{11te} - \varepsilon_{11tp} = f(\varphi, \nu)$ (see Eqs. (3.48)–(3.50), (12.23)), presented for both solutions (8.4), (8.5) in the part B of Section 6.4.1 (see Eqs. (6.214)–(6.231)), regarding the boundary condition given by Eq. (1.7).

8.1 Mathematical techniques 7, 8

With regard to Eq. (3.35), assuming $U_{1q} = r^{\lambda_q}$, we get

$$U_{1q} = \frac{\partial^2 u_{1q}}{\partial \varphi^2} + \frac{\partial^2 u_{1q}}{\partial \nu^2} = C_q\, r^{c_{3q}}, \tag{8.1}$$

where C_q is an integration constant. Considering Eq. (8.1) and $r\,[\partial \text{Eq.}(8.1)/\partial r]$, the differential equations (3.36) and (5.12) are transformed to the forms

$$\frac{\partial^2 u_{1q}}{\partial r^2} + \frac{2}{r}\frac{\partial u_{1q}}{\partial r} - \frac{2 u_{1q}}{r^2} = -C_q\, a_q^{(3)}\, r^{c_{3q}-2}, \tag{8.2}$$

$$\frac{\partial^3 u_{1q}}{\partial r^3} + \frac{4}{r}\frac{\partial^2 u_{1q}}{\partial r^2} = -C_q\, a_q^{(2)}\, r^{c_{3q}-3}, \tag{8.3}$$

representing linear differential equations of the second and third rank regarding the variable r with non-zero right sides, where the elastic modulus s_{44q} and the coefficients c_{3q}, $a_q^{(2)}$, $a_q^{(3)}$ are given by Eqs. (3.25) and (3.40), (5.14), (5.19), respectively.

Using the Wronskian's method [5] presented in Section 12.4.5, representing a short introduction to problems of linear differential equations with non-zero right sides, the solutions of the differential equations (8.2), (8.3) for the radial displacement u_{1q} of the infinitesimal spherical cap (see Figs. 3.2, 3.3) are derived as

$$u_{1q} = C_q \, a_q^{(1)} \, C_q^{(1)} \, r^{c_{3q}}, \tag{8.4}$$

$$u_{1q} = C_q \, a_q^{(2)} \, C_q^{(2)} \, r^{c_{3q}}, \tag{8.5}$$

and the absolute value $|u_{1q}|$ related to Eqs. (8.4), (8.5) represents decreasing functions of the variable r due to $c_{3q} < 0$ (see Eq. (3.40)), where the coefficients $a_q^{(1)}$, $C_q^{(1)}$, $C_q^{(2)}$ are given by Eqs. (5.6), (6.5), (7.4), respectively.

Chapter 9

Isotropic multi- and one-particle-(envelope)-matrix systems. Solution 9

9.1 Analysis of conditions $\beta_p \neq \beta_e = \beta_m$ and $\varepsilon_{11te} - \varepsilon_{11tp} \neq f(\varphi, \nu)$

The different mathematical techniques presented in Section 9.2 and applied to Eq. (3.36) transformed to the form given by Eq. (9.2) are considered for the determination of the thermal stresses and the corresponding quantities, the latter represented by w_q, W_q, σ_{iq}, w_{iq} $(i = 1,2,3)$ (see Eqs. (3.41), (3.46), (3.47)), for the spherical particle and for the spherical envelope (see Sections 9.4.1, 9.4.2) related to the multi-particle-envelope-matrix system

1. on the conditions $\beta_p \neq \beta_e = \beta_m$ and $\varepsilon_{11tp} = \varepsilon_{11te} = \varepsilon_{11tm}$ (see Eqs. (3.48)–(3.50), (12.23)), where the phase-transformation induced radial strain ε_{11tq} $(q = p, e, m)$ is a function of the variables φ, ν, and accordingly $\varepsilon_{11te} - \varepsilon_{11tp} = 0 \neq f(\varphi, \nu)$,
2. on the condition $\beta_p \neq \beta_e = \beta_m$, considering ε_{11tq} to be φ, ν-independent with regard to simplification represented by Eq. (12.24), and accordingly $\varepsilon_{11te} - \varepsilon_{11tp} \neq f(\varphi, \nu)$.

Provided that $\beta_p \neq \beta_e = \beta_m$, the boundary conditions (4.3), (4.5) and (4.7) for the determination of the thermal stresses and the corresponding quantities in the spherical particle and in the spherical envelope, respectively, are not connected with the φ, ν-dependent distance r_c (see Section 2.3). Consequently, the coefficients ϱ_{1p} and ϱ_{1e}^{me} resulting from the tangential strains $(\varepsilon_{22p})_{r=R_1}$ and $(\varepsilon_{22e})_{r=R_1}$, respectively, representing constant functions regarding the variables φ, ν, and included in Eq. (3.63) for the determination of the radial stress p_1 acting on the particle-envelope boundary, are also φ, ν-independent. Additionally, disregarding the φ, ν-dependent radial strains ε_{11tp}, ε_{11tp} (see Eq. (12.23)), the condition $\varepsilon_{11tp} = \varepsilon_{11te} = \varepsilon_{11tm}$ (see Item 1), strictly speaking $\varepsilon_{11te} - \varepsilon_{11tp} = 0$, results in the φ, ν-independent difference $\beta_e - \beta_p$. Finally, on the condition $\varepsilon_{11tq} \neq f(\varphi, \nu)$ $(q = p, e, m)$ (see Item 2) to result in $\varepsilon_{11te} - \varepsilon_{11tp} \neq f(\varphi, \nu)$, the difference $\beta_e - \beta_p$ is also φ, ν-independent. Accordingly, Items 1, 2 are expressed by the conditions $\beta_p \neq \beta_e = \beta_m$ and $\varepsilon_{11te} - \varepsilon_{11tp} \neq f(\varphi, \nu)$, the latter resulting in the φ, ν-independent difference $\beta_e - \beta_p$.

Accordingly, the radial stress $p_1 = (\beta_e - \beta_p) / (\varrho_{1p} - \varrho_{1e}^{me})$ (see Eq. (10.4)) resulting from Eq. (3.63), represents a constant function regarding the variables φ, ν, resulting in the φ, ν-independent radial displacements u_{1p} and u_{1e} of the infinitesimal spherical cap in the spherical particle and envelope, and then $\partial u_{1q}/\partial \eta = \partial^2 u_{1q}/\partial \eta^2 = 0$ $(q = p, e;\ \eta = \varphi, \nu)$ resulting in $U_{1q} = 0$ (see Eq. (3.37)).

With regard to the conditions $\beta_p \neq \beta_e = \beta_m$ and $\varepsilon_{11te} - \varepsilon_{11tp} \neq f(\varphi, \nu)$ resulting in the φ, ν-independent difference $\beta_e - \beta_p$ and consequently in the mathematical techniques presented in Section 9.2, the coefficients ϱ_{1p} and ϱ_{1e}^{me} are given by Eqs. (9.50) and (9.63), respectively. Provided that $\beta_p \neq \beta_e = \beta_m$ and $\varepsilon_{11te} - \varepsilon_{11tp} = f(\varphi, \nu)$, resulting in the φ, ν-dependent difference $\beta_e - \beta_p$, the φ, ν-independent coefficients ϱ_{1p} and ϱ_{1e}^{me} are given by Eqs. (5.36) and (5.106), (6.207), (6.228), (6.246), (7.230), respectively, considering the mathematical techniques presented in Sections 5.1–5.3, 6.1, 7.1, 7.2, 8.1, and resulting in the solutions (5.7), (5.15), (5.20), (6.4), (7.3), (7.12), (8.4), (8.5) for the radial displacement u_{1q} $(q = p, e, m)$.

With regard to the different mathematical techniques presented in Section 9.2, the radial displacement u_{1q} related to the conditions $\beta_p \neq \beta_e = \beta_m$ and $\varepsilon_{11te} - \varepsilon_{11tp} \neq f(\varphi, \nu)$ has the form

$$u_{1q} = \sum_{i=1}^{n} C_{iq}\, r^{\lambda_i}, \tag{9.1}$$

representing a linear combination of the particular solutions $u_{1r} = r^{\lambda_1}, \ldots,$ $u_{nr} = r^{\lambda_n}$, where $\lambda_1, \ldots, \lambda_n$ are real roots of one of the characteristic equations (9.3), (9.7), (9.11), (9.14), (9.18), (9.22), (9.25), (9.29), (9.33), (9.37) related to one of the linear differential equations (9.2), (9.6), (9.10), (9.13), (9.17), (9.21), (9.24), (9.28), (9.32), (9.36) with zero right sides, respectively. The integration constants $C_{1q}, \ldots, C_{nq}$ are determined from the boundary conditions for the spherical particle and for the spherical envelope (see Sections 4.1.1, 4.1.2), regarding $u_{ir} = r^{\lambda_i}$ as a decreasing, constant and increasing function of the variable r for $\lambda_i < 0$, $\lambda_i = 0$ and $\lambda_i > 0$, respectively.

Additionally, with regard to $\partial u_{1q}/\partial \eta = \partial^2 u_{1q}/\partial \eta^2 = 0$ $(q = p, e;\ \eta = \varphi, \nu)$ resulting in $U_{1q} = 0$ (see Eq. (3.37)), the differential equations (9.2), (9.6), (9.10), (9.13), (9.17), (9.21), (9.24), (9.28), (9.32), (9.36) are suitable, provided that $\varepsilon_{11tm} - \varepsilon_{11tp} \neq f(\varphi, \nu)$ and $\varepsilon_{11te} - \varepsilon_{11tp} \neq f(\varphi, \nu)$, $\varepsilon_{11tm} - \varepsilon_{11te} \neq f(\varphi, \nu)$ (see Eqs. (12.23), (12.24)), for the determination of the thermal stresses and the corresponding quantities in the spherical particle $(q = p)$, in the spherical envelope $(q = e)$, in the infinite matrix $(q = m)$, representing components of the one-particle-matrix and one-particle-envelope-matrix systems (see Section 9.5), respectively, containing, in contrast to the multi-particle-(envelope)-matrix system, one spherical particle only. Considering Section 2.2.2, the analytical model of the thermal stresses in the one-particle-(envelope)-matrix system are required to be derived with regard to the thermal-stress induced elastic energy $W_q^{(1)}$ $(q=p,e,m)$ in each of the components.

9.2 Mathematical techniques 9

Section 9.2 presents different mathematical techniques to be applied to Eq. 9.2, as a transformation of Eq. (3.36), and to result in the different linear differential equations (9.6), (9.10), (9.13), (9.17), (9.21), (9.24), (9.28), (9.32), (9.36) with zero right sides. Consequently, with regard to $\partial u_{1q}/\partial \eta = \partial^2 u_{1q}/\partial \eta^2 = 0$ $(\eta = \varphi, \nu)$ resulting in $U_{1q} = 0$ (see Eq. (3.37)), the system of the differential equations (3.35) and (3.36) is consequently represented by Eq. (3.36) only, transformed to the form

$$r^2 \frac{\partial^2 u_{1q}}{\partial r^2} = 2 \left(u_{1q} - r \frac{\partial u_{1q}}{\partial r} \right). \tag{9.2}$$

Assuming $u_{1q} = r^{\lambda}$ and considering Eqs. (12.82), (12.83), the characteristic equation resulting from the linear differential equation (9.2) with zero right side has the form

$$(\lambda - 1)(\lambda + 2) = 0, \tag{9.3}$$

where λ_1, λ_2 representing roots of the quadratic algebraic equation (9.3) are derived as

$$\lambda_1 = 1, \tag{9.4}$$

$$\lambda_2 = -2. \tag{9.5}$$

Performing $r\,[\partial \text{Eq.}(9.2)/\partial r]$, the differential equation (9.2) is transformed to the form

$$r \frac{\partial^3 u_{1q}}{\partial r^3} = -4 \frac{\partial^2 u_{1q}}{\partial r^2}. \tag{9.6}$$

Similarly, assuming $u_{1q} = r^{\lambda}$, the characteristic equation resulting from the linear differential equation (9.6) with zero right side has the form

$$\lambda (\lambda - 1)(\lambda + 2) = 0, \tag{9.7}$$

with the real roots λ_1, λ_2 and λ_3 given by Eqs. (9.4), (9.5), respectively, and derived as

$$\lambda_3 = 0. \tag{9.8}$$

Performing $r^2\times$Eq. (9.6), we get

$$r^3 \frac{\partial^3 u_{1q}}{\partial r^3} = -4\, r^2 \frac{\partial^2 u_{1q}}{\partial r^2}, \tag{9.9}$$

and consequently substituting Eq. (9.2) to Eq. (9.9), the differential equation (9.9) is transformed to the form

$$r^3 \frac{\partial^3 u_{1q}}{\partial r^3} = 8 \left(u_{1q} - r \frac{\partial u_{1q}}{\partial r} \right), \tag{9.10}$$

exhibiting the characteristic equation derived as

$$(\lambda - 1)\,(\lambda + 2)\,(\lambda - 4) = 0, \tag{9.11}$$

with the real roots λ_1, λ_2 and λ_4 given by Eqs. (9.4), (9.5), respectively, and derived as

$$\lambda_4 = 4. \tag{9.12}$$

Performing $r\,[\partial\text{Eq.}(9.6)/\partial r]$, the differential equation (9.6) is transformed to the form

$$r \frac{\partial^4 u_{1q}}{\partial r^4} = -5 \frac{\partial^3 u_{1q}}{\partial r^3}, \tag{9.13}$$

exhibiting the characteristic equation derived as

$$\lambda\,(\lambda - 1)\,(\lambda + 2)\,(\lambda - 2) = 0, \tag{9.14}$$

with the real roots λ_1, λ_2, λ_3 and λ_5 given by Eqs. (9.4), (9.5), (9.8), respectively, and derived as

$$\lambda_5 = 2. \tag{9.15}$$

Performing $r\times$Eq. (9.13), we get

$$r^2 \frac{\partial^4 u_{1q}}{\partial r^4} = -5\, r \frac{\partial^3 u_{1q}}{\partial r^3}, \tag{9.16}$$

and consequently substituting Eq. (9.6) to Eq. (9.16), the differential equation (9.16) is transformed to the form

$$r^2 \frac{\partial^4 u_{1q}}{\partial r^4} = 20 \frac{\partial^2 u_{1q}}{\partial r^2}, \tag{9.17}$$

exhibiting the characteristic equation derived as

$$\lambda\,(\lambda - 1)\,(\lambda + 2)\,(\lambda - 7) = 0, \tag{9.18}$$

with the real roots λ_1, λ_2, λ_3 and λ_6 given by Eqs. (9.4), (9.5), (9.8), respectively, and derived as

$$\lambda_6 = 7. \tag{9.19}$$

Performing $r^2 \times$Eq. (9.17), we get

$$r^4 \frac{\partial^4 u_{1q}}{\partial r^4} = 20\, r^2 \frac{\partial^2 u_{1q}}{\partial r^2}, \tag{9.20}$$

and consequently substituting Eq. (9.2) to Eq. (9.20), the differential equation (9.20) is transformed to the form

$$r^4 \frac{\partial^4 u_{1q}}{\partial r^4} = 40 \left(u_{1q} + r \frac{\partial u_{1q}}{\partial r} \right), \tag{9.21}$$

exhibiting the characteristic equation derived as

$$(\lambda - 1)(\lambda + 2)\left(\lambda^2 - 7\lambda + 20\right) = 0, \tag{9.22}$$

with the real roots λ_1, λ_2 given by Eqs. (9.4), (9.5), respectively. Considering the discriminant $D = -31 < 0$ (see Eqs. (12.51)–(12.53)), the quadratic algebraic equation $\lambda^2 - 7\lambda + 20 = 0$ has imaginary roots $\lambda_1^{(I)}$, $\lambda_2^{(I)}$ in the form

$$\lambda_{1,2}^{(I)} = \frac{1}{2}\left(7 \pm i\sqrt{31}\right). \tag{9.23}$$

Performing $r\,[\partial\text{Eq.}(9.13)/\partial r]$, the differential equation (9.13) is transformed to the form

$$r \frac{\partial^5 u_{1q}}{\partial r^5} = -6 \frac{\partial^4 u_{1q}}{\partial r^4}, \tag{9.24}$$

exhibiting the characteristic equation derived as

$$\lambda(\lambda - 1)(\lambda + 2)(\lambda - 2)(\lambda - 3) = 0, \tag{9.25}$$

with the real roots λ_1, λ_2, λ_3, λ_5 and λ_7 given by Eqs. (9.4), (9.5), (9.8), (9.15), respectively, and derived as

$$\lambda_7 = 3. \tag{9.26}$$

Performing $r \times$Eq. (9.24), we get

$$r^2 \frac{\partial^5 u_{1q}}{\partial r^5} = -6\, r \frac{\partial^4 u_{1q}}{\partial r^4}, \tag{9.27}$$

and consequently substituting Eq. (9.13) to Eq. (9.27), the differential equation (9.27) is transformed to the form

$$r^2 \frac{\partial^5 u_{1q}}{\partial r^5} = 30 \frac{\partial^3 u_{1q}}{\partial r^3}, \tag{9.28}$$

exhibiting the characteristic equation derived as

$$\lambda (\lambda - 1) (\lambda + 2) (\lambda - 2) (\lambda - 9) = 0, \tag{9.29}$$

with the real roots λ_1, λ_2, λ_3, λ_5 and λ_8 given by Eqs. (9.4), (9.5), (9.8), (9.15), respectively, and derived as

$$\lambda_8 = 9. \tag{9.30}$$

Performing $r\times$Eq. (9.28), we get

$$r^3 \frac{\partial^5 u_{1q}}{\partial r^5} = 30\, r \frac{\partial^3 u_{1q}}{\partial r^3}, \tag{9.31}$$

and consequently substituting Eq. (9.6) to Eq. (9.31), the differential equation (9.31) is transformed to the form

$$r^3 \frac{\partial^5 u_{1q}}{\partial r^5} = -120 \frac{\partial^2 u_{1q}}{\partial r^2}, \tag{9.32}$$

exhibiting the characteristic equation derived as

$$\lambda (\lambda - 1) (\lambda + 2) \left(\lambda^2 - 11\lambda + 48\right) = 0, \tag{9.33}$$

with the real roots λ_1, λ_2, λ_3 given by Eqs. (9.4), (9.5), (9.8), respectively. Considering the discriminant $D = -71 < 0$ (see Eqs. (12.51)–(12.53)), the quadratic algebraic equation $\lambda^2 - 11\lambda + 48 = 0$ has imaginary roots $\lambda_3^{(I)}$, $\lambda_4^{(I)}$ in the form

$$\lambda_{3,4}^{(I)} = \frac{1}{2} \left(11 \pm i \sqrt{71}\right). \tag{9.34}$$

Performing $r^2\times$Eq. (9.32), we get

$$r^5 \frac{\partial^5 u_{1q}}{\partial r^5} = -120\, r^2 \frac{\partial^2 u_{1q}}{\partial r^2}, \tag{9.35}$$

and consequently substituting Eq. (9.2) to Eq. (9.35), the differential equation (9.35) is transformed to the form

$$r^5 \frac{\partial^5 u_{1q}}{\partial r^5} = -240 \left(u_{1q} - r \frac{\partial u_{1q}}{\partial r}\right), \tag{9.36}$$

exhibiting the characteristic equation derived as

$$(\lambda - 1) (\lambda + 2) \left(\lambda^3 - 11\lambda^2 + 48\lambda - 120\right) = 0, \tag{9.37}$$

with the real roots λ_1, λ_2 given by Eqs. (9.4), (9.5), respectively. The cubic algebraic equation $\lambda^3 - 11\lambda^2 + 48\lambda - 120 = 0$ has one real and two imaginary roots, λ_9 and $\lambda_5^{(I)}$, $\lambda_6^{(I)}$, respectively, in the forms[1]

$$\lambda_9 = \frac{1}{3}\left\{11 - \left(\frac{23^2}{25+18\sqrt{2}}\right)^{1/3} + \left[23\left(25+18\sqrt{2}\right)\right]^{1/3}\right\} = 6.43999. \qquad (9.38)$$

$$\lambda_{5,6}^{(I)} = \frac{1}{6}\left\{22 + \left(\frac{23^2}{25+18\sqrt{2}}\right)^{1/3}\left(1 \pm i\sqrt{3}\right) - \left[23\left(25+18\sqrt{2}\right)\right]^{1/3}\left(1 \mp i\sqrt{3}\right)\right\}$$
$$= 2.28001 \pm 3.6654\, i. \qquad (9.39)$$

With regard to the mathematical techniques presented in Section 9.2, additional real exponents can be accordingly derived.

9.3 Analysis of exponent λ

9.3.1 Spherical particle

As presented in Sections 4.1.1, 5.4.1, the absolute value $|u_{1p}|$ of the radial displacement $u_{1p} = C_p r^{\lambda_p}$ (see Eq. (9.1)) in the spherical particle is required to represent an increasing function of the variable $r \in \langle 0, R_1 \rangle$ provided that $\beta_p \neq \beta_m$, $\beta_p \neq \beta_e \neq \beta_m$, $\beta_p \neq \beta_e = \beta_m$, $\beta_p = \beta_e \neq \beta_m$ for $\varepsilon_{11tm} - \varepsilon_{11tp} \neq f(\varphi, \nu)$, $\varepsilon_{11te} - \varepsilon_{11tp} \neq f(\varphi, \nu)$, $\varepsilon_{11tm} - \varepsilon_{11te} \neq f(\varphi, \nu)$ (see Eqs. (3.48)–(3.50), (12.23), (12.24)). Consequently, the radial displacement u_{1p} for the real exponent $\lambda_p \geq 1$ fulfils the boundary conditions (4.3)–(4.5), and accordingly $\lambda_p \subset \{\lambda_1, \lambda_4, \ldots, \lambda_9\}$ (see Eqs. (9.4), (9.12), (9.15), (9.19), (9.26), (9.30), (9.38)). Provided that $\lambda_p < 1$, the stress $(\sigma_{iip})_{r\to 0} \propto \left(r^{\lambda_p - 1}\right)_{r\to 0} \to \infty$ ($i = 1,2,3$) (see Eqs. (3.31)–(3.34)) and the radial displacement $(u_{1p})_{r\to 0} \to \infty$, respectively, what is physically unacceptable. Additionally, with regard to the Castigliano's theorem (see Section 3.1.9), such real exponent λ_p from the set $\{\lambda_1, \lambda_4, \ldots, \lambda_9\} = \{1, 4, 2, 7, 3, 9, 6.43999\}$ is considered to result in minimal thermal-stress induced elastic energy, $W_c = W_p + W_e + W_m$ and $W^{(1)} = W_p^{(1)} + (W_e^{(1)}) + W_m^{(1)}$ (see Section 2.2.2.1), related to the cell of the multi-particle-envelope-matrix system and to the one-particle-(envelope)-matrix system, respectively.

9.3.2 Spherical envelope

With regard to the one-particle-envelope-matrix system provided that $\beta_p \neq \beta_e \neq \beta_m$ for $\varepsilon_{11te} - \varepsilon_{11tp} \neq f(\varphi, \nu)$ and $\varepsilon_{11tm} - \varepsilon_{11te} \neq f(\varphi, \nu)$, (see Eqs. (3.48)–(3.50), (12.23),

[1]The exponents λ_9, $\lambda_{5,6}^{(I)}$ are derived using a software product.

(12.24)), considering the analyses presented in Sections 4.1.2, 5.4.2 and the boundary conditions (4.7), (4.8), the exponents $\lambda_{1e} \neq \lambda_{2e}$ included in $u_{1e} = C_{1e}r^{\lambda_{1e}} + C_{2e}r^{\lambda_{2e}}$ (see Eq. (9.1)) are required to represent real roots of one of the characteristic equations (9.3), (9.7), (9.11), (9.14), (9.18), (9.22), (9.25), (9.29), (9.33), and consequently λ_{1e}, λ_{2e} are both from one of the sets $\{\lambda_1, \lambda_2, \lambda_3, \lambda_5\}$, $\{\lambda_1, \lambda_2, \lambda_3, \lambda_6\}$, $\{\lambda_1, \lambda_2, \lambda_4\}$, $\{\lambda_1, \lambda_2, \lambda_3, \lambda_5, \lambda_7\}$, $\{\lambda_1, \lambda_2, \lambda_3, \lambda_5, \lambda_8\}$ to result in the 19 combinations $\{-2, 0\}$, $\{-2, 1\}$, $\{-2, 2\}$, $\{-2, 3\}$, $\{-2, 4\}$, $\{-2, 7\}$, $\{-2, 9\}$, $\{0, 1\}$, $\{0, 2\}$, $\{0, 3\}$, $\{0, 7\}$, $\{0, 9\}$, $\{1, 2\}$, $\{1, 3\}$, $\{1, 4\}$, $\{1, 7\}$, $\{1, 9\}$, $\{2, 3\}$, $\{2, 9\}$ for $\lambda_{1e} \neq \lambda_{2e}$. Similarly, such combination of λ_{1e}, λ_{2e} is considered to result in minimal thermal-stress induced elastic energy $W^{(1)} = W_p^{(1)} + (W_e^{(1)}) + W_m^{(1)}$ (see Section 2.2.2.1).

Finally, with regard to the multi- and one-particle-envelope-matrix systems provided that $\beta_p \neq \beta_e = \beta_m$ for $\varepsilon_{11te} - \varepsilon_{11tp} \neq f(\varphi, \nu)$ or $\beta_p = \beta_e \neq \beta_m$ for $\varepsilon_{11tm} - \varepsilon_{11te} \neq f(\varphi, \nu)$, considering the boundary condition (4.7) or (4.8), the absolute value $|u_{1e}|$ of the radial displacement $u_{1e} = C_e r^{\lambda_e}$ (see Eq. (9.1)) in the spherical envelope is required to represent a decreasing or increasing function of the variable $r \in \langle R_1, R_2 \rangle$, at the very most, a constant function regarding the variable r, and accordingly $\lambda_e \leq 0$ or $\lambda_e \geq 0$, respectively. Although the assumption for the radial displacement u_{1e} as a constant function regarding the variable r may be improbable, nevertheless it is considered, with regard to completeness of the presented topic, in Sections 9.4.2, 9.5.2.3 for $\lambda_e = 0$. Finally, the exponent λ_e is from the sets $\{\lambda_2, \lambda_3\} = \{-2, 0\}$ or $\{\lambda_1, \lambda_3, \ldots, \lambda_9\} = \{1, 0, 4, 2, 7, 3, 9, 6.43999\}$ for $\beta_p \neq \beta_e = \beta_m$ or $\beta_p = \beta_e \neq \beta_m$, respectively, considering the Castigliano's theorem (see Section 3.1.9).

9.3.3 Infinite matrix

With regard to the one-particle-matrix and one-particle-envelope-matrix systems provided that $\beta_p \neq \beta_m$ and $\beta_p \neq \beta_e \neq \beta_m$, $\beta_p \neq \beta_e = \beta_m$, $\beta_p = \beta_e \neq \beta_m$ for $\varepsilon_{11tm} - \varepsilon_{11tp} \neq f(\varphi, \nu)$ and $\varepsilon_{11te} - \varepsilon_{11tp} \neq f(\varphi, \nu)$, $\varepsilon_{11tm} - \varepsilon_{11te} \neq f(\varphi, \nu)$ (see Eqs. (3.48)–(3.50), (12.23), (12.24)), the absolute values $|u_{1m}|$, $|\sigma_{11m}|$ represent decreasing functions of the variable $r \in \langle R_1, \infty)$ and $r \in \langle R_2, \infty)$ (see Section 4.2), respectively, and consequently the exponent $\lambda_m = \lambda_2 = -2$ (see Eq. (9.5)) regarding $u_{1m} = C_m r^{\lambda_m}$, $\sigma_{11m} \propto C_m r^{\lambda_m - 1}$ (see Eqs. (9.1), (3.31)).

9.4 Multi-particle-envelope-matrix system. Conditions $\beta_p \neq \beta_e = \beta_m$ and $\varepsilon_{11te} - \varepsilon_{11tp} \neq f(\varphi, \nu)$

9.4.1 Thermal stresses in spherical particle

The conditions presented in Section 9.4.1, concerning the coefficient β_q $(q = p, e, m)$ (see Eqs. (3.48)–(3.50)) and the phase-transformation induced radial strain ε_{11tq} dependent or independent on the variables φ, ν (see Eqs. (12.23) or (12.24)), are defined in Items 1, 2 in Section 9.1, resulting in the φ, ν-independent difference $\beta_e - \beta_p$ and consequently in the φ, ν-independent radial stress $p_1 = (\beta_e - \beta_p) / (\varrho_{1p} - \varrho_{1e}^{me})$ (see Eq. (10.4)) acting on the particle-envelope boundary, the latter derived from Eq. (3.63) and included in the formulae in Section 9.4.1, where the φ, ν-independent coefficients ϱ_{1p} and ϱ_{1e}^{me} are given by Eqs. (9.50) and (9.63) presented below, respectively, considering the mathematical techniques in Section 9.2 to result in solutions exhibiting the form given by Eq. (9.1).

With regard to the boundary conditions (4.3)–(4.5) to determine one integration constant only, considering the analysis presented in Section 9.3.1 to define u_{1p} as an increasing function of the variable $r \in \langle 0, R_1 \rangle$, the radial displacement u_{1p} of the infinitesimal spherical cap in the spherical particle, regarding Eq. (9.1), has the form

$$u_{1p} = -p_1 \, \varrho_{1p} \, r \left(\frac{r}{R_1} \right)^{\lambda_p - 1}, \tag{9.40}$$

where the coefficient ϱ_{1p} is given by Eq. (9.50), and the exponent λ_p is from the set $\{1, 4, 2, 7, 3, 9, 6.43999\}$ (see Section 9.3.1).

Consequently, the radial and tangential and stresses, σ_{11p} and σ_{22p}, σ_{33p}, acting in the spherical particle, respectively, regarding Eqs. (3.31), (3.32), (9.40), are derived as

$$\sigma_{11p} = -p_1 \left(\frac{r}{R_1} \right)^{\lambda_p - 1}, \tag{9.41}$$

$$\sigma_{22p} = \sigma_{33p} = -p_1 \, \varrho_{1p} \left(c_{1p} - \lambda_p \, c_{2p} \right) \left(\frac{r}{R_1} \right)^{\lambda_p - 1}, \tag{9.42}$$

where the coefficients c_{1p}, c_{2p} are given by Eqs. (3.38), (3.39), respectively, and the shear stresses $\sigma_{12p} = \sigma_{13p} = 0$ (see Eqs. (3.33), (3.34)) due to $\partial u_{1p} / \partial \eta = 0$ $(\eta = \varphi, \nu)$ as presented in Section 9.1.

Finally, the radial and tangential strains, ε_{11p} and ε_{22p}, ε_{33p}, in the spherical particle, respectively, regarding Eqs. (3.1)–(3.5), (9.40), have the forms

$$\varepsilon_{11p} = -p_1 \, \varrho_{1p} \, \lambda_p \left(\frac{r}{R_1} \right)^{\lambda_p - 1}, \tag{9.43}$$

$$\varepsilon_{22p} = \varepsilon_{33p} = -p_1 \, \varrho_{1p} \left(\frac{r}{R_1} \right)^{\lambda_p - 1}, \tag{9.44}$$

where the shear strains $\varepsilon_{12p} = \varepsilon_{13p} = 0$ (see Eqs. (3.4), (3.5)) due to $\partial u_{1p} / \partial \eta = 0$ $(\eta = \varphi, \nu)$ as presented in Section 9.1.

With regard to the determination of the compressive or tensile radial stress $p_1 > 0$ or $p_1 < 0$, acting on the particle-matrix boundary, thus for $r = R_1$, respectively, considering Eq. (3.58), the tangential strain on the particle-matrix boundary, $(\varepsilon_{22p})_{r=R_1}$, is required to be derived in the form

$$(\varepsilon_{22p})_{r=R_1} = -p_1\, \varrho_{1p}. \tag{9.45}$$

Considering Eqs. (3.41), (3.43), (9.41)–(9.44), regarding $\sigma_{12+ip} = 0$, $\varepsilon_{12+ip} = 0$ ($i = 0{,}1$), the thermal-stress induced elastic energy density w_p and the thermal-stress induced elastic energy W_p, accumulated in an arbitrary point of the spherical particle and in the volume of the spherical particle, respectively, are derived as

$$w_p = \frac{\omega_{1p}\, \varrho_{1p}^2}{2} \left(\frac{r}{R_1}\right)^{2(\lambda_p - 1)}, \tag{9.46}$$

$$W_p = 4\pi \int_0^{R_1} w_p\, r^2\, dr = \frac{2\pi\, \omega_{1p}\, \varrho_{1p}^2\, R_1^3}{2\lambda_p + 1}, \tag{9.47}$$

where the coefficient ω_{1p} is given by Eq. (9.51).

With regard to the determination of the thermal-stress induced strengthening presented in Volume III, the stress σ_{ip} acting in a direction of the axis x_i and the elastic energy density w_{ip} induced by the stress σ_{ip} ($i = 1{,}2{,}3$) (see Eqs. (3.46), (3.47), (9.41), (9.42)), both related to an arbitrary point of the spherical particle, have the forms

$$\sigma_{ip} = -\xi_{i1p}\, \varrho_{1p} \left(\frac{r}{R_1}\right)^{\lambda_p - 1}, \quad i = 1, 2, 3, \tag{9.48}$$

$$w_{ip} = \frac{s_{11p}\, (\xi_{i1p}\, \varrho_{1p})^2}{2} \left(\frac{r}{R_1}\right)^{2(\lambda_p - 1)}, \quad i = 1, 2, 3, \tag{9.49}$$

where the elastic modulus s_{11p} and the coefficient ξ_{i1p} ($i = 1{,}2{,}3$) are given by Eqs. (3.23) and (5.39), respectively.

Appendix. Considering the coefficients c_{1p}, c_{2p}, a'_{1i}, a'_{2i}, a'_{3i} ($i = 1{,}2{,}3$) (see Eqs. (3.38), (3.39), (12.26)–(12.34)) and the exponent $\lambda_p \subset \{1, 4, 2, 7, 3, 9, 6.43999\}$ (see Section 9.3.1), the coefficients ϱ_{1p}, ω_{1+ip} ($i = 0{,}1$), ξ_{i1+jp} ($i = 1{,}2{,}3$; $j = 0{,}1$) have the forms

$$\varrho_{1p} = \frac{1}{\lambda_p\,(c_{1p} + c_{2p}) - 2c_{2p}}, \tag{9.50}$$

$$\omega_{1+ip} = p_{1+i}^2 \left[\lambda_p^2\,(c_{1p} + c_{2p}) - 4\lambda_p\, c_{2p} + 2c_{1p}\right], \quad i = 0, 1, \tag{9.51}$$

$$\xi_{i\,1+jp} = p_{1+j}\left[\frac{a'_{1i}}{\varrho_{1p}} + \left(a'_{2i} + a'_{3i}\right)\left(c_{1p} - \lambda_p\, c_{2p}\right)\right], \quad i = 1, 2, 3; \quad j = 0, 1. \tag{9.52}$$

9.4.2 Thermal stresses in spherical envelope

The conditions presented in Section 9.4.2, concerning the coefficient β_q $(q = p, e, m)$ (see Eqs. (3.48)–(3.50)) and the phase-transformation induced radial strain ε_{11tq} dependent or independent on the variables φ, ν (see Eqs. (12.23) or (12.24)), are defined in Items 1, 2 in Section 9.1, resulting in the φ, ν-independent difference $\beta_e - \beta_p$ and consequently in the φ, ν-independent radial stress $p_1 = (\beta_e - \beta_p) / (\varrho_{1p} - \varrho_{1e}^{me})$ (see Eq. (10.4)) acting on the particle-envelope boundary, the latter derived from Eq. (3.63) and included in the formulae in Section 9.4.1, where the φ, ν-independent coefficients ϱ_{1p} and ϱ_{1e}^{me} are given by Eqs. (9.50) and (9.63) presented below, respectively, considering the mathematical techniques in Section 9.2 to result in solutions exhibiting the form given by Eq. (9.1).

With regard to the boundary conditions (4.3)–(4.5) to determine one integration constant only, considering the analysis presented in Section 9.3.2 to define u_{1e} as a decreasing function of the variable $r \in \langle R_1, R_2 \rangle$, the radial displacement u_{1e} of the infinitesimal spherical cap in the spherical envelope, regarding Eq. (9.1), has the form

$$u_{1e} = -p_1\, \varrho_{1e}^{me}\, r \left(\frac{r}{R_1}\right)^{\lambda_e - 1}, \tag{9.53}$$

where the coefficient ϱ_{1e}^{me} is given by Eq. (9.63), and the exponent λ_e is from the set $\{-2, 0\}$ (see Section 9.3.2).

Consequently, the radial and tangential and stresses, σ_{11e} and σ_{22e}, σ_{33e}, acting in the spherical envelope, respectively, regarding Eqs. (3.31), (3.32), (9.53), are derived as

$$\sigma_{11e} = -p_1 \left(\frac{r}{R_1}\right)^{\lambda_e - 1}, \tag{9.54}$$

$$\sigma_{22e} = \sigma_{33e} = -p_1\, \varrho_{1e}^{me} \left(c_{1e} - \lambda_e\, c_{2e}\right) \left(\frac{r}{R_1}\right)^{\lambda_e - 1}, \tag{9.55}$$

where the coefficients c_{1e}, c_{2e} are given by Eqs. (3.38), (3.39), respectively, and the shear stresses $\sigma_{12e} = \sigma_{13e} = 0$ (see Eqs. (3.33), (3.34)) due to $\partial u_{1e}/\partial \eta = 0$ $(\eta = \varphi, \nu)$ as presented in Section 9.1.

Finally, the radial and tangential strains, ε_{11e} and ε_{22e}, ε_{33e}, in the spherical envelope, respectively, regarding Eqs. (3.1)–(3.5), (9.53), have the forms

$$\varepsilon_{11e} = -p_1\, \varrho_{1e}^{me}\, \lambda_e \left(\frac{r}{R_1}\right)^{\lambda_e - 1}, \tag{9.56}$$

$$\varepsilon_{22e} = \varepsilon_{33e} = -p_1\, \varrho_{1e}^{me} \left(\frac{r}{R_1}\right)^{\lambda_e - 1}, \tag{9.57}$$

where the shear strains $\varepsilon_{12e} = \varepsilon_{13e} = 0$ (see Eqs. (3.4), (3.5)) due to $\partial u_{1e}/\partial\eta = 0$ ($\eta = \varphi, \nu$) as presented in Section 9.1.

With regard to the determination of the compressive or tensile radial stress $p_1 > 0$ or $p_1 < 0$ acting on the particle-envelope boundary, thus for $r = R_1$, respectively, considering Eq. (3.63), and to the boundary condition for the cell matrix provided that $\beta_p \neq \beta_e = \beta_m$ (see Eq. (4.12)), the tangential strains on the particle-envelope and matrix-envelope boundaries, $(\varepsilon_{22e})_{r=R_1}$ and $(\varepsilon_{22e})_{r=R_2}$, respectively, are required to be derived in the form

$$(\varepsilon_{22e})_{r=R_{1+i}} = -p_1\, \varrho_{1+ie}^{me}, \quad i = 0, 1, \tag{9.58}$$

where the coefficient ϱ_{1+ie}^{me} ($i=0,1$) is given by Eq. (9.63) for $\lambda_e \subset \{-2, 0\}$ (see Section 9.3.2).

Considering Eqs. (3.41), (3.43), (9.54)–(9.57), regarding $\sigma_{12+ie} = 0$, $\varepsilon_{12+ie} = 0$ ($i=0,1$), the thermal-stress induced elastic energy density w_e and the thermal-stress induced elastic energy W_e, accumulated in an arbitrary point of the spherical envelope and in the volume of the spherical envelope, respectively, are derived as

$$w_e = \frac{\omega_{1e} \left(\varrho_{1e}^{me}\right)^2}{2} \left(\frac{r}{R_1}\right)^{2(\lambda_e - 1)}, \tag{9.59}$$

$$W_e = 4\pi \int_{R_1}^{R_2} w_e\, r^2\, dr = \frac{2\pi\, \omega_{1e} \left(\varrho_{1e}^{me}\right)^2 R_1^3}{2\lambda_e + 1} \left[\left(\frac{R_2}{R_1}\right)^{2\lambda_e + 1} - 1\right], \tag{9.60}$$

where the coefficient ω_{1e} is given by Eq. (9.64) for $\lambda_e \subset \{-2, 0\}$ (see Section 9.3.2).

With regard to the determination of the thermal-stress induced strengthening presented in Volume III, the stress σ_{ie} acting in a direction of the axis x_i and the elastic energy density w_{ie} induced by the stress σ_{ie} ($i=1,2,3$) (see Eqs. (3.46), (3.47), (9.54), (9.55)), both related to an arbitrary point of the spherical envelope, are derived as

$$\sigma_{ie} = -\xi_{i1e}\, \varrho_{1e}^{me} \left(\frac{r}{R_1}\right)^{\lambda_e - 1}, \quad i = 1, 2, 3, \tag{9.61}$$

$$w_{ie} = \frac{s_{11e} \left(\xi_{i1e}\, \varrho_{1e}^{me}\right)^2}{2} \left(\frac{r}{R_1}\right)^{2(\lambda_e - 1)}, \quad i = 1, 2, 3, \tag{9.62}$$

where the elastic modulus s_{11e} and the coefficient ξ_{i1e} ($i=1,2,3$) are given by Eqs. (3.23) and (9.65), respectively, the latter for $\lambda_e \subset \{-2, 0\}$ (see Section 9.3.2).

Appendix. Considering the coefficients c_{1e}, c_{2e}, a'_{1i}, a'_{2i}, a'_{3i} ($i=1,2,3$) (see Eqs. (3.38), (3.39), (12.26)–(12.34)), with regard to the exponent $\lambda_e \subset \{-2,0\}$ in Eq. (9.63) corresponding to the subscript $i=0,1$; with regard to the exponent $\lambda_e \subset \{-2,0\}$ and $\lambda_e \subset \{1,0,4,2,7,3,9,6.43999\}$ in Eqs. (9.64), (9.65) (see Section 9.3.2) corresponding to the subscripts $j=0$ and $j=1$ related to Sections 9.4.2 and 9.5.2.3, respectively; the coefficients ϱ_{1+ie}^{me} ($i=0,1$), ω_{1+je}, $\xi_{i\,1+je}$ ($i=1,2,3$; $j=0,1$) have the forms

$$\varrho_{1+ie}^{me} = \frac{1}{\lambda_e\left(c_{1e}+c_{2e}\right)-2c_{2e}}\left(\frac{R_{1+i}}{R_1}\right)^{\lambda_e-1}, \quad i=0,1, \tag{9.63}$$

$$\omega_{1+je} = p_{1+j}^2\left[\lambda_e^2\left(c_{1e}+c_{2e}\right)-4\lambda_e c_{2e}+2c_{1e}\right], \quad j=0,1, \tag{9.64}$$

$$\xi_{i\,1+je} = p_{1+j}\left\{a'_{1i}\left[\lambda_e\left(c_{1e}+c_{2e}\right)-2c_{2e}\right]+\left(a'_{2i}+a'_{3i}\right)\left(c_{1e}-\lambda_e c_{2e}\right)\right\}, \quad i=1,2,3; \quad j=0,1. \tag{9.65}$$

9.5 One-particle-(envelope)-matrix system. Conditions $\varepsilon_{11tm}-\varepsilon_{11tp} \neq f(\varphi,\nu)$, $\varepsilon_{11te}-\varepsilon_{11tp} \neq f(\varphi,\nu)$, $\varepsilon_{11tm}-\varepsilon_{11te} \neq f(\varphi,\nu)$

9.5.1 Thermal stresses in spherical particle

9.5.1.1 Conditions $\beta_p \neq \beta_m$, $\beta_p \neq \beta_e \neq \beta_m$, $\beta_p \neq \beta_e = \beta_m$

Formulae for the thermal stresses and for the corresponding quantities related to the spherical particle in the one-particle-matrix and one-particle-envelope-matrix systems provided that $\beta_p \neq \beta_m$ and $\beta_p \neq \beta_e \neq \beta_m$, $\beta_p \neq \beta_e = \beta_m$ are identical with those presented in Section 9.4.1 to result from the mathematical techniques in Section 9.2, considering the radial stress p_1 acting on the particle-matrix and particle-envelope boundaries given by Eq. (10.1), (10.2), (10.4), where the coefficients ϱ_{1m} and ϱ_{2m} in Eqs. (10.1) and (10.2), respectively, related to the infinite matrix are given by Eq. (9.110) (see Section 10.1.2).

9.5.1.2 Condition $\beta_p = \beta_e \neq \beta_m$

With regard to the boundary conditions (4.3), (4.4), (4.6) to determine one integration constant only, considering the analysis presented in Section 9.3.1 to define u_{1p} as an increasing function of the variable $r \in \langle 0, R_1 \rangle$, the radial displacement u_{1p} of the infinitesimal spherical cap in the spherical particle, regarding Eq. (9.1), has the form

$$u_{1p} = -p_2\,\varrho_{1e}^{pe}\,r\left(\frac{r}{R_1}\right)^{\lambda_p-1}, \tag{9.66}$$

where the coefficient ϱ_{1e}^{pe} is given by Eq. (9.99), the exponent λ_p is from the set $\{1, 4, 2, 7, 3, 9, 6.43999\}$ (see Section 9.3.1), and the radial stress p_2 acting on the matrix-envelope boundary is given by Eq. (10.5) regarding Section 10.1.2.

Consequently, the radial and tangential stresses, σ_{11p} and σ_{22p}, σ_{33p}, acting in the spherical particle, respectively, regarding Eqs. (3.31), (3.32), (9.66), are derived as

$$\sigma_{11p} = -\frac{p_2\,\varrho_{1e}^{pe}}{\varrho_{1p}}\left(\frac{r}{R_1}\right)^{\lambda_p-1}, \tag{9.67}$$

$$\sigma_{22p} = \sigma_{33p} = -p_2\,\varrho_{1e}^{pe}\left(c_{1p} - \lambda_p\,c_{2p}\right)\left(\frac{r}{R_1}\right)^{\lambda_p-1}, \tag{9.68}$$

where the coefficients c_{1p}, c_{2p}, ϱ_{1p} are given by Eqs. (3.38), (3.39), (9.50), respectively, and the shear stresses $\sigma_{12p} = \sigma_{13p} = 0$ (see Eqs. (3.33), (3.34)) due to $\partial u_{1p}/\partial\eta = 0$ $(\eta = \varphi, \nu)$ as presented in Section 9.1.

Finally, the radial and tangential strains, ε_{11p} and ε_{22p}, ε_{33p}, in the spherical particle, respectively, regarding Eqs. (3.1)–(3.5), (9.40), have the forms

$$\varepsilon_{11p} = -p_2\,\varrho_{1e}^{pe}\,\lambda_p\left(\frac{r}{R_1}\right)^{\lambda_p-1}, \tag{9.69}$$

$$\varepsilon_{22p} = \varepsilon_{33p} = -p_2\,\varrho_{1e}^{pe}\left(\frac{r}{R_1}\right)^{\lambda_p-1}, \tag{9.70}$$

where the shear strains $\varepsilon_{12p} = \varepsilon_{13p} = 0$ (see Eqs. (3.4), (3.5)) due to $\partial u_{1p}/\partial\eta = 0$ $(\eta = \varphi, \nu)$ as presented in Section 9.1.

Considering Eqs. (3.41), (3.43), (9.67)–(9.70), regarding $\sigma_{12+ip} = 0$, $\varepsilon_{12+ip} = 0$ $(i = 0,1)$, the thermal-stress induced elastic energy density w_p and the thermal-stress induced elastic energy $W_p^{(1)}$ [2], accumulated in an arbitrary point of the spherical particle and in the volume of the spherical particle, respectively, are derived as

$$w_p = \frac{\omega_{2p}\left(\varrho_{1e}^{pe}\right)^2}{2}\left(\frac{r}{R_1}\right)^{2(\lambda_p-1)}, \tag{9.71}$$

$$W_p^{(1)} = 4\pi\int_0^{R_1} w_p\,r^2\,dr = \frac{2\pi\,\omega_{2p}\left(\varrho_{1e}^{pe}\right)^2 R_1^3}{2\lambda_p+1}, \tag{9.72}$$

where the coefficient ω_{2p} is given by Eq. (9.51).

With regard to the determination of the thermal-stress induced strengthening presented in Volume III, the stress σ_{ip} acting in a direction of the axis x_i and the elastic energy density w_{ip} induced by the stress σ_{ip} $(i = 1,2,3)$ (see Eqs. (3.46), (3.47),

[2] The superscript (1) is related to the *one*-particle-(envelope)-matrix system.

(9.67), (9.68)), both related to an arbitrary point of the spherical particle, have the forms

$$\sigma_{ip} = -\xi_{\,i2p}\,\varrho_{\,1p}\left(\frac{r}{R_1}\right)^{\lambda_{1p}-1}, \quad i = 1, 2, 3, \tag{9.73}$$

$$w_{ip} = \frac{s_{11p}\left(\xi_{\,i2p}\,\varrho_{\,1p}\right)^2}{2}\left(\frac{r}{R_1}\right)^{2(\lambda_{1p}-1)}, \quad i = 1, 2, 3, \tag{9.74}$$

where the elastic modulus s_{11p} and the coefficient $\xi_{\,i1p}$ (i=1,2,3) are given by Eqs. (3.23) and (9.52), respectively.

9.5.2 Thermal stresses in spherical envelope

9.5.2.1 Condition $\beta_p \neq \beta_e \neq \beta_m$

With regard to the boundary conditions (4.7), (4.8) to determine two integration constants, considering the analysis presented in Section 9.3.2 to define u_{1e} as a function of the variable $r \in \langle R_1, R_2\rangle$, respectively, the radial displacement u_{1e} of the infinitesimal spherical cap in the spherical envelope, regarding Eq. (9.1), has the form

$$u_{1e} = -\sum_{i,j=0}^{1} p_{1+i}\,\kappa^{(1)}_{1+i+2je}\left(\delta_{0i} - \delta_{1i}\right) r^{\lambda_{1+je}}, \tag{9.75}$$

where the exponents $\lambda_{1e} \neq \lambda_{2e}$ are from one of the sets $\{-2, 0\}$, $\{-2, 1\}$, $\{-2, 2\}$, $\{-2, 3\}$, $\{-2, 4\}$, $\{-2, 7\}$, $\{-2, 9\}$, $\{0, 1\}$, $\{0, 2\}$, $\{0, 3\}$, $\{0, 7\}$, $\{0, 9\}$, $\{1, 2\}$, $\{1, 3\}$, $\{1, 4\}$, $\{1, 7\}$, $\{1, 9\}$, $\{2, 3\}$, $\{2, 9\}$ (see Section 9.3.2), and consequently, with regard to the Castigliano's theorem (see Section 3.1.9), such combination of λ_{1e}, λ_{2e} is considered to result in minimal thermal-stress induced elastic energy $W^{(1)} = W_p^{(1)} + (W_e^{(1)}) + W_m^{(1)}$ (see Section 2.2.2.1). Considering Section 10.1.2, the radial stresses p_1 and p_2 acting on the particle-envelope and matrix-envelope boundaries are given by Eqs. (10.2) and (10.3), respectively.

With regard to Section 5.6.2.1, the coefficients $\varrho_{\,1+i\;1+je}$, $\omega_{1+i\;1+je}$ (i, j = 0,1), $\xi_{\,i\,1+je}$ (i=1,2,3; j=0,1) mentioned below (see Eqs. (9.80), (9.81), (9.83), (9.84)) depend on $\kappa^{(1)}_{1e}, \ldots, \kappa^{(1)}_{4e}$ (see Eq. (12.35)) representing functions of $\chi_{1+j\;1+ie}$ (i, j = 0,1) given by Eq. (9.85).

Consequently, the radial and tangential stresses, σ_{11e} and σ_{22e}, σ_{33e}, acting in the spherical envelope, respectively, regarding Eqs. (3.31), (3.32), (9.75), are derived as

$$\sigma_{11e} = -\sum_{i,j=0}^{1} p_{1+i}\,\kappa^{(1)}_{1+i+2je}\left(\delta_{0i} - \delta_{1i}\right)\left[\lambda_{1+je}\left(c_{1e} + c_{2e}\right) - 2c_{2e}\right] r^{\lambda_{1+je}-1}, \tag{9.76}$$

$$\sigma_{22e} = \sigma_{33e} = -\sum_{i,j=0}^{1} p_{1+i}\, \kappa_{1+i+2je}^{(1)} \left(\delta_{0i} - \delta_{1i}\right) \left(c_{1e} - \lambda_{1+je} c_{2e}\right) r^{\lambda_{1+je}-1}, \tag{9.77}$$

where the coefficients c_{1e}, c_{2e} are given by Eqs. (3.38), (3.39), respectively, and the shear stresses $\sigma_{12e} = \sigma_{13e} = 0$ (see Eqs. (3.33), (3.34)) due to $\partial u_{1p}/\partial\eta = 0$ $(\eta = \varphi, \nu)$ as presented in Section 9.1.

Finally, the radial and tangential strains, ε_{11e} and ε_{22e}, ε_{33e}, in the spherical envelope, respectively, regarding Eqs. (3.1)–(3.5), (9.75), have the forms

$$\varepsilon_{11e} = -\sum_{i,j=0}^{1} p_{1+i}\, \kappa_{1+i+2je}^{(1)}\, \lambda_{1+je} \left(\delta_{0i} - \delta_{1i}\right) r^{\lambda_{1+je}-1}, \tag{9.78}$$

$$\varepsilon_{22e} = \varepsilon_{33e} = -\sum_{i,j=0}^{1} p_{1+i}\, \kappa_{1+i+2je}^{(1)} \left(\delta_{0i} - \delta_{1i}\right) r^{\lambda_{1+je}-1}, \tag{9.79}$$

where the shear strains $\varepsilon_{12e} = \varepsilon_{13e} = 0$ (see Eqs. (3.4), (3.5)) due to $\partial u_{1e}/\partial\eta = 0$ $(\eta = \varphi, \nu)$ as presented in Section 9.1.

With regard to the determination of the compressive or tensile radial stresses $p_1 > 0$ or $p_1 < 0$ and $p_2 > 0$ or $p_2 < 0$ acting on the particle-envelope and matrix-envelope boundaries, thus for $r = R_1$ and $r = R_2$, respectively, considering Eqs. (3.63), (3.64), the tangential strains on the particle-envelope and matrix-envelope boundaries, $(\varepsilon_{22e})_{r=R_1}$ and $(\varepsilon_{22e})_{r=R_2}$, respectively, are required to be derived in the form

$$\left(\varepsilon_{22e}\right)_{r=R_{1+i}} = -\sum_{j=0}^{1} p_{1+j}\, \varrho_{1+i\;1+je}, \quad i = 0, 1, \tag{9.80}$$

where the coefficient $\varrho_{1+i\;1+je}$ $(i, j = 0,1)$ is given by Eq. (9.86).

Considering Eqs. (3.41), (3.43), (9.76)–(9.79), the thermal-stress induced elastic energy density w_e and the thermal-stress induced elastic energy $W_e^{(1)}$[3], accumulated in an arbitrary point of the spherical envelope and in the volume of the spherical envelope, respectively, have the forms

[3]The superscript (1) is related to the *one*-particle-(envelope)-matrix system.

$$w_e = \frac{1}{2} \sum_{i,j=0}^{1} \omega_{1+i\,1+je}\, r^{\lambda_{1+ie}+\lambda_{1+je}-2}, \tag{9.81}$$

$$W_e^{(1)} = 4\pi \int_0^{R_1} w_e\, r^2\, dr$$
$$= 2\pi \sum_{i,j=0}^{1} \frac{\omega_{1+i\,1+je}}{\lambda_{1+ie}+\lambda_{1+je}+1} \left(R_2^{\lambda_{1+ie}+\lambda_{1+je}+1} - R_1^{\lambda_{1+ie}+\lambda_{1+je}+1} \right), \tag{9.82}$$

where the coefficient $\omega_{1+i\,1+je}$ $(i,j=0,1)$ is given by Eq. (9.87).

With regard to the determination of the thermal-stress induced strengthening presented in Volume III, the stress σ_{ie} acting in a direction of the axis x_i and the elastic energy density w_{ie} induced by the stress σ_{ie} $(i=1,2,3)$ (see Eqs. (3.46), (3.47), (9.76), (9.77)), both related to an arbitrary point of the spherical envelope, are derived as

$$\sigma_{ie} = -\left(\xi_{i\,1e}\, r^{\lambda_{1e}-1} + \xi_{i\,2e}\, r^{\lambda_{2e}-1} \right), \quad i = 1,2,3, \tag{9.83}$$

$$w_{ie} = \frac{s_{11e}}{2} \left[\xi_{i\,1e}^2\, r^{2(\lambda_{1e}-1)} + \xi_{i\,2e}^2\, r^{2(\lambda_{2e}-1)} + 2\,\xi_{i\,1e}\,\xi_{i\,2e}\, r^{\lambda_{1e}+\lambda_{2e}-2} \right],$$
$$i = 1,2,3, \tag{9.84}$$

where the elastic modulus s_{11e} and the coefficient $\xi_{i\,1+je}$ $(i=1,2,3;\ j=0,1)$ are given by Eqs. (3.23) and (9.88), respectively.

Appendix. Considering the coefficients c_{1e}, c_{2e}, a'_{1i}, a'_{2i}, a'_{3i} $(i=1,2,3)$ (see Eqs. (3.38), (3.39), (12.26)–(12.34)), and the exponents $\lambda_{1e} \neq \lambda_{2e}$ from one of the sets $\{-2,0\}$, $\{-2,1\}$, $\{-2,2\}$, $\{-2,3\}$, $\{-2,4\}$, $\{-2,7\}$, $\{-2,9\}$, $\{0,1\}$, $\{0,2\}$, $\{0,3\}$, $\{0,7\}$, $\{0,9\}$, $\{1,2\}$, $\{1,3\}$, $\{1,4\}$, $\{1,7\}$, $\{1,9\}$, $\{2,3\}$, $\{2,9\}$ (see Section 9.3.2), the coefficients $\chi_{1+j\,1+ie}$, $\varrho_{1+i\,1+je}$, $\omega_{1+i\,1+je}$ $(i,j=0,1)$, ξ_{i1+je} $(i=1,2,3;\ j=0,1)$ have the forms

$$\chi_{1+j\,1+ie} = R_{1+j}^{\lambda_{1+ie}-1} \left[\lambda_{1+ie} \left(c_{1e} + c_{2e} \right) - 2c_{2e} \right], \quad i,j = 0,1, \tag{9.85}$$

$$\varrho_{1+i\,1+je} = \sum_{k=0}^{1} \kappa_{1+j+2ke}^{(1)} \left(\delta_{0j} - \delta_{1j} \right) R_{1+i}^{\lambda_{1+ke}-1}, \quad i,j = 0,1, \tag{9.86}$$

$$\omega_{1+i\,1+je} = \sum_{k,l=0}^{1} p_{1+k}\, p_{1+l}\, \kappa_{1+2i+ke}^{(1)}\, \kappa_{1+2j+le}^{(1)} \left(\delta_{0k} - \delta_{1k} \right) \left(\delta_{0l} - \delta_{1l} \right)$$
$$\times \left[\lambda_{1+ie}\lambda_{1+je} \left(c_{1e} + c_{2e} \right) - 4\lambda_{1+ie} c_{2e} + 2c_{1e} \right], \quad i,j = 0,1, \tag{9.87}$$

$$\xi_{i\,1+je} = \sum_{k=0}^{1} p_{1+k}\, \kappa_{1+2j+ke}^{(1)} \left(\delta_{0k} - \delta_{1k}\right)$$
$$\times \left\{a'_{1i} \left[\lambda_{1+je} \left(c_{1e} + c_{2e}\right) - 2c_{2e}\right] + \left(a'_{2i} + a'_{3i}\right) \left(c_{1e} - \lambda_{1+je} c_{2e}\right)\right\},$$
$$i = 1, 2, 3; \quad j = 0, 1. \tag{9.88}$$

9.5.2.2 Condition $\beta_p \neq \beta_e = \beta_m$

Formulae for the thermal stresses and for the corresponding quantities related to the spherical envelope in the one-particle-envelope-matrix system provided that $\beta_p \neq \beta_e = \beta_m$ are identical with those presented in Section 9.4.2, considering the radial stress p_1 acting on the particle-envelope boundary (see Section 10.1.2; see Eq. (10.4)).

9.5.2.3 Condition $\beta_p = \beta_e \neq \beta_m$

With regard to the boundary condition (4.8) to determine one integration constant only, considering the analysis presented in Section 9.3.2 to define u_{1e} as an increasing function of the variable $r \in \langle R_1, R_2 \rangle$, the radial displacement u_{1e} of the infinitesimal spherical cap in the spherical envelope, regarding Eqs. (9.1), has the form

$$u_{1e} = -p_2\, \varrho_{2e}^{pe}\, r \left(\frac{r}{R_2}\right)^{\lambda_e - 1}, \tag{9.89}$$

where the coefficient ϱ_{2e}^{pe} is given by Eq. (9.99), the exponent λ_e is from the set $\{1, 0, 4, 2, 7, 3, 9, 6.43999\}$ (see Section 9.3.2), and the radial stress p_2 is given by Eq. (10.5) regarding Section 10.1.2.

Consequently, the radial and tangential stresses σ_{11e} and σ_{22e}, σ_{33e}, acting in the spherical envelope, respectively, regarding Eqs. (3.31)–(3.34), (9.89), are derived as

$$\sigma_{11e} = -p_2 \left(\frac{r}{R_2}\right)^{\lambda_e - 1}, \tag{9.90}$$

$$\sigma_{22e} = \sigma_{33e} = -p_2\, \varrho_{2e}^{pe} \left(c_{1e} - \lambda_e\, c_{2e}\right) \left(\frac{r}{R_2}\right)^{\lambda_e - 1}, \tag{9.91}$$

where the coefficients c_{1e}, c_{2e} are given by Eqs. (3.38), (3.39), respectively, and the shear stresses $\sigma_{12e} = \sigma_{13e} = 0$ (see Eqs. (3.33), (3.34)) due to $\partial u_{1e}/\partial \eta = 0$ $(\eta = \varphi, \nu)$ as presented in Section 9.1.

Finally, the radial and tangential strains, ε_{11e} and ε_{22e}, ε_{33e}, in the spherical envelope, respectively, regarding Eqs. (3.1)–(3.5), (9.89), have the forms

$$\varepsilon_{11e} = -p_2\, \varrho_{2e}^{pe}\, \lambda_e \left(\frac{r}{R_2}\right)^{\lambda_e - 1}, \tag{9.92}$$

$$\varepsilon_{22e} = \varepsilon_{33e} = -p_2\,\varrho_{2e}^{pe}\left(\frac{r}{R_2}\right)^{\lambda_e - 1}, \tag{9.93}$$

where the shear strains $\varepsilon_{12e} = \varepsilon_{13e} = 0$ (see Eqs. (3.4), (3.5)) due to $\partial u_{1e}/\partial\eta = 0$ $(\eta = \varphi, \nu)$ as presented in Section 9.1.

With regard to the determination of the compressive or tensile radial stress $p_2 > 0$ or $p_2 < 0$ acting on the matrix-envelope boundary, thus for $r = R_2$, respectively, considering Eq. (3.64), and to the boundary condition for the spherical particle provided that $\beta_p = \beta_e \neq \beta_m$ (see Eq. (4.6)), the tangential strains on the particle-envelope and matrix-envelope boundaries, $(\varepsilon_{22e})_{r=R_1}$ and $(\varepsilon_{22e})_{r=R_2}$, respectively, are required to be derived in the form

$$(\varepsilon_{22e})_{r=R_{1+i}} = -p_2\,\varrho_{1+ie}^{pe}, \quad i = 0, 1, \tag{9.94}$$

where the coefficient ϱ_{1+ie}^{pe} $(i = 0,1)$ is given by Eq. (9.99) for $\lambda_e \subset \{1, 0, 4, 2, 7, 3, 9, 6.43999\}$ (see Section 9.3.2).

Considering Eqs. (3.41), (3.43), (9.90)–(9.93), the thermal-stress induced elastic energy density w_e and the thermal-stress induced elastic energy $W_e^{(1)}$ [4], accumulated in an arbitrary point of the spherical envelope and in the volume of the spherical envelope, respectively, have the forms

$$w_e = \frac{\omega_{2e}\left(\varrho_{2e}^{pe}\right)^2}{2}\left(\frac{r}{R_2}\right)^{2(\lambda_e - 1)}, \tag{9.95}$$

$$W_e^{(1)} = 4\pi\int_{R_1}^{R_2} w_e\, r^2\, dr = \frac{2\pi\,\omega_{2e}\left(\varrho_{2e}^{pe}\right)^2 R_2^3}{2\lambda_e + 1}\left[1 - \left(\frac{R_1}{R_2}\right)^{2\lambda_e + 1}\right], \tag{9.96}$$

where the coefficient ω_{2e} is given by Eq. (9.64) for $\lambda_e \subset \{1, 0, 4, 2, 7, 3, 9, 6.43999\}$ (see Section 9.3.2).

With regard to the determination of the thermal-stress induced strengthening presented in Volume III, the stress σ_{ie} acting in a direction of the axis x_i and the elastic energy density w_{ie} induced by the stress σ_{ie} $(i = 1,2,3)$ (see Eqs. (3.46), (3.47), (5.111), (5.112)), both related to an arbitrary point of the spherical envelope, are derived as

$$\sigma_{ie} = -\xi_{i2e}\,\varrho_{2e}^{pe}\left(\frac{r}{R_2}\right)^{\lambda_e - 1}, \quad i = 1, 2, 3, \tag{9.97}$$

$$w_{ie} = \frac{s_{11e}\left(\xi_{i2e}\,\varrho_{2e}^{pe}\right)^2}{2}\left(\frac{r}{R_2}\right)^{2(\lambda_e - 1)}, \quad i = 1, 2, 3, \tag{9.98}$$

[4] The superscript (1) is related to the *one*-particle-(envelope)-matrix system.

where the elastic modulus s_{11e} and the coefficient ξ_{i2e} $(i=1,2,3)$ are given by Eqs. (3.23) and (9.65), respectively, the latter for $\lambda_e \subset \{1, 0, 4, 2, 7, 3, 9, 6.43999\}$ (see Section 9.3.2).

Appendix. Considering the coefficients c_{1e}, c_{2e} (see Eqs. (3.38), (3.39)), and the exponent $\lambda_e \subset \{1, 0, 4, 2, 7, 3, 9, 6.43999\}$ (see Section 9.3.2), the coefficient ϱ_{1+ie}^{pe} $(i=0,1)$ has the form

$$\varrho_{1+ie}^{pe} = \frac{1}{\lambda_e \left(c_{1e} + c_{2e}\right) - 2c_{2e}} \left(\frac{R_{1+i}}{R_2}\right)^{\lambda_e - 1}, \quad i = 0, 1. \tag{9.99}$$

9.5.3 Thermal stresses in infinite matrix

9.5.3.1 Conditions $\beta_p \neq \beta_m$, $\beta_p \neq \beta_e \neq \beta_m$, $\beta_p = \beta_e \neq \beta_m$

Considering Eqs. (9.100)–(9.110) containing the subscript $i=0,1$, the thermal stresses and the corresponding quantities for the infinite matrix presented in Section 9.5.3.1 are derived

1. with regard to the condition $\beta_p \neq \beta_m$ related to the one-particle-matrix system, and consequently the subscript $i = 0$,

2. with regard to the conditions $\beta_p \neq \beta_e \neq \beta_m$, $\beta_p = \beta_e \neq \beta_m$ related to the one-particle-envelope-matrix system, and consequently the subscript $i = 1$.

With regard to the boundary conditions (4.9), (4.16) to determine one integration constant only, considering the analysis presented in Section 9.3.3 to define u_{1m} as a decreasing function of the variable $r \in \langle R_{1+i}, \infty)$ $(i=0,1)$, the radial displacement u_{1m} of the infinitesimal spherical cap in the infinite matrix, regarding Eq. (9.1), has the form

$$u_{1m} = -p_{1+i}\, \varrho_{1+im}\, r \left(\frac{R_{1+i}}{r}\right)^3, \quad i = 0, 1, \tag{9.100}$$

where the coefficient ϱ_{1+im} $(i=0,1)$ is given by Eqs. (9.110). Considering Section 10.1.2, provided that $\beta_p \neq \beta_m$ and $\beta_p \neq \beta_e \neq \beta_m$ or $\beta_p = \beta_e \neq \beta_m$, the radial stresses p_1, p_2 are given by Eqs. (10.1) and (10.2), (10.3) or (10.5), respectively.

Consequently, the radial and tangential stresses, σ_{11m} and σ_{22m}, σ_{33m}, acting in the infinite matrix, respectively, regarding Eqs. (3.31)–(3.34), (9.100), are derived as

$$\sigma_{11m} = -p_{1+i} \left(\frac{R_{1+i}}{r}\right)^3, \quad i = 0, 1, \tag{9.101}$$

$$\sigma_{22m} = \sigma_{33m} = \frac{p_{1+i}}{2} \left(\frac{R_{1+i}}{r}\right)^3, \quad i = 0, 1, \tag{9.102}$$

where the shear stresses $\sigma_{12e} = \sigma_{13e} = 0$ (see Eqs. (3.33), (3.34)) due to $\partial u_{1e}/\partial \eta = 0$ $(\eta = \varphi, \nu)$ as presented in Section 9.1.

Finally, the radial and tangential strains, ε_{11m} and ε_{22m}, ε_{33m}, in the infinite matrix, respectively, regarding Eqs. (3.1)–(3.5), (9.100), have the forms[5]

$$\varepsilon_{11m} = 2\, p_{1+i}\, \varrho_{1+im} \left(\frac{R_{1+i}}{r} \right)^3, \quad i = 0, 1, \tag{9.103}$$

$$\varepsilon_{22m} = \varepsilon_{33m} = -p_{1+i} \varrho_{1+im} \left(\frac{R_{1+i}}{r} \right)^3, \quad i = 0, 1, \tag{9.104}$$

where the shear strains $\varepsilon_{12m} = \varepsilon_{13m} = 0$ (see Eqs. (3.4), (3.5)) due to $\partial u_{1m}/\partial \eta = 0$ $(\eta = \varphi, \nu)$ as presented in Section 9.1.

With regard to the determination of the compressive or tensile radial stress $p_{1+i} > 0$ or $p_{1+i} < 0$ acting on the particle-matrix, matrix-envelope boundary, thus for $r = R_1$, $r = R_2$, respectively, considering Eqs. (3.58), (3.63), (3.64), the tangential strains on the particle-matrix and matrix-envelope boundaries, $(\varepsilon_{22m})_{r=R_1}$ and $(\varepsilon_{22m})_{r=R_2}$, respectively, are required to be derived in the form

$$(\varepsilon_{22m})_{r=R_{1+i}} = -p_{1+i}\, \varrho_{1+im}, \quad i = 0, 1, \tag{9.105}$$

where the coefficient ϱ_{1+im} $(i = 0,1)$ is given by Eq. (9.110).

Considering Eqs. (3.41), (3.43), (9.101)–(9.104), the thermal-stress induced elastic energy density w_m and the thermal-stress induced elastic energy $W_m^{(1)}$[6], accumulated in an arbitrary point of the infinite matrix and in the volume of the infinite matrix, respectively, have the forms

$$w_m = \frac{3\, p_{1+i}^2}{4} \left(s_{11m} - s_{12m} \right) \left(\frac{R_{1+i}}{r} \right)^6, \quad i = 0, 1, \tag{9.106}$$

$$W_m^{(1)} = 4\pi \int_{R_{1+i}}^{\infty} w_m\, r^2\, dr = \pi\, p_{1+i}^2\, R_{1+i}^3 \left(s_{11m} - s_{12m} \right), \quad i = 0, 1, \tag{9.107}$$

where the elastic moduli s_{11m}, s_{12m} are given by Eqs. (3.23), (3.24), respectively.

With regard to the determination of the thermal-stress induced strengthening presented in Volume III, the stress σ_{jm} acting in a direction of the axis x_j and the elastic energy density w_{jm} induced by the stress σ_{jm} $(j = 1,2,3)$ (see Eqs. (3.46), (3.47), (9.101), (9.102)), both related to an arbitrary point of the infinite matrix, are derived as[7]

[5]To avoid misunderstanding, signs on the right sides in Eqs. (9.102), (9.103) are really plus, in contrast to minus signs on the right sides in Eqs. (9.101), (9.104).

[6]The superscript (1) is related to the *one*-particle-(envelope)-matrix system.

[7]To avoid misunderstanding, a sign on the right sides in Eq. (9.108) is really plus.

$$\sigma_{jm} = \frac{p_{1+i}}{2} \left(a'_{2j} + a'_{3j} - 2\, a'_{1j} \right) \left(\frac{R_{1+i}}{r} \right)^3, \quad i = 0, 1; \quad j = 1, 2, 3, \tag{9.108}$$

$$w_{jm} = \frac{s_{11m}\, p_{1+i}^2}{8} \left(a'_{2j} + a'_{3j} - 2\, a'_{1j} \right)^2 \left(\frac{R_{1+i}}{r} \right)^6, \quad i = 0, 1; \quad j = 1, 2, 3, \tag{9.109}$$

where the coefficients a'_{1j}, a'_{2j}, a'_{3j} ($j = 1,2,3$) are given by Eqs. (12.26)–(12.34).

Appendix. Considering the elastic moduli s_{11m}, s_{12m} (see Eqs. (3.23), (3.24)), the coefficient ϱ_{1+im} [8] has the form

$$\varrho_{1+im} = -\frac{s_{11m} - s_{12m}}{2}, \quad i = 0, 1. \tag{9.110}$$

9.5.3.2 Condition $\beta_p \neq \beta_e = \beta_m$

With regard to Section 10.1.2, the formulae in Section 9.5.3.2 consider the φ, ν-independent radial stress $p_1 = (\beta_e - \beta_p) / (\varrho_{1p} - \varrho_{1e}^{me})$ (see Eq. (10.4)) acting on the particle-envelope boundary and derived from Eq. (3.63), where the φ, ν-independent coefficients ϱ_{1p} and ϱ_{1e}^{me}, along with ϱ_{2e}^{me} included in the formulae in Section 9.5.3.2, are given by Eqs. (9.50) and (9.63), respectively. The φ, ν-independence of the radial stress p_1 results from the φ, ν-independent difference $\varepsilon_{11te} - \varepsilon_{11tp} \neq f(\varphi, \nu)$ (see Eq. (12.23), (12.24)), resulting in the φ, ν-independent difference $\beta_e - \beta_p$ (see Eqs. (3.48)–(3.50)).

With regard to the boundary conditions (4.12), (4.16) to determine one integration constant only, considering the analysis presented in Section 9.3.3 to define u_{1m} as a decreasing function of the variable $r \in \langle R_2, \infty)$, the radial displacement u_{1m} of the infinitesimal spherical cap in the infinite matrix, regarding Eq. (9.1), has the form

$$u_{1m} = -p_1\, \varrho_{2e}^{me}\, \varrho_{1m}\, r \left(\frac{R_2}{r} \right)^3, \quad i = 0, 1, \tag{9.111}$$

where the coefficient ϱ_{1m} is given by Eq. (9.110).

Consequently, the radial and tangential stresses, σ_{11m} and σ_{22m}, σ_{33m}, acting in the infinite matrix, respectively, regarding Eqs. (3.31)–(3.34), (9.111), are derived as

[8] The subscript $1 + i$ in Eq. (9.110) is used due to uniform notations applied in formulae for the tangential strain $(\varepsilon_{22q})_{r=R_{1+i}}$ ($i = 0,1$; $q = p,e,m$), e.g. as presented in Eq. (9.105). Additionally, to avoid misunderstanding, a sign on the right side in Eq. (9.110) is really minus.

$$\sigma_{11m} = -p_1\, \varrho_{2e}^{me} \left(\frac{R_2}{r}\right)^3, \tag{9.112}$$

$$\sigma_{22m} = \sigma_{33m} = \frac{p_1\, \varrho_{2e}^{me}}{2} \left(\frac{R_2}{r}\right)^3, \tag{9.113}$$

where the shear stresses $\sigma_{12e} = \sigma_{13e} = 0$ (see Eqs. (3.33), (3.34)) due to $\partial u_{1e}/\partial \eta = 0$ $(\eta = \varphi, \nu)$ as presented in Section 9.1.

Finally, the radial and tangential strains, ε_{11m} and ε_{22m}, ε_{33m}, in the infinite matrix, respectively, regarding Eqs. (3.1)–(3.5), (9.111), have the forms[9]

$$\varepsilon_{11m} = 2\, p_1\, \varrho_{2e}^{me}\, \varrho_{1m} \left(\frac{R_2}{r}\right)^3, \tag{9.114}$$

$$\varepsilon_{22m} = \varepsilon_{33m} = -p_1\, \varrho_{2e}^{me}\, \varrho_{1m} \left(\frac{R_2}{r}\right)^3, \tag{9.115}$$

where the shear strains $\varepsilon_{12m} = \varepsilon_{13m} = 0$ (see Eqs. (3.4), (3.5)) due to $\partial u_{1m}/\partial \eta = 0$ $(\eta = \varphi, \nu)$ as presented in Section 9.1.

Considering Eqs. (3.41), (3.43), (9.112)–(9.115), the thermal-stress induced elastic energy density w_m and the thermal-stress induced elastic energy $W_m^{(1)}$[10], accumulated in an arbitrary point of the infinite matrix and in the volume of the infinite matrix, respectively, have the forms

$$w_m = \frac{3\left(p_1\, \varrho_{2e}^{me}\right)^2}{4} \left(s_{11m} - s_{12m}\right) \left(\frac{R_2}{r}\right)^6, \tag{9.116}$$

$$W_m^{(1)} = 4\pi \int_{R_2}^{\infty} w_m\, r^2\, dr = \pi \left(p_1\, \varrho_{2e}^{me}\right)^2 R_2^3 \left(s_{11m} - s_{12m}\right), \tag{9.117}$$

where the elastic moduli s_{11m}, s_{12m} are given by Eqs. (3.23), (3.24), respectively.

With regard to the determination of the thermal-stress induced strengthening presented in Volume III, the stress σ_{im} acting in a direction of the axis x_i and the elastic energy density w_{im} induced by the stress σ_{im} $(i=1,2,3)$ (see Eqs. (3.46), (3.47), (9.112), (9.113)), both related to an arbitrary point of the infinite matrix, are derived as[11]

$$\sigma_{im} = \frac{p_1\, \varrho_{2e}^{me}}{2} \left(a'_{2i} + a'_{3i} - 2\, a'_{1i}\right) \left(\frac{R_2}{r}\right)^3, \quad i = 1, 2, 3, \tag{9.118}$$

[9]To avoid misunderstanding, signs on the right sides in Eqs. (9.113), (9.114) are really plus, in contrast to minus signs on the right sides in Eqs. (9.112), (9.115).

[10]The superscript (1) is related to the *one*-particle-(envelope)-matrix system.

[11]To avoid misunderstanding, a sign on the right sides in Eq. (9.118) is really plus.

$$w_{im} = \frac{s_{11m}\left(p_1\,\varrho_{2e}^{me}\right)^2}{8}\left(a'_{2i} + a'_{3i} - 2\,a'_{1i}\right)^2\left(\frac{R_2}{r}\right)^6, \quad i = 1, 2, 3, \tag{9.119}$$

where the coefficients a'_{1i}, a'_{2i}, a'_{3i} $(i = 1,2,3)$ are given by Eqs. (12.26)–(12.34).

9.6 Supplement

9.6.1 Mathematical techniques 10

In contrast to Section 9.2, Section 9.6.1 presents the mathematical techniques to lead to different linear differential equations for σ_{11q} with zero right sides, and consequently the radial stress σ_{11q} has the form

$$\sigma_{11q} = \sum_{i=0}^{n} B_{iq}\, r^{\zeta_i}, \tag{9.120}$$

representing a linear combination of the particular solutions $\sigma_{1r} = r^{\zeta_1}, \ldots,$ $\sigma_{nr} = r^{\zeta_n}$ related to the integration constants $B_{1q}, \ldots, B_{nq}$, respectively, where $\zeta_1, \ldots, \zeta_n$ are real roots of one of the characteristic equations (9.127), (9.131), (9.135), (9.138), (9.142), (9.145), (9.148), (9.152), (9.156), (9.160), (9.163), (9.167), (9.171), (9.175), (9.179). With regard to $\sigma_{22q} = \sigma_{33q}$ and $\sigma_{12q} = \sigma_{13q} = 0$ resulting from $\varepsilon_{22q} = \varepsilon_{33q}$ and $\partial u_{1q}/\partial\eta = 0$ $(\eta = \varphi, \nu)$ (see Eqs. (3.2), (3.3), (3.19)–(3.22)), the equilibrium equation (3.15) is transformed to the form

$$\sigma_{22q} = \sigma_{33q} = \sigma_{11q} + \frac{r}{2}\frac{\partial\sigma_{11q}}{\partial r}, \tag{9.121}$$

and considering Eq. (9.120), (9.121), we get

$$\sigma_{22q} = \sigma_{33q} = \sum_{i=0}^{n} B_{iq}\left(1 + \frac{\zeta_i}{2}\right) r^{\zeta_i}. \tag{9.122}$$

Consequently, with regard to Eqs. (3.27), (9.120), (9.122), the radial displacement u_{1q}, the radial and tangential strains, ε_{11q} and ε_{22q}, ε_{33q}, respectively, have the forms

$$u_{1q} = \sum_{i=0}^{n} B_{iq}\left[\left(1 + \frac{\zeta_i}{2}\right)(s_{11q} + s_{12q}) + s_{12q}\right] r^{\zeta_i + 1}, \tag{9.123}$$

$$\varepsilon_{11q} = \sum_{i=0}^{n} B_{iq}\,\zeta_i\left[\left(1 + \frac{\zeta_i}{2}\right)(s_{11q} + s_{12q}) + s_{12q}\right] r^{\zeta_i}, \tag{9.124}$$

$$\varepsilon_{22q} = \varepsilon_{33q} = \sum_{i=0}^{n} B_{iq} \left[\left(1 + \frac{\zeta_i}{2} \right) (s_{11q} + s_{12q}) + s_{12q} \right] r^{\zeta_i}, \tag{9.125}$$

where the shear strains $\varepsilon_{12q} = \varepsilon_{13q} = 0$ (see Eqs. (3.4), (3.5)) due to $\partial u_{1p}/\partial \eta = 0$ ($\eta = \varphi, \nu$) as presented in Section 9.1.

Substituting Eqs. (3.19), (3.20), (9.121) to the Saint-Venant's equation (3.6), we get

$$r \frac{\partial^2 \sigma_{11q}}{\partial r^2} = -4 \frac{\partial \sigma_{11q}}{\partial r}, \tag{9.126}$$

and consequently assuming $\sigma_{11q} = r^{\zeta}$, the characteristic equation resulting from the linear differential equation (9.126) with zero right side has the form

$$\zeta (\zeta + 3) = 0, \tag{9.127}$$

where ζ_1, ζ_2 representing roots of the quadratic algebraic equation 9.127 are derived as

$$\zeta_1 = 0, \tag{9.128}$$

$$\zeta_2 = -3. \tag{9.129}$$

Performing $r\,[\partial \text{Eq.}(9.126)/\partial r]$, the differential equation (9.126) is transformed to the form

$$r \frac{\partial^3 \sigma_{11q}}{\partial r^3} = -5 \frac{\partial^2 \sigma_{11q}}{\partial r^2}. \tag{9.130}$$

Similarly, assuming $\sigma_{11q} = r^{\zeta}$, the characteristic equation resulting from the linear differential equation (9.130) with zero right side has the form

$$\zeta (\zeta + 3)(\zeta - 1) = 0, \tag{9.131}$$

with the real roots ζ_1, ζ_2 and ζ_3 given by Eqs. (9.128), (9.129), respectively, and derived as

$$\zeta_3 = 1. \tag{9.132}$$

Performing $r \times$Eq. (9.130), we get

$$r^2 \frac{\partial^3 \sigma_{11q}}{\partial r^3} = -5\, r \frac{\partial^2 \sigma_{11q}}{\partial r^2}, \tag{9.133}$$

and consequently substituting Eq. (9.126) to Eq. (9.133), the differential equation (9.133) is transformed to the form

$$r^2 \frac{\partial^3 \sigma_{11q}}{\partial r^3} = 20 \frac{\partial \sigma_{11q}}{\partial r}, \tag{9.134}$$

exhibiting the characteristic equation derived as

$$\zeta (\zeta + 3) (\zeta - 6) = 0, \tag{9.135}$$

with the real roots ζ_1, ζ_2 and ζ_4 given by Eqs. (9.128), (9.129), respectively, and derived as

$$\zeta_4 = 6. \tag{9.136}$$

Performing $r\,[\partial \text{Eq.}(9.130)/\partial r]$, the differential equation (9.130) is transformed to the form

$$r \frac{\partial^4 \sigma_{11q}}{\partial r^4} = -\,6 \frac{\partial^3 \sigma_{11q}}{\partial r^3}, \tag{9.137}$$

exhibiting the characteristic equation derived as

$$\zeta (\zeta + 3) (\zeta - 1) (\zeta - 2) = 0, \tag{9.138}$$

with the real roots ζ_1, ζ_2, ζ_3 and ζ_5 given by Eqs. (9.128), (9.129), (9.132), respectively, and derived as

$$\zeta_5 = 2. \tag{9.139}$$

Performing $r \times$Eq. (9.137), we get

$$r^2 \frac{\partial^4 \sigma_{11q}}{\partial r^4} = -\,6\, r \frac{\partial^3 \sigma_{11q}}{\partial r^3}, \tag{9.140}$$

and consequently substituting Eq. (9.130) to Eq. (9.140), the differential equation (9.140) is transformed to the form

$$r^2 \frac{\partial^4 \sigma_{11q}}{\partial r^4} = 30 \frac{\partial^2 \sigma_{11q}}{\partial r^2}, \tag{9.141}$$

exhibiting the characteristic equation derived as

$$\zeta (\zeta + 3) (\zeta - 1) (\zeta - 8) = 0, \tag{9.142}$$

with the real roots ζ_1, ζ_2, ζ_3 and ζ_6 given by Eqs. (9.128), (9.129), (9.132), respectively, and derived as

$$\zeta_6 = 8. \tag{9.143}$$

Substituting Eq. (9.126) to Eq. (9.141), the differential equation (9.141) is transformed to the form

$$r^3 \frac{\partial^4 \sigma_{11q}}{\partial r^4} = -120 \frac{\partial \sigma_{11q}}{\partial r}, \tag{9.144}$$

exhibiting the characteristic equation derived as

$$\zeta\,(\zeta+3)\left(\zeta^2 - 9\,\zeta + 38\right) = 0, \tag{9.145}$$

with the real roots ζ_1, ζ_2 given by Eqs. (9.128), (9.129), respectively. Considering the discriminant $D = -71 < 0$ (see Eqs. (12.51)–(12.53)), the quadratic algebraic equation $\zeta^2 - 9\,\zeta + 38 = 0$ has imaginary roots $\zeta_1^{(I)}$, $\zeta_2^{(I)}$ in the form

$$\zeta_{1,2}^{(I)} = \frac{1}{2}\left(9 \pm i\,\sqrt{71}\right). \tag{9.146}$$

Performing $r\,[\partial\text{Eq.}(9.137)/\partial r]$, the differential equation (9.137) is transformed to the form

$$r \frac{\partial^5 \sigma_{11q}}{\partial r^5} = -7 \frac{\partial^4 \sigma_{11q}}{\partial r^4}, \tag{9.147}$$

exhibiting the characteristic equation derived as

$$\zeta\,(\zeta+3)\,(\zeta-1)\,(\zeta-2)\,(\zeta-3) = 0, \tag{9.148}$$

with the real roots ζ_1, ζ_2, ζ_3, ζ_5 and ζ_7 given by Eqs. (9.128), (9.129), (9.132), (9.139), respectively, and derived as

$$\zeta_7 = 3. \tag{9.149}$$

Performing $r\times$Eq. (9.147), we get

$$r^2 \frac{\partial^5 \sigma_{11q}}{\partial r^5} = -7\,r \frac{\partial^4 \sigma_{11q}}{\partial r^4}, \tag{9.150}$$

and consequently substituting Eq. (9.137) to Eq. (9.150), the differential equation (9.150) is transformed to the form

$$r^2 \frac{\partial^5 \sigma_{11q}}{\partial r^5} = 42 \frac{\partial^3 \sigma_{11q}}{\partial r^3}, \tag{9.151}$$

exhibiting the characteristic equation derived as

$$\zeta\,(\zeta+3)\,(\zeta-1)\,(\zeta-2)\,(\zeta-10) = 0, \tag{9.152}$$

with the real roots ζ_1, ζ_2, ζ_3, ζ_5 and ζ_8 given by Eqs. (9.128), (9.129), (9.132), (9.139), respectively, and derived as

$$\zeta_8 = 10. \tag{9.153}$$

Performing $r\times$Eq. (9.151), we get

$$r^3\,\frac{\partial^5\sigma_{11q}}{\partial r^5} = 42\,r\,\frac{\partial^3\sigma_{11q}}{\partial r^3}, \tag{9.154}$$

and consequently substituting Eq. (9.130) to Eq. (9.154), the differential equation (9.154) is transformed to the form

$$r^3\,\frac{\partial^5\sigma_{11q}}{\partial r^5} = -210\,\frac{\partial^2\sigma_{11q}}{\partial r^2}, \tag{9.155}$$

exhibiting the characteristic equation derived as

$$\zeta\,(\zeta+3)\,(\zeta-1)\left(\zeta^2-12\,\zeta+62\right)=0, \tag{9.156}$$

with the real roots ζ_1, ζ_2, ζ_3 given by Eqs. (9.128), (9.129), (9.132), respectively. Considering the discriminant $D=-104<0$ (see Eqs. (12.51)–(12.53)), the quadratic algebraic equation $\zeta^2-12\,\zeta+62=0$ has imaginary roots $\zeta_3^{(I)}$, $\zeta_4^{(I)}$ in the form

$$\zeta_{3,4}^{(I)} = 6 \pm i\,\sqrt{26}. \tag{9.157}$$

Performing $r\times$Eq. (9.155), we get

$$r^4\,\frac{\partial^5\sigma_{11q}}{\partial r^5} = -210\,r\,\frac{\partial^2\sigma_{11q}}{\partial r^2}, \tag{9.158}$$

and consequently substituting Eq. (9.126) to Eq. (9.158), the differential equation (9.158) is transformed to the form

$$r^4\,\frac{\partial^5\sigma_{11q}}{\partial r^5} = 840\,\frac{\partial\sigma_{11q}}{\partial r}, \tag{9.159}$$

exhibiting the characteristic equation derived as

$$\zeta\,(\zeta+3)\,(\zeta-8)\left(\zeta^2-5\,\zeta+34\right)=0, \tag{9.160}$$

with the real roots ζ_1, ζ_2, ζ_6 given by Eqs. (9.128), (9.129), (9.143), respectively. Considering the discriminant $D=-111<0$ (see Eqs. (12.51)–(12.53)), the quadratic algebraic equation $\zeta^2-5\,\zeta+34=0$ has imaginary roots $\zeta_5^{(I)}$, $\zeta_6^{(I)}$ in the form

$$\zeta_{5,6}^{(I)} = \frac{1}{2}\left(5 \pm i\,\sqrt{111}\right). \tag{9.161}$$

Performing $r\,[\partial\text{Eq.}(9.147)/\partial r]$, the differential equation (9.147) is transformed to the form

$$r\,\frac{\partial^6\sigma_{11q}}{\partial r^6} = -8\,\frac{\partial^5\sigma_{11q}}{\partial r^5}, \tag{9.162}$$

exhibiting the characteristic equation derived as

$$\zeta\,(\zeta+3)\,(\zeta-1)\,(\zeta-2)\,(\zeta-3)\,(\zeta-4)=0, \tag{9.163}$$

with the real roots ζ_1, ζ_2, ζ_3, ζ_5, ζ_7 and ζ_9 given by Eqs. (9.128), (9.129),(9.132), (9.139), (9.149), respectively, and derived as

$$\zeta_9 = 4. \tag{9.164}$$

Performing $r\times$Eq. (9.162), we get

$$r^2\,\frac{\partial^6\sigma_{11q}}{\partial r^6} = -8\,r\,\frac{\partial^5\sigma_{11q}}{\partial r^5}, \tag{9.165}$$

and consequently substituting Eq. (9.147) to Eq. (9.165), the differential equation (9.165) is transformed to the form

$$r^2\,\frac{\partial^6\sigma_{11q}}{\partial r^6} = 56\,\frac{\partial^4\sigma_{11q}}{\partial r^4}, \tag{9.166}$$

exhibiting the characteristic equation derived as

$$\zeta\,(\zeta+3)\,(\zeta-1)\,(\zeta-2)\,(\zeta-3)\,(\zeta-12)=0, \tag{9.167}$$

with the real roots ζ_1, ζ_2, ζ_3, ζ_5, ζ_7 and ζ_{10} given by Eqs. (9.128), (9.129),(9.132), (9.139), (9.149), respectively, and derived as

$$\zeta_{10} = 12. \tag{9.168}$$

Performing $r\times$Eq. (9.166), we get

$$r^3\,\frac{\partial^6\sigma_{11q}}{\partial r^6} = 56\,r\,\frac{\partial^4\sigma_{11q}}{\partial r^4}, \tag{9.169}$$

and consequently substituting Eq. (9.137) to Eq. (9.169), the differential equation (9.169) is transformed to the form

$$r^3\,\frac{\partial^6\sigma_{11q}}{\partial r^6} = -336\,\frac{\partial^3\sigma_{11q}}{\partial r^3}, \tag{9.170}$$

exhibiting the characteristic equation derived as

$$\zeta\,(\zeta+3)\,(\zeta-1)\,(\zeta-2)\left(\zeta^2-15\,\zeta+92\right)=0, \tag{9.171}$$

with the real roots ζ_1, ζ_2, ζ_3, ζ_5 given by Eqs. (9.128), (9.129), (9.132), (9.139), respectively. Considering the discriminant $D=-143<0$ (see Eqs. (12.51)–(12.53)), the quadratic algebraic equation $\zeta^2-15\,\zeta+92=0$ has imaginary roots $\zeta_7^{(I)}$, $\zeta_8^{(I)}$ in the form

$$\zeta_{7,8}^{(I)} = \frac{1}{2}\left(15\pm i\,\sqrt{143}\right). \tag{9.172}$$

Performing $r\times$Eq. (9.170), we get

$$r^4\,\frac{\partial^6\sigma_{11q}}{\partial r^6} = -336\,r\,\frac{\partial^3\sigma_{11q}}{\partial r^3}, \tag{9.173}$$

and consequently substituting Eq. (9.130) to Eq. (9.173), the differential equation (9.173) is transformed to the form

$$r^4\,\frac{\partial^6\sigma_{11q}}{\partial r^6} = 1680\,\frac{\partial^2\sigma_{11q}}{\partial r^2}, \tag{9.174}$$

exhibiting the characteristic equation derived as

$$\zeta\,(\zeta+3)\,(\zeta-1)\,(\zeta-10)\left(\zeta^2-7\,\zeta+52\right)=0, \tag{9.175}$$

with the real roots ζ_1, ζ_2, ζ_3, ζ_8 given by Eqs. (9.128), (9.129), (9.132), (9.153), respectively. Considering the discriminant $D=-159<0$ (see Eqs. (12.51)–(12.53)), the quadratic algebraic equation $\zeta^2-7\,\zeta+52=0$ has imaginary roots $\zeta_9^{(I)}$, $\zeta_{10}^{(I)}$ in the form

$$\zeta_{9,10}^{(I)} = \frac{1}{2}\left(7\pm i\,\sqrt{159}\right). \tag{9.176}$$

Performing $r\times$Eq. (9.174), we get

$$r^5\,\frac{\partial^6\sigma_{11q}}{\partial r^6} = 1680\,r\,\frac{\partial^2\sigma_{11q}}{\partial r^2}, \tag{9.177}$$

and consequently substituting Eq. (9.126) to Eq. (9.177), the differential equation (9.177) is transformed to the form

$$r^5\,\frac{\partial^6\sigma_{11q}}{\partial r^6} = -6720\,\frac{\partial\sigma_{11q}}{\partial r}, \tag{9.178}$$

exhibiting the characteristic equation derived as

$$\zeta\,(\zeta+3)\left(\zeta^4-18\,\zeta^3+139\,\zeta^2-642\,\zeta+2200\right)=0, \tag{9.179}$$

with the real roots ζ_1, ζ_2 given by Eqs. (9.128), (9.129), respectively. The algebraic equation $\zeta^4-18\,\zeta^3+139\,\zeta^2-642\,\zeta+2200=0$ has imaginary roots $\zeta_{11}^{(I)}$, $\zeta_{12}^{(I)}$, $\zeta_{13}^{(I)}$, $\zeta_{14}^{(I)}$ in the form [12]

$$\zeta_{11,12}^{(I)} = 1.14703\pm 5.37956\,i, \tag{9.180}$$

$$\zeta_{13,14}^{(I)} = 7.85297\pm 3.32342\,i. \tag{9.181}$$

With regard to the mathematical techniques presented in Section 9.6.1, additional real exponents can be accordingly derived.

[12] The exponents $\zeta_{11,12}^{(I)}$, $\zeta_{13,14}^{(I)}$ are derived using a software product.

9.6.2 Analysis of exponent ζ

With regard to the boundary conditions in Sections 4.1.1, 4.1.2, 4.2, formulae for the thermal stresses and for the corresponding quantities in the spherical particle, in the spherical envelope and in the infinite matrix, using the solutions to result from the mathematical techniques presented in Section 9.6.1, can be easily derived, being similar to those in Sections 9.4, 9.5.

9.6.2.1 Spherical particle

As presented in Sections 4.1.1, 5.4.1, the absolute value $|u_{1p}|$ of the radial displacement $u_{1p} \propto B_p r^{\zeta_p + 1}$ (see Eq. (9.123)) in the spherical particle is required to represent an increasing function of the variable $r \in \langle 0, R_1 \rangle$ provided that $\beta_p \neq \beta_m$, $\beta_p \neq \beta_e \neq \beta_m$, $\beta_p \neq \beta_e = \beta_m$, $\beta_p = \beta_e \neq \beta_m$ for $\varepsilon_{11tm} - \varepsilon_{11tp} \neq f(\varphi, \nu)$, $\varepsilon_{11te} - \varepsilon_{11tp} \neq f(\varphi, \nu)$, $\varepsilon_{11tm} - \varepsilon_{11te} \neq f(\varphi, \nu)$. Consequently, the radial displacement u_{1p} for the real exponent $\zeta_p \geq 0$ fulfils the boundary conditions (4.3)–(4.5), and accordingly $\zeta_p \subset \{\zeta_1, \zeta_3, \ldots, \zeta_{10}\}$ (see Eqs. (9.128), (9.132), (9.136), (9.139), (9.143), (9.149), (9.153), (9.164), (9.168)). Provided that $\zeta_p < 0$, the stress $(\sigma_{iip})_{r \to 0} \propto \left(r^{\zeta_p}\right)_{r \to 0} \to \infty$ $(i = 1,2,3)$ (see Eqs. (9.120), (9.122)), what is physically unacceptable. Additionally, with regard to the Castigliano's theorem (see Section 3.1.9), such real exponent ζ_p from the set $\{\zeta_1, \zeta_3, \ldots, \zeta_{10}\} = \{0, 1, 6, 2, 8, 3, 10, 4, 12\}$ is considered to result in minimal thermal-stress induced elastic energy, $W_c = W_p + W_e + W_m$ and $W^{(1)} = W_p^{(1)} + (W_e^{(1)}) + W_m^{(1)}$ (see Section 2.2.2.1), related to the cell of the multi-particle-envelope-matrix system and to the one-particle-(envelope)-matrix system, respectively. The analysis presented in Section 9.6.2.1 is also valid for the multi-particle-envelope-matrix system provided that $\beta_p \neq \beta_e = \beta_m$ for $\varepsilon_{11te} - \varepsilon_{11tp} \neq f(\varphi, \nu)$ (see Section 9.4.1).

9.6.2.2 Spherical envelope

With regard to the one-particle-envelope-matrix system provided that $\beta_p \neq \beta_e \neq \beta_m$ for $\varepsilon_{11te} - \varepsilon_{11tp} \neq f(\varphi, \nu)$ and $\varepsilon_{11tm} - \varepsilon_{11te} \neq f(\varphi, \nu)$, (see Eqs. (3.48)–(3.50), (12.23), (12.24)), considering the analyses presented in Sections 4.1.2, 5.4.2 and the boundary conditions (4.7), (4.8), the exponents $\zeta_{1e} \neq \zeta_{2e}$ included in $\sigma_{11e} = B_{1e} r^{\zeta_{1e}} + B_{2e} r^{\zeta_{2e}}$ (see Eq. (9.120)) are required to represent real roots of one of the characteristic equations (9.127), (9.131), (9.135), (9.138), (9.145), (9.148), (9.152), (9.156), (9.160), (9.163), (9.167), (9.171), (9.175), (9.179), and consequently ζ_{1e}, ζ_{2e} are both from one of the sets $\{\zeta_{1e}, \zeta_{2e}, \zeta_{4e}\}$, $\{\zeta_{1e}, \zeta_{2e}, \zeta_{3e}, \zeta_{6e}\}$, $\{\zeta_{1e}, \zeta_{2e}, \zeta_{3e}, \zeta_{5e}, \zeta_{8e}\}$, $\{\zeta_{1e}, \zeta_{2e}, \zeta_{3e}, \zeta_{5e}, \zeta_{7e}, \zeta_{9e}\}$, $\{\zeta_{1e}, \zeta_{2e}, \zeta_{3e}, \zeta_{5e}, \zeta_{7e}, \zeta_{10e}\}$ to result in the 29 combinations $\{-3, 0\}$, $\{-3, 1\}$, $\{-3, 2\}$, $\{-3, 3\}$, $\{-3, 4\}$, $\{-3, 6\}$, $\{-3, 8\}$, $\{-3, 10\}$, $\{-3, 12\}$, $\{0, 1\}$, $\{0, 2\}$, $\{0, 3\}$, $\{0, 4\}$, $\{0, 6\}$, $\{0, 8\}$, $\{0, 10\}$, $\{0, 12\}$, $\{1, 2\}$, $\{1, 3\}$, $\{1, 4\}$, $\{1, 8\}$, $\{1, 10\}$, $\{1, 12\}$, $\{2, 3\}$, $\{2, 4\}$, $\{2, 10\}$, $\{2, 12\}$, $\{3, 4\}$, $\{3, 12\}$ for $\zeta_{1e} \neq \zeta_{2e}$. Similarly, such combination of ζ_{1e}, ζ_{2e} is considered to result in minimal thermal-stress induced elastic energy $W^{(1)} = W_p^{(1)} + (W_e^{(1)}) + W_m^{(1)}$ (see Section 2.2.2.1).

Finally, with regard to the one-particle-envelope-matrix system provided that $\beta_p \neq \beta_e = \beta_m$ for $\varepsilon_{11te} - \varepsilon_{11tp} \neq f(\varphi, \nu)$ or $\beta_p = \beta_e \neq \beta_m$ for $\varepsilon_{11tm} - \varepsilon_{11te} \neq f(\varphi, \nu)$, considering the boundary condition (4.7) or (4.8), the absolute value $|u_{1e}|$ of the radial displacement $u_{1e} \propto B_e \, r^{\zeta_e + 1}$ (see Eq. (9.123)) in the spherical envelope is required to represent a decreasing or increasing function of the variable $r \in \langle R_1, R_2 \rangle$, at the very most, a constant function regarding the variable r, and accordingly $\zeta_e \leq -1$ or $\zeta_e \geq -1$, respectively. Although the assumption for the radial displacement u_{1e} as a constant function regarding the variable r for $\zeta_e = -1$ may be improbable, nevertheless it is considered with regard to completeness of the presented topic. Finally, the exponent $\zeta_e = \zeta_2 = -3$ (see Eq. (9.129)) or ζ_e is from the set $\{0, 1, 6, 2, 8, 3, 10, 4, 12\}$ for $\beta_p \neq \beta_e = \beta_m$ or $\beta_p = \beta_e \neq \beta_m$, respectively, considering the Castigliano's theorem (see Section 3.1.9). The analysis presented in Section 9.6.2.2 is also valid for the multi-particle-envelope-matrix system provided that $\beta_p \neq \beta_e = \beta_m$ for $\varepsilon_{11te} - \varepsilon_{11tp} \neq f(\varphi, \nu)$ (see Section 9.4.2).

9.6.2.3 Infinite matrix

With regard to the one-particle-matrix and one-particle-envelope-matrix systems provided that $\beta_p \neq \beta_m$ and $\beta_p \neq \beta_e \neq \beta_m$, $\beta_p \neq \beta_e = \beta_m$, $\beta_p = \beta_e \neq \beta_m$ for $\varepsilon_{11tm} - \varepsilon_{11tp} \neq f(\varphi, \nu)$ and $\varepsilon_{11te} - \varepsilon_{11tp} \neq f(\varphi, \nu)$, $\varepsilon_{11tm} - \varepsilon_{11te} \neq f(\varphi, \nu)$ (see Eqs. (3.48)–(3.50), (12.23), (12.24)), the absolute values $|u_{1m}|$, $|\sigma_{11m}|$ represent decreasing functions of the variable $r \in \langle R_1, \infty)$ and $r \in \langle R_2, \infty)$ (see Section 4.2), respectively, and consequently the exponent $\zeta_m = \zeta_2 = -3$ (see Eq. (9.129)) regarding $\sigma_{11m} = B_m \, r^{\zeta_m}$, $u_{1m} \propto B_m \, r^{\zeta_m + 1}$ (see Eqs. (9.120), (9.123)).

Chapter 10

Radial stresses p_1, p_2 and temperature range

10.1 Radial stresses p_1, p_2

10.1.1 Multi-particle-(envelope)-matrix system

Resulting from $\beta_p \neq \beta_m$ and $\beta_p \neq \beta_e$, $\beta_m \neq \beta_e$ (see Eqs. (3.48)–(3.50)), the radial stresses p_1 and p_1, p_2 determined from the conditions given by Eqs. (3.58) and (3.63), (3.64), acting on the particle-matrix and particle-envelope, matrix-envelope boundaries related to the multi-particle-matrix and multi-particle-envelope-matrix systems, respectively, represent common parameters included in all formulae for the thermal stresses and for the corresponding quantities derived by the different mathematical techniques presented in Sections 5.1–5.3, 6.1, 7.1, 7.2, 8.1, 9.2, where the conditions (3.63), (3.64) are both considered provided that $\beta_p \neq \beta_e \neq \beta_m$ (see Eqs. (3.48)–(3.50)). Finally, the conditions (3.63) and (3.64) are considered provided that $\beta_p \neq \beta_e = \beta_m$ and $\beta_p = \beta_e \neq \beta_m$, respectively. Equations (3.58), (3.63), (3.64) include the tangential strains $(\varepsilon_{22p})_{r=R_1}$, $(\varepsilon_{22e})_{r=R_{1+i}}$, $(\varepsilon_{22m})_{r=R_{1+i}}$ $(i=0,1)$ on the particle-matrix, particle-envelope and matrix envelope boundaries, thus for $r = R_1$ and $r = R_2$, respectively.

Included in Eqs. (3.58), (3.63), the tangential strain $(\varepsilon_{22p})_{r=R_1} = -p_1\,\varrho_{1p}$ in the spherical particle on the particle-matrix and particle-envelope boundaries, the latter related to the condition $\beta_p \neq \beta_e \neq \beta_m$, is derived by one of the mathematical techniques presented in Sections 5.1–5.3 to result in the φ, ν-independent coefficient ϱ_{1p} given by Eq. (5.36). Equation (5.36) is also considered on the conditions $\beta_p \neq \beta_e = \beta_m$ and $\varepsilon_{11te} - \varepsilon_{11tp} = f(\varphi, \nu)$ (see Eqs. (12.23), (12.24)) resulting in the φ, ν-dependent difference $\beta_e - \beta_p$.

Provided that $\beta_p \neq \beta_e = \beta_m$ and $\varepsilon_{11te} - \varepsilon_{11tp} \neq f(\varphi, \nu)$, resulting in the φ, ν-independent difference $\beta_e - \beta_p$, the tangential strain $(\varepsilon_{22p})_{r=R_1} = -p_1\,\varrho_{1p}$ (see Eq. (9.45)) in the spherical particle on the particle-envelope boundary is derived by the mathematical techniques presented in Section 9.2, where the φ, ν-independent coefficient ϱ_{1p} is given by Eq. (9.50) to include the exponent λ_p from the set $\{1, 4, 2, 7, 3, 9, 6.43999\}$ (see Section 9.3.1).

Resulting from the mathematical techniques presented in Sections 5.1–5.3, the solutions (5.7), (5.15), (5.20) for u_{1q} $(q = p, e, m)$ result in the same formula (5.31) for the radial displacement u_{1p} in the spherical particle. Conversely, the mathematical techniques presented in Sections 6.1, 7.1, 7.2, 8.1 result in solutions (6.4), (7.3), (7.12) not to be suitable for the determination of the thermal stresses and the corresponding quantities in the spherical particle, due to $(u_{1p})_{r\to 0} \to \infty$ or $(u_{1p})_{r=0} \neq 0$ what is physically unacceptable (see Section 4.1.1).

Similarly, with regard to the spherical envelope, provided that $\beta_p \neq \beta_e \neq \beta_m$, the coefficient $\varrho_{1+i\,1+je}$ in the formula $(\varepsilon_{22e})_{r=R_{1+i}} = -\sum_{j=0}^{1} p_{1+j}\varrho_{1+i\,1+je}$ $(i, j = 0{,}1)$ is given by Eqs. (5.82), (6.132), (6.159), (6.180), (7.148), (7.174), (7.199), where the coefficients $\Lambda_{1+ke} = \Lambda_{1+ke}^{(l)}$ $(k = 0{,}1;\ l = 1{,}2{,}3)$ and C_e included in Eqs. (5.88) and (7.155), (7.205) are substituted by $\Lambda_{1+ke}^{(1)}$, $\Lambda_{1+ke}^{(2)}$, $\Lambda_{1+ke}^{(3)}$ and $C_e^{(1)}$, $C_e^{(2)}$ given by Eqs. (5.8), (5.16), (5.21) and (6.5), (7.4), respectively.

Provided that $\beta_p \neq \beta_e = \beta_m$ and $\beta_p = \beta_e \neq \beta_m$, the coefficients ϱ_{1e}^{me} and ϱ_{2e}^{pe} in the tangential strains $(\varepsilon_{22e})_{r=R_1} = -p_1\,\varrho_{1e}^{me}$ and $(\varepsilon_{22e})_{r=R_2} = -p_2\,\varrho_{2e}^{pe}$ are given by Eqs. (5.106), (6.207), (6.228), (6.246), (7.230), (9.63) and (5.134), (6.264), (6.279), (7.251), (7.266), respectively. Considering $\varepsilon_{11te} - \varepsilon_{11tp} \neq f(\varphi, \nu)$, Equation (9.58) is related to the φ, ν-independent difference $\beta_e - \beta_p$, consequently resulting in the φ, ν-independent radial stress p_1 (see Eq. (10.4)), regarding the φ, ν-independent coefficients ϱ_{1p} and ϱ_{1e}^{me} given by Eqs. (9.50) and (9.63) to include the exponents λ_p and λ_e from the sets $\{1, 4, 2, 7, 3, 9, 6.43999\}$ and $\{-2, 0\}$, respectively.

With regard to the tangential strains in the cell matrix, $(\varepsilon_{22m})_{r=R_1} = -p_1\,\varrho_{1m}$ and $(\varepsilon_{22m})_{r=R_2} = -p_2\,\varrho_{2m}$, the latter provided that $\beta_p \neq \beta_e \neq \beta_m$ or $\beta_p = \beta_e \neq \beta_m$, the coefficient ϱ_{1+im} $(i = 0{,}1)$ is given by Eqs. (5.56), (6.32), (6.59), (6.81), (6.107), (7.34), (7.60), (7.88), (7.118), where the coefficients $\Lambda_{1+jm} = \Lambda_{1+jm}^{(k)}$ $(j = 0{,}1;$ $k = 1{,}2{,}3)$ and C_m included in Eqs. (5.56) and (7.34), (7.88), (7.118) are substituted by $\Lambda_{1+jm}^{(1)}$, $\Lambda_{1+jm}^{(2)}$, $\Lambda_{1+jm}^{(3)}$ and $C_m^{(1)}$, $C_m^{(2)}$ given by Eqs. (5.8), (5.16), (5.21) and (6.5), (7.4), respectively.

With regard to the tangential strains $(\varepsilon_{22p})_{r=R_1}$, $(\varepsilon_{22m})_{r=R_1}$, and considering Eq. (3.58), the radial stress p_1 related to the multi-particle-matrix system has the form

$$p_1 = \frac{\beta_m - \beta_p}{\varrho_{1p} - \varrho_{1m}}. \tag{10.1}$$

Similarly, with regard to the tangential strains $(\varepsilon_{22p})_{r=R_1}$, $(\varepsilon_{22e})_{r=R_{1+i}}$ $(i = 0{,}1)$, $(\varepsilon_{22m})_{r=R_2}$, and considering Eqs. (3.63), (3.64), the radial stresses p_1 and p_2 related to the multi-particle-envelope-matrix system provided that $\beta_p \neq \beta_e \neq \beta_m$ are derived as

$$p_1 = \frac{\varrho_{12e}\,(\beta_m - \beta_e) - (\varrho_{2m} - \varrho_{22e})\,(\beta_e - \beta_p)}{\varrho_{12e}\varrho_{21e} - (\varrho_{1p} - \varrho_{11e})\,(\varrho_{2m} - \varrho_{22e})}, \tag{10.2}$$

$$p_2 = \frac{(\varrho_{1p} - \varrho_{11e})(\beta_m - \beta_e) - \varrho_{21e}(\beta_e - \beta_p)}{\varrho_{12e}\varrho_{21e} - (\varrho_{1p} - \varrho_{11e})(\varrho_{2m} - \varrho_{22e})}. \tag{10.3}$$

Moreover, with regard to the tangential strains $(\varepsilon_{22p})_{r=R_1}$, $(\varepsilon_{22e})_{r=R_1}$, and considering Eq. (3.63), the radial stress p_1 related to the multi-particle-envelope-matrix system provided $\beta_p \neq \beta_e = \beta_m$ has the form

$$p_1 = \frac{\beta_e - \beta_p}{\varrho_{1p} - \varrho_{1e}^{me}}. \tag{10.4}$$

As presented above, with regard to the φ, ν-dependent and φ, ν-independent difference $\beta_e - \beta_p$ resulting from $\varepsilon_{11te} - \varepsilon_{11tp} = f(\varphi, \nu)$ and $\varepsilon_{11te} - \varepsilon_{11tp} \neq f(\varphi, \nu)$ (see Eqs. (3.48)–(3.50), (12.23), (12.24)), the coefficients ϱ_{1p}; ϱ_{1e}^{me} are given by Eqs. (5.36); (5.106), (6.207), (6.228), (6.246), (7.230) and (9.50); (9.63), respectively.

Finally, with regard to the tangential strains $(\varepsilon_{22m})_{r=R_2}$, $(\varepsilon_{22e})_{r=R_2}$, and considering Eq. (3.64), the radial stress p_2 related to the multi-particle-envelope-matrix system provided $\beta_p = \beta_e \neq \beta_m$ is derived as

$$p_2 = \frac{\beta_m - \beta_e}{\varrho_{2e}^{pe} - \varrho_{2m}}. \tag{10.5}$$

In contrast to the φ, ν-dependent coefficient ϱ_{1+im} $(i=0,1)$, as a function of the φ, ν-dependent distance r_c to include the experimentally determined parameters v, R_{1ex}, d_{1ex}, d_{2ex}, d_{3ex} (see Section 2.2.1, 2.3), and consequently resulting in the φ, ν-dependent the radial stresses p_1, p_2 given by Eqs. (10.1)–(10.3), (10.5), the radial stress p_1 given by Eq. (10.4) represents a constant function regarding the variables φ, ν, resulting from the φ, ν-independent coefficients ϱ_{1p}, ϱ_{1+ie}^{me} $(i=0,1)$ (see Eqs. (9.50), (9.63)) and provided that $\varepsilon_{11te} - \varepsilon_{11tp} \neq f(\varphi, \nu)$, otherwise p_1 is a function of φ, ν provided that $\varepsilon_{11te} - \varepsilon_{11tp} = f(\varphi, \nu)$.

With regard to the Castigliano's theorem (see Section 3.1.9), provided that the coefficient ϱ, represented by one of the coefficients ϱ_{1p}, $\varrho_{1+i\,1+je}$, ϱ_{1e}^{me}, ϱ_{2e}^{pe}, ϱ_{1+im} $(i,j=0,1)$, is determined by several formulae (see Sections 1.4, 1.8.1), such combination of the coefficients ϱ_{1m}, ϱ_{1e}^{me}, ϱ_{2e}^{pe} in Eqs. (10.1), (10.4), (10.5) and such combination of the coefficients $\varrho_{1+i\,1+je}$ $(i,j=0,1)$, ϱ_{2m} in Eqs. (10.2), (10.3) are considered to result in minimal thermal-stress induced elastic energy of the cell, $W_c = W_p + W_m$ and $W_c = W_p + W_e + W_m$ (see Eqs. (3.44), (3.45)), related to the multi-particle-matrix and multi-particle-envelope-matrix systems, respectively [1].

Finally, with regard to the conditions $\beta_p = \beta_m$ and $\beta_p = \beta_e = \beta_m$, the thermal stresses in the multi-particle-matrix and multi-particle-envelope-matrix systems do not occur (see Eqs. (10.1)–(10.5)), respectively.

[1] To avoid misunderstanding, on the one hand, the coefficients ϱ_{1p}, ϱ_{1+ie}^{me} $(i=0,1)$ related to the multi-particle-envelope-matrix system provided that $\beta_p \neq \beta_e = \beta_m$ and $\varepsilon_{11te} - \varepsilon_{11tp} \neq f(\varphi, \nu)$ (see Eq. (12.23)) are given by Eqs. (9.50), (9.63) only, respectively. On the other hand, the exponents λ_p and λ_e included in Eqs. (9.50) and (9.63) are from the sets $\{1, 2, 3, 4, 6.43999, 7, 9\}$ and $\{-2, 0\}$, respectively.

10.1.2 One-particle-(envelope)-matrix system

With regard to the one-particle-(envelope)-matrix system provided that $\beta_p \neq \beta_m$, $\beta_p \neq \beta_e \neq \beta_m$, $\beta_p \neq \beta_e = \beta_m$, $\beta_p = \beta_e \neq \beta_m$ and for $\varepsilon_{11tm} - \varepsilon_{11tp} = f(\varphi, \nu)$, $\varepsilon_{11te} - \varepsilon_{11tp} = f(\varphi, \nu)$, $\varepsilon_{11tm} - \varepsilon_{11te} = f(\varphi, \nu)$ (see Sections 5.7, 6.5), the radial stresses p_1, p_2 are given by Eqs. (10.1)–(10.5) to include the coefficients ϱ_{1p}, $\varrho_{1+i\,1+je}$ $(i, j = 0{,}1)$, ϱ_{1e}^{me}, ϱ_{2e}^{pe} identical with those related to the multi-particle-(envelope)-matrix system, and additionally the coefficient ϱ_{1+im} $(i = 0{,}1)$ is given by Eqs. (5.178), (6.476).

Considering $\varepsilon_{11tm} - \varepsilon_{11tp} = f(\varphi, \nu)$, $\varepsilon_{11te} - \varepsilon_{11tp} = f(\varphi, \nu)$, $\varepsilon_{11tm} - \varepsilon_{11te} = f(\varphi, \nu)$ (see Section 9.5), the coefficients ϱ_{1p}, $\varrho_{1+i\,1+je}$, ϱ_{1e}^{me}, ϱ_{2e}^{pe}, ϱ_{1+im} $(i, j = 0{,}1)$ in Eqs. (10.1)–(10.5) are given by Eqs. (9.50), (9.86), (9.63), (9.99), (9.110), where the exponents λ_p and λ_e included in Eqs. (9.50) and (9.63), (9.99) are from the sets $\{1, 4, 2, 7, 3, 9, 6.43999\}$ and $\{-2, 0\}$, $\{1, 0, 4, 2, 7, 3, 9, 6.43999\}$ (see Sections 9.3.1, 9.3.2), respectively. Finally, the exponents $\lambda_{1e} \neq \lambda_{2e}$ included in Eq. (9.86) are from one of the sets $\{-2, 0\}$, $\{-2, 1\}$, $\{-2, 2\}$, $\{-2, 3\}$, $\{-2, 4\}$, $\{-2, 7\}$, $\{-2, 9\}$, $\{0, 1\}$, $\{0, 2\}$, $\{0, 3\}$, $\{0, 7\}$, $\{0, 9\}$, $\{1, 2\}$, $\{1, 3\}$, $\{1, 4\}$, $\{1, 7\}$, $\{1, 9\}$, $\{2, 3\}$, $\{2, 9\}$ (see Section 9.3.2).

10.2 Dependencies $p_1 = p_1(v, R_1, R_2)$, $p_2 = p_2(v, R_1, R_2)$

As a function of the particle volume fraction v, the compressive or tensile radial stresses $p_1 > 0$ or $p_1 < 0$ and $p_2 > 0$ or $p_2 < 0$, acting on the particle-matrix, particle-envelope and matrix-envelope boundaries, given by Eqs. (10.1), (10.4) and (10.5), correspond to the conditions $\beta_m > \beta_p$, $\beta_m > \beta_e$ or $\beta_m < \beta_p$, $\beta_m < \beta_e$ and $\beta_m > \beta_e$ or $\beta_m < \beta_e$ (see Eqs. (3.56), (3.59), (3.60)), respectively.

However, considering the coefficients ϱ_{1p} and ϱ_{1e}^{me} given by Eqs. (5.36) and (5.106), (6.228), (6.246), (9.63), respectively, the radial stress p_1 given by Eq. (10.4), dependent on ϱ_{1p}, ϱ_{1e}^{me} and derived on the condition $\beta_p \neq \beta_e = \beta_m$ for the multi-particle-envelope-matrix system, is paradoxically independent on v, R_1, R_2 representing parameters of composite materials. Accordingly, the radial displacements u_{1p}, u_{1e} and consequently the thermal stresses in the spherical particle, in the spherical envelope are not functions of v, R_1, R_2 for $r \in \langle 0, R_1 \rangle$, $r \in \langle R_1, R_2 \rangle$, respectively, as also valid for $(u_{1m})_{r=R_2} = (u_{1e})_{r=R_2}$ resulting from the condition $\beta_e = \beta_m$. Finally, the thermal stresses in the cell matrix can be dependent on v, R_1, R_2 for $r \in \langle R_2, r_c \rangle$, resulting from the function $r_c = r_c(v, R_1, R_2)$ included in the coefficients $\kappa_{1m}^{(1)}$, $\kappa_{3m}^{(1)}$ and $\kappa_{4m}^{(2)}$, $\kappa_{5m}^{(2)}$, $\kappa_{6m}^{(2)}$ (see Eqs. (12.35), (12.37)–(12.39)) as functions of χ_{21m}, χ_{22m} and χ_{21m}, χ_{22m}, χ_{23m}, χ_{31m}, χ_{32m}, χ_{33m} related to the boundary conditions (4.11) and (4.11), (4.14), respectively, where the φ, ν-dependent distance r_c as a function of the experimentally determined parameters v, R_{1ex}, d_{1ex}, d_{2ex}, d_{3ex} (see Section 2.2.1) is derived in Section 2.3.

Conversely, considering the multi-particle-envelope-matrix system, the radial stresses $p_1 \gtrless 0$, $p_2 \gtrless 0$ can or need not correspond to the conditions $\beta_e \gtrless \beta_p$, $\beta_m \gtrless \beta_e$ (see Eqs. (3.59), (3.60)), respectively, resulting from dependencies of p_1, p_2 on the parameters v, R_1, R_2 included in the coefficients ϱ_{1+im}, $\varrho_{1+i\,1+je}$ $(i, j = 0{,}1)$ (see Eqs. (10.2), (10.3)). Consequently, one of the radial stresses p_1, p_2 or both p_1, p_2 can exhibit zero values at the parameters v_c, R_{1c}, R_{2c} at which one of the radial stresses p_1, p_2 or both p_1, p_2 are transformed from compressive to tensile or vice versa.

The compressive or tensile radial stresses $p_1 > 0$ or $p_1 < 0$ and $p_2 > 0$ or $p_2 < 0$ result in compressive or tensile thermal stresses in the spherical particle and the cell matrix, accordingly representing resistance against compressive or tensile mechanical loading. The parameters v_c, R_{1c}, R_{2c} represent critical values at which the thermal stresses change from positive to negative or vice versa, and a resistive effect against the mechanical loading is transformed to a contributory effect or vice versa, respectively. Consequently, the resistive and contributory effects result in the thermal-stress induced strengthening and weakening, respectively, derived by mathematical techniques presented in Volume III.

With regard to the critical parameters v_c, R_{1c}, R_{2c}, determined from numerically generated dependencies of the radial stresses p_1, p_2 on the variables v, R_1, R_2 and related to a real composite material included in one of Items 1–4 presented in Section 2.1, materials scientists are able to design and develop microstructure of the real composite material with such material parameters v, R_1, R_2 resulting in the thermal-stress induced strengthening against required mechanical loading. Additionally, provided that the time-dependent development of the microstructure to be tailored regarding the required mechanical loading is defined analytically or experimentally, the time when the real composite material exhibits the critical parameters v_c, R_{1c}, R_{2c} can be consequently determined.

As an example, microstructure to exhibit usually the unstable equilibrium regarding the enthalpy of a real composite material is significantly transformed at the temperature T higher than the room temperature, 20°C [1, p 38–40]. On the condition of mechanical loading at $T \gg 20$°C, the time τ_c is required to be determined, accordingly representing the lifetime of the real composite material with regard to a transformation of the resistive effect to the contributory effect resulting from a transformation of the microstructure with the critical parameters v_c, R_{1c}, R_{2c}.

Finally, the similar analysis concerning the radial stresses p_1, p_2 (see Eqs. (10.2)–(10.5)) related to the multi-particle-envelope-matrix system is considered for the radial stress p_1 as a function of v, R_1 (see Eq. (10.1)), acting on the particle-matrix boundary related to the multi-particle-matrix system, and regarding the critical parameters v_c, R_{1c} at which p_1 is extremal. Additionally, the radial stress p_1 is not assumed to be transformed from positive to negative or vice versa.

With regard to the φ, ν-dependent radial stresses p_1, p_2 given by Eqs. (10.1)–(10.3), (10.5), the average values $\overline{p_1}$ and $\overline{p_2}$ defined as [6, p 474]

$$\overline{p_{1+i}} = \left(\frac{2}{\pi}\right)^2 \int_0^{\pi/2}\int_0^{\pi/2} p_{1+i}\, d\varphi\, d\nu, \quad i = 0, 1, \tag{10.6}$$

are considered for the evaluation of the elastic thermal-stress loading and for the determination of the critical parameters v_c, R_{1c}, R_{2c}. The integrals in Eq. (10.6) can be derived numerically for a real composite material, considering Eq. (1.7).

Due to symmetry of the multi-particle-(envelope)-matrix system, resulting from the matrix infinity and from the periodical distribution of the spherical particles shown in Fig. 2.1a,b, the radial stresses p_1, p_2 are sufficient to be integrated in the interval $\varphi, \nu \in \langle 0, \pi/2\rangle$ regarding the spatial polar coordinates $[r, \varphi, \nu]$ (see Fig. 3.1).

The same analysis as presented in Section 10.2 is also valid for the one-particle-(envelope)-matrix system.

10.3 Temperature range

As presented in Section 3.2.1, the thermal stresses, induced as a consequence of the difference in the thermal expansion coefficients α_p, α_e, α_m, originate in the temperature interval $\langle T_f, T_r\rangle$, considering a temperature dependence of the coefficient β_q $(q = p,e,m)$ (see Eqs. (3.48)–(3.50)), where T_f is final temperature of a cooling process, T_r is minimal temperature from the sets $\{T_{rp}, T_{rm}\}$ and $\{T_{rp}, T_{re}, T_{rm}\}$ related to the multi-particle-matrix and multi-particle-envelope-matrix systems, respectively. Additionally, the recovery temperature T_{rq}, below that the stress relaxation as a consequence of thermal-activated processes does not occur in a particle, an envelope, a matrix, thus for $q = p,e,m$, respectively, is determined approximately by the relationship $T_{rq} = (0.35 - 0.4) \times T_{mq}$ [1, p 39] and exactly by an experiment, where T_{mq} is the melting temperature. Generally, the thermal stresses originating at the temperature $T > T_r$ are completely released by plastic deformation during a *slow* cooling process [1, p 38–40].

On the conditions $\alpha_p \neq \alpha_m$ and $\alpha_p \neq \alpha_e \neq \alpha_m$ or $\alpha_p \neq \alpha_e = \alpha_m$ or $\alpha_p = \alpha_e \neq \alpha_m$ related to the multi-particle-matrix and multi-particle-envelope-matrix systems, respectively, temperature dependencies of the thermal stresses results from the integrals regarding the variable T, included in the coefficient β_q (see Eqs. (3.48)–(3.50)).

As presented in Section *Preface*, the analytical models presented in this book are derived regarding elasticity of an isotropic solid continuum represented by the isotropic multi-particle-(envelope)-matrix system acted by the thermal stresses.

As presented in Section 10.1.1, absolute values of the thermal stresses in the spherical particle and in the cell matrix of the multi-particle-matrix system are maximal on the particle-matrix boundary, equal to the value $|p_1|$ as a function of the temperature range $\langle T_f, T_r\rangle$. Consequently, regarding the elastic thermal-stress loading, and provided that $\alpha_p \neq \alpha_m$, the analytical models of the thermal stresses are considered up to the critical final temperature of a cooling process, T_{ys}, derived from the condition $|p_{1max}| = \sigma_{ys}$, where $\sigma_{ys} > 0$, representing a yield stress in compression or in tension on the conditions $p_1 > 0$ or $p_1 < 0$, is the minimum from the sets $\{|\sigma_{yscp}|, |\sigma_{yscm}|\}$ or $\{|\sigma_{ystp}|, |\sigma_{ystm}|\}$, respectively; σ_{yscp}, σ_{ystp} and σ_{yscm}, σ_{ystm} are yield stresses in compression, in tension, for the spherical particle and for the cell matrix respectively; $|p_{1max}|$ is a maximal value of the φ, ν-dependence, $|p_1| = |p_1(\varphi, \nu)|$, given by Eq. (10.1).

Similarly, with regard to the multi-particle-envelope-matrix system provided that $\alpha_p \neq \alpha_e \neq \alpha_m$ (see Eqs. (3.48)–(3.50)), the critical final temperature T_{ys} is derived from the conditions $|p_{1max}| = \sigma_{ys}$ or $|p_{2max}| = \sigma_{ys}$, where $\sigma_{ys} > 0$, representing a yield stress in compression, in tension on the conditions $p_1 > 0$, $p_1 < 0$ or $p_2 > 0$, $p_2 < 0$, is the minimum from the sets $\{|\sigma_{yscp}|, |\sigma_{ysce}|\}$, $\{|\sigma_{ystp}|, |\sigma_{yste}|\}$ or $\{|\sigma_{yscm}|, |\sigma_{yscm}|\}$, $\{|\sigma_{ystm}|, |\sigma_{ystm}|\}$, related to the conditions $|p_{1max}| > |p_{2max}|$ or $|p_{1max}| < |p_{2max}|$, respectively; $|p_{1max}|$ and $|p_{2max}|$ are maximal values of the φ, ν-dependencies, $|p_1| = |p_1(\varphi, \nu)|$ and $|p_2| = |p_2(\varphi, \nu)|$, given by Eqs. (10.2) and (10.3), respectively; σ_{ysce} and σ_{ysce} are yield stresses in compression and in tension for the spherical envelope, respectively.

Consequently, provided that $\alpha_p \neq \alpha_e = \alpha_m$, the critical final temperature T_{ys} is derived from the condition $|p_1| = \sigma_{ys}$, where $\sigma_{ys} > 0$, representing a yield stress in compression or in tension on the conditions $p_1 > 0$ or $p_1 < 0$, is the minimum from the sets $\{|\sigma_{yscp}|, |\sigma_{ysce}|\}$ or $\{|\sigma_{ystp}|, |\sigma_{yste}|\}$, respectively, and the radial stress p_1, as a constant function regarding the variables φ, ν, is given by Eq. (10.4).

Finally, provided that $\alpha_p = \alpha_e \neq \alpha_m$, the critical final temperature T_c is derived from the condition $|p_{2max}| = \sigma_{ys}$, where $\sigma_{ys} > 0$, representing a yield stress in compression or in tension on the conditions $p_2 > 0$ or $p_2 < 0$, respectively, is the minimum from the sets $\{|\sigma_{yscm}|, |\sigma_{ysce}|\}$ or $\{|\sigma_{ystm}|, |\sigma_{yste}|\}$, respectively, and $|p_{2max}|$ is a maximal value of the φ, ν-dependence, $|p_2| = |p_2(\varphi, \nu)|$, given by Eq. (10.5).

The particle-matrix and particle-envelope, matrix-envelope boundary bonds can be characterized by the adhesion strength σ_{as}^{pm} and σ_{as}^{pe}, σ_{as}^{me}, respectively. Provided that the particle-matrix and particle-envelope, matrix-envelope boundaries are acted by the tensile radial stresses, $p_1 < 0$ and $p_1 < 0$, $p_2 < 0$, the conditions for the determination of the critical temperature T_{ys} related to the yield stress σ_{ys} are used for the determination of the critical temperature T_{as} related to the adhesion strength σ_{as}, replacing the yield stress σ_{ys} by the corresponding adhesion strength σ_{as}^{pm} and σ_{as}^{pe}, σ_{as}^{me}, respectively. Accordingly, on the conditions $p_{1+i} < 0$ $(i = 0{,}1)$ and $T_a > T_c$, the latter resulting also from $\sigma_{as} < \sigma_{ys}$, the analytical models presented in this book, regarding the elastic thermal-stress loading, are considered up to the critical temperature T_{as}.

Provided that $\alpha_p = \alpha_m$ and $\alpha_p = \alpha_e = \alpha_m$, a phase transformation originating at the temperature T_{tq} $(q = p, e, m)$ at least in one of components of the multi-particle-(envelope)-matrix system is a reason of the thermal stresses originating at the temperature $T_t \in \langle T_f, T_r \rangle$, where T_t is maximal temperature from the sets $\{T_{tp}, T_{tm}\}$

and $\{T_{tp}, T_{te}, T_{tm}\}$ related to the multi-particle-matrix and multi-particle-envelope-matrix systems, respectively. Accordingly, temperature dependencies of the thermal stresses results only from temperature dependencies of the elastic moduli s_{11q}, s_{12q}, s_{44q} (see Eqs. (3.23)–(3.25)) in the temperature interval $\langle T_f, T_t \rangle$, and T_{ys} or T_{as} are derived from the conditions presented above related to the yield stress σ_{ys} or to the adhesion strength σ_{as}, respectively, where $\langle T_f, T_t \rangle \subset \langle T_f, T_r \rangle$. Additionally, in case of the temperature independent elastic moduli s_{11q}, s_{12q}, s_{44q} included in the coefficients ϱ_{1p}, ϱ_{1+im}, $\varrho_{1+i\,1+je}$ $(i, j = 0{,}1)$ (see Eqs. (10.1)–(10.5)), the thermal stresses are temperature-independent, and consequently the analytical models presented in this book, considering the elastic thermal-stress loading, are derived on the conditions $|p_{1+imax}| \leq \sigma_{ys}$ or $|p_{1+imax}| \leq \sigma_{as}$ $(i = 0{,}1)$.

Finally, applying parameters of a real composite material, numerical values of T_{ys}, T_{as} can be derived from the presented conditions by a numerical method.

The same analysis as presented in Section 10.3 is also valid for the one-particle-(envelope)-matrix system.

Chapter 11

Related phenomenon

The analytical models of the thermal stresses and the corresponding quantities presented in this book are transformable to analytical models of stresses originating in an isotropic crystalline lattice with a central substitutive atom. The isotropic crystalline lattice is represented by the primitive, body-centered (bcc) and face-centered cubic crystalline lattices with the dimensions $d \times d \times d$, denoted as K_6, K_8, K_{12} [1, p 5–8], respectively. In case of the cubic crystalline lattice, the φ, ν-dependent distance r_c, as a function of the dimension d, is given by Eqs. (2.17), (2.18). The radial stress p_1 acting on the particle-matrix boundary boundary and the particle radius R_1, both included in the formulae for the thermal stresses presented in Sections 5.5, 6.3, 7.4, are replaced by those given by Eq. (11.3) and derived from Eq. (11.5) by numerical methods for the determination of roots of transcendental equations [7, p 970–971], respectively.

Provided that the inter-particle distance $d_i = d$ $(i = 1,2,3)$, the infinite matrix with the cubic cells (see Fig. 2.1a) is considered to represent an infinite matrix with cubic crystalline lattices with central spherical holes with the radius R_{1m}. Consequently, considering the spherical particle to represent a substitutive atom with the radius R_{1p} in the centre O of the cubic crystalline lattice (see Figs. 2.1a, 3.1), the stresses originate on the condition $R_{1p} \neq R_{1m}$, where the parameters d, R_{1p}, R_{1m} are taken at the same temperature. The compressive or tensile radial stress, $p_1 > 0$ or $p_1 < 0$, induced due to $R_{1p} - R_{1m} > 0$ or $R_{1p} - R_{1m} < 0$, respectively, results in the radial displacements $(u_{1p})_{r=R_1}$, $(u_{1m})_{r=R_1}$, and consequently the distance R_{1T} from the centre O to a point on "the particle-matrix boundary boundary" in a radial direction, determined by the angles φ, ν and represented by the axis x_1' (see Fig. 3.1), is derived as

$$R_{1T} = R_{1p} + (u_{1p})_{r=R_1} = R_{1m} + (u_{1m})_{r=R_1} \, . \tag{11.1}$$

Considering the Cauchy's equation $u_1 = r\,\varepsilon_{22}$ (see Eq. (3.2)), using the uniform notation for the tangential strains, $(\varepsilon_{22p})_{r=R_1} = -p_1\,\varrho_{1p}$ and $(\varepsilon_{22m})_{r=R_1} = -p_1\,\varrho_{1m}$, given by Eqs. (5.31) and (5.49), (6.23), (6.52), (6.74), (6.97), (7.25), (7.53), (7.81), (7.108), respectively, we get

$$R_{1T} = R_{1p} - R_1 p_1\,\varrho_{1p} = R_{1m} - R_1 p_1\,\varrho_{1m}, \tag{11.2}$$

where the coefficients ϱ_{1p} and ϱ_{1m} are given by Eqs. (5.36) and (5.56), (6.32), (6.59), (6.81), (6.107), (7.34), (7.60), (7.88), (7.118), respectively, resulting from the different mathematical techniques (see Sections 5.1–5.3, 6.1, 7.1), 7.2).

Consequently, the radial stress p_1 has the form

$$p_1 = \frac{R_{1p} - R_{1m}}{R_1 \left(\varrho_{1p} - \varrho_{1m}\right)}, \tag{11.3}$$

and substituting Eq. (11.3) to Eq. (11.2), we get

$$R_{1T} = \frac{R_{1m}\varrho_{1p} - R_{1p}\varrho_{1m}}{\varrho_{1p} - \varrho_{1m}}. \tag{11.4}$$

The parameter R_1, representing an average value of the φ, ν-dependent distance R_{1T}, is derived from the equation

$$R_1 = \left(\frac{2}{\pi}\right)^2 \int_0^{\pi/2}\int_0^{\pi/2} R_{1T}\, d\varphi\, d\nu = \left(\frac{2}{\pi}\right)^2 \int_0^{\pi/2}\int_0^{\pi/2} \frac{R_{1m}\varrho_{1p} - R_{1p}\varrho_{1m}}{\varrho_{1p} - \varrho_{1m}}\, d\varphi\, d\nu, \tag{11.5}$$

using numerical methods for the determination of roots of transcendental equations [7, p 970–971]. Provided that R_{1T} is R_1-independent due to the reduction of the term R_1 in fractions in the coefficients ϱ_{1p}, ϱ_{1m}, the integral in Eq. (11.5) can be derived numerically for a real composite material, using the same method as included in Eqs. (1.5)–(1.7). Finally, the parameter d_{ex} in Eqs. (2.17), (2.18) is replaced by the parameter d, and consequently the distance r_c included in formulae for the stresses in the isotropic crystalline lattice is a function of d, φ, ν only (see Eqs. (2.17), (2.18)).

Chapter 12

Appendix

12.1 Phase-transformation induced radial displacement and radial strain

As presented in Section 3.1.2, the infinitesimal spherical cap in the spherical particle, in the spherical envelope and in the cell matrix is shifted in a direction of a normal to the particle-matrix, particle-envelope and matrix-envelope boundaries, and consequently the normal in the arbitrary point P on the boundaries is identical with the axis x_1' of the Cartesian system $(Px_1'x_2'x_3')$ (see Fig. 3.1), where $x_1' = \overline{OP}$ represents a radial direction.

The determination of the radial displacement $(u_{1qt})_{r=R_{1+i}}$ $(i=0,1)$ on the particle-matrix, particle-envelope and matrix envelope boundaries, thus for $r = R_1$ $(i=0)$ and $r = R_2$ $(i=1)$, respectively, and consequently the radial strain ε_{11tq}, both induced by a phase transformation originating at the temperature $T_{tq} \in \langle T_f, T_r \rangle$ at least in one of components represented by the spherical particle $(q=p)$ with the radius R_1, by the spherical envelope $(q=e)$ with the radii $R_1 < R_2$ and by the cell matrix $(q=m)$, results from the change of dimensions of mutually transforming cubic crystalline latices in a radial direction represented by the axis x_1' and given by the angles φ, ν, considering general angle positions of the cubic crystalline latices regarding the Cartesian system $(Ox_1x_2x_3)$ (see Figs. 2.1, 12.1), where T_r is relaxation temperature and T_f is final temperature of a cooling process (see Section 10.3).

Accordingly, the determination of the phase-transformation induced displacement and strain, $(u_{1qt})_{r=R_{1+i}}$ $(i=0,1)$ and ε_{11tq}, respectively, is considered provided that the axis x_1' as a radial direction goes through the point O, representing the beginning of the mutually transforming cubic crystalline lattices (see Fig. 12.1). On the one hand, with regard to a general position of the axis x_1', the determination of $(u_{1qt})_{r=R_{1+i}}$ $(i=0,1)$ and ε_{11tq}, considering length of abscissae between points of intersection of the axis x_1' with the surfaces *0142*, *0361*, *0273* and *1456*, *2754*, *3657*, is general, but, on the other hand, such determination is complicated with respect to a position of the mutually transforming cubic crystalline lattices given by the spatial polar coordinates $[r, \varphi, \nu]$ (see Fig. 3.1).

As presented in Section 2.1, due to symmetry of the multi-particle-(envelope)-matrix system resulting from the matrix infinity and from the periodical distribution of the spherical particles shown in Fig. 2.1a,b, thermal stresses are sufficient to be investigated in the interval $\varphi, \nu \in \langle 0, \pi/2 \rangle$ regarding the spatial polar coordinates $[r, \varphi, \nu]$ (see Fig. 3.1).

Finally, with regard to the angle positions of the cubic crystalline latices, a short introduction to the vector algebra is presented in Section 12.1.1 [3, p 11–13].

12.1.1 Transformation in Cartesian systems

Cooling down the isotropic multi-particle-(envelope)-matrix system from the relaxation temperature T_r to the final temperature T_f, the cubic crystalline lattice, e.g. $K6$, $K8$, $K12$ (see Chapter 11), in the Cartesian system $\left(Ox''_{1qI}x''_{2qI}x''_{3qI}\right)$ at the temperature $T = T_{tq} + dT$ is transformed to a cubic crystalline lattice in the Cartesian system $\left(Ox''_{1qII}x''_{2qII}x''_{3qII}\right)$ at the temperature $T = T_{tq} - dT$ (see Fig. 12.1).

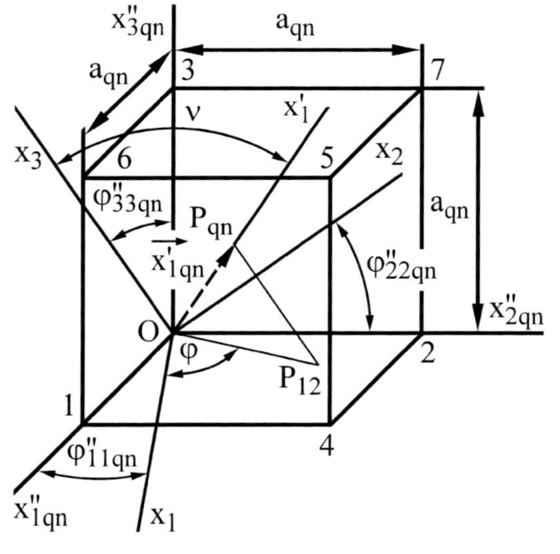

Figure 12.1: The cubic crystalline lattice in the Cartesian system $(Ox''_{1qn}x''_{2qn}x''_{3qn})$ with a position determined by the nine coefficients a''_{11qn}, $a''_{12qn} \ldots, a''_{32qn}$, a''_{33qn} (see Eq. (12.1)) related to the Cartesian system $(Ox_1x_2x_3)$ (see Fig. 3.1) and with dimension a_{qn} along the axis x''_{iqn} at the temperature $T_{tq} \in \langle T_f, T_r \rangle$ ($i = 1,2,3$; $q = p,e,m$; $n = I,II$). The points P_{qn} and P_{12} represent an intersection of the axis x'_1 with one of the surfaces *1456*, *2754*, *3657* and a projection of the point P_{qn} onto the plane x_1x_2, respectively. As an example, the angles φ''_{11qn}, φ''_{22qn}, φ''_{33qn} are presented (see Eq. (12.1)).

Strictly speaking, with regard to $dT \to 0$, the cubic crystalline lattice with a position determined by the nine coefficients a''_{11qI}, $a''_{12qI}, \ldots, a''_{32qI}$, a''_{33qI} (see Eq. (12.1)) related to the Cartesian system $(Ox_1x_2x_3)$ (see Fig. 3.1) and with dimension a_{qI} along the axis x''_{iqI} at the temperature $T_{tq} \in \langle T_f, T_r \rangle$ is transformed during the cooling process at the temperature $T_{tq} \in \langle T_f, T_r \rangle$ to the cubic crystalline lattice with a position given by the nine coefficients a''_{11qII}, $a''_{12qII}, \ldots, a''_{32qII}$, a''_{33qII} (see Eq. (12.1)) related to the Cartesian system $(Ox_1x_2x_3)$ and with dimension a_{qII} at the temperature $T_{tq} \in \langle T_f, T_r \rangle$ along the axis x''_{iqI} ($i = 1,2,3$), and the coefficient derived as

$$a''_{ijqn} = \cos \varphi''_{ijqn} = \cos \left[\angle \left(x''_{iqn}, x_j \right) \right], \tag{12.1}$$

is the direction cosine of the angle φ''_{ijqn} formed by the axes x''_{iqn} and x_j $(i,j=1,2,3;$ $n=I,II)$.

Considering the coordinates $[x_1,x_2,x_3]$ and $\left[x''_{1qn},x''_{2qn},x''_{3qn}\right]$ of the point P_{qn} in the Cartesian systems $(Ox_1x_2x_3)$ and $\left(Ox''_{1qn}x''_{2qn}x''_{3qn}\right)$, respectively, the transformation formulae have the forms [3, p 11]

$$x''_{1qn} = a''_{11qn}x_1 + a''_{12qn}x_2 + a''_{13qn}x_3, \tag{12.2}$$

$$x''_{2qn} = a''_{21qn}x_1 + a''_{22qn}x_2 + a''_{23qn}x_3, \tag{12.3}$$

$$x''_{3qn} = a''_{31qn}x_1 + a''_{32qn}x_2 + a''_{33qn}x_3, \tag{12.4}$$

derived by one formula

$$x''_{iqn} = \sum_{j=1}^{3} a''_{ijqn}x_j, \quad i=1,2,3. \tag{12.5}$$

Conversely, the formula for a transformation of the coordinates $[x_1,x_2,x_3]$ to $\left[x''_{1qn},x''_{2qn},x''_{3qn}\right]$ is derived as

$$x_i = \sum_{j=1}^{3} a''_{ijqn}x''_{jqn}, \quad i=1,2,3. \tag{12.6}$$

With regard to Eqs. (12.2)–(12.4), the condition

$$|OP_{qn}| = \sqrt{x_1^2+x_2^2+x_3^2} = \sqrt{\left(x''_{1qn}\right)^2+\left(x''_{2qn}\right)^2+\left(x''_{3qn}\right)^2}, \tag{12.7}$$

is transformed to the form

$$\begin{aligned}
\left(x''_{1qn}\right)^2+\left(x''_{2qn}\right)^2+\left(x''_{3qn}\right)^2 &= x_1^2\left[\left(a''_{11qn}\right)^2+\left(a''_{21qn}\right)^2+\left(a''_{31qn}\right)^2\right]\\
&+ x_2^2\left[\left(a''_{12qn}\right)^2+\left(a''_{22qn}\right)^2+\left(a''_{32qn}\right)^2\right]\\
&+ x_3^2\left[\left(a''_{13qn}\right)^2+\left(a''_{23qn}\right)^2+\left(a''_{33qn}\right)^2\right]\\
&+ 2x_1x_2\left(a''_{11qn}a''_{12qn}+a''_{21qn}a''_{22qn}+a''_{31qn}a''_{32qn}\right)\\
&+ 2x_1x_3\left(a''_{11qn}a''_{13qn}+a''_{21qn}a''_{23qn}+a''_{31qn}a''_{33qn}\right)\\
&+ 2x_2x_3\left(a''_{12qn}a''_{13qn}+a''_{22qn}a''_{23qn}+a''_{32qn}a''_{33qn}\right)\\
&= x_1^2+x_2^2+x_3^2.
\end{aligned} \tag{12.8}$$

The comparison of the coefficients at the term $x''_{iqn}x''_{jqn}$ $(i,j=1,2,3)$ in the middle and on the right side of Eq. (12.8) results in the conditions

$$\left(a''_{11qn}\right)^2 + \left(a''_{21qn}\right)^2 + \left(a''_{31qn}\right)^2 = 1, \tag{12.9}$$

$$\left(a''_{12qn}\right)^2 + \left(a''_{22qn}\right)^2 + \left(a''_{32qn}\right)^2 = 1, \tag{12.10}$$

$$\left(a''_{13qn}\right)^2 + \left(a''_{23qn}\right)^2 + \left(a''_{33qn}\right)^2 = 1, \tag{12.11}$$

$$a''_{11qn}a''_{12qn} + a''_{21qn}a''_{22qn} + a''_{31qn}a''_{32qn} = 0, \tag{12.12}$$

$$a''_{11qn}a''_{13qn} + a''_{21qn}a''_{23qn} + a''_{31qn}a''_{33qn} = 0, \tag{12.13}$$

$$a''_{12qn}a''_{13qn} + a''_{22qn}a''_{23qn} + a''_{32qn}a''_{33qn} = 0, \tag{12.14}$$

derived by one formula

$$\sum_{i=1}^{3} a''_{ijqn}a''_{ikqn} = \delta_{jk}, \quad j,k = 1,2,3, \tag{12.15}$$

where $\delta_{jk} = 0$ and $\delta_{jk} = 1$ for $j \neq k$ and $j = k$ [3, p 13], respectively.

To determine a position of the Cartesian system $\left(Ox''_{1qn}x''_{2qn}x''_{3qn}\right)$ $(n = I,II)$ related to the Cartesian system $(Ox_1x_2x_3)$, three of the nine coefficients a''_{11qn}, $a''_{12qn}, \ldots, a''_{32qn}$, a''_{33qn} are required to be given by numerical values obtained from experimental methods, and six coefficients can be derived from the six conditions (12.9)–(12.14). Equations (12.2)–(12.6) and the conditions (12.9)–(12.15) represent formulae to be considered for a transformation of coordinates derived in two different Cartesian systems [3, p 11–13], where the axes x_1, x_2, x_3 and x''_{1qn}, x''_{21qn}, x''_{1qn} of the systems $[x_1, x_2, x_3]$ and $\left[x''_{1qn}, x''_{2qn}, x''_{3qn}\right]$ are mutually perpendicular, respectively, and thus $\cos\left[\angle\left(x_i, x_j\right)\right] = \cos\left[\angle\left(x''_{iqn}, x''_{jqn}\right)\right] = \pi\left(1-\delta_{ij}\right)/2$ for $\angle\left(x_i, x_j\right) = \angle\left(x''_{iqn}, x''_{jqn}\right) = \pi\left(1-\delta_{ij}\right)/2$, regarding $i,j = 1,2,3$.

As an example, the condition $x_i = x''_{iqn}$ results in $\varphi''_{ijqn} = \pi\left(1-\delta_{ij}\right)/2$, and accordingly $a''_{ijqn} = \cos\varphi''_{ijqn} = \delta_{ij}$, regarding $(i,j = 1,2,3)$.

12.1.2 Radial displacement u_{1qt} and radial strain ε_{11tq}

With regard to the point P_{qn} as an intersection of the axis x'_1 with the surface *1456*, the length $x'_{1qn} = |OP_{qn}|$ of the abscissa OP_{qn} is a dimension of the cubic crystalline lattice in a radial direction.

Considering the unit vector[1] $\vec{e_1}'$ in a radial direction derived in the Cartesian system $(Ox_1x_2x_3)$ as

$$\vec{e_1}' = \sum_{i=1}^{3} x_i\vec{e_i} = \vec{e_1}\cos\varphi\sin\nu + \vec{e_2}\sin\varphi\sin\nu + \vec{e_3}\cos\nu, \tag{12.16}$$

is transformed by the transformation formulae (12.2)–(12.4) related to the Cartesian system $\left(Ox''_{1qn}x''_{2qn}x''_{3qn}\right)$ to the form

$$\vec{e_1}' = \sum_{i=1}^{3} c_{iqn}\vec{e_i}'', \tag{12.17}$$

and the coefficient c_{iqn} is derived as

$$c_{iqn} = \left(a''_{i1qn}\cos\varphi + a''_{i2qn}\sin\varphi\right)\sin\nu + a''_{i3qn}\cos\nu,$$
$$i = 1, 2, 3; \quad q = p, e, m; \quad n = I, II, \tag{12.18}$$

where $\vec{e_i}$ and $\vec{e_i}''$ are unit vectors along the axes x_i and x''_{iqn} $(i=1,2,3;\ q=p,e,m;\ n=I,II)$, respectively. Accordingly, coordinates of the point P_{qn} are equal to the coordinates x'_{11qn}, x'_{12qn}, x'_{13qn} of the vector $\vec{x'_{1qn}} = x'_{1qn}\vec{e_1}' = |OP_{qn}|\,\vec{e_1}'$ derived in the Cartesian system $\left(Ox''_{1qn}x''_{2qn}x''_{3qn}\right)$ as

$$x'_{1iqn} = x'_{1qn}c_{iqn}. \tag{12.19}$$

A position of the point of intersection P_{qn} on the surface *1456* with the normal x''_{1qn} (see Fig. 12.1) results in the conditions $x'_{11qn} = a_{qn}$, $x'_{12qn} \le a_{qn}$, $x'_{13qn} \le a_{qn}$ transformed, regarding Eq. (12.18), to the forms

$$x'_{1qn} = \frac{a_{qn}}{c_{1qn}} \quad \text{on the conditions} \quad \frac{c_{2qn}}{c_{1qn}} \le 1 \quad \text{and} \quad \frac{c_{3qn}}{c_{1qn}} \le 1. \tag{12.20}$$

Similarly, a position of the point of intersection P_{qn} on the surfaces *2754* and *3657* with the normals x''_{2qn} and x''_{3qn} results in the conditions $x'_{12qn} = a_{qn}$, $x'_{11qn} \le a_{qn}$, $x'_{13qn} \le a_{qn}$ and $x'_{13qn} = a_{qn}$, $x'_{11qn} \le a_{qn}$, $x'_{12qn} \le a_{qn}$, respectively, transformed, regarding Eq. (12.18) and along with Eq. (12.19), to the forms

$$x'_{1qn} = \frac{a_{qn}}{c_{iqn}} \quad \text{on the conditions} \quad \frac{c_{jqn}}{c_{iqn}} \le 1 \quad \text{and} \quad \frac{c_{kqn}}{c_{iqn}} \le 1,$$
$$i, j, k = 1, 2, 3; \quad i \ne j \ne k; \quad q = p, e, m; \quad n = I, II. \tag{12.21}$$

With regard to the subscript i in Eq. (12.21), the point P_{qn} represents an intersection of the axis x'_1 with the surface of the cubic crystalline lattice with the normal

[1] The unit vector is a vector with length equal to 1.

x''_{iqn} $(i=1,2,3)$. The determination of the subscript i results from the conditions in Eq. (12.21), required to be simultaneously fulfilled.

The difference $u_{1qt} = x'_{1qII} - x'_{1qI}$ induced by a phase transformation and the fraction r/x'_{1qI} represent a radial displacement related to one transforming cubic crystalline lattice and a number of the transforming cubic crystalline lattices in a radial direction related to the distance r, respectively, both at the temperature $T = T_{tq} + dT$ for $dT \to 0$. Accordingly, the radial displacement and strain, $(u_{1qt})_{r=R_{1+i}}$ and ε_{11tq}, respectively, in the spherical particle $(q=p;\ i=0)$, in the spherical envelope $(q=e;\ i=0,1)$ and in the cell matrix $(q=m;\ i=0,1)$ are derived as

$$(u_{1qt})_{r=R_{1+i}} = \frac{R_{1+i}}{x'_{1qI}} \left(x'_{1qII} - x'_{1qI}\right) = R_{1+i}\,\varepsilon_{11tq}, \tag{12.22}$$

$$\varepsilon_{11tq} = \frac{|OP_{qII}| - |OP_{qI}|}{|OP_{qI}|} = \frac{x'_{1qII} - x'_{1qI}}{x'_{1qI}}, \tag{12.23}$$

representing, with regard to the coefficient c_{iqn} (see Eqs. (12.18), (12.21)), functions of the variables $\varphi, \nu \in \langle 0, \pi/2 \rangle$. Considering simplification, Equation (12.23) can be transformed to the form

$$\varepsilon_{11tq} = \frac{a_{qII} - a_{qI}}{a_{qI}}. \tag{12.24}$$

12.2 Coefficient a'_{ji}

With regard to the determination of the elastic energy density w_i induced by the stress σ_i (see Eqs. (3.46), (3.47)) acting along the axis x_i, the direction cosine $a'_{ij} = \cos\varphi'_{ij}$ of the angle φ'_{ij} formed by the axes x'_i, x_j $(i,j=1,2,3)$ (see Fig. 3.1) is required to be derived.

Considering the unit vectors $\vec{e_i}' = [e'_{i1}, e'_{i2}, e'_{i3}]$, $\vec{e_j} = [e_{j1}, e_{j2}, e_{j3}]$ in directions of the axes x'_i, x_j of the Cartesian systems $(Px'_1x'_2x'_3)$, $(Ox_1x_2x_3)$ (see Fig. 3.1), respectively, the direction cosine a'_{ij}, representing a scalar product of the vector $\vec{e'_j}$, $\vec{e_i}$ [6, p 146,150], has the form

$$\begin{aligned} a'_{ij} &= \left|\vec{e_i}'\right| \left|\vec{e_j}\right| \cos\left[\angle\left(x'_i, x_j\right)\right] = \cos\left[\angle\left(x'_i, x_j\right)\right] \\ &= \vec{e_i}'\vec{e_i} = e'_{i1}e_{j1} + e'_{i2}e_{j2} + e'_{i3}e_{j3}, \end{aligned} \tag{12.25}$$

where e'_{i1}, e'_{i2}, e'_{i3} and e_{j1}, e_{j2}, e_{j3} are coordinates of the vectors $\vec{e_i}'$ and $\vec{e_j}$ along the axes x_1, x_2, x_3, respectively, and the length $\left|\vec{e_i}'\right| = \left|\vec{e_j}\right| = 1$ $(i,j=1,2,3)$.

With regard to $\vec{e_1}' = [\cos\varphi\sin\nu,\ \sin\varphi\sin\nu,\ \cos\nu]$, $\vec{e_2}' = [-\sin\varphi,\ \cos\varphi,\ 0]$, $\vec{e_3}' = [-\cos\varphi\cos\nu,\ -\sin\varphi\cos\nu,\ \sin\nu]$ and $\vec{e_1} = [1,0,0]$, $\vec{e_2} = [0,1,0]$, $\vec{e_3} = [0,0,1]$, we get

$$a'_{11} = \cos\varphi \sin\nu, \tag{12.26}$$

$$a'_{12} = \sin\varphi \sin\nu, \tag{12.27}$$

$$a'_{13} = \cos\nu, \tag{12.28}$$

$$a'_{21} = -\sin\varphi, \tag{12.29}$$

$$a'_{22} = \cos\varphi, \tag{12.30}$$

$$a'_{23} = 0, \tag{12.31}$$

$$a'_{31} = -\cos\varphi \cos\nu, \tag{12.32}$$

$$a'_{32} = -\sin\varphi \cos\nu, \tag{12.33}$$

$$a'_{33} = \sin\nu. \tag{12.34}$$

12.3 Coefficients κ and ψ

The coefficients $\kappa^{(1)}_{1+i+2jq}$, $\psi^{(1)}_{1+i\ 1+j\,\eta q}$ $(i,j=0,1;\ \eta=\varphi,\nu)$ have the forms

$$\kappa^{(1)}_{1+i+2jq} = \frac{\chi_{2-i\ 2-jq}\left(\delta_{0j}-\delta_{1j}\right)}{\chi_{11q}\chi_{22q}-\chi_{12q}\chi_{21q}}, \quad i,j=0,1, \tag{12.35}$$

$$\psi^{(1)}_{1+i\ 1+j\,\eta q} = p_{1+i}\frac{\partial\kappa^{(1)}_{1+2jq}}{\partial\eta} + \kappa^{(1)}_{1+2jq}\frac{\partial p_{1+i}}{\partial\eta}, \quad i,j=0,1; \quad \eta=\varphi,\nu. \tag{12.36}$$

With regard to the coefficients $\kappa^{(1)}_{1+i+2jq}$, $\chi_{1+i\ 1+jq}$, the subscript q is related to material parameters of the spherical envelope and the cell matrix, thus for $q=e$ and $q=m$, respectively. With regard to the coefficient $\psi^{(1)}_{1+i\ 1+j\,\eta q}$, the subscript q is related to material parameters of the cell matrix, thus for $q=m$.

The coefficients $\kappa^{(2)}_{1+iq}$ $(i=0,1)$, $\kappa^{(2)}_{3q}$, $\kappa^{(2)}_{3+iq}$ $(i=1,2,3)$, $\psi^{(2)}_{1+ij\,\eta q}$ $(i=0,1;\ j=1,2,3;$ $\eta=\varphi,\nu)$ have the forms

$$\kappa^{(2)}_{1+iq} = \frac{\chi_{2+i\ 3-i}\,\chi_{3-i\,1} - \chi_{2+i\,1}\,\chi_{3-i\ 3-i}}{\chi_{22q}\chi_{33q}-\chi_{23q}\chi_{32q}}, \quad i=0,1, \tag{12.37}$$

$$\kappa_{3q}^{(2)} = \chi_{11q} + \sum_{i=0}^{1} \chi_{12+iq}\kappa_{1+iq}^{(2)}, \tag{12.38}$$

$$\kappa_{3+iq}^{(2)} = \frac{1}{\kappa_{3q}^{(2)}}\left[\delta_{1i} + \delta_{2i}\kappa_{1q}^{(2)} + \delta_{3i}\kappa_{2q}^{(2)}\right], \quad i = 1, 2, 3, \tag{12.39}$$

$$\psi_{1+i\ j\ \eta q}^{(2)} = p_{1+i}\frac{\partial \kappa_{3+jq}^{(2)}}{\partial \eta} + \kappa_{3+jq}^{(2)}\frac{\partial p_{1+i}}{\partial \eta}, \quad i = 0, 1; \quad j = 1, 2, 3; \quad \eta = \varphi, \nu, \tag{12.40}$$

and the subscript q in Eqs. (12.37)–(12.40) is related to material parameters of the cell matrix, thus for $q = m$.

The coefficient $\chi_{1+i\ 1+jq}$ $(i, j = 0,1,2)$ is presented in subsections *Appendix* on pages 72, 78, 86, 89, 105, 110, 113, 117, 123, 127, 130, 146, 148, 151, 154, 157, 162, 166, 170, 185, 189, 194, 199, 206, 210, 214, 227, 230, 232, 235, 239, 244, 249, 255.

12.4 Selected topics of Applied Mathematics

The selected topics of the scientific branch Applied Mathematics presented in Section 12.4 represents the mathematical knowledge to be required for the study of the book.

12.4.1 Definition of determinant. Cramer's rule

With regard to $n \geq 3$, the formula for a calculation of the determinant $D^{(n)}$ with n rows and n columns has the form [6, p 76]

$$D^{(n)} = \begin{vmatrix} a_{11}, & a_{12}, & \dots & a_{1n} \\ a_{21}, & a_{22}, & \dots & a_{2n} \\ \vdots & \vdots & \vdots & \vdots \\ a_{n1}, & a_{n2}, & \dots & a_{nn} \end{vmatrix} = \sum_{i=1}^{n} (-1)^{1+i} a_{1i} D_{1i}^{(n-1)}$$

$$= \sum_{i=1}^{n} (-1)^{1+i} a_{i1} D_{i1}^{(n-1)}, \tag{12.41}$$

and, in general, the sub-determinant $D_{ij}^{(n-1)}$ $(i, j = 1, \dots, \mathrm{n})$ with $(n-1)$ rows and $(n-1)$ columns is created from the determinant $D^{(n)}$, omitting the row i and the column j, considering the determinant $D^{(2)}$ derived as

$$D^{(2)} = \begin{vmatrix} b_{11}, & b_{12} \\ b_{21}, & b_{22} \end{vmatrix} = b_{11}b_{22} - b_{12}b_{21}. \tag{12.42}$$

As an example, considering $n = 4$ and Eqs. (12.41), (12.42), we get

$$D^{(4)} = \begin{vmatrix} a_{11}, & a_{12}, & a_{13}, & a_{14} \\ a_{21}, & a_{22}, & a_{23}, & a_{24} \\ a_{31}, & a_{32}, & a_{33}, & a_{34} \\ a_{41}, & a_{42}, & a_{43}, & a_{44} \end{vmatrix} = a_{11} \begin{vmatrix} a_{22}, & a_{23}, & a_{24} \\ a_{32}, & a_{33}, & a_{34} \\ a_{42}, & a_{43}, & a_{44} \end{vmatrix}$$

$$- a_{12} \begin{vmatrix} a_{21}, & a_{23}, & a_{24} \\ a_{31}, & a_{33}, & a_{34} \\ a_{41}, & a_{43}, & a_{44} \end{vmatrix} + a_{13} \begin{vmatrix} a_{21}, & a_{22}, & a_{24} \\ a_{31}, & a_{32}, & a_{34} \\ a_{41}, & a_{42}, & a_{44} \end{vmatrix} - a_{14} \begin{vmatrix} a_{21}, & a_{22}, & a_{23} \\ a_{31}, & a_{32}, & a_{33} \\ a_{41}, & a_{42}, & a_{43} \end{vmatrix}$$

$$= a_{11} \left(a_{22} \begin{vmatrix} a_{33}, & a_{34} \\ a_{43}, & a_{44} \end{vmatrix} - a_{23} \begin{vmatrix} a_{32}, & a_{34} \\ a_{42}, & a_{44} \end{vmatrix} + a_{24} \begin{vmatrix} a_{32}, & a_{33} \\ a_{42}, & a_{43} \end{vmatrix} \right)$$

$$- a_{12} \left(a_{21} \begin{vmatrix} a_{33}, & a_{34} \\ a_{43}, & a_{44} \end{vmatrix} - a_{23} \begin{vmatrix} a_{31}, & a_{34} \\ a_{41}, & a_{44} \end{vmatrix} + a_{24} \begin{vmatrix} a_{31}, & a_{33} \\ a_{41}, & a_{43} \end{vmatrix} \right)$$

$$+ a_{13} \left(a_{21} \begin{vmatrix} a_{32}, & a_{34} \\ a_{42}, & a_{44} \end{vmatrix} - a_{22} \begin{vmatrix} a_{31}, & a_{34} \\ a_{41}, & a_{44} \end{vmatrix} + a_{24} \begin{vmatrix} a_{31}, & a_{32} \\ a_{41}, & a_{42} \end{vmatrix} \right)$$

$$- a_{14} \left(a_{21} \begin{vmatrix} a_{32}, & a_{33} \\ a_{42}, & a_{43} \end{vmatrix} - a_{22} \begin{vmatrix} a_{31}, & a_{33} \\ a_{41}, & a_{43} \end{vmatrix} + a_{23} \begin{vmatrix} a_{31}, & a_{32} \\ a_{41}, & a_{42} \end{vmatrix} \right)$$

$$\begin{aligned} &= (a_{11}a_{22} - a_{12}a_{21})(a_{33}a_{44} - a_{34}a_{43}) + (a_{14}a_{22} - a_{12}a_{24})(a_{31}a_{43} - a_{33}a_{41}) \\ &+ (a_{13}a_{24} - a_{14}a_{23})(a_{31}a_{42} - a_{32}a_{41}) + (a_{12}a_{23} - a_{13}a_{22})(a_{31}a_{44} - a_{34}a_{41}) \\ &+ (a_{11}a_{24} - a_{14}a_{21})(a_{32}a_{43} - a_{33}a_{42}) \\ &+ (a_{13}a_{21} - a_{11}a_{23})(a_{32}a_{44} - a_{34}a_{42}) \,. \end{aligned} \tag{12.43}$$

Determinants are considered for a calculation of roots of a system of linear algebraic equations, using the Cramer's rule represented by Eq. (12.45). Provided that $D^{(n)} \neq 0$, the roots $x_1, \ldots, x_n$ of the system of the linear algebraic equations [6, p 87,88]

$$\begin{aligned} a_{11}\,x_1 + a_{12}\,x_2 + \,\ldots\, + a_{1n}\,x_n &= b_1, \\ a_{11}\,x_1 + a_{12}\,x_2 + \,\ldots\, + a_{1n}\,x_n &= b_2, \\ \vdots \quad\quad \vdots \quad\quad \vdots \quad\quad \vdots \quad & \quad \vdots \\ a_{n1}\,x_1 + a_{n2}\,x_2 + \,\ldots\, + a_{nn}\,x_n &= b_n, \end{aligned} \tag{12.44}$$

are derived as

$$x_i = \frac{D_i^{(n)}}{D^{(n)}}, \quad i = 1, \ldots, n, \tag{12.45}$$

where the determinant $D_i^{(n)}$ $(i = 1, \ldots, \mathrm{n})$ is created from the determinant $D^{(n)}$ (see Eq. (12.41)) by replacing the column i by the column with n rows in the form

$$\left.\begin{matrix} b_1 \\ b_2 \\ \vdots \\ b_n \end{matrix}\right\} n \text{ rows.} \tag{12.46}$$

As an example, with regard to the solution (5.7) and to Eq. (3.31), considering the boundary conditions (4.9), (4.11) for the cell matrix related to the isotropic multi-particle-matrix system, the integration constants C_{1m}, C_{2m} representing roots of the system of the linear algebraic equations

$$\chi_{11m}C_{1m} + \chi_{12m}C_{2m} = -\frac{p_1}{a_m^{(1)}}, \tag{12.47}$$

$$\chi_{21m}C_{1m} + \chi_{22m}C_{2m} = 0, \tag{12.48}$$

are derived in the forms

$$C_{1m} = \begin{vmatrix} -p_1/a_m^{(1)}, & \chi_{12m} \\ 0, & \chi_{22m} \end{vmatrix} \left(\begin{vmatrix} \chi_{11m}, & \chi_{12m} \\ \chi_{21m}, & \chi_{22m} \end{vmatrix} \right)^{-1} = -\frac{p_1 \kappa_{1m}^{(1)}}{a_m^{(1)}}, \tag{12.49}$$

$$C_{2m} = \begin{vmatrix} -p_1/a_m^{(1)}, & \chi_{12m} \\ 0, & \chi_{22m} \end{vmatrix} \left(\begin{vmatrix} \chi_{11m}, & \chi_{12m} \\ \chi_{21m}, & \chi_{22m} \end{vmatrix} \right)^{-1} = -\frac{p_1 \kappa_{3m}^{(1)}}{a_m^{(1)}}, \tag{12.50}$$

where $a_m^{(1)}$, χ_{11+im}, χ_{21+im} and $\kappa_{1+2im}^{(1)}$ $(i=0,1)$ are given by Eqs. (5.6), (5.54), (5.55) and (12.35), respectively.

12.4.2 Quadratic algebraic equation

With regard to the coefficients $a_1 \neq 0$, a_2, a_3 representing real numbers, and considering the discriminant

$$D = a_2^2 - 4a_1a_3, \tag{12.51}$$

the roots λ_1, λ_2 of the quadratic algebraic equation

$$a_1\lambda^2 + a_2\lambda + a_3 = 0, \tag{12.52}$$

determined using the formula

$$\lambda_{1,2} = \frac{1}{2a_1}\left(-a_2 \pm \sqrt{D}\right), \tag{12.53}$$

are real or imaginary for the discriminant $D \geq 0$ or $D < 0$ [6, p 118], respectively. Provided that $D < 0$, the term $\sqrt{D}$ is transformed to $i\sqrt{-D}$, where $i = \sqrt{-1}$ is the imaginary unit [6, p 50].

12.4.3 Monotonous functions

With regard to $x_1 < x_2$ from a definition range of the function $f = f(x)$, the conditions (12.54) and (12.55) derived in the forms [6, p 218]

$$f(x_1) \leq f(x_2), \tag{12.54}$$

$$f(x_1) < f(x_2), \tag{12.55}$$

define $f = f(x)$ as a non-decreasing and increasing function of the variable x, respectively. Similarly, the conditions (12.56) and (12.57) derived in the forms [6, p 218]

$$f(x_1) \geq f(x_2), \tag{12.56}$$

$$f(x_1) > f(x_2), \tag{12.57}$$

define $f = f(x)$ as a non-increasing and decreasing function of the variable x, respectively. Finally, non-decreasing, non-increasing and increasing, decreasing functions are monotonous and pure monotonous, respectively.

12.4.4 Indefinite integrals

With regard to the determination of the elastic energy W (see Eq. (3.42)), the following indefinite integrals are required to be presented.

The derivations of the constant function C regarding the variable r, and of the functions $f = r^\lambda$, $f = \ln r$ with the variable r applied in this book are as follows [6, p 340–344]

$$C' = 0, \tag{12.58}$$

$$\left(r^\lambda\right)' = \lambda\, r^{\lambda-1}, \tag{12.59}$$

$$(\ln r)' = \frac{1}{r}. \tag{12.60}$$

Similarly, the indefinite integrals of the constant function C regarding the variable r, and of the functions $f = r^\lambda$, $f = 1/r$ with the variable r applied in this book are as follows [6, p 340–344]

$$\int C\, dr = Cr, \tag{12.61}$$

$$\int r^\lambda\, dr = \frac{r^\lambda + 1}{\lambda + 1}, \quad \lambda \neq -1, \tag{12.62}$$

$$\int \frac{dr}{r} = \ln r. \tag{12.63}$$

Considering the derivation of the product fg of the functions $f(r)$, $g(r)$ with the variable r in the form

$$(fg)' = f'g + fg', \tag{12.64}$$

the formula representing the method Per Partes [6, p 420] for the solving of indefinite integrals has the form

$$\int f'g\, dr = fg - \int fg'\, dr. \tag{12.65}$$

As an example of an application of the method Per Partes, with regard to Eq. (12.65), considering $f' = r^{\lambda}$, $g = \ln r$, the indefinite integral $\int r^{\lambda} \ln r\, dr$ for $\lambda \neq -1$ is solved, regarding Eqs. (12.60), (12.62), in the following way

$$\begin{aligned} \int r^{\lambda} \ln r\, dr &= \frac{r^{\lambda+1}}{\lambda+1} \ln r - \int \frac{r^{\lambda+1}}{\lambda+1} \frac{1}{r}\, dr = \frac{r^{\lambda+1}}{\lambda+1} \ln r - \frac{1}{\lambda+1} \int r^{\lambda}\, dr \\ &= \frac{r^{\lambda+1}}{\lambda+1} \left(\ln r - \frac{1}{\lambda+1} \right), \quad \lambda \neq -1. \end{aligned} \tag{12.66}$$

Similarly, with regard to the indefinite integral $\int \ln r\, dr$ equal to that given by Eq. (12.66) for $\lambda = 0$, considering $f' = 1$, $g = \ln r$, $C = 1$ (see Eq. (12.61)), we get

$$\begin{aligned} \int \ln r\, dr &= \int 1 \times \ln r\, dr = r \ln r - \int r \frac{1}{r}\, dr \\ &= r \ln r - \int 1 \times dr = r\left(\ln r - 1\right). \end{aligned} \tag{12.67}$$

Other indefinite integrals, required for the determination of the elastic energy W (see Eq. (3.42)) and solved by the method Per Partes, are as follows

$$\int \frac{(\ln r)^{n}}{r}\, dr = \frac{(\ln r)^{n+1}}{n+1}, \tag{12.68}$$

$$\int r^{\lambda} \ln^2 r\, dr = \frac{r^{\lambda+1}}{\lambda+1} \left[\left(\ln r - \frac{1}{\lambda+1} \right)^2 + \frac{1}{(\lambda+1)^2} \right], \quad \lambda \neq -1. \tag{12.69}$$

Finally, with regard to Eqs. (3.42), (3.43), a definite integral is required to be defined. Provided that $F(r)$ is a primitive function of $f(r)$ in the interval $r \in \langle a, b \rangle$, thus $f = dF/dr$, the definite integral $\int_a^b f\,dr$ is defined by the Newton-Leibniz's formula [6, p 471]

$$\int_a^b f\,dr = F(b) - F(a)\,. \tag{12.70}$$

12.4.5 Wronskian's method

As a short introduction to problems of linear differential equations with non-zero right sides, Equations (5.2) and (5.5) are solved by the following mathematical techniques including the Wronskian's method [5].

Assuming the term $U_{1q} = \left[\left(\partial^2 u_{1q}/\partial\varphi^2\right) + \left(\partial^2 u_{1q}/\partial\nu^2\right)\right]$ $(q = p, e, m)$ in the form

$$U_{1q} = r^{\lambda_q}, \tag{12.71}$$

and considering the second derivation regarding the variable r

$$\frac{\partial^2 U_{1q}}{\partial r^2} = \lambda_q\left(\lambda_q - 1\right) r^{\lambda_q - 2}, \tag{12.72}$$

the differential equation (5.2)

$$r^2\,\frac{\partial^2 U_{1q}}{\partial r^2} + c_{3q}\left(1 - c_{3q}\right) U_{1q} = 0, \tag{12.73}$$

is transformed to the characteristic equation, regarding the exponent λ_q, derived as

$$\lambda_q^2 - \lambda_q + c_{3q}\left(1 - c_{3q}\right) = 0. \tag{12.74}$$

where the subscripts p, e and m are related to the spherical particle, the spherical envelope and the cell matrix, respectively. Accordingly, we get the differential equation (5.3)

$$U_{1q} = \frac{\partial^2 u_{1q}}{\partial\varphi^2} + \frac{\partial^2 u_{1q}}{\partial\nu^2} = \sum_{i=0}^{1} C_{1+iq}\, r^{\lambda_{1+iq}}, \tag{12.75}$$

where the exponent λ_{1+iq} $(i = 0{,}1)$, representing a root of the quadratic algebraic equation (12.74) (see Section 12.4.2), has the form

$$\lambda_{1+iq} = \frac{1}{2}\left[1 + \left(\delta_{0i} - \delta_{1i}\right)\sqrt{1 - 4c_{3q}\left(1 - c_{3q}\right)}\right], \quad i = 0, 1. \tag{12.76}$$

Considering Eq. (12.75), the differential equation (3.36)

$$r^2\,\frac{\partial^2 u_{1q}}{\partial r^2} + 2r\,\frac{\partial u_{1q}}{\partial r} - 2u_{1q} + \frac{U_{1q}}{s_{44q}\left(c_{1q} + c_{2q}\right)} = 0, \tag{12.77}$$

is transformed to the form

$$\frac{\partial^2 u_1}{\partial r^2} + \frac{2}{r}\frac{\partial u_1}{\partial r} - \frac{2u_1}{r^2} = g, \tag{12.78}$$

representing a linear differential equation of the n-th rank regarding the variable r, where $n = 2$, with a non-zero right side represented by the function g in the form

$$g = -3a_q^{(1)} \sum_{i=0}^{1} C_{1+iq}\, r^{\lambda_{1+iq}-2} \tag{12.79}$$

solvable by the Wronskian's method [5] presented consecutively, and the coefficient $a_q^{(1)}$ is given by Eq. (5.6). With regard to an application of the Wronskian's method, a linear differential equation of the n-th rank regarding the variable r, with a non-zero right side, is required to be generally derived in the form

$$\frac{\partial^n u_{1q}}{\partial r^n} + f_1 \frac{\partial^{n-1} u_{1q}}{\partial\, r^{n-1}} + \ldots + f_i \frac{\partial^{n-i} u_{1q}}{\partial\, r^{n-i}} + \ldots + f_n\, u_{1q} = g, \tag{12.80}$$

where f_i $(i = 1, \ldots, \mathrm{n})$ and g are functions of the variable r or constant functions regarding the variable r.

Assuming

$$u_{1q} = r^{\lambda}, \tag{12.81}$$

considering the first and second derivations regarding the variable r

$$\frac{\partial u_{1q}}{\partial r} = \lambda\, r^{\lambda-1}, \tag{12.82}$$

$$\frac{\partial^2 u_{1q}}{\partial r^2} = \lambda\,(\lambda - 1)\, r^{\lambda-2}, \tag{12.83}$$

the linear differential equation with zero right side, thus for $g = 0$,

$$\frac{\partial^2 u_{1q}}{\partial r^2} + \frac{2}{r}\frac{\partial u_{1q}}{\partial r} - \frac{2u_{1q}}{r^2} = 0, \tag{12.84}$$

is accordingly transformed to the characteristic equation, regarding the exponent λ, derived as

$$\lambda^2 + \lambda - 2 = 0, \tag{12.85}$$

where $\lambda_1 = 1$ and $\lambda_2 = -2$ are roots of the quadratic algebraic equation (12.85), using the formula (12.53). Accordingly, the functions

$$u_{11} = r^{\lambda_1} = r, \tag{12.86}$$

$$u_{12} = r^{\lambda_2} = \frac{1}{r^2}, \tag{12.87}$$

represent solutions of the linear differential equation (12.84) with zero right side. With regard to the analytical models of the thermal stresses in composite materials, **real** roots of a characteristic equation are recommended to be considered.

Generally, when $u_{11}, \ldots, u_{1n}$, representing solutions of a linear differential equation of the n-th rank with zero right side, exhibit continuous derivations to the $(n-1)$-th degree regarding the variable r, the solution of a linear differential equation with the right side $g \neq 0$, representing a linear combination of the terms $a_{1q}u_{11}, \ldots, a_{nq}u_{1n}$, is derived as

$$u_1 = \sum_{i=1}^{n} a_{iq} u_{1i}, \tag{12.88}$$

considering the coefficient a_i $(i = 1, \ldots, \mathrm{n})$ derived as

$$a_{iq} = \int \frac{W_i^{(n)}}{W^{(n)}} dr, \tag{12.89}$$

where the Wronskian's determinant $W^{(n)}$ with n rows and n columns has the form

$$W^{(n)} = \begin{vmatrix} u_{11}, & u_{12}, & \cdots & u_{1n} \\ \dfrac{\partial u_{11}}{\partial r}, & \dfrac{\partial u_{12}}{\partial r}, & \cdots & \dfrac{\partial u_{1n}}{\partial r} \\ \vdots & \vdots & \vdots & \vdots \\ \dfrac{\partial^{n-1} u_{11}}{\partial r^{n-1}}, & \dfrac{\partial^{n-1} u_{12}}{\partial r^{n-1}}, & \cdots & \dfrac{\partial^{n-1} u_{1n}}{\partial r^{n-1}} \end{vmatrix}, \tag{12.90}$$

and the determinant $W_i^{(n)}$ $(i = 1, \ldots, \mathrm{n})$ is created from the determinant $W^{(n)}$ by replacing the column i by the column with n rows in the form

$$\left.\begin{matrix} 0 \\ 0 \\ \vdots \\ g \end{matrix}\right\} n \text{ rows}. \tag{12.91}$$

With regard to the right side g and the solutions u_{11}, u_{12} (see Eqs. (12.79), (12.86), (12.87)), considering the first derivations, with respect to $n = 2$, derived as

$$\frac{\partial u_{11}}{\partial r} = 1, \tag{12.92}$$

$$\frac{\partial u_{12}}{\partial r} = -\frac{2}{r^3}, \tag{12.93}$$

we get

$$W^{(2)} = \begin{vmatrix} u_{11}, & u_{12} \\ \frac{\partial u_{11}}{\partial r}, & \frac{\partial u_{12}}{\partial r} \end{vmatrix} = u_{11}\frac{\partial u_{11}}{\partial r} - u_{12}\frac{\partial u_{12}}{\partial r} = -\frac{3}{r^2}, \tag{12.94}$$

$$W_1^{(2)} = \begin{vmatrix} 0, & u_{12} \\ g, & \frac{\partial u_{12}}{\partial r} \end{vmatrix} = -g\, u_{12} = a_q^{(1)} \sum_{i=0}^{1} C_{1+iq}\, r^{\lambda_{1+iq}-4}, \tag{12.95}$$

$$W_2^{(2)} = \begin{vmatrix} u_{11}, & 0 \\ \frac{\partial u_{11}}{\partial r}, & g \end{vmatrix} = g\, u_{11} = -a_q^{(1)} \sum_{i=0}^{1} C_{1+iq}\, r^{\lambda_{1+iq}-1}, \tag{12.96}$$

and consequently, considering Eq. (12.62), the coefficients a_{1q}, a_{2q} (see Eq. (12.89)) have the forms

$$a_{1q} = -a_q^{(1)} \sum_{i=0}^{1} \frac{C_{1+iq}\, r^{\lambda_{1+iq}-1}}{\lambda_{1+iq} - 1}, \tag{12.97}$$

$$a_{2q} = a_q^{(1)} \sum_{i=0}^{1} \frac{C_{1+iq}\, r^{\lambda_{1+iq}+2}}{\lambda_{1+iq} + 2}. \tag{12.98}$$

Considering Eq. (12.88) to result in

$$u_{1q} = a_{1q}\, u_{11} + a_{2q}\, u_{12}, \tag{12.99}$$

the solution of the linear differential equation (5.5) with a non-zero right side, as presented in Section 5.1 (see Eq. (5.7)), is derived as

$$u_{1q} = a_q^{(1)} \sum_{i=0}^{1} C_{1+iq}\, \Lambda_{1+iq}^{(1)}\, r^{\lambda_{1+iq}}, \tag{12.100}$$

where the coefficient $\Lambda_{1+iq}^{(1)}$ $(i = 0{,}1)$ is given by Eq. (5.8).

With regard to Eq. (12.89), formulae for a calculation of some integrals are presented in Section 12.4.4.

Bibliography

[1] Skočovský, P.; Bokůvka, O.; Palček, P. Materials Science; ISBN 80-7100-303-4; EDIS-University in Žilina: Žilina, SK, 1996; p 5–15, 38–40, 59 (in Slovak).

[2] Kuba, F. Theory of Elasticity and Selected Applications; 04-215-82; SNTL/Alfa: Prague, CZ, 1982; p 12–14, 125–126 (in Czech).

[3] Hearmon, R.F.S. Introduction to Applied Anisotropic Elasticity; 04-013-65; SNTL: Prague, CZ, 1965; p 11–14, 17, 24, 30 (in Czech); translation from English: The Clarendon Press Oxford, 1st edition, Oxford, 1961.

[4] Brdička, M.; Samek, L.; Sopko, B. Mechanics of Continuum; ISBN 80-200-0772-5; Academia: Prague, CZ, 2000; p 162, 219 (in Czech).

[5] Kluvánek, I.; Mišík, L.; Švec, M. Mathematics; 3rd edition; 63-001-70; Alfa: Bratislava, SK, 1970; Vol. 2, p 265–310 (in Slovak).

[6] Kluvánek, I.; Mišík, L.; Švec, M. Mathematics; 4th edition; 63-551-71; Alfa: Bratislava, SK, 1971; Vol. 1, p 50, 76, 87, 88, 118, 146, 150, 218, 247, 340–344, 368, 420, 471, 474 (in Slovak).

[7] Rektorys, K. Review of Applied Mathematics; 3rd edition; 04-001-73; SNTL: Prague, CZ, 1973; p 117, 970–971 (in Czech).

INDEX

G

H

I

M

N

O

P

Q

R

S

T

W